CW01335788

ENGLISH EPISCOPAL ACTA
IX

WINCHESTER 1205–1238

ENGLISH EPISCOPAL ACTA

I. LINCOLN 1067–1185. Edited by David M. Smith. 1980.

II. CANTERBURY 1162–1190. Edited by C. R. Cheney and Bridget E. A. Jones. 1986.

III. CANTERBURY 1193–1205. Edited by C. R. Cheney and E. John. 1986.

IV. LINCOLN 1186–1206. Edited by David M. Smith. 1986.

V. YORK 1070–1154. Edited by Janet E. Burton. 1988.

VI. NORWICH 1070–1214. Edited by Christopher Harper-Bill. 1990.

VII. HEREFORD 1079–1234. Edited by Julia Barrow. 1993.

VIII. WINCHESTER 1070–1204. Edited by M. J. Franklin. 1993.

IX. WINCHESTER 1205–1238. Edited by Nicholas Vincent. 1994.

ENGLISH EPISCOPAL ACTA IX

WINCHESTER 1205–1238

EDITED BY
NICHOLAS VINCENT

Published *for* THE BRITISH ACADEMY
by OXFORD UNIVERSITY PRESS

Oxford University Press, Walton Street, Oxford OX2 6DP
Oxford New York Toronto
Delhi Bombay Calcutta Madras Karachi
Kuala Lumpur Singapore Hong Kong Tokyo
Nairobi Dar es Salaam Cape Town
Melbourne Auckland Madrid

and associated companies in
Berlin Ibadan

Published in the United States
by Oxford University Press Inc., New York

© The British Academy 1994

All rights reserved. No part of this publication may be reproduced,
stored in a retrieval system, or transmitted, in any form or by any means,
electronic, mechanical, photocopying, recording, or otherwise, without
the prior permission of the British Academy

British Library Cataloguing in Publication Data

English episcopal acta.
9 Winchester 1205-1238
I. England. Christian church. Dioceses.
Administration. Decisions of bishops 1066-1300—Documents
—Latin texts
I. Vincent, Nicholas.   II. British Academy
262.3

ISBN 0-19-726130-2

Typeset by Alden Press Limited
Printed in Great Britain
on acid-free paper by
The Cromwell Press Limited
Melksham, Wiltshire

For Olivia (1964–1986)

CONTENTS

LIST OF PLATES	viii
ACKNOWLEDGEMENTS	ix
MANUSCRIPT SOURCES CITED	xi
ABBREVIATED REFERENCES TO MANUSCRIPTS, PRINTED BOOKS AND ARTICLES	xiii
OTHER ABBREVIATIONS	xxv

INTRODUCTION
- Peter des Roches, Bishop of Winchester 1205–1238 — xxvii
- The Bishop in his Diocese — xxxi
- Auxiliary Bishops, Archdeacons and Officials — xxxiv
- The Wider Household: Clerks, Bailiffs and Knights — xxxviii
- The Bishop's Chancery — xliii
- The Acta: Sources and Content — xlviii
- The Bishop and the Monks — liv
- The Bishop and Reform — lvi
- The Acta: External Features — lix
- The Acta: Internal Features — lxiv
- Editorial Method — lxxv

THE ACTA
1. Diocesan Affairs nos.1–85 — 1
2. The Bishop in Politics nos.86–133 — 77
3. Writs cited in the Winchester Pipe Rolls nos.134–343 — 105

APPENDICES
1. The Records of the Bishop's Years as Chief Justiciar and Regent 1213–1215 — 125
2. References to uses of the Bishop's Seal and to Actions which may have involved the issue of Episcopal Letters — 133
3. The Bishop's Itinerary — 143
4. The Bishop's Household — 163

INDEX OF PERSONS AND PLACES	217
INDEX OF SUBJECTS	253

LIST OF PLATES

(*between page* lxiv *and page* lxv)

I. SCRIBES 1 AND 2, ACTA nos. 51, 48 (A^2)
II. SCRIBE 3, ACTA no. 48 (A^1)
III. SCRIBES 4, 5 AND 6, ACTA nos. 119, 16, 3
IV. SEAL AND COUNTERSEAL OF BISHOP PETER

ACKNOWLEDGEMENTS

The present collection began life as part of a wider study of Peter des Roches, much of which will appear in due course as a political biography, to be published by the Cambridge University Press. Des Roches has dogged my footsteps for the past eight years. But for the support and encouragement afforded me by countless colleagues and friends, I doubt that I should ever have finished with him, delightful as his company has been. As it is, my chief debt of gratitude lies with the editor and the chairman of this series, David Smith and Christopher Brooke. Their criticism and advice have been splendid beyond compare. Whatever errors remain are, needless to say, entirely of my own devising. Preliminary versions were read by my research supervisor, John Maddicott and by my undergraduate tutor, Henry Mayr-Harting. To their teaching this work owes most of its merits and none of its faults. At an early stage Barbara Harvey and Susan Reynolds delivered criticisms for which I owe a special burden of thanks. Michael Franklin most kindly showed me a preliminary version of his collection of the charters of the bishops of Winchester prior to des Roches. Amongst the many archivists from whose help I have benefited, I should particularly like to pay tribute to those of the Hampshire Record Office, Winchester Cathedral Library and Winchester College; Roger Custance, Rosemary Dunhill, John Hardacre, Caroline Humphreys, Jennifer Thorp and Keith Walker. For advice on particular texts and problems I am grateful to Martin Brett, David Carpenter, Henry Chadwick, Mary Cheney, Michael Clanchy, H.E.J. Cowdrey, David Crook, David Crouch, Frances Davies (now Ramsey), Suzanne Eward, Michael Franklin, Joan Greatrex, Diana Greenway, the late Canon Hill, Sir James Holt, David Johnson, Derek Keene, Deirdre Le Faye, Roger Lovatt, Robert Patterson, J.O. Prestwich, Nigel Ramsay, Huw Ridgeway, Jane Sayers, Martin Snape, Benjamin Thompson, George Watts, Brian Wormald and Patrick Zutshi. The Rev. Robert Ferguson very generously loaned me his house in Winchester for a period of several months, whilst Jonathon Bayntun, David Clement-Davies, James Cope, Jeremy Davies, Julian Large, Lucien Miers and Tom Price served, albeit unwittingly, to improve my acquaintance with Hampshire and its churches. For other practical assistance I am grateful to the Alikhani family, the British Academy, Marcel M. Mille, the Master and Fellows of Peterhouse and above all to my parents.

Peterhouse, Cambridge   NICHOLAS VINCENT

MANUSCRIPT SOURCES

ORIGINAL CHARTERS OF PETER DES ROCHES

Canterbury Cathedral, Dean and Chapter Library: Chartae Antiquae M247: *100*
London, British Library: Additional Charter 47850: *40*
London, Public Record Office:
- Ancient Correspondence: SC1/1/20: *114*; SC1/1/21: *132*; SC1/1/198: *112*; SC1/1/199: *113*; SC1/1/200: *111*; SC1/2/13: *118*; SC1/2/14: *119*; SC1/4/167: *127*; SC1/6/24: *110*; SC1/6/26: *116*
- Ancient Deeds: E326/12579: *3*
- Ecclesiastical Documents: E135/5/35: *85*
- Significations of Excommunication: C85/153/1: *15*; C85/153/2: *16*

Oxford, Magdalen College: Basing Charters 34: *32*; 35: *48*; Basingstoke Charter 5: *48*; Petersfield Charters 147: *9*; 152: *51*; Selborne Charters 6: *49*; 97: *50*; 233: *52*; 250: *52*; 257: *53*; 269: *54*; 273: *54*
Salisbury, Cathedral Library: Dean and Chapter Muniments, Press III, Hurstbourne and Burbage Box: *18*
Westminster Abbey: Muniments 1846: *74*; 12753: *73*
Winchester Cathedral Library, Alchin Scrapbook II no.2: *82*
Untraced: *107*

COPIES, TRANSCRIPTS AND MENTIONS OF CHARTERS OF PETER DES ROCHES

Cambridge, Trinity College: Muniments Box 22 Chesterton no.6: *107*
Lichfield Cathedral: Dean and Chapter Library ms.28: *App.1 no.1*
London, British Library:
- Additional Charters 44694: *63*; 47398: *37, 38, 39, 40*
- Additional Manuscripts 6040: *App.2 no.28*; 29436: *81*
- Additional Roll 47861*: *38*
- Cotton Manuscripts: Tiberius C ix: *72*; Tiberius D vi, part i: *8*; Nero C iii: *36*; Domitian A xiv: *19, 20*; Vespasian A xvi: *129*; Vespasian F xv: *26*; Cleopatra C vii: *App.2 no.28*; Faustina A iii: *73*; Appendix xxi: *35*
- Egerton Manuscripts 2104a: *75, 76, 77*; 3667: *27, 28, 29*
- Loans Manuscripts: 29/55: *67*; 29/330: *2*
- Stowe Manuscript: 942: *56, 57, 58, 59, 60, 61*; *App.2 no.44*
- Wolley Charter XI.25: *24*
London, Public Record Office:
- Ancient Deeds E40/14233: *26*; E40/15431: *26*

- Bench Plea Rolls CP40/348: *21*
- Chancery, Ancient Correspondence SC1/47/53: *25*
- Chancery Ecclesiastical Miscellanea C270/36/21: *23*
- Charter Rolls C53/10: *94;* C53/124: *83, 84*; C53/131: *63*
- Close Rolls C54/12: *98*; C54/13: *98*; C54/16: *105, 106*; C54/17: *105, 106*
- Exch. KR misc. books E164/2: *97*; E164/20; *11*; E164/25: *5, 6*
- Justices Itinerant Plea Rolls JUST1/775: *22*
- KR Memoranda Rolls E159/2: *102*; E159/3: *115*
- LTR Memoranda Rolls E368/1: *103*; E368/4: *120*
- Papal Bulls SC7/18/3: *104*
- Patent Rolls C66/14: *99*; C66/20: *108*; C66/21: *109*; C66/35: *128*; C66/131: *63*; C66/148: *71*; C66/149: *68*; C66/162: *31*; C66/328: *63*; C66/448: *63*; C66/507: *63*
Oxford, Bodleian Library: Rawlinson Manuscripts B408: *11*; D763: *56, 57, 58, 59, 60, 61, 62*; *App.2 nos.29, 44*

Oxford, Magdalen College: Selborne Charters 100: *48*; 245: *54*
Salisbury, Cathedral Library, Dean and Chapter Muniments: Liber Evidentiarum C: *18*, *45*
Southampton, City Archives Office: Southampton Borough Council SC 15/4: *2*
Trowbridge, Wiltshire County Record Office: Bishop of Salisbury Manuscripts D1/1/1: *45*, *46*, *47*; D1/1/2: *18*; D1/1/3: *18*
Vatican, Archivio Segreto: Reg.Vat.12 part 2: *107*
Wells Cathedral, Dean and Chapter Library: Charter 39: *86*, *87*; Charter 40: *86*; Liber Albus I: *86*, *87*
Westminster Abbey Muniments: Domesday: *73*, *74*
Winchester Cathedral Library:
– Alchin Scrapbook II: *App.2 no.26*
– Baigent Papers, Cathedral Records: *82*
– Book of Endowments: *69*, *70*
– Book of John Chase: *81*
– Libellus Basyng: *App.2 no.26*
– Cartulary: *14*, *21*, *50*, *54*, *69*, *70*

Winchester College: Muniments 4222: *14*; 4227: *14*; 12094: *21*
Winchester, Hampshire County Record Office: 46M48/109: *67*; 5M53/998: *67*, *67a*, *71*; 1M54/2: *64*, *65*; 1M54/3: *64*, *65*; 1M63/1: *App.2 no.34*; 1M63/2: *34*; 21M65/A/1/2: *14*, *31*; 21M65/A/1/5: *31*, *56*, *57*, *59*, *60*, *61*, *62*; 21M65/A/1/11: *1*; Eccles. II 159270: *134–9*; Eccles. II 159270a: *140–52*; Eccles. II 159271: *153–63*; Eccles. II 159272: *164–74*; Eccles. II 159273: *175–86*; Eccles. II 159274: *187–93*; Eccles. II 159275: *194–210*; Eccles. II 159276: *211–28*; Eccles. II 159277: *229–51*; Eccles. II 159278: *252–68*; Eccles. II 159279: *269–93*; Eccles. II 159280: *294–304*; Eccles. II 159281: *305–14*; Eccles. II 159282: *315–28*; Eccles. II 159283: *329–35*; Eccles. II 159284: *336–39*; Eccles. II 159285: *340–43*.
Winchester, St Cross Hospital: Liber Secundus: *80*
York Minster Library: M2(3)2: *33*
Untraced: *13*

ABBREVIATED REFERENCES TO MANUSCRIPTS, PRINTED BOOKS AND ARTICLES

Acta Langton	*Acta Stephani Langton Cantuariensis Archiepiscopi A.D.1207–1228*, ed. K.Major, Canterbury and York Society l (Oxford 1950)
Act.Phil.Auguste	*Catalogue des actes de Philippe-Auguste*, ed. L.Delisle (Paris 1856)
AM	*Annales Monastici*, ed. H.R.Luard, 5 vols., Rolls Series (London 1864–9)
'Annals of Southwark'	'The Annals of Southwark and Merton', ed. M. Tyson, *Surrey Archaeological Collections* xxxv (1925), pp.24–57
Antient Kalendars	*Antient Kalendars and Inventories of His Majesty's Exchequer*, ed. F. Palgrave, 3 vols. (London 1836)
Baigent, F.J. and Millard, J.E.	*A History of the Town and Manor of Basingstoke* (Basingstoke 1889)
Bath Cartulary	*Two Chartularies of the Priory of St Peter at Bath*, ed. W. Hunt, Somerset Record Society vii (1893)
Beauchamp Cart.	*The Beauchamp Cartulary: charters, 1100–1268*, ed. E. Mason, Pipe Roll Society new series xliii (1980)
Beaulieu Cartulary	*The Beaulieu Cartulary*, ed. S.F. Hockey, Southampton Record Series xvii (1974)
Becket Materials	*Materials for the history of Thomas Becket, archbishop of Canterbury*, ed. J.C. Robertson and J.B. Sheppard, 7 vols., Rolls Series (London 1875–85)
Berkeley Charters	*Descriptive Catalogue of the Charters and Muniments in the possession of Lord Fitzhardinge at Berkeley Castle*, ed. I.H. Jeayes (Bristol 1892)
Berkshire Eyre	*The Roll and Writ File of the Berkshire Eyre of 1248*, ed. M.T. Clanchy, Selden Society xc (1973)
BF	*Liber Feodorum. The Book of Fees commonly called Testa de Nevill*, 3 vols.(London 1920–31)
BIHR	*Bulletin of the Institute of Historical Research*
BNB	*Bracton's Note Book*, ed. F.W.Maitland, 3 vols. (London 1887)
Bouquet, M. ed.	*Recueil des Historiens des Gaules et de la France: Rerum Gallicarum et Francicarum Scriptores*, 24 vols. (Paris 1738–1904)
Bradenstoke Cart.	*The Cartulary of Bradenstoke Priory*, ed. V.C.M. London, Wiltshire Record Society xxxv (1979)
Brecon Cartulary	'Cartularium Prioratus S. Johannis Evang. de Brecon', ed. R.W. Banks, *Archaeologia Cambrensis* 4th series xiii (1882), pp.275–308, xiv (1883), pp.18–49, 137–68,221–36,274–311
CACW	*Calendar of Ancient Correspondance concerning Wales*, ed. J.G.Edwards, Board of Celtic Studies (Cardiff 1935)
Cal.Chart.R.	*Calendar of the Charter Rolls*,6 vols.(London 1903–27)
Cal.Inq.Misc.	*Calendar of Miscellaneous Inquisitions I: Henry III and Edward I* (London 1916)

BOOKS AND ARTICLES

Cal.Lib.R.	*Calendar of Liberate Rolls*, 6 vols. (London 1917–64)
Cal.Pap.Reg. i	*Calendar of entries in the Papal Registers relating to Great Britain and Ireland 1198–1304*, ed. W.H. Bliss (London 1893)
Cambridge Portion	*The Cambridgeshire Portion of the Chartulary of priory of St Pancras of Lewes*, ed. J.H. Bullock and W.M. Palmer, Cambridge Antiquarian Society and Sussex Record Society (Cambridge 1938)
Cant.Professions	*Canterbury Professions*, ed. M.Richter, Canterbury and York Society lxvii (1973)
CAR	*Cartae Antiquae Rolls*, ed. L.Landon and J. Conway Davies, 2 vols., Pipe Roll Soc. new series xvii, xxxiii (1939–60)
Carisbrooke Cart.	*The Cartulary of Carisbrooke Priory*, ed. S.F.Hockey, Isle of Wight Record Series ii (1981)
Carpenter, D.A.	*The Minority of Henry III* (London 1990)
Cart.Château-du-Loir	*Cartulaire de Château-du-Loir*, ed. E. Vallée, Archives Historiques du Maine vi (Le Mans 1905)
Carte Nativorum	*Carte Nativorum: a Peterborough Abbey cartulary of the Fourteenth Century*, ed. C.N.L. Brooke and M.M. Postan, Northamptonshire Record Society xx (1960)
Cart. Gyseburne	*Cartularium Prioratus de Gyseburne*, ed. W. Brown, 2 vols., Surtees Society lxxxvi, lxxxix (1889–94)
Cart. Normand	*Cartulaire Normand de Philippe-Auguste, Louis VIII, Saint-Louis et Philippe-le-Hardi*, ed. L. Delisle (Caen 1882)
Cart. Saint-Pierre	*Cartulaire des Abbayes de Saint-Pierre de la Couture et de Saint-Pierre de Solesmes*, ed. P. d'Albert duc de Chaulnes (Le Mans 1881)
Cart. St Bartholomew's	*Cartulary of St Bartholomew's Hospital*, ed. J.M. Kerling (London 1973)
Cart. St Werburgh	*The Chartulary or Register of the Abbey of St Werburgh, Chester*, ed. J. Tait, 2 vols., Chetham Society new series lxxix, lxxxii (1920–3)
Cart.Treas.York	*The Cartulary of the Treasurer of York Minster and related documents*, ed. J.E. Burton, Borthwick Texts and Calendars v (York 1978)
Cart. Vendôme	*Cartulaire de l'abbaye cardinale de la Trinité de Vendôme*, ed. C. Métais, 6 vols. in 5, Société Archéologique du Vendômois (Paris/Chartres 1893–1904)
Cat.Anc.Deeds	*A Descriptive Catalogue of Ancient Deeds in the Public Record Office*, 6 vols. (London 1890–1915)
CCR	*Calendar of Close Rolls: Edward I*, 5 vols. (London 1900–1908)
CFR	*Calendar of Fine Rolls*, 22 vols. (London 1911–63)
Chart. Dieulacres	*Chartulary of Dieulacres Abbey*, ed. G. Wrottesley, Collections for a History of Staffordshire edited by the William Salt Archaeological Society, new series ix (1906)
Charters of Finchale	*The Priory of Finchale. The Charters of endowment, inventories and account rolls of the priory of Finchale in the county of Durham*, ed. J. Raine, Surtees Society vi (1837)
Chart.Pontefract	*The Chartulary of St John of Pontefract*, ed. R. Holmes, 2 vols., Yorkshire Archaeological Society Record Series xxv, xxx (1899–1902)
Cheney, C.R.	*English Bishops' Chanceries 1100–1250* (Manchester 1950)
	From Becket to Langton: English Church Government 1170–1213 (Manchester 1956)
	Hubert Walter (London 1967)

BOOKS AND ARTICLES

	Notaries Public in England in the Thirteenth and Fourteenth Centuries (Oxford 1972)
	Pope Innocent III and England (Stuttgart 1976)
Cheney, M.G.	'Master Geoffrey de Lucy, an early chancellor of the University of Oxford', *English Historical Review* lxxxii (1967), pp.750–63
Chertsey Cart.	*Chertsey Abbey Cartularies*, ed. M.S. Giuseppi, C.A.F. Meekings and P. Barnes, 3 vols. in 5 parts, Surrey Record Society xii (1915–63)
Chichester Cart.	*The Chartulary of the High Church of Chichester*, ed. W.D.Peckham, Sussex Record Society xlvi (1946)
Chobham *Summa*	Thomas de Chobham, *Summa Confessorum*, ed. F. Broomfield, Analecta Medievalia Namurcensia xxv (1968)
Chron. de Touraine	*Recueil des Chroniques de Touraine*, ed. A.Salmon (Tours 1854)
Chronicle of Hugh	*The Chronicle of the Election of Hugh Abbot of Bury St Edmunds and later Bishop of Ely*, ed. R.M. Thomson (Oxford Medieval Texts 1974)
Chron. Saint-Martial	*Chroniques de Saint-Martial de Limoges*, ed. H. Duplès-Agier, Société de l'Histoire de France (Paris 1874)
CIPM	*Calendar of Inquisitions Post Mortem*, 14 vols.(London 1904–52)
Cirencester Cart.	*The Cartulary of Cirencester Abbey*, ed. C.D. Ross, 3 vols. (London 1964–77)
Cl.R.	*Close Rolls of the Reign of Henry III*, 14 vols. (London 1902–38)
Coggeshall	*Radulphi de Coggeshall Chronicon Anglicanum*, ed. J. Stevenson, Rolls Series (London 1875)
Colchester Cart.	*Cartularium Monasterii Sancti Johannis Baptiste de Colecestria*, ed. S.A. Moore, 2 vols. (Roxburghe Club 1897)
Cole *Documents*	*Documents illustrative of English History in the Thirteenth and Fourteenth Centuries*, ed. H. Cole (London 1844)
Colvin, H.M.	*The White Canons in England* (Oxford 1951)
CP	*The Complete Peerage*, ed. G.E. Cockayne, revised by V.Gibbs, H.E. Doubleday, Lord Howard de Walden and G.H. White, 12 vols in 13 (London 1910–57)
CPR	*Calendar of Patent Rolls* (London 1891–)
Crook *Eyre*	D.Crook, *Records of the General Eyre*, Public Record Office Handbooks xx (1982)
Crouch, D.B.	*The Beaumont Twins: The Roots and Branches of Power in the Twelfth Century* (Cambridge 1986)
CRR	*Curia Regis Rolls of the Reigns of Richard I, John and Henry III preserved in the Public Record Office*, 17 vols. (London 1922–92)
C. & S.	*Councils and Synods with other documents relating to the English Church II 1205–1313*, ed. F.M. Powicke and C.R. Cheney, 2 vols. (Oxford 1964), part i (1205–1265)
Daventry Cartulary	*The Cartulary of Daventry Priory*, ed. M.J. Franklin, Northamptonshire Record Society xxxv (1988)
DD	*Diplomatic Documents preserved in the Public Record Office 1101–1272*, ed. P. Chaplais (London 1964)
Derbyshire Charters	*Descriptive Catalogue of Derbyshire Charters*, ed. I.H. Jeayes (London 1906)
Diceto	*Radulfi de Diceto decani Lundoniensis opera historica*, ed. W. Stubbs, 2 vols., Rolls Series (London 1876)
Domerham	*Adami de Domerham Historia de Rebus Glastoniensibus*, ed. T. Hearne, 2 vols. (Oxford 1727)
Dunstable Cartulary	*A Digest of the Charters preserved in the cartulary of the priory of*

	Dunstable, ed. G.H. Fowler, Bedfordshire Historical Record Society x (1926)
Early Bucks.Charters	*Early Buckinghamshire Charters*, ed. G.H. Fowler and J.G. Jenkins, Buckinghamshire Archaeological Society iii (1939)
EEA	*English Episcopal Acta*; i *Lincoln 1067–1185*, ed. D.M. Smith; ii *Canterbury 1162–1190*, ed. C.R. Cheney and B.E.A. Jones; iii *Canterbury 1193–1205*, ed. C.R. Cheney and E. John; iv *Lincoln 1186–1206*, ed. D.M. Smith; v *York 1070–1154*, ed. J.E. Burton; vi *Norwich 1070–1214*, ed. C. Harper-Bill; vii *Hereford 1079–1234*, ed. J. Barrow; viii *Winchester 1070–1204*, ed. M.J. Franklin (British Academy 1980–)
EHR	*English Historical Review*
Emden, A.B.	*A Biographical Register of the University of Oxford to A.D.1500*, 3 vols. (Oxford 1957–9)
Epistolae Cant.	*Chronicles and Memorials of the Reign of Richard I*, ed. W. Stubbs, 2 vols., Rolls Series (London 1864–5), vol.2: *Epistolae Cantuarienses*
E.Rot.Fin.	*Excerpta e Rotulis Finium in Turri Londinensi Asservatis ... A.D.1216–72*, ed. C.Roberts, 2 vols. (London 1835–6)
Essex Fines	*Feet of Fines for Essex i: 1182–1272*, ed. R.E.G. Kirk, Essex Archaeological Society (1899–1910)
EYC	*Early Yorkshire Charters*, vols. i–iii ed. W.Farrer (Edinburgh 1914–16), vols. iv–xii ed. C.T. Clay, Yorkshire Archaeological Society Record Series, extra series (1935–65)
Eynsham Cart.	*The Cartulary of the Abbey of Eynsham*, ed. H.E. Salter, 2 vols., Oxford Historical Society xlix, li (1906–8)
Farrer, W.	*Honors and Knights Fees*, 3 vols. (London 1923–5)
Fasti	John Le Neve, *Fasti Ecclesiae Anglicanae 1066–1300*, revised edition by D.E.Greenway, 4 vols. (London 1968–)
Feod. Dunelmensis	*Feodarium prioratus Dunelmensis*, ed. W. Greenwell, Surtees Society lviii (1872)
FF Henry II	*Feet of Fines of the reign of Henry II and of the first seven years of the reign of Richard I, A.D.1182 to A.D.1196*, Pipe Roll Society xvii (1894)
FF 9 Richard I	*Feet of Fines of the ninth year of the reign of king Richard I, A.D.1197 to A.D.1198*, Pipe Roll Society xxiii (1898)
Fitznells Cartulary	*Fitznells Cartulary: a calendar*, ed. C.A.F. Meekings and P. Shearman, Surrey Record Society xxvi (1968)
Flores Hist.	*Flores Historiarum*, ed. H.R.Luard, 3 vols., Rolls Series (London 1890)
Foedera	*Foedera, Conventiones, Litterae et cujuscumque generis Acta Publica*, ed. T.Rymer, new edition, vol.I part i, ed. A. Clark and F. Holbrooke (London 1816)
Foreign Accounts	*Roll of Divers accounts for the early years of the Reign of Henry III*, ed. F.A. Cazel Jr., Pipe Roll Society new series xliv (1974–5)
Furness Coucher Bk.	*The Coucher Book of Furness Abbey*, ed. J.C. Atkinson, 3 vols., Chetham Society new series ix,xi,xiv (1886–8)
Galbraith, V.H.	*Studies in the Public Records* (London 1948)
Gallia Christiana	*Gallia Christiana in Provincias Ecclesiasticas distributa*, 16 vols. (Paris 1715–1865)
Gams, P.B.	*Series Episcoporum Ecclesiae Catholicae* (Ratisbon 1873)
Gervase	*The Historical Works of Gervase of Canterbury*, ed. W. Stubbs, 2 vols., Rolls Series (London 1879–80)

BOOKS AND ARTICLES

Gesta Abbatum	*Gesta Abbatum Monasterii Sancti Albani*, ed. H.T. Riley, 3 vols., Rolls Series (London 1867)
Giles, K.R.	'Two English Bishops in the Holy Land', *Nottingham Medieval Studies* xxxi (1987), pp. 46–57
Glastonbury Cart.	*The Great Chartulary of Glastonbury*, ed. A. Watkin, 3 vols., Somerset Record Society lix, lxiii-iv (1947–52)
Gloucester Charters	*Earldom of Gloucester Charters: The Charters and Scribes of the Earls and Countesses of Gloucester to A.D.1217*, ed. R.B. Patterson (Oxford 1973)
God's House Cart.	*The Cartulary of God's House, Southampton*, ed. J.M. Kaye, 2 vols., Southampton Records Series xix–xx (1976)
Hatton *Seals*	*Sir Christopher Hatton's Book of Seals*, ed. L.C. Loyd and D.M. Stenton (Oxford 1950)
Haughmond Cart.	*The Cartulary of Haughmond Abbey*, ed. U. Rees, Shropshire Archaeological Society (Cardiff 1985)
HDF	*Historia Diplomatica Friderici Secundi*, ed. J.H.A. Huillard-Bréholles, 6 vols. in 11 (Paris 1852–61)
Heales, A., ed.	*Records of Merton Priory* (Oxford 1898)
Henry of Avranches	*The Shorter Latin Poems of Master Henry of Avranches relating to England*, ed. J. Cox Russell and J.P. Heironimus, Medieval Academy of America Studies and Documents i (Cambridge Mass. 1935)
Hist. Gloucestriae	*Historia et Cartularium Monasterii Sancti Petri Gloucestriae*, ed. W.H. Hart, 3 vols., Rolls Series (London 1863–7)
Hist.Maréchal	*L'Histoire de Guillaume le Maréchal*, ed. P. Meyer, 3 vols., Société de l'Histoire de France (Paris 1891–1901)
HMC	Reports of the Royal Commission on Historical Manuscripts
HMC Wells	*Calendar of the Manuscripts of the Dean and Chapter of Wells Vol.I*, Historical Manuscripts Commission (London 1907)
Hockey, S.F.	*Insula Vecta: The Isle of Wight in the Middle Ages* (Chichester 1982)
Holm Cultram Cart.	*The Register and Records of Holm Cultram*, ed. F. Grainger and W.G. Collingwood, Cumberland and Westmorland Antiquarian and Archaeological Society, record series vii (1929)
Holt, J.C.	*Magna Carta* (2nd edition, Cambridge 1992)
Interdict Docs.	*Interdict Documents*, ed. P.M. Barnes and W.R. Powell. Pipe Roll Society new series xxxiv (1958)
Jolliffe, J.E.A.	'The Chamber and the Castle Treasures under King John', *Studies in Medieval History presented to F.M. Powicke*, ed. R.W. Hunt and others (Oxford 1948), pp. 117–32
Keene *Survey*	D. Keene, *Survey of Medieval Winchester*, Winchester Studies II, 2 vols. (Oxford 1985)
Lacock Charters	*Lacock Abbey Charters*, ed. K.H. Rogers, Wiltshire Record Society xxxv (1979)
Layettes	*Layettes du Trésor des Chartes*, ed. A. Teulet, H.-F. Delaborde and E Berger, 5 vols. (Paris 1863–1909)
Letters of Innocent III	*The Letters of Pope Innocent III (1198–1216) concerning England and Wales*, ed. C.R.Cheney and M.G.Cheney (Oxford 1967)
Lincs. Assize Rolls	*The Earliest Lincolnshire assize rolls, A.D.1202–1209*, ed. D.M. Stenton, Lincoln Record Society xxii (1926)
Longueville Charters	*Chartes du Prieuré de Longueville de l'ordre de Cluny au diocèse de Rouen antérieures à 1204*, ed. P. le Cacheux, Société de l'Histoire de Normandie (Rouen/Paris 1934)
Loyd, L.C.	*The Origins of some Anglo-Norman Families*, ed. C.T. Clay and D.C.

	Douglas, Harleian Society ciii (1951)
Luffield Charters	*Luffield Priory Charters*, ed. G.R. Elvey, 2 vols., Northamptonshire Record Society xxii,xxvi (1968–75)
Lyttelton Charters	*A Descriptive Catalogue of the Charters and Documents of the Lyttelton Family at Hagley Hall*, ed. I.H. Jeayes (London 1893)
McKechnie, W.S.	*Magna Carta* (Glasgow 1914)
Magnum Reg.Album	*The Great Register of Lichfield Cathedral known as Magnum Registrum Album*, ed. H.E. Savage, Collections for a History of Staffordshire edited by the William Salt Archaeological Society (1926 for 1924)
Malmesbury Cart.	*Registrum Malmesburiense*, ed. J.S. Brewer and C. Trice Martin, 2 vols., Rolls Series (London 1879–80)
Martyrologe de Tours	*Martyrologe Obituaire de l'Église Métropolitaine de Tours*, ed. J.J. Bourassé, Mémoires de la Société Archéologique de Touraine xviii (1865)
Meekings, C.A.F.	'The Early Years of Netley Abbey', ed. R.F. Hunnisett, in Meekings, *Studies in Thirteenth Century Justice and Administration* (London 1981), chapter 17
Memoranda 1 John	*The Memoranda Roll for the Michaelmas Term of the first year of the reign of King John, 1199–1200*, ed. H.G. Richardson, Pipe Roll Society new series xxi (1943)
Memoranda 10 John	*The Memoranda Roll for the tenth year of the reign of King John, 1207–8*, ed. R. Allen Brown, Pipe Roll Society new series xxxi (1956)
Memoranda 14 Henry III	*The Memoranda Roll of the King's remembrancer for Michaelmas 1230–Trinity 1231*, ed. C. Robinson, Pipe Roll Society new series xi (1933)
MGH	*Monumenta Germaniae Historica*
Migne *PL*	*Patrologiae Latinae cursus completus*, ed. J.-P. Migne (Paris 1844–64)
Misae Roll 11 John	Misae Roll 11 John in *Rotuli de Liberate ac de Misis et Praestitis*, ed. T.Duffus Hardy (London 1844)
Misae Roll 14 John	Misae Roll 14 John in *Documents Illustrative of English History in the Thirteenth and Fourteenth Centuries*, ed. H.Cole (London 1844)
Missenden Cart.	*The Cartulary of Missenden Abbey*, ed. J.G. Jenkins, 3 vols., Buckinghamshire Record Society ii,x,xii (1939–62)
Monasticon	Sir William Dugdale and Roger Dodsworth, *Monasticon Anglicanum*, ed. J.Caley, H.Ellis and B.Bandinel, 6 vols. (London 1846)
Mon. Exoniensis	G.Oliver, *Monasticon diocesis Exoniensis: records illustrating the ancient foundations in Cornwall and Devon* (Exeter 1846)
Montacute Cart.	*Two Cartularies of the Augustinian priory of Bruton and the Cluniac priory of Montacute in the county of Somerset*, ed. H.C. Maxwell-Lyte and others, Somerset Record Society viii (1894)
Ms. 4DR	Winchester, Hampshire Record Office ms. Eccl.II 159270 (Pipe Roll of the Bishopric of Winchester for the year 4 Des Roches [Michaelmas 1208–09])
Ms. 6DR	Winchester, Hampshire Record Office ms. Eccl.II 159270a (Pipe Roll of the Bishopric of Winchester for the year 6 Des Roches [Michaelmas 1210–11])
Ms. 7DR	Winchester, Hampshire Record Office ms. Eccl.II 159271 (Pipe Roll of the Bishopric of Winchester for the year 7 Des Roches [Michaelmas 1211–12])
Ms. 9DR	Winchester, Hampshire Record Office ms. Eccl.II. 159272 (Pipe Roll

	of the Bishopric of Winchester for the year 9 Des Roches [Michaelmas 1213–14])
Ms. 11DR	Winchester, Hampshire Record Office ms. Eccl.II 159273 (Pipe Roll of the Bishopric of Winchester for the year 11 Des Roches [Michaelmas 1215–16])
Ms. 13DR	Winchester, Hampshire Record Office ms. Eccl.II 159274 (Pipe Roll of the Bishopric of Winchester for the year 13 Des Roches [Michaelmas 1217–18])
Ms. 14DR	Winchester, Hampshire Record Office ms. Eccl.II 159275 (Pipe Roll of the Bishopric of Winchester for the year 14 Des Roches [Michaelmas 1218–19])
Ms. 15DR	Winchester, Hampshire Record Office ms. Eccl.II 159276 (Pipe Roll of the Bishopric of Winchester for the year 15 Des Roches [Michaelmas 1219–20])
Ms. 16DR	Winchester, Hampshire Record Office ms. Eccl.II 159277 (Pipe Roll of the Bishopric of Winchester for the year 16 Des Roches [Michaelmas 1220–21])
Ms. 19DR	Winchester, Hampshire Record Office ms. Eccl.II 159278 (Pipe Roll of the Bishopric of Winchester for the year 7 Des Roches [Michaelmas 1223–24])
Ms. 20DR	Winchester, Hampshire Record Office ms. Eccl.II 159279 (Pipe Roll of the Bishopric of Winchester for the year 20 Des Roches [Michaelmas 1224–25])
Ms. 21DR	Winchester, Hampshire Record Office ms. Eccl.II 159280 (Pipe Roll of the Bishopric of Winchester for the year 21 Des Roches [Michaelmas 1225–26])
Ms. 22DR	Winchester, Hampshire Record Office ms. Eccl.II 159281 (Pipe Roll of the Bishopric of Winchester for the year 22 Des Roches [Michaelmas 1226–27])
Ms. 27DR	Winchester, Hampshire Record Office ms. Eccl.II 159282 (Pipe Roll of the Bishopric of Winchester for the year 27 Des Roches [Michaelmas 1231–32])
Ms. 28DR	Winchester, Hampshire Record Office ms. Eccl.II 159283 (Pipe Roll of the Bishopric of Winchester for the year 28 Des Roches [Michaelmas 1232–33])
Ms. 31DR	Winchester, Hampshire Record Office ms. Eccl.II 159284 (Pipe Roll of the Bishopric of Winchester for the year 31 Des Roches[Michaelmas 1235–36])
Ms. 32DR	Winchester, Hampshire Record Office ms. Eccl.II 159285 (Pipe Roll of the Bishopric of Winchester for the year 32 Des Roches [Michaelmas 1236–37])
Mss. Windsor	*The Manuscripts of St George's Chapel Windsor Castle*, ed. J.N. Dalton (Windsor 1957)
Nash,T.R.	*Collections for the History of Worcestershire*, 2 vols. (London 1781–2)
New Forest Docs.	*A Calendar of New Forest Documents 1244–1334*, ed. D.J. Stagg, Hampshire Record Series iii (1979)
Norfolk Fines	*Feet of Fines for the County of Norfolk ... 1201–1215, for the County of Suffolk ... 1199–1214*, ed. B. Dodwell, Pipe Roll Society new series xxxii (1958)
Norgate, K.	*The Minority of Henry III* (London 1912)
Norwich Charters	*The Charters of Norwich Cathedral Priory*, ed. B. Dodwell, 2 vols.,

	Pipe Roll Society new series xl, xlvi (1974–85)
'Nutley Cartulary'	J.G. Jenkins, 'The Lost Cartulary of Nutley Abbey', *The Huntington Library Quarterly* xvii (1953–4), pp.379–96
Oseney Cartulary	*Cartulary of Oseney Abbey*, ed. H.E. Salter, 6 vols., Oxford Historical Society lxxxix–xci, xcvii–viii, ci (1929–36)
Painter, S.	*The Reign of King John* (Baltimore 1949)
Paris *CM*	*Matthaei Parisiensis, Monachi Sancti Albani,Chronica Majora*, ed. H.R.Luard, 7 vols.,Rolls Series (London 1884–9)
Paris *Hist.Ang.Min.*	*Matthaei Parisiensis Historia Anglorum*, ed. F.Madden, 3 vols., Rolls Series (London 1866–9)
Pat.R.	*Patent Rolls*, 2 vols. (London 1901–3)
PBK	*Pleas Before the King or his Justices*, ed. D.M. Stenton, 4 vols., Selden Society lxvii–viii, lxxxiii–iv (1952–67)
Pinchbeck Register	*The Pinchbeck Register*, ed. F. Hervey, 2 vols. (Brighton 1925)
Plac.Quo Warrant.	*Placita de Quo Warranto*, ed. W. Illingworth and J. Caley (London 1818)
Political Songs	*The Political Songs of England from the reign of John to Edward II*, ed. J.T. Wright, Camden Society 1st series vi (1839)
Potthast, A.	*Regesta Pontificum Romanorum A.D. 1198–1304*, 2 vols. (Berlin 1874–5)
Powicke, F.M.	*King Henry III and the Lord Edward*, 2 vols. (Oxford 1947)
PR	*Pipe Rolls 1 John–14 Henry III*, ed. D.M. Stenton, P.M. Barnes and others, Pipe Roll Society (London 1933–90)
PR4DR	*The Pipe Roll of the Bishopric of Winchester 1208–9*, ed. H. Hall (London 1903)
PR6DR	*The Pipe Roll of the Bishopric of Winchester 1210–11*, ed. N.R. Holt (Manchester 1964)
PR 26 Henry III	*Pipe Roll 26 Henry III*, ed. H.L. Cannon (Yale 1916)
Prynne *Records* iii	W. Prynne, *The Third Tome of an Exact Chronological Vindication . . . of the Supreme Ecclesiastical Jurisdiction of our . . . English Kings* (London 1668)
Quarr Charters	*The Charters of Quarr Abbey*, ed. S.F. Hockey, Isle of Wight Record Series iii (1991)
Ramsey Cartulary	*Cartularium Monasterium de Rameseia*, ed. W.H. Hart and P.A. Lyons, 3 vols., Rolls Series (London 1884–93)
Rawlinson,R.	*The History and Antiquities of Hereford* (London 1717)
Reading Cartularies	*Reading Abbey Cartularies*, ed. B.R. Kemp, 2 vols., Camden Society 4th series xxxi–ii (1986–7)
Records of St Barth.	E.A. Webb, *The Records of St Bartholomew's Priory*, 2 vols. (Oxford 1921)
Red Book	*The Red Book of the Exchequer*, ed. H. Hall, 3 vols., Rolls Series (London 1896)
Reg.Antiq.Lincs.	*The Registrum Antiquissimum of the Cathedral Church of Lincoln*, ed. C.W.Foster and K.Major, 12 vols.,Lincoln Record Society xxvii–ix,xxxii,xxxiv, xli–ii,xlvi,li,lxii,lxvii–viii (1931–73)
Reg. Common Seal	*The Register of the Common Seal of the priory of St Swithun, Winchester 1345–1497*, ed. J. Greatrex, Hampshire Record Series ii (1978)
Regesta	*Regesta Regum Anglo-Normannorum 1066–1154*, ed. H.W.C. Davis, C.Johnson, H.A. Cronne and R.H.C. Davis, 4 vols. (Oxford 1913–69)
Reg.Cantilupo	*Registrum Thome de Cantilupo, episcopi Herefordensis, A.D.*

	MCCLXXV–MCCLXXXII, ed. R.G. Griffiths and W.W. Capes, Canterbury and York Society ii (1907)
Reg.Greg.IX	*Les Registres de Grégoire IX (1227–41)*, ed. L. Auvray, 4 vols. (École française de Rome 1896–1955)
Reg.Hamonis Hethe	*Registrum Hamonis Hethe, diocesis Roffensis A.D. 1319–1352*, ed. C. Johnson, 2 vols., Canterbury and York Society xlviii–ix (1948)
Reg.Hon.III	*Regesta Honorii Papae III*, ed. P. Pressutti, 2 vols. (Rome 1888–95)
Reg.Pal.Dun.	*Registrum Palatinum Dunelmense: The register of Richard de Kellawe, lord palatine and Bishop of Durham, 1314–1316*, ed. T.D. Hardy, 4 vols., Rolls Series (London 1873–8)
Reg.Pont.	*Registrum Johannis de Pontissara, episcopi Wintoniensis, A.D. MCCLXXII–MCCCIV*, ed. C. Deedes, 2 vols., Canterbury and York Society xix, xxx (London 1915–24)
Reg.St Osmund	*The Register of St Osmund*, ed. W.H.Rich Jones, 2 vols., Rolls Series (London 1883–4)
Reg. Walter Gray	*The Register or Rolls of Walter Gray, lord archbishop of York*, ed. J. Raine, Surtees Society lvi (1872)
Reg. Winchelsey	*Registrum Roberti Winchelsey Cantuariensis Archiepiscopi A.D. 1294–1313*, ed. R. Graham, 2 vols., Canterbury and York Society li–ii (1952–6)
Reg.Woodlock	*Registrum Henrici Woodlock, diocesis Wintoniensis, A.D.1305–1316*, ed. A.W. Goodman, 2 vols., Canterbury and York Society xliii–iv (1940–1)
RL	*Royal and other Historical Letters illustrative of the Reign of Henry III*, ed. W.W. Shirley, 2 vols., Rolls Series (London 1862–6)
RLC	*Rotuli Litterarum Clausarum in Turri Londinensi asservati*, ed. T.Duffus Hardy, 2 vols.(London 1833–4)
RLP	*Rotuli Litterarum Patentium in Turri Londinensi asservati*, ed. T.Duffus Hardy (London 1835)
Rot.Chart.	*Rotuli Chartarum in Turri Londinensi asservati*, ed. T.Duffus Hardy (London 1837)
Rot.Cur.Reg.	*Rotuli Curiae Regis rolls and records of the court held before the king's justiciars or justices, 6 Richard I–1 John*, ed. F. Palgrave, 2 vols. (London 1835)
Rot.Grosseteste	*Rotuli Roberti Grosseteste*, ed. F.N.Davis, Canterbury and York Society x (London 1910–13)
Rot.Hugh Welles	*Rotuli Hugonis de Welles episcopi Lincolniensis A.D. MCCIX–MCCXXXV*, ed. W.P.W. Phillimore and F.N. Davis, 3 vols., Canterbury and York Society i, iii,iv (London 1907–9)
Rot.Lib.	*Rotuli de Liberate ac de Misis et Praestitis*, ed. T.Duffus Hardy (London 1844)
Rot.Norm.	*Rotuli Normanniae in Turri Londinensi asservati*, ed. T.Duffus Hardy (London 1835)
Rot.Ob.	*Rotuli de Oblatis et Finibus in Turri Londinensi asservati*, ed. T.Duffus Hardy (London 1835)
Round, J.H.	*The King's Serjeants and Officers of State, with their Coronation Services* (London 1911)
St Denys Cartulary	*The Cartulary of the Priory of St Denys near Southampton*, ed. E.O. Blake, 2 vols., Southampton Records Series xxiv–v (1981)
St Paul's Charters	*The Early Charters of the Cathedral Church of St Paul London*, ed. M. Gibbs, Camden Society 3rd series lvii (1939)
St Thomas' Cartulary	*Chartulary of the Hospital of St Thomas the Martyr, Southwark*, ed.

	L. Drucker (Southwark 1932)
Sanders, I.J.	*English Baronies: a Study of their Origin and Descent 1086–1327* (Oxford 1960)
San Germano	The Chronicle of Richard of San Germano in *Rerum Italicarum Scriptores* viii part ii, ed. C.A. Garufi and L.A. Muratori (Bologna 1937–8)
Sarum Charters	*Charters and other Documents illustrating the history of the Cathedral . . . of Salisbury*, ed. W.H. Rich Jones and W.D. Macray, Rolls Series (London 1891)
Sayers, J.	*Papal Judges Delegate in the Province of Canterbury 1198–1254* (Oxford 1971)
	Papal Government and England during the Pontificate of Honorius III (1216–1227) (Cambridge 1984)
Selborne Charters	*Calendar of Charters and Documents relating to Selborne and its Priory*, ed. W.D.Macray, 2 vols., Hampshire Record Society (1891–4)
SLI	*Selected Letters of Innocent III*, ed. C.R. Cheney and W.H. Semple, Nelson's Medieval Texts (London 1953)
Southwick Cart.	*The Cartularies of Southwick Priory*, ed. K.A.Hanna, 2 vols., Hampshire Record Series ix–x (1988–9)
Stapleton *MRN*	*Magni Rotuli Scaccarii Normanniae sub Regibus Anglie*, ed. T. Stapleton, 2 vols. (London 1840–44)
Statutes	*Statutes of the Realm*, ed. A. Luders, T.E. Tomlins and others, 11 vols. (London 1810–1828)
Stoke by Clare Cart.	*Stoke by Clare Cartulary*, ed. C.Harper-Bill and R.Mortimer, 3 vols., Suffolk Record Series Suffolk Charters iv–vi (1982–4)
Stubbs, W.	*Select Charters and other illustrations of English Constitutional History*, 9th edition, ed. H.W.C Davis (Oxford 1921)
Surrey Eyre	*The 1235 Surrey Eyre*, ed. C.A.F. Meekings and D. Crook, 2 vols., Surrey Record Society xxxi–ii (1979–83)
Surrey Fines	*Pedes Finium or Fines relating to the county of Surrey*, ed. F.B. Lewis, Surrey Archaeological Society, extra series i (Guildford 1894)
'Surrey Portion'	'The Surrey Portion of the Lewes Cartulary', ed. D. Harrison, *Surrey Archaeological Collections* xliii (1935), pp. 84–112
Sussex Fines	*An Abstract of Feet of Fines relating to the county of Sussex from 2 Richard I to 33 Henry III*, ed. L.F. Salzmann, Sussex Record Society ii (1903)
Swift, E.	'The Machinery of Manorial Administration: with special reference to the lands of the bishopric of Winchester 1208–1454', Unpublished M.A. Thesis, University of London (1929)
Taxatio 1291	*Taxatio Ecclesiastica Angliae et Walliae, auctoritate Papae Nicholai IV, circa 1291*, ed. S.Ayscough and J.Caley (London 1802)
Thorpe, J.	*Registrum Roffense* (London 1769)
Tout, T.F.	*Chapters in the Administrative History of Medieval England: the wardrobe, the chamber and the small seals*, 6 vols. (Manchester 1920–33)
TRHS	*Transactions of the Royal Historical Society*
Trivet	*Annales F. Nicholai de Triveti*, ed. T.Hog, English Historical Society (London 1845)
Tropenell Cartulary	*The Tropenell Cartulary, being the contents of an old Wiltshire muniment chest*, ed. J. Silvester, 2 vols., Wiltshire Archaeological and Natural History Society iv (1908)

Turner, R.V.	*Men Raised from the Dust: administrative service and upward mobility in Angevin England* (Philadelphia 1988)
VCH	*Victoria History of the Counties of England*, ed. H.A. Doubleday and others (London 1900–)
Vincent, N.C.	'Peter des Roches, Bishop of Winchester: an Alien in English Politics 1205–1231', Unpublished Oxford D.Phil. Thesis (1991)
	'Jews, Poitevins and the Bishop of Winchester 1231–1234', in *Christianity and Judaism: Studies in Church History* xxix, ed. D. Wood (Oxford 1992), pp.119–132
	'The Origins of the Chancellorship of the Exchequer', *English Historical Review* cviii (1993), pp.105–121
	'Simon de Montfort's First Quarrel with King Henry III', in *Thirteenth Century England IV*, ed. P.Coss and S.D. Lloyd (Woodbridge 1992), pp.167–77
Vincent *Guala*	*The Acta of the Legate Guala 1216–1218*, ed. N.C. Vincent, Canterbury and York Society (forthcoming)
WAC	*Westminster Abbey Charters 1066–c.1214*, ed. E. Mason, London Record Society xxv (1988)
Walter of Coventry	*Memoriale fratris Walteri de Coventria. The historical collections of Walter of Coventry*, ed. W. Stubbs, 2 vols., Rolls Series (London 1872–3)
Waltham Charters	*The Early Charters of Waltham Abbey 1062–1230*, ed. R.Ransford (Woodbridge 1989)
WC	*The Chartulary of Winchester Cathedral*, ed. A.W.Goodman (Winchester 1927)
WCM	*Winchester College Muniments: A Descriptive List*, ed. S. Himsworth, P. Gwyn, J. Harvey and W. Graham, 3 vols. (Chichester 1976–84)
Wendover *Flores*	*Rogeri de Wendover Chronica sive Flores Historiarum*, ed. H.O.Coxe, 5 vols., English Historical Society (London 1841–4)
West, F.	*The Justiciarship in England 1066–1232* (Cambridge 1966)
'Wherwell Kalendar'	M.R. James, *A Descriptive Catalogue of the Manuscripts in the library of St John's College Cambridge* (Cambridge 1913), pp.90–2, analysing Cambridge, St John's College ms.68 fos.1r–7v
White *Selborne*	Gilbert White, *The Natural History and Antiquities of Selborne*, ed. T. Bell, 2 vols. (London 1877)
Wilkins *Concilia*	David Wilkins, *Concilia Magnae Britanniae et Hiberniae*, 4 vols. (London 1737)
Worcester Cartulary	*The Cartulary of Worcester Cathedral Priory*, ed. R.R. Darlington, Pipe Roll Society new series xxxviii (1968)
York Minster Fasti	*York Minster Fasti*, ed. C.T.Clay, 2 vols., Yorkshire Archaeological Society Record Series cxxiii–iv (1958–9)

OTHER ABBREVIATIONS

Add.	Additional
archbp(s)	archbishop(s)
archdn(s)	archdeacon(s)
BL	British Library
Bodl.	Bodleian Library, Oxford
bp(s)	bishop(s)
Ch.	Charter
CUL	Cambridge University Library
D.& C.	Dean and Chapter
fo(s).	folio(s)
m.	membrane
ms(s).	manuscript(s)
pb(d)e	*per breve (domini) episcopi*
pd	printed
PRO	Public Record Office, London
s.-ex.	late – century
s.-in.	early – century
s.-med.	middle – century
transl.	translated

INTRODUCTION

PETER DES ROCHES, BISHOP OF WINCHESTER 1205-1238

The present collection covers the episcopate of a single bishop of Winchester: Peter des Roches, or *de Rupibus*, one of the most influential politicians of his day. Des Roches was a native of the Touraine, sprung from the region between the rivers Loir and Loire in north west France.[1] Almost certainly he was a kinsman of William des Roches, lord of Château-du-Loir, a key figure in the government of northern France. Peter began his career in the service of king Richard I, according to contemporary legend being better versed in how to lay siege to a castle than in preaching the gospels.[2] He makes his first recorded appearance at Richard's court in April 1197, rising thereafter to become dean of the colleges of Loches and St Martin, Angers, and treasurer, *de facto* head, of St Hilaire, the great collegiate establishment at Poitiers. Meanwhile, at the Plantagenet court he had come to occupy a leading position within financial administration, serving as one of the more important officers of the royal chamber.

This budding career came close to collapse in 1204, following king John's defeat by Philip Augustus. Peter might have followed his kinsman, William des Roches, into service with the Capetians. Instead, he opted for exile in England; an exile, though he can have had little idea of this at the time, destined to last for the remainder of his life. Showered with ecclesiastical preferment including the precentorship at Lincoln, by the end of his first year in England he had emerged as the king's favoured candidate for the see of Winchester left vacant by the death of bishop Godfrey de Lucy (d.11 September 1204). Des Roches first appears with the title bishop-elect on 5

---

[1] Vincent 'Peter des Roches' provides a full-scale political biography of the bishop to 1231. It is hoped that this may shortly be published, together with a concluding part, already in typescript, continuing the study to 1238. The evidence for des Roches' birth and family is assembled in chapter 1, pp.1-35. For published sketches see *Dictionary of National Biography*, ed. L. Stephens and others, reissued in 21 vols. (London 1908-9) xv 938-42 *sub* Peter (by W.H. Rhodes); M. Lecointre-Dupont, 'Pierre des Roches, Trésorier de Saint-Hilaire de Poitiers, Evêque de Winchester', *Mémoires de la Société des Antiquaires de l'Ouest*, series 1 xxxii part i (1868) 3-16, both of which labour under the misapprehension that des Roches was a Poitevin. Individual aspects of the bishop's career are ably dealt with by West *The Justiciarship* 178-211, and Carpenter *The Minority, passim* esp. 134-40.

[2] Wendover *Flores* iii 181, iv 19, 327.

February 1205.[3] But although a majority of the Winchester monks appears to have supported him, a minority, backed by the see's two archdeacons, sought to promote Richard Poer, dean of Salisbury, son of a former bishop of Winchester, Richard of Ilchester. The disputed election was carried to Rome, the king providing financial and diplomatic support for des Roches. At some time before 21 June 1205 pope Innocent III quashed the elections both of des Roches and of Richard Poer. Thereafter he prevailed upon the Winchester monks to proceed to a fresh election, admitting some degree of participation by the archdeacons.[4]

Des Roches proved the unanimous choice of both parties, his election being confirmed by the pope on 21 June. He was consecrated by Innocent at St Peter's Rome on 25 September 1205. Shortly afterwards he received papal letters, exempting him from suspension or excommunication by any ecclesiastical authority save that of the pope.[5] In the immediate term these letters appear to have been intended to further a commission, entrusted to des Roches, to secure payment of Peter's Pence.[6] However, coming so soon after his personal consecration by the pope and coinciding with a prolonged and acrimonious vacancy in the see of Canterbury, his special privileges afforded him a degree of independence from metropolitan jurisdiction almost unique amongst his contemporaries, greater certainly than that which the pope had intended. There is no evidence that des Roches ever made formal profession of obedience to Canterbury. For his first eight years as bishop, Canterbury was either vacant or filled by a candidate at odds with the court. Thereafter, despite the admission to England of Stephen Langton, des Roches appears to have avoided issuing the customary profession.[7] As late as 1233 he was to claim that the papal privilege of 1205 exempted him from the threat of excommunication or discipline from the suffragans of Canterbury.[8] The peculiarity of his standing in relation to his metropolitan is highly significant. Combined with the vast wealth of the see of Winchester and the degree of Peter's political influence at court, it afforded him a freedom of action denied to virtually any other English bishop of his day.

Returning to England in March 1206, des Roches was formally enthroned

---

[3] *RLC* i 18b,19; RLP 49–49b. The election dispute is described in Cheney *Innocent III* 144–7.
[4] Migne *PL* ccxv cols.671–3; *Letters of Innocent III* no.631.
[5] *Flores Hist.* ii 129; *AM* ii (Winchester) 79; *Letters of Innocent III* no.664.
[6] *Letters of Innocent III* no.673; *AM* ii (Waverley) 257; Cheney *Innocent III and England* 38–9.
[7] Bps consecrated at Rome were not normally dispensed from professing obedience to Canterbury. See for example *Cant.Professions* nos.31,191,195,201,241,246. John de Pontoise, provided to the see of Winchester by pope Martin IV and consecrated by him in Rome, nonetheless professed obedience to archbp Pecham (*Ibid.* no.228a).
[8] Wendover *Flores* iv 276; Paris *CM* iii 252.

in Winchester cathedral on Palm Sunday, 26 March. The temporalities of the see, worth over £2000 a year in net, cash receipts, had been formally restored to him on March 24, but in practice it seems that the restoration was backdated to the time of his election in Rome, the account for the king's vacancy receipts ending on 24 June 1205.[9] The course of his political career thereafter can be sketched here only in the briefest detail. It falls into three basic phases, the first stretching from 1206 to 1224, a period during which des Roches exercised enormous power at court, to begin with as one of the closest confidants of king John, and then, after John's death, as guardian of the boy-king Henry III. His loyalty was most clearly demonstrated during the years of papal interdict, 1208-1213, when as one of only two English bishops to remain at the excommunicate court, he was castigated in satire as "the warrior of Winchester, up at the Exchequer; good at accountancy, slack at the gospels".[10] In 1213 he served briefly as *de facto* royal chancellor, although accorded no official title.[11] In the following year he emerged as chief justiciar and regent during the king's absence on the ill-fated Poitevin expedition, a controversial appointment which was overturned on the field of Runnymede when des Roches was replaced by the Englishman, Hubert de Burgh.[12] In the ensuing civil war Peter reverted to type as the warrior of Winchester, masterminding the royalist victory at the battle of Lincoln.[13]

Twice during this period, he came close to obtaining an archbishopric: in 1214 when he was postulated unsuccessfully to the see of York, and in 1221 when he is said to have been promoted to Damietta, an appointment that lapsed almost immediately with the disintegration of the fifth crusade.[14] In 1215 he may have refused an offer of the see of Durham.[15] In the same year, as

---

[9] *RLP* 60b,62; *RLC* i 52; *AM* ii (Winchester) 79. For the vacancy receipts to 24 June 1205 see *PR 7 John* 11-13, In addition the bp received over 1500 marks in outright gifts from the crown in 1205-6, besides being advanced considerable sums, in theory re-payable but in practice transformed into gifts: *PR 8 John* 54,155; *Rot.Ob.* 358; *RLC* i 71,101.

[10] *Political Songs* 10-11: *Wintoniensis armiger praesidet ad scaccarium, ad computandum impiger, piger ad evangelium, regis revolvens rotulum*. The translation given above is adapted from that by Michael Clanchy.

[11] *Rot.Chart.* 194b-196b; *Foedera* 118; *Norwich Charters* i no.39. Discussed by Painter *King John* 80; West *The Justiciarship* 190; Galbraith *Studies in the Public Records* 129-30.

[12] West *The Justiciarship* 178-211; Vincent 'Des Roches' 91-126. The exact date of his appointment is unknown. Geoffrey fitz Peter died on 14 October 1213. Des Roches' commission is implied in royal letters of 12 January 1214 and explicitly stated on 1 February: *RLC* i 160; *RLP* 110; *Foedera* 118.

[13] Carpenter *The Minority* 36-40; Vincent 'Des Roches' 159-64.

[14] For York see Cheney *Innocent III and England* 162-5. For Damietta see *Coggeshall* 190, discussed by Giles 'Two English Bishops' 48n.

[15] Galbraith *Studies in the Public Records* 136-7,161-2, from PRO SC1/1/6.

papal commissioner, it was des Roches who suspended Stephen Langton from office as archbishop of Canterbury.[16] One of the half dozen or so most influential figures at court, des Roches was nonetheless dogged by controversy. Firmly committed to the reconquest of the Plantagenet lands in France, he was derided by the native baronage as an alien. His close association with king John, his personal profiteering as virtual vice-regent of Hampshire and his links to a close-knit and volatile group of alien constables and sheriffs made him an increasingly isolated figure at court. In 1221 he was deprived of personal custody of the king. Two years later, despite an attempt to wrest control from de Burgh by having the king's minority revoked, des Roches was toppled from power.[17]

The second phase of his career, from 1224 to 1231, saw him overshadowed at court by de Burgh, Langton and those others, mostly Englishmen, who now controlled royal policy. To escape, des Roches looked to adventures abroad. Between 1227 and 1229 he led the English contingent on the sixth crusade. Here, once again, he found himself at odds with the church, supporting the emperor Frederick II in the face of papal excommunication and attacks from the patriarch of Jerusalem.[18] His military skills in the re-fortification of the Holy Land won him praise even from the generally hostile English chroniclers.[19] In the spring of 1229 he joined Frederick II at the head of the first Christian army to enter Jerusalem since the city's fall to the Saracens half a century earlier. Back in Europe, in the summer of 1229 he helped bring about peace between pope and emperor, remaining abroad, in all likelihood at the papal curia, for the next two years.[20] Only in August 1231 did he return to England, very much the hero of the hour, a victorious warrior and peacemaker, fêted by king Henry III who in the intervening years had grown increasingly disenchanted with the administration headed by Hubert de Burgh.

The third and final phase of the bishop's career, from 1231 to 1238, began with the ousting of de Burgh from court and the accession of des Roches and his satellites to unprecedented power. The bishop's nephew, Peter de Rivallis, was granted an extraordinary array of offices and lands. With des Roches as *éminence grise* working from behind the scenes, the king was persuaded to devote vast resources to diplomatic and military initiatives in Poitou and Brittany. At the same time, des Roches and his supporters, many of them

---

[16] Below no.100, appendix 1 no.6.
[17] Carpenter *The Minority* 239-43, 301-42.
[18] Giles 'Two English Bishops' 46-57; Vincent 'Des Roches' 294-337.
[19] *AM* iii (Dunstable) 112, 126-7; Paris *CM* iii 490; *Hist.Ang.Min.* ii 304.
[20] *AM* i (Tewkesbury) 76, iii (Dunstable) 126; below no.130.

aliens, veterans of John's reign, set about overturning the landed settlement put in place by de Burgh since 1224. In the process they appeared to threaten various of the legal principles enshrined in Magna Carta, trespassing on the interests of the English church and of a powerful constituency of barons alarmed by the prospect of a return to the arbitrary style of government favoured by king John. The outcome was baronial rebellion in the Welsh marches and in Ireland, headed by the earl of Pembroke, Richard Marshal. Fiscal and administrative reforms initiated by des Roches and de Rivallis led only to chaos. Enormous sums were shipped overseas to allies in Brittany and Poitou, but the supply of money soon ran short. The king faced bankruptcy. Church and baronage were in uproar. In the spring of 1234 des Roches was dismissed. Shortly afterwards baronial rebellion was ended with the death of Richard Marshal. The king's continental alliances collapsed for want of further supply, a decisive turning point for the history of Plantagenet rule in France.[21]

Despite threats of retribution, des Roches was permitted to retire relatively unscathed. An old man by now, once again he sought consolation for political failure in adventures abroad. In 1235 he joined pope Gregory IX, lending military aid in the pope's struggle with the Italian communes. Broken in health, he returned to England late in 1236, finding a limited degree of re-acceptance at court.[22] He died on 9 June 1238 at his manor of Farnham, probably aged seventy or more. His body was buried in Winchester cathedral, his heart at the Cistercian abbey of Waverley.[23]

THE BISHOP IN HIS DIOCESE

For a period of more than thirty years Peter des Roches exercised an influence over the Plantagenet court greater, arguably, than that of any other single figure save king John. From the thirteenth-century onwards, he has been portrayed as a warrior and financier first and foremost, a bishop in little more than name. Certainly, as the itinerary given below (appendix 3) bears witness, his political career involved him in prolonged absences from the diocese of Winchester. In 1206 he joined the king's expedition to Poitou. He was

---

[21] There is as yet no satisfactory published study of the bp's regime of 1232-4. Two articles by D.A. Carpenter, 'The Fall of Hubert de Burgh', *Journal of British Studies* xix no.2 (1980) 1-17, and 'The Decline of the Curial Sheriff in England 1194-1258', *EHR* (1976) 1-32, provide essential revisions to the outdated account given by Powicke *Henry III* 74-83,123-47. See also N.C. Vincent, 'Jews, Poitevins and the Bishop of Winchester' 119-32; 'Simon de Montfort's First Quarrel with King Henry III' 167-77, and references there cited.
[22] Paris *CM* iii 304,309,378,393,478.
[23] Paris *CM* iii 489-90; *AM* ii (Winchester/Waverley) 87-8, 319.

overseas on pilgrimage for several months in 1221 and spent more than four years abroad, on crusade and at the papal curia between July 1227 and August 1231. His last exile took him out of the country for nearly eighteen months, from the spring of 1235 to the autumn of 1236. In all his thirty four years as bishop, he was abroad for the whole or parts of twelve years, and in a further twelve travelled with the king north of Trent or into the marches with Wales.

However, we should not be misled by this into assuming that des Roches was an absentee bishop. His recorded itinerary is supplied for the most part from attestations to royal writs or charters, evidence which by its very nature is loaded in favour of his political rather than his pastoral concerns. Even then, and despite the fact that des Roches' is the most comprehensive itinerary yet compiled for any thirteenth-century bishop, we know of his precise whereabouts on an average of only forty days in each of his thirty four years. There are vast disparities between our knowledge of those years when he was at court and those in which he languished in the political wilderness. In 1219, for example, at the height of his influence in the minority council of Henry III, he can be located more or less precisely on 152 days of the year, in 1220 on 104 days, in 1233 following his return to power, on 142 days. By contrast, in 1224, the year of his political downfall, he can be located only once, and between 1236 and his death in 1238 on only twenty occasions, ten of them within the diocese of Winchester. Even in the years of his greatest activity at court he was never wholly absent from his see; in 1219, for example, he is found in the diocese or on his own manors on thirteen separate dates, whilst on average, nine in each of the forty days for which he can be located in any particular year were spent within the see of Winchester or on his outlying manors in Oxfordshire, Wiltshire and Somerset. Westminster, the centre of the court's activities, lay only a few hundred yards across the Thames from des Roches' own diocese.[1]

Taken all in all, the bishop's itinerary suggests that even at the height of his political career des Roches was never entirely cut off from diocesan affairs. During his prolonged absence abroad after 1227 he is known to have maintained communications by letter, reporting on the progress of the crusade, or announcing the death overseas of one of his nephews (no.129; appendix 2 no.31). Other letters, from the time of Henry III's minority, suggest that the bishop was particularly keen to be in Winchester in early

---

[1] The Winchester pipe rolls provide numerous references to undated visits by des Roches to his manors *PR 4DR* – Ms.32DR *passim*. It has proved impossible to include many of these references, of which there are several hundred, in the itinerary given below.

September, the time of the great Winchester fair and of the harvest on his manors (no.119). The Winchester pipe rolls provide ample evidence of his personal intervention in matters financial, his respiting of arrears, request for further enquiry or demand for accounts. Even so it is unlikely that he was present at the Wolvesey exchequer for the main session of account, between October and December each year.[2] Des Roches was also a devotee of the chase, a passion which he shared with king John and one amply illustrated by the Winchester pipe rolls, with their lists of the expenses of hawks, hounds, warreners and huntsmen. The late thirteenth-century chronicle of Lanercost describes des Roches as delighting more in the suffering of wild beasts than in the salvation of men's souls.[3] By 1238 several new hunting parks had been created on the bishopric estates, and it is inconceivable that des Roches did not make regular use of them for sport. The Winchester rolls record the regular stream of correspondence passing between the bishop and his estates, noted as warranty in individual manorial accounts (below nos.134–343), ordering everything from the provision of ginger and spices to the purchase of livestock and building materials.

Hunting and estate management are secular affairs: they tell us nothing of des Roches' concerns as pastor or churchmen, save incidentally that the bishop was very much a man of the world; a soldier as much as a preacher, his piety overshadowed by his political career. Adam of Eynsham, the biographer of St Hugh, has a story to tell of a young bishop of exceptional strength, who rather than dismount to offer his blessing to bystanders, sprinkled them with chrism from horseback, so that the children howled and screamed amidst the hooves of his fiery and kicking horses.[4] Could this have been an early glimpse of bishop Peter at work? As we shall see, there is good reason to suppose that des Roches was a far more conscientious pastor than his posthumous reputation would allow. Nonetheless, some means had to be devised to enable the ordinary business of the diocese to function when des Roches' was absent overseas or with the king.

---

[2] A statement based upon comparison between the bishop's itinerary and the accounts preserved in the Winchester pipe rolls, which from the year 16 des Roches are precisely dated.

[3] *Chronicon de Lanercost*, ed. J. Stevenson (Edinburgh 1839) 23. For king John as huntsman see *Political Songs* 4.

[4] *Magna Vita Sancti Hugonis*, ed. D.L. Douie and D.H. Farmer, 2 vols., revised ed., Oxford Medieval Texts (Oxford 1985) i 128.

AUXILIARY BISHOPS, ARCHDEACONS AND OFFICIALS

For the most part, a diocesan absent for long periods from his see could entrust the management of his affairs to subordinates in lower orders. But there were certain functions, such as ordinations, confirmations, consecrations and the issuing of indulgences, which only a bishop could perform. In this respect des Roches was by no means unique in calling upon the services of auxiliary bishops, most of them drawn from the Celtic fringe. During the vacancy of 1204-5, Herlewin bishop of Leighlin consecrated an altar at Southwick priory.[1] The Scottish bishops, Walter of Whithorn and Nicholas of the Isles, are to be found performing similar services at Christchurch Twinham in 1214 and 1221.[2] Albinus, bishop of Ferns in Ireland, consecrated altars at Waverley in 1201 and 1214, and was commemorated with obit celebrations in St Swithun's priory.[3] Another Irish bishop, named Godfrey, is said to have carried out ordinations at Winchester before 1222, although a deacon he had ordained was subsequently found to be ignorant of Latin and incapable of intoning the offices.[4]

The most important of des Roches' suffragans was a man named John, consecrated bishop of Ardfert before July 1219, but deprived of his see following a lengthy dispute. In February 1225 John was granted papal licence to exercise the office of suffragan.[5] He is already to be found acting as an auxiliary under archbishop Langton in October 1222.[6] Probably in 1226 he issued an indulgence for those visiting the shrine of St William at York.[7] Active in the diocese of Winchester from before 1222, in January 1226 he is found in des Roches' presence, consecrating altars at Waverley.[8] Thereafter he appears to have acted as auxiliary during the bishop's absence on crusade. Indulgences issued by him survive for Andwell priory and for the hospital of St Thomas at Southwark.[9] In June 1231 he was back at Waverley,

---

[1] *Southwick Cart.* vol.2 part iii no.964.
[2] BL ms. Cotton Tiberius D vi part ii (Christchurch cartulary) fo.149v.
[3] *AM* ii (Waverley) 253,282; BL ms. Add. 29436 (St Swithun's cartulary) fo.46r.
[4] *Reg. St Osmund* i 306.
[5] *Cal.Pap.Reg.* i 68,98,100; *Pat.R.1216-25* 323.
[6] *Les Chartes de Saint-Bertin*, ed. D. Haigneré, 4 vols. (St Omer 1886-99) i no.632; *Acta Langton* p.xx, and see BL ms. Cotton Claudius D x (Cartulary of St Augustine's Canterbury) fos.258v-59r.
[7] BL ms. Cotton Claudius B iii (York Minster cartulary) fo.34r.
[8] *AM* ii (Waverley) 301-2; BL ms. Cotton Cleopatra C vii (Merton cartulary) fo.89v, mistranslated in Heales *Records of Merton* 77 as J. bp of Arles.
[9] *WCM* no.2786; Bodl. ms. Rawlinson D763 (St Thomas' cartulary) p.2, where he is styled merely *Iohannes episcopus*, without his title as bp of Ardfert. He is probably the *Iohannes Hibernensis quondam episcopus* who assisted in the blessing of an abbot of Gloucester in 1228: *Hist.Gloucestriae* i 28.

consecrating further altars.[10] Following des Roches' return, it was bishop John who consecrated the king's new house for Jewish converts in London.[11] Later he appears as a suffragan of Edmund, archbishop of Canterbury.[12] Before his death in 1245 he retired to St Albans, acting as a source of information to the chronicler, Matthew Paris.[13]

Ranked below the auxiliary bishops, but arguably the most important figure in the diocese, was the bishop's official, a dignitary said to have been introduced to the diocese of Winchester during the episcopate of bishop Richard of Ilchester (1173/4–1188), although in practice there is no evidence of an official at Winchester before the election of des Roches[14] Four such officials are found in succession during des Roches' time, all of them men of considerable learning.[15] Master John of London appears to have come to des Roches from the household of Philip of Poitiers, bishop of Durham. He may well be identified as a scholar said to have taught the grammarian master John of Garland. His successor, master Alan of Stokes, is virtually unrecorded outside the diocese of Winchester, but his surviving charters bear witness to considerable skills as lawyer and commissioner in canon law. It is master Alan to whom routine papal judge delegacy business was directed. When the pope instructed des Roches to grant a cemetery to the hospital of St Thomas Southwark, the bishop delayed for some considerable time, and only after prompting by the bishops of Rochester and Chichester and a further reminder from archbishop Langton, was he persuaded to act, as master Alan phrases it, *ne de contemptu et inobediencia posset redargui post tot mandata et commonitiones.* Even then, being unable to attend to the matter in person, des Roches deputed it to master Alan, reserving the right to alter any ordinance that Alan might make.[16] Elsewhere, in a tithe settlement first heard before des Roches, master Alan describes himself as acting *vices domini Petri Wintoniensis*

---

[10] *AM* ii (Waverley) 309.
[11] PRO E372/76 m.8d; and see Ms. 28DR m.8 for "bp John" at Farnham, 9 March 1233, returning from *Kenitun'* to Winchester.
[12] BL ms. Harley 2110 (Castle Acre cartulary) fo.130r-v, where he appears simply as bp J(ohn), without diocesan title.
[13] Paris *CM* iii 243, iv 324,501-2, v 2, vi 385-6; BL ms.Cotton Nero D vii (List of St Albans' benefactors) fo.87v; and see *Gesta Abbatum* i 286, where a bp John is said to have consecrated a bell at St Albans, used in the celebration of the mass of the Virgin, 1214 × 1235.
[14] Cheney *English Bishops' Chanceries* 20. R.W. Southern, 'Some New Letters of Peter of Blois', *EHR* liii (1938) 412n., cites the claim by Peter of Blois, almost certainly incorrect, that bp Richard of Winchester was the first to introduce the office to England.
[15] Appendix 4 nos. 1-4.
[16] BL ms. Stowe 942 (St Thomas' cartulary) fos.2v-3r; Bodl. ms. Rawlinson D763 (St Thomas' cartulary) pp.2-3, and see below appendix 3 no.29.

episcopi iure ordinario.[17] During the bishop's absence on crusade after 1227, master Alan was addressed by the crown as if he had powers to institute clergy, suggesting a greater degree of delegated authority than that allowed to the official of the diocese of Lincoln, who during the absence of bishop Hugh of Wells, 1209–1213, appears to have possessed power merely to grant custody of churches, requiring special licence from the bishop before institution could take place.[18]

Master Alan was undoubtedly the single most important officer within des Roches' administration. Even after his retirement as official, he continued to rank in the witness lists to the bishop's charters above all other dignitaries save heads of religious houses; appearing above the archdeacons and even above his successor as official, master William de Ste Mère Église, a canon lawyer, active in much papal judge delegate work, promoted dean of St Paul's in 1241. If we were to look for the man most likely to have drafted des Roches' diocesan statutes, issued after 1222, there is no more plausible candidate than master Alan of Stokes. The statutes themselves, although modelled to a large extent upon those of Langton for Canterbury and Richard Poer for Salisbury, are composed in a distinctively terse Latin, very different from the prolix phraseology favoured by des Roches' contemporaries.[19] Throughout, they bear witness to the burden of responsibility carried by the bishop's official. It was to his official that des Roches deputed the licensing of mendicant preachers, and in company with the archdeacons, it was the bishop's *officiales* (an ambiguous term but one which must surely include the *officialis* proper) who were entrusted with the disciplining of clergy.[20] Elsewhere, the official is charged by des Roches' rule for the college of chaplains at Marwell, with maintaining discipline, appointing the college's prior when the chaplains were unable to agree on an election, and with the licencing of chaplains who sought absence from the college for more than a week (no.31). As we shall see, des Roches was lax in his endowment of vicarages. Nonetheless, of his two surviving endowments, one is specifically said to have been drawn up in the presence of master Alan of Stokes, whilst the other, composed on the bishop's deathbed, is said to have been undertaken *arbitrio officialis nostri* (nos. 70,84), the official in question being master Humphrey de Millières, the fourth and final man to hold such office under des Roches, and like his predecessors clearly a man of some learning, associated with the schools of Oxford. Besides

---

[17] Winchester, Hampshire Record Office ms. 13M63/1 (Mottisfont cartulary I) fos.69v–70r.
[18] D. Smith, 'The Rolls of Hugh of Wells, Bishop of Lincoln 1209–35', *BIHR* xlv (1972) 161–3.
[19] *C.& S.* 125–37; C.R. Cheney, *English Synodalia of the Thirteenth Century*, revised ed. (Oxford 1968) 75–6.
[20] *C.& S.* 128 no.11, 129 no.19.

their learning, one other factor unites at least three of des Roches' four officials: master John, master Alan and master Humphrey all seem to have served as keeper of St Cross hospital outside Winchester; a lucrative official perquisite.[21]

Of the archdeacons of Winchester and Surrey, little need be said here.[22] At des Roches' election the offices were held by adherents of his rival for the see, Richard Poer. Roger, archdeacon of Winchester, died shortly afterwards. Master Amicius, archdeacon of Surrey, may possibly have defected to the rebel cause in 1215. Thereafter, for the majority of des Roches' episcopate, the two archdeaconries were held by his own kinsmen, Bartholomew, Hugh and Luke des Roches. All served as papal judges delegate, whilst the bishop's diocesan legislation makes plain their duties, or at least their theoretical duties, in respect to the disciplining of clergy and the holding of archidiaconal synods.[23] Hugh and Luke played a controversial role in the election which followed bishop Peter's death. Luke was deposed or resigned in 1243, but Hugh survived in office for the whole of the episcopate of bishop William de Raleigh and for part of that of Aymer de Valence. Archdeacons' officials appear in both archdeaconries under des Roches, the bishop's diocesan statutes explicitly associating these officials with the archdeacons in a way that marks a departure from the statutes of Langton and Richard Poer, perhaps because des Roches' archdeacons, as wealthy aliens, were absent from their archdeaconries for long periods.[24] Master Luke des Roches seems to have been a man of some learning; his will bestowed alms upon the religious houses of the see and what appears to have been a considerable collection of books upon the bishop's posthumous foundation, Netley abbey.[25]

---

[21] Appendix 4 nos. 1,2,4. All of the bp's officials appear regularly in the witness lists to his charters. However, it is clear that for the long stretches of the year they were not actually resident in the bp's household, bearing out the general conclusions of M.Burger, '*Officiales* and the *familias* of the Bishops of Lincoln, 1258-99', *Journal of Medieval History* xvi (1990) 39-53.
[22] Appendix 4 nos. 5-13.
[23] *C.& S.* 125-37, esp. nos.13, 57, 76.
[24] *C.& S.* 126-9 nos.5,6,13,18. For officials in the archdeaconry of Winchester, master Walter de Langton, master Adam of Ebbesbourne, master Roger, master Ralph of Southwick and master Alberic de Vitriaco see appendix 4 nos.5,10; *Reg.St Osmund* ii 76. For master Amicius, Martin of Canterbury and Eudo dean of Ewell, officials of Surrey, see appendix 4 no.11; 'Surrey Portion' nos.51-2; *CRR* xvii no.32.
[25] PRO E210/11304.

THE WIDER HOUSEHOLD: CLERKS, BAILIFFS AND KNIGHTS

The proliferation of written records and above all the survival of the Winchester pipe rolls from 1208 onwards, enable us to plot the biographies of the members of des Roches' household in a detail quite impossible with the households of most twelfth-century bishops. Much information of this sort is presented below as an appendix, providing individual biographies of more than forty of des Roches' *familiares*.[1] The men chosen have been selected for their prominence in the witness lists to the bishop's charters. They make up only a small fraction of the bishop's wider household. Over the course of his thirty four years at Winchester des Roches employed or was associated with an enormous range of clerks, bailiffs and knights. The Winchester pipe rolls provide the names of more than two hundred men in regular service in estate administration, besides listing the entertainment and expenses of many others, clerks and laymen, who flitted in and out of the bishop's circle. The evidence suggests a considerable degree of flexibility and flux in office holding and administrative duties. The title of treasurer, familiar in later Winchester sources, appears to have been unknown in des Roches' day, although a succession of men, styled chamberlain, dean of Wolvesey, or keeper and receiver of Wolvesey, were clearly performing the function of treasurer throughout des Roches' time as bishop.[2] The non-hereditary office of marshal, held under des Roches by two men, father and son, Achard and Nicholas Achard, is accounted in the Winchester pipe rolls indiscriminately as 'scutage', 'the marshalsy' or 'Achard's account'.[3] Likewise, with the bishop's stewards, the see appears to have been managed by two or three stewards holding office simultaneously, laymen or clerks, with a further officer, in later sources styled bailiff of the bishop's liberties, supervising relations with the crown.[4] Of the bishop's clerks, his official could be entrusted with the custody of castles, or the archdeacon of Winchester with a role at the episcopal exchequer.

Des Roches was a Frenchman, and a remarkably high proportion of his

---

[1] Appendix 4 below.
[2] Appendix 4 nos.14–7, and in general see Swift 'Manorial Administration' 26–58.
[3] For the marshals and their office see *CIPM* i nos.477,899; *Selborne Charters* ii 82; *Cl.R.1259–61* 424; *WC* nos. 76–7; Swift 'Manorial Administration' 47. For their accounts see *PR 6DR* 184–5; Mss. 9DR m.10d; 13DR m. 15; 14DR m.12d; 16DR m.12; 20DR m.15; 21DR m.14d; 31DR m.15; 32DR m.15. The pipe rolls also reveal the activities of lesser marshals, charged with the humbler functions of attending to the bp's horses and stables.
[4] Six of the bp's stewards are treated in detail below appendix 4 nos.21–6. In general see Swift 'Manorial Administration' 103–4.

inner household, both clerks and knights, were drawn from France. They include Poitevins, such as master John of Limoges, and an even larger number of Normans or Anglo-Normans, including master Humphrey de Millières, the bishop's official; Roger Wacelin, des Roches' steward; the clerk Robert de Clinchamps and the knight Eustace de Greinville.[5] The bishop's fellow Tourangeaux are represented by his nephews, his chamber clerk, Denis de Bourgueil, and Peter de Chanceaux, who appears to have been active in the bishop's chancery.[6] Most of these men had no previous connection to England. Even amongst the Normans, most appear to have held little or no English land prior to 1200. The estates of master Humphrey de Millières, Eustace de Greinville and Geoffrey de Luverez were all later subject to confiscation by the crown as *Terre Normannorum*.[7] As England broke free from the patterns of patronage established under Henry II and his Angevin successors, so des Roches' household came to serve as a haven for a large number of aliens, increasingly isolated from the sympathies of the English church and baronage. Such men had a vital role to perform in what was still regarded as the imminent reconquest of the Plantagenet lands across the Channel. Several of des Roches' clerks served in diplomatic initiatives, to Rome and to France.[8]

After 1204 there were many in Normandy, excluded by the Capetian conquest from office-holding or advancement, who preferred to seek their fortunes in England.[9] The household of the Frenchman, des Roches, with its close connections to the court, was a natural point of embarkation for such men. At the time of des Roches' election to Winchester, aliens filled the vast majority of English sees; nine Frenchmen and one Italian, set against only four Englishmen, of whom one, Giles de Braose, might best be styled Anglo-Norman.[10] Within thirty years all of this had changed. By the time of his regime of the 1230s des Roches was the only bishop of an English see known to have been born in France. On occasion it is possible to observe a direct transferral of service from the older generation of alien bishops to des Roches. From the household of Philip of Poitiers, bishop of Durham, des Roches appears to have recruited at least three men; the English scholar, master John

---

[5] Appendix 4 nos.4,17,25,33,38.
[6] Appendix 4 nos.7,10,13,15,19,41-4.
[7] Appendix 4 nos. 4,38-9.
[8] Appendix 4 nos.18,22,27,29,30,33,36,41.
[9] In general see L.Musset, 'Quelques Problèmes posés par l'annexation de la Normandie au domaine royal français', in *La France de Philippe-Auguste: le temps des mutations*, ed. R.H. Bautier (Paris 1982) 291-309.
[10] Vincent 'Des Roches' 28-9.

of London; the Norman clerk, Robert de Clinchamps, and the Tourangeau knight, Andrew de Chanceaux.[11]

It was his patronage of aliens that formed one of the chief charges brought against des Roches by political rivals, from as early as 1214 right up to the time of his death. In one other respect his household stands apart from the general norm, in its heavy bias towards knights. Little of this emerges from the witness lists to the bishop's charters which are dominated by clerks. However, throughout his career des Roches was to be alternately pilloried or praised for his skills as a soldier.[12] The Winchester pipe rolls and the records of the royal court show that he consistently patronised knights, crossbowmen and other fighting men.[13] Few of these were drawn from the upper level of the bishopric's tenantry by knight service.[14] By the early thirteenth-century, great magnates had ceased for the most part to draw their inner households from amongst the principal tenants of their fees. Instead they looked to an affinity built up from office-holding, recruitment from outside and the creation of new tenancies.[15] At Winchester, although the bishop continued to profit from the scutages and feudal incidents levied from the anciently established military tenants of the see, his military following consisted for the most part of men he himself had introduced: Normans such as Roger Wacelin and Eustace de Greinville, or the bishop's Tourangeau nephew, Geoffrey des Roches, for whom new tenancies were created.[16]

Beyond the men who migrated from other episcopal households, such as those drawn from Durham, des Roches is known to have retained at least part of the household of bishop Godfrey de Lucy, his predecessor at Winchester. Master Eustace de Fauconberg, master Philip de Fauconberg, master Philip

---

[11] Appendix 4 nos. 1,17. For Andrew de Chanceaux see Vincent 'Des Roches' 29–30,34; *PR 4DR* 17; Mss. 14DR mm.7,7d,8,10d,11; 15DR m.2; 16DR mm.1,2,4d,5,6d,9–11; 21DR mm.1d,5d,9d; *Martyrologe de Tours* 37; R. Blair, 'A Number of Interesting Documents', *Proc. Soc. Antiquaries of Newcastle upon Tyne*, series 2 x (1902) 304.

[12] Wendover *Flores* iii 181, iv 19, 327; *AM* (Dunstable) 126–7; Paris *CM* iii 304,309; *Hist.Maréchal* lines 16313–8.

[13] Vincent 'Des Roches' 53–6, and see below no.118.

[14] A full list of the bp's contingent for the Montgomery campaign of 1223 is supplied by PRO C72/3. Of twenty two knights listed, fifteen are named, of whom only three appear to have held knight's fees of the bpric. For the bp's tenantry by knight service see *Red Book* 204–7; *Reg.Pont.* 387–90,592–7.

[15] See especially D. Crouch, *William Marshal: Court, Career and Chivalry in the Angevin Empire 1147–1219* (London 1990) 133–49,195–204; G.G. Simpson, 'The *Familia* of Roger de Quincy, Earl of Winchester and Constable of Scotland' in *Essays on the Nobility of Medieval Scotland*, ed. K.J. Stringer (Edinburgh 1985) 102–29; H. Ridgeway, 'William de Valence and his *Familiares*, 1247–72', *Historical Research* lxv (1992) 240–57.

[16] Appendix 4 nos.25,38,43.

de Lucy and the knight, Maurice de Turville, all spanned the service of bishops Godfrey and Peter.[17] On his appointment as justiciar in succession to Geoffrey fitz Peter, des Roches recruited at least two of Geoffrey's clerks; Richard of Barking and Richard of Stapleford.[18] For the rest, the bishop's contacts with the king and his court ensured that there was a constant flitting to and fro of clerks between service at Winchester and service under the crown. Master William de St Maixent and master Philip de Lucy served under des Roches both at Winchester and in the *camera regis*.[19] Richard of Barking acted as bishop's attorney and as a clerk of the royal exchequer.[20] Geoffrey de Caux is described in a letter from king John to des Roches as *clericus noster et vester*.[21] Others of the bishop's familiars were drawn in on occasion to serve as ambassadors overseas or on minor commissions at court. From 1217 to 1224 and again between 1232 and 1234 the bishop's stewards, William of Shorwell and Roger Wacelin, served both as episcopal stewards and as sheriffs of Hampshire, accounting at the royal exchequer as the bishop's deputies, providing des Roches with a stranglehold over the affairs of Hampshire that does much to explain his unpopularity with the local gentry.[22]

The close connection between des Roches and the court ensured that many of his *familiares* went on to obtain high office in church and state. Eustace de Greinville served as steward of the royal household (1217-1225), Peter de Rivallis rose to become clerk of the king's wardrobe (1217-23) and later royal treasurer, with custody over a bewildering variety of court offices during des Roches' regime of 1232-34. William de St Mère Église migrated from a position as bishop's official under des Roches to become dean of St Paul's. None of des Roches' familiars rose higher than master Eustace de Fauconberg, already a royal justice at the time of des Roches' election, subsequently promoted royal treasurer (1217-1228) and bishop of London (1221-1228). On the fringes of the bishop's circle, master Robert of Bingham and master Giles of Bridport, future bishops of Salisbury, both appear early in their careers, recorded in the Winchester pipe rolls as guests on the Winchester estates.[23] A tantalising series of references from the bishopric pipe roll for 1213-14 shows des Roches employing a steward named Ralph de Neville.[24] It

---

[17] Appendix 4 nos.30,31,34,40.
[18] Appendix 4 nos.24,28.
[19] Appendix 4 nos. 34,35.
[20] Appendix 4 no.28.
[21] *RLP* 118b, and see below appendix 4 no.29.
[22] Appendix 4 nos.23,25, and in general see Vincent 'Des Roches' 230-44.
[23] Mss. 7DR m.1d; 31DR m.15, and see below no.335.
[24] Ms. 9DR mm.1d,2,4d,5d,6d-8,9.

is possible but by no means certain that this is the same Ralph later promoted bishop of Chichester and royal chancellor. Certainly, the future chancellor's first introduction to royal service appears to have come in December 1213 when as des Roches' deputy he received custody of the king's great seal.[25]

Service under des Roches could do much to advance a man at court or in the church. But by no means all of the men promoted in this way were hard-bitten curialists. The bishop's wealth was a powerful magnet, attracting scholars and artists as much as mercenaries and knights. There were undoubtedly schools of some sort at Winchester in the early thirteenth-century.[26] The poet master Henry of Avranches composed verses at des Roches' request. In 1233 he addressed a panegyric to the pope in support of master John Blund, the Aristotelian scholar of Oxford and Paris whose candidacy for the archbishopric of Canterbury was enthusiastically promoted by des Roches.[27] Master John Blund is only one of several of des Roches' protégés who enjoyed contact with the schools of Oxford. Master Humphrey de Millières, des Roches' official, owned houses there to which he retired in 1231 to renew his studies.[28] Bishop John of Ardfert, des Roches' suffragan, was in residence at the Oxford schools in 1229.[29] Masters Eustace and Philip de Fauconberg and master Philip de Lucy were all close associates of master Geoffrey de Lucy, one of the earliest chancellors of Oxford.[30] In 1211–12 the Winchester pipe rolls show des Roches supplying firewood and provisions to a nephew resident at

---

[25] *RLP* 107; *RLC* i 158b. For most of 1214 Ralph de Neville was absent with the king in Poitou which greatly reduces the likelihood of his identification with des Roches' steward.

[26] *RLC* i 27b, an order that a man named Geoffrey attend the schools there. Northampton, Northamptonshire Record Office, ms. Stopford-Sackville 2175, records Stephen prior of St Swithun's and O. prior of Hyde acting on a papal mandate addressed to them and to A., late master of the schools of Winchester (*A. magistro scolar' Winton'...sublato de medio*) (1199 × 1214). The (bp's) official, the dean and the *magister scolarum* of Winchester, all of them un-named, served as arbiters in a dispute between master Robert de Forde and the canons of Notley, 10 February 1231 (Oxford, Christ Church Archives ms. Notley charter roll m.12). In 1241 M. the wife of Stephen the goldsmith took a case before the master of the schools of Winchester, complaining that Robert de Hibernia, clerk, had called her a dirty whore (*coram magistro scolar' de Winton' traxisset in causam super eo quod dictus clericus ipsam meretriciam fecadam vocaverat*): Canterbury Cathedral D.& C. Library, ms. Cart.Antiq. M364/8. Master Andrew, *rector scolar' Wynt'*, appears as investigator in a tithe dispute, December 1268: Salisbury Cathedral D. & C. Library, ms. Press I, Box 5 D-F (Fordingbridge) no.3. For the city schools in the 1150s see *The Letters of John of Salisbury* i, ed. W.J. Millor, H.E. Butler and C.N.L. Brooke (London 1955) 95–6 no.56.

[27] *Henry of Avranches* 123–4,127–36; Paris *CM* iii 243; *Hist.Ang.Min.* ii 355; *AM* iv (Osney) 73–4; D.A. Callus, 'Introduction of Aristotelian Learning to Oxford', *Proceedings of the British Academy* xxxix (1943) 241–9.

[28] Appendix 4 no.4.

[29] *Cl.R.1227–31* 265.

[30] M.G. Cheney 'Master Geoffrey de Lucy' 750–63.

Oxford, in all likelihood to be identified as Peter de Rivallis.[31] Master John of London and master John of Limoges both have claims to be considered as authors or scholars. Master Luke des Roches owned what appears to have been a substantial library.[32] Master Thomas of Ebbesbourne, active in the bishop's circle from 1208 to 1224, was almost certainly the author of a medical receipt.[33] After 1232 we find the ubiquitous master Elias of Dereham, a man long associated with building projects and artistic commissions, appearing as witness to the charters of des Roches' foundation at Selborne.[34] The Winchester pipe rolls show Elias visiting Titchfield, or the nearby manor of Fareham, on at least fourteen separate occasions between 1232 and 1237, possibly to supervise building work at des Roches' new abbey.[35] In 1236 he was named amongst des Roches' executors, a service he performed for at least half a dozen other bishops and magnates.[36] Elsewhere we know him to have been involved in the making of monumental effigies commissioned by the king. At least one of his attractions as executor may have lain in his ability to supervise the construction of bishops' tombs.[37]

THE BISHOP'S CHANCERY

By contrast to other dioceses where a dignitary styled bishop's chancellor appears regularly from the twelfth century, at Winchester there is no evidence of a titular chancellor between the episcopate of Richard of Ilchester and the early fourteenth century.[1] The majority of bishop Godfrey de Lucy's acta conclude with a *per manum* clause, naming one of several men, Roger, Reginald, Godfrey, William and Philip, all styled clerk, by whose hand the bishop's charters are said to have been given. De Lucy appears to have been the first English bishop to date his acta routinely, generally specifying place of issue, day and month in the Roman calendar and pontifical year. Something

---

[31] Appendix 4 no.41.
[32] Appendix 4 no.13.
[33] Below no.26.
[34] Below nos.51,53; *Selborne Charters* i 6,16,20,31,36, ii 6,65. For master Elias in general see A.Hamilton Thompson, 'Master Elias of Dereham and the King's Works', *Archaeological Journal* xcviii (1942) 1–35; *Fasti* iv p.xxixn.
[35] Mss.28DR mm.3,4,6d,7,14; 31DR mm.5,6d,8,8–9,11,12d; 32DR m.2d.
[36] *CPR 1232–47* 166; below no.43. Elias served as executor to Hubert Walter, Stephen Langton and Richard Grant, successive archbps of Canterbury; Richard Poer, bp of Durham; William Marshal I, earl of Pembroke, and was appointed as executor of the first will of Hugh of Wells, bp of Lincoln.
[37] See *EEA* iii p.307; *Cal.Lib.R.1226–40* 316.
[1] Cheney *English Bishops' Chanceries* 38,42.

of this routine may have survived into the early years of des Roches. Before 1210 we find four charters with a place/date clause. Of these one original and two cartulary copies are said to have been issued by Peter Russinol *per manum* (nos. 32,34,81). In the original and one of the copies Peter is styled *sigilli nostri custodis*, in the other copy he appears merely as 'our clerk'. Whatever Russinol's precise duties as keeper of the bishop's seal there is no consistency to the eschatocol of the three texts said to have been drawn up by his hand; one carries a place/date according to the Roman calendar, with no year (15 October [1205]); another a place/date in the Roman calendar with the pontifical year of des Roches (7 May 1208), and the third neither place nor date (1206 × 1211). Elsewhere Russinol appears as witness to charters, with no mention of his title or function as seal-keeper. Another early charter with a *per manum* clause is said to have been given by the hand of *domini Gregorii capellani nostri*, with place/date clause by day in the English rather than the Roman form, month and pontifical year (no.21, 28 May 1209). Gregory appears elsewhere as bishop's chaplain in the witness lists to half a dozen of des Roches' acta (nos.1,11,33,56,72,83), all to be dated before 1227 and probably prior to 1220. In the Winchester pipe rolls he appears performing routine commissions associated with the bishop's estates.[2] It is tempting to identify him as Gregory the chaplain, witness to more than twenty charters of bishop Godfrey de Lucy from *c*.1190 onwards.

Between 1209 and 1234 the *per manum* clause vanishes from des Roches' charters. Several of the acta of this period are precisely dated, although in general undated preponderate over dated acta. Thereafter, between January 1234 and March 1237, we find three charters, all originals associated with Selborne priory, said to have been issued by Peter de Chanceaux *per manum*, with place/date according to the Roman or the church calendar, and the year of the incarnation (nos.48–9, 52). Throughout this period the bishop continued to issue charters, some of them carrying place/date, without the clause *per manum*. On his deathbed, on 4 June 1238, des Roches granted a final series of charters, two of them issued by Peter Russinol *per manum*, the third witnessed by William the clerk, parson of Baughurst, *presentium scriptore* (nos.53–4, 84).

None of this evidence is particularly illuminating. Peter Russinol I, the bishop's seal keeper before 1210, died *c*.1219 and must therefore be distinguished from a namesake, Peter Russinol II, who is said to have issued charters in 1238. Were it not for their appearances in the dating clauses to the bishop's charters, we would have no indication from external sources that any

---

[2] *PR 4DR* 6,8; *6DR* 65,70–1,95,97,124,155,157,186.

of the men named above was active in the bishop's chancery. William, parson of Baughurst (*Bagehurst*), is otherwise unknown. Peter Russinol I and II, and Peter de Chanceaux all appear regularly in royal or episcopal service, but as administrators, bailiffs or clerks involved with matters financial, rather than as scribes or seal keepers.[3] The clause *per manum* is itself notoriously ambiguous. There seems little likelihood that it was used to designate the draftsman or scribe of a charter. Probably it implies no more than general oversight. A comparison between the originals issued by Peter de Chanceaux *per manum*, suggests that at least two and possibly three different scribes were actually employed in the writing. Between August 1213 and February 1214 we find a large number of royal charters said to have been issued by bishop des Roches *per manum*.[4] A charter of Thomas Basset of much the same period, concerning the marriage settlement of Reginald de Vautort, the bishop's ward, is witnessed *domino Petro Wintoniensi episcopo per cuius manum hec facta sunt*.[5] It seems certain that the bishop played little or no part in the drafting or writing of these documents. His name was singled out merely as that of the courtier entrusted with general oversight of affairs at the royal chancery, or in the case of the Basset charter as that of the man who had specified the basic terms of agreement.[6]

Equally mysterious is the title 'keeper of the bishop's seal' applied before 1210 to Peter Russinol I. Russinol was a Frenchman, like his namesake, Peter Russinol II, and like Peter de Chanceaux. It is possible that his title is also French, although no parallel has yet been found in the household of any contemporary bishop, English or French. The first text issued by Russinol as seal keeper is dated at Rome, in October 1205 (no.81). No original survives, though one would dearly like to know how the document was sealed. Had des Roches already commissioned a seal as bishop elect before leaving England;

---

[3] Appendix 4 nos.18–20. The Latin version, *Cancellis*, of the place name Chanceaux, has in the past been misread as if it were derived from *cancellaria* or chancery: Swift 'Manorial Administration' 81–2.

[4] *Rot.Chart.* 194b-96b; *Foedera* 118; *Norwich Charters* i no.39.

[5] Exeter, Devon County Record Office ms. Courtenay of Powderham 1508M/TD51 (Courtenay cartulary) pp.155–6, noticed in *HMC 9th Report* part ii (1884) 404b. Des Roches' name and the *per manum* clause appear at the head of the witness list. The charter is to be dated between 13 October 1206, the date of des Roches' acquistition of the Vautort wardship, and May 1221, the death of Thomas Basset, whose daughter Joan was married to the bp's ward. It probably dates before Vautort's coming of age c.1215 (*RLC* i 74b; *RLP* 159b,162b; *E.Rot.Fin.* i 48).

[6] Walter de Gray, the titular chancellor, was pre-occupied in the last months of 1213 with missions to Flanders. Richard Marsh, the senior chancery clerk, was absent in Rome, defending his part in the papal interdict: *RLP* 103,105; *Foedera* 113; Cheney *Innocent III and England* 69, 347. For the clause *per manum*, papal in origin, borrowed after 1189 by the royal chancery, see Galbraith *Studies in the Public Records* 126–30; Cheney *English Bishops' Chanceries* 84–90.

was Russinol keeper of an episcopal seal especially made in Rome, or of the private seal des Roches had used before consecration?

Beyond those clerks credited with the issue of particular charters, the witness lists to des Roches' acta record the names of a large number of men styled chaplain or clerk, any one of whom might have been involved in the drafting or writing of letters. Some of these men merit particular attention. Four charters, to be dated 1206 × 1224, are witnessed by the same three clerks, Richard of Barking, Denis (de Bourgueil) and Richard of Elmham, clumped together as a group but in no particular order (nos. 26,38,45,83). Of the three, Richard of Barking has been identified elsewhere as a scribe of the royal exchequer, responsible, between 1217 and 1228, for the writing of the royal pipe rolls. Almost certainly he was a satellite of master Eustace de Fauconberg, himself a protégé of des Roches.[7] Master Eustace appears as witness to three of the four charters of des Roches witnessed by Barking, whose activities at the royal exchequer coincide with Eustace's term of office as royal treasurer. After 1221 Barking also witnesses Eustace's acta as bishop of London. In the present context it is particulary interesting to note his appearance in des Roches' charters alongside Denis the clerk. Active as an officer of the bishop's chamber from at least 1208, supervising matters financial, Denis served between 1218 and 1226 as *de facto* treasurer of the bishop's exchequer at Wolvesey. As such he is likely to have been entrusted with oversight of the episcopal pipe rolls, though not apparently with the duty of writing them.[8]

Any one of these men would have been capable of writing writs or rolls, although their chief sphere of activity appears to have lain in the bishop's chamber and exchequer rather than in anything resembling an episcopal chancery. The association between Denis and Richard of Barking, officers respectively of the royal and episcopal exchequers, does much to explain the speed with which reforms introduced to the Westminster exchequer were transmitted to Wolvesey. In the year 1220-1, for example, a system was introduced more or less simultaneously, at Westminster to date the *communia* section of the memoranda rolls, and at Wolvesey to date the individual manorial accounts audited by the bishop's exchequer.[9] Once again we are reminded of the innumerable points of contact between des Roches and the court. As chief justiciar in 1214-15 the bishop inherited various clerks who had worked under his predecessor, Geoffrey fitz Peter. It seems likely too that

---

[7] Appendix 4 no.28; *PR 5 Henry III* pp.xxiv-vi.
[8] Appendix 4 no.15.
[9] Carpenter *The Minority* 226; *PR 5 Henry III* pp.vi-vii; Ms.16DR *passim*.

he inherited a branch of the royal chancery specially appointed to serve the justiciar in the king's absence. The surviving records of des Roches' justiciarship, the close, originalia and scutage rolls, are composed in styles and hands indistinguishable from those of the royal chancery. The thousand or more writs which we can assume the justiciar to have issued in 1214 alone, must surely have required labour far beyond the capacity of his episcopal clerks and chaplains (below appendix 1). Likewise, after 1217, the official correspondence issued by des Roches as a member of the minority council (nos. 110–120), could have been written either by the bishop's clerks or, on occasion, by officers attached to the royal household.

The Winchester pipe rolls provide only incidental glimpses of the men and materials most significant for a study of the bishop's chapel and chancery. The rolls were never intended to be records of the bishop's household expenses. However there is one item of expenditure which may have a direct link to the chancery; the purchase of wax. In 1224–5, for example, we find the bishop's bailiffs at Southwark buying ten pounds of wax 'for the spigurnel by the bishop's order'. In the same account, a further £21 was spent at Southwark on one thousand units of wax (*i. mill(aria) de cere*)[10]. Five hundred units were bought at Southwark the following year at a cost of £10 and fifteen shillings.[11] Assuming these units to have been pounds, the cost works out at roughly five pence the pound by weight; sixpence a pound in the case of the wax bought for the spigurnel in 1224–5. Such prices tally closely with the cost of sealing wax recorded in the *liberate* rolls of the court, which during the period in question was using more than three pounds of wax a week to seal writs in chancery.[12] Unfortunately, the cost of wax for sealing appears to be almost indistinguishable from that of high-grade candle-wax.[13] There can be no guarantee that the bulk of the wax purchased at Southwark was for sealing rather than for lights. In 1223–4, for example, we find one hundred pounds of wax being carried from Southwark to Bedford, presumably during the siege of Bedford castle, but they are far more likely to have been for candles and tapers than for sealing.[14] Only with the ten pounds of wax intended for the spigurnel in 1224–5 can we be fairly confident that we are dealing with chancery expenses, and even then there is a possibility that the spigurnel in question was not an episcopal officer but the spigurnel of the royal chancery, entrusted with the

---

[10] Ms. 20DR m.15.
[11] Ms. 21DR m.14.
[12] *Cal.Lib.R.1226–40 passim*, tabulated by M.T. Clanchy, *From Memory to Written Record* (London 1979) 43–6, 58–9.
[13] For the cost of candle-wax see *Cal.Lib.R.1226–40* 442, 449.
[14] Ms.19DR m.12.

appending and pressing of the royal seal. In the following year, 1225–6, the Southwark account records a sum of twenty two shillings and four pence paid to Walkelin *de Scriptoria* (*sic*), by William fitz Humphrey, a clerk of des Roches who regularly witnesses the bishop's charters.[15] But a lack of any further reference to Walkelin makes it impossible to judge whether his name implies service in a writing office, episcopal, royal or monastic. All in all, and despite the great proliferation in written records during des Roches' lifetime, we know as little about the chancery of des Roches as we do of the chanceries of most twelfth-century bishops.

THE ACTA: SOURCES AND CONTENT

The bishopric of Winchester possesses records richer than those of many other dioceses. Above all, the Winchester pipe rolls, apparently an innovation of des Roches' years as bishop, provide us with an incomparable wealth of evidence on the manorial economy and the bishop's household.[1] Sixteen such rolls survive from des Roches' thirty four years, the earliest covering the period Michaelmas 1208 to 1209. As a supplement to the records of royal and monastic administration the pipe rolls are invaluable. However, they were originally only one aspect of record-keeping at Winchester, intended as a means by which the bishop could keep a check on his financial affairs. Unlike the royal pipe rolls, those at Winchester were not used for the enrolment of deeds and settlements save in one isolated instance where the roll for 1235–6 recites a quit-claim to bishop Peter from John, son of Robert de la Bere.[2] Beyond the pipe rolls, other capitular and episcopal records at Winchester have suffered more harshly from the ravages of time. By contrast to the situation at Lincoln and York, there is no evidence of registration of episcopal

---

[15] Ms.21DR m.14. For William fitz Humphrey see appendix 4 no.32.

[1] For the rolls see the full bibliography provided by T.W. Mayberry, *Estate Records of the Bishops of Winchester in the Hampshire Record Office* (Hampshire Record Office Pamphlet 1988) 4–6. The question of their origin is dealt with in my paper 'The Origins of the Winchester Pipe Rolls', *Archives* (forthcoming 1994).

[2] Ms.31DR m.10, between the accounts for Hambledon and Farnham. John quitclaims to the bishop Ralph de Wippeley *cum omnibus catallis suis et tota sequela sua*, the bishop paying John 24 shillings. The deed is witnessed by dom. Robert de Walth(am), seneschal, and dom. William de Ho, knights, Matthew de Donemede, William clerk, Roger son of Adam, Robert clerk, William de Ranvill' and Nicholas de Langeress'.

THE ACTA: SOURCES AND CONTENT   xlix

records at Winchester until the years of bishop Nicholas of Ely (1268-1280).[3] At one time there must have been an enormous accumulation of documentary evidence, including charters, indentures, and the mass of individual manorial accounts, court rolls and schedules which underlie the order of the Winchester pipe rolls. Some of this material was later copied into the bishopric registers, or survived together with the pipe rolls as part of the bishop's exchequer archive at Wolvesey, but the vast majority of it has been lost.[4] Today the bishopric charter collection contains less than fifty items earlier than 1500, the earliest dating to the 1250s.[5] Similarly, with the capitular archive at St Swithun's, although several thousand medieval charters survived there into the 1640s, an enormous number were lost or irrevocably damaged during two attacks by parliamentary troops, in 1642 and 1646. An inventory compiled after the first of these attacks, by the then chapter clerk, John Chase, provides some indication of the extent of our losses.[6]

The collection presented here has been divided into three sections, numbered consecutively. The first section, dealing with the documents issued by des Roches as diocesan or landholder, is presented in accordance with the conventions of other volumes in this series, in alphabetical order according to beneficiary or addressee. Eighty five such documents or mentions of documents now lost, have been assembled here, drawn for the most part from monastic cartularies or charter collections. It is these diocesan acta that form

---

[3] Cheney *English Bishops' Chanceries* 148-9; D.M. Smith, *Guide to Bishops' Registers of England and Wales* (London 1981) 203. The reference in *VCH Hampshire* ii 203 to a register of des Roches, is based on a mistranslation. Gilbert White (*Selborne* i 352-3, letter xxiii) claims to have examined a register of Peter des Roches for evidence on Selborne Priory, but it seems likely that the record he consulted, probably at Winchester cathedral, was the register of a later bishop or possibly one of the Winchester cartularies.

[4] The earliest list of the bishopric muniments, entered amongst business for August 1335 in the register of bp Adam Orleton, deserves to be quoted in full: Winchester, Hampshire Record Office ms. 21M65/A/1/6 fo.24v: *Quoddam memorandum munimentorum episcopi Wyntoniensis: Ista sunt munimenta domini episcopi Wyntoniensis, videlicet una carta reg(ia) de castris, hundredis et maneriis totius episcopatus. Item una carta libertatum episcopatus. Item in uno hamperio xxi carte reg(ie), in alio hamperio vii carte reg(ie) cum uno scripto ducis Normannie. Item in tercio hamperio xv carte reg(ie) et una carta de quadam terra in Kynggesclere. In quarto hamperio xviii scripta de compositione inter episcopum et priorem et conventum. In quadam pixide iiii munimenta tangentia terram in Estthorpnatele infra manerium de Farnhame. In quinto hamperio iiii^{xx} viii scripte et munimenta et v fines. Item in quodam coffro xvii carte reg(ie) confirmat(ionum).* The total of 178 documents listed here is surprisingly small. It is odd, for example, that there is no mention of papal letters or privileges. In all, its seems possible that the list deals not with the bishop's archive as a whole, dispersed amongst his various manors or stored at Wolvesey, but with the charters, for the most part kept in baskets, which accompanied him on a day to day basis as he travelled the diocese.

[5] Winchester, Hampshire Record Office mss. Eccles.II 159692,159695-6.

[6] Winchester Cathedral Library ms. The Book of John Chase.

the chief subject of the diplomatic analysis which follows. But beyond his role as diocesan, des Roches was also involved in the issue of letters and writs of a political or general administrative nature, and it is these which form the second of our three sections; fifty or so acta, chiefly drawn from the records of the royal chancery, almost all of them to be dated within fairly narrow limits, and so presented here in chronological rather than alphabetical order. The third section, drawn exclusively from the Winchester pipe rolls, lists the two hundred or more mentions of episcopal writs, now lost, noted as warranty for expenditure within the accounts of the bishop's manors. These too are presented in chronological order, covering a bewildering variety of subjects from the entertainment of huntsmen and clerks, to the purchase of luxuries such as spices and gold, or the garrisoning of the bishop's castles in time of civil war. Thereafter, an appendix lists the records of des Roches' time as justiciar in 1214–15, covering several hundred individual writs, virtually all of them already printed, and in any case, too numerous to give here in detail. A second appendix lists those of the bishop's actions, institutions, presentations, foundations of religious houses, and other such activities likely to have involved the issue of written orders, besides occasional uses of the bishop's seal.

Amongst the diocesan acta we find a typical cross-section of mandates, confirmations and grants. As one might expect of a thirteenth-century bishop, a large number of des Roches' charters involved the renewal of grants and settlements by his twelfth-century predecessors or by private individuals, including the king (nos. 1,2,8,9,11,14,18,21,32,34,36,38,45,51, 53,56,61,64,66,75–7,81,83). Nearly a third of the eighty five acta involve the confirmation of previous awards, occasionally with some degree of revision or re-settlement by des Roches, demonstrating the eagerness with which parties resorted to the diocesan as a guarantor. Of a similar nature are the various letters testifying to the legitimacy of marriages or of ecclesiastical orders (nos.12,22,23), intended to serve as evidence in the law courts or before the pope. Likewise we can assume that letters of institution, admission or appointment (nos.27,28,37,39–41,72,74,80,84–5) were preserved by their beneficiaries chiefly as a means of proving title to churches or other offices.

Of awards and grants newly made in the time of des Roches, we find gifts of pensions or tithes to the religious, occasionally specifying particular uses; wax for lights at Chichester for example, or money for the infirmary of Southwick priory (nos.7,19,35–6,48,59–60,65,74,81). A considerable number of acta concern the bishop's own religious foundations; at Halesowen (no.13), Selborne (nos.48–54) and Titchfield (nos.67–71), or the houses des Roches' helped to refound at Marwell (no.31) and St Thomas' Southwark (nos.56–62).

Several of these (nos.50,54,69) involve the appropriation of churches. The bishop is also known to have appropriated churches to the convents of Lewes (no.26), Newark (no.35) and Wintney (no.83). Here and elsewhere, for example in his letters of institution, des Roches was keen to emphasise the reservation of a vicar's portion. However, in only a few cases (nos.28,70,84) do we have any specific evidence that such vicarages were endowed. In part this may be because the majority of acta survive in monastic cartularies, the religious being less anxious to preserve documents relating to their obligations to vicars than they were to record their own title to churches and tithes. Des Roches' diocesan statutes stress the need for adequate vicarages where a rector was non-resident, and at the time of his election in 1205 the bishop had secured papal letters empowering him to intervene in churches where the vicarage was farmed to a sub-vicar, incapable of serving the parish, or where the patron had established a token payment of pepper or wax to the parson in lieu of a proper stipend.[7] Nonetheless, by contrast to other dioceses, most prominently Lincoln, where the early thirteenth-century witnessed a great spate of vicarage endowments, it appears that Winchester lagged behind the vanguard of reform.[8] Of the appropriations licenced or obtained by des Roches, the grant of the church of Halesowen to the bishop's foundation at Hales preceded the establishment of any vicar's portion by more than fifty years.[9] At St Olave's Southwark, no vicarage was established until after des Roches' death (no.26), whilst the vicarage the bishop ordained at Titchfield seems to have proved inadequate, being augmented twice before the end of the century (no.70).

Judicial business is another area in which a reforming bishop might be expected to make his mark, and from which des Roches is yet again conspicuous by his absence. We have a single tithe settlement devised by des Roches as ordinary (no.6), whilst a mere handful of his surviving diocesan acta were issued in response to papal commissions. Of these only one is a straightforward decision as papal judge delegate (no.5). Of the others, two represent major political settlements involving the see of London (nos.30,73), whilst the third transmits papal instructions for the preaching of the crusade

---

[7] *Letters of Innocent III* nos.646,649.
[8] For the system in general see Cheney *From Becket to Langton* 131-6.
[9] The church was appropriated in 1215 by bp Walter de Gray of Worcester, but there is no record of any vicarage being endowed until 1270 (Nash *Worcestershire* ii appendix xxix; *Monasticon* vi 928 no.5). Walsall church, granted to Halesowen before 1233, was officially appropriated in 1248 by the bp of Coventry and Lichfield who at the same time endowed a vicarage (*Monasticon* vi 927 no.4; *Cal.Pap.Reg.* i 147; *Magnum Reg.Album* nos.595-7; Bodl. ms. Ashmole 1527 (Lichfield cartulary) fos.93v-94r).

(no.47). This is not to say that des Roches was seldom commissioned by the papacy. He was called upon by Innocent III to act against the rebel barons in 1215 (no.100), and to intervene in various political cases, including the Canterbury election and a dispute over the patronage of Glastonbury abbey.[10] Honorius III commissioned him to preach the crusade and addressed several mandates to him on the government of the realm.[11] His relations with Gregory IX appear to have been particularly close, no doubt as a result of des Roches' role as peacemaker between Gregory and Frederick II (below no.130). During the 1230s the bishop was called upon to supervise peace negotiations with France, to punish attacks against Roman clergy in England, to settle a dispute involving the earl of Chester and the convent of St Nicholas Angers, and to preach the crusade.[12] Nonetheless, only a handful of such papal commissions concern des Roches as diocesan rather than as diplomat or politician.[13] The average run of delegated hearings would appear to have been committed not to the bishop but to his subordinates, the episcopal officials and the archdeacons. In general, both at court and in his diocese, des Roches appears to have had no great experience as a justice, nor much of an inclination for judicial work.

Of the other diocesan acta given below, several concern des Roches as private landholder, involved in transactions over wardships or in ransom settlements following the civil war (nos. 3,4,10,25,78). With reference to the system of record-keeping at Winchester, it may be significant that the one surviving indenture known to have involved the bishop, a settlement between des Roches, the Winchester monks and William Brewer, was drawn up as a bipartite cyrograph, one part presumably entrusted to Brewer, the other to the archive of St Swithun's priory. There appears to have been no provision for a permanent record to remain with the bishop, suggesting that episcopal archive keeping had still to attain any degree of sophistication. Amongst the few surviving routine administrative writs, we find three significations of excommunication addressed to the crown (nos.15–17), orders to the bishop's official

---

[10] *Letters of Innocent III* nos.673,703,834,949,1016,1020.

[11] *Reg.Hon.III* nos.50,142,248,705,1105,1325,1612,3297,3312,6157,6160; Sayers *Honorius III* 228 no.20, 238 no.31. This last mandate, although superficially concerned with a matrimonial suit, had wide political ramifications since the marriage in question was that of William de Breauté, brother of Fawkes.

[12] *Reg.Greg.IX* nos.621,1311,1313,1562,1801–3,1804–5; BL ms. Add.35296 (Spalding cartulary) fos. 13r-14r; Wendover *Flores* iv 241; *Pat.R.1225–32* 498. The bp's commission to defend alien clergy underlies *Reg.Greg.IX* no.1234, and probably too the cases involving William Clinton and the church of Massingham referred to in *Cl.R.1231–4* 164,322.

[13] For example PRO SC7/15/34, on Quarr abbey; *Cal.Pap.Reg.* i 160, on the church of Damerham, both dated to 1237.

on the collection of a royal tax (no.46) and letters patent for annual submission at the Wolvesey exchequer, authorising the bishop's keeper there to make payments to Hyde abbey (no.20). Rather surprisingly, des Roches is recorded as having issued a mere two indulgences (nos.24,58), of which only one, a particularly grandiloquent example, survives. Finally, we have evidence of the bishop's will, now lost, composed and revised over the course of more than twenty years (nos.42-3).

The acta touching upon des Roches' role as politician and royal confidant need not detain us here. The majority were issued jointly with other bishops or magnates. They include letters of supplication or testimony addressed to the papacy (nos.86-8,104,121,132), pledges of support for political allies (nos.93-4), writs of account at the royal exchequer (nos.89,102,109,115,120), and a large number of letters involved with domestic politics: the papal interdict (nos.91-2,95-6), Magna Carta and the civil war (nos. 96a-100), and the bishop's part in the minority council of Henry III (nos.105-26).

What does the subject matter of the bishop's surviving acta teach us of des Roches' diocesan administration? There are certain obvious omissions. Nowhere, for example, do we read of the bishop conducting systematic visitations of the religious houses of his see. We have already seen that he may have been lax in his endowment of vicarages, and that personally he took little part in the definition or enforcement of canon law. Despite his establishment of two new cemeteries (nos.5,33), there is virtually nothing on the consecration of churches, or the definition of parish boundaries, suggesting either that such matters were of little concern to des Roches, or, more likely, that the parish network within the diocese had been established some time before his election. Rather surprisingly, in a period which witnessed the spread of manorial chapels, there is nothing in his acta to testify to the establishment of private oratories or the first stirrings of the chantry system.[14] Above all, and in stark contrast both to the acta collections of his predecessors at Winchester and to those of his contemporaries in other monastic cathedrals, we have only one charter of bishop Peter addressed directly to his monastic chapter at St Swithun's (no.81): a confirmation of tithes, issued within a few weeks of his election. This, together with a charter of the 1220s in favour of Chertsey abbey (no.6), is one of only two of the bishop's charters to have been witnessed by monks of St Swithun's.

[14] In November 1231 des Roches was commissioned by the pope to investigate a request by Thomas, lord of Warblington, that he have a private chapel and chaplain for his own estate (*HMC Various Collections* iv 160). Des Roches himself ordered the construction of a chapel on the episcopal manor of Brightwell in 1225-6, at a cost of more than £3: Ms.21DR m.6d.

THE BISHOP AND THE MONKS

The chroniclers tell us that the Winchester monks considered des Roches a hard task-master; 'hard as rocks' (a pun upon Peter's name).[1] Various papal mandates suggest that relations between monks and bishop were by no means entirely peaceful.[2] In 1224 and again a decade later, the monks faced attacks upon their property as a result of the political vicissitudes which befell des Roches.[3] On the first of these occasions the monks went so far as to obtain a confirmation of their estates from des Roches' political rival, archbishop Langton; an intervention by the metropolitan which is unlikely to have been welcome to bishop Peter.[4] Elsewhere, we learn that des Roches attended, perhaps that he supervised, the profession of monks of St Swithun's, and that he commanded the appointment to various lay offices within the convent.[5] Both of these subjects, profession and appointment to conventual offices, were to prove the subject of serious dispute between St Swithun's and the bishops later in the thirteenth-century.[6]

At the same time, it would be foolish to regard des Roches as an outright enemy of the monks. At Winchester he is known to have carried on the building of the cathedral retrochoir initiated by bishop Godfrey de Lucy.[7] He appears to have done much to foster the cults of Swithun and Birinus, the cathedral's two principal saints.[8] Although there are signs of strain in his relations with St Swithun's, these are as nothing compared to the hostilities between bishops and monks that had marred the pontificates of his predecessors, Henry of Blois and Godfrey de Lucy, and which were to break out again at Winchester in the 1260s. Whenever a secular bishop and his household came up against an entrenched, conservative community of monks, there was almost certain to be friction, often violent friction, over the distribution of estates, over jurisdiction and over the patronage of churches.

---

[1] *AM* i (Tewkesbury) 110.
[2] *Letters of Innocent III* no.1096; *Reg.Hon.III* no.2022; *Reg.Greg.IX* no.2686.
[3] Below appendix 2 nos.25,39.
[4] Winchester Cathedral Library ms. Book of John Chase (Muniment list *c.*1640) fo.4v: 1224, *Nota carta confirmationis archiepiscopi Cant' priori Winton' edit' cum munibus (sic.)*.
[5] Cambridge, Gonville and Caius College ms.123/60 (Formulary) fo.96v: *Ego frater Gervasius subdiaconus promitto stabilitatem meam et conversionem morum meorum et obedientiam secundum regulam sancti Benedicti coram domino et omnibus sanctis eius in hoc monasterio quod constructum est in honore Petri et Pauli et sancti Swythuni confessoris, in presentia domini Petri episcopi*. For appointments by des Roches see below no. 41, appendix 2 nos.46-9.
[6] *Cal.Chart.R.1257-1300* 287-8.
[7] P. Draper, 'The Retrochoir of Winchester Cathedral', *Journal of the Society of Architectural Historians of Great Britain* xxi (1978) 1-16; Vincent 'Des Roches' 80.
[8] *C.& S.* 127-8; *Henry of Avranches* 123-5; Vincent 'Des Roches' 315-20.

Yet des Roches' time at Winchester saw nothing to compare to the struggles which afflicted his contemporaries at Canterbury, Coventry and Durham.

Elsewhere, away from Winchester itself, he showed himself a fervent patron of the religious orders, particularly of the Cistercians and Premonstratensians. Altogether ten houses were founded or principally sponsored by des Roches: the Premonstratensian abbeys of Halesowen (1215) and Titchfield (1233), the Augustinian priory of Selborne (1234), the two Cistercian abbeys of Netley and Clarté Dieu founded under the terms of his will; the Dominican house at Winchester (before 1225); the college of chaplains at Marwell (re-endowed and provided with new rules 1227); and the hospitals of Portsmouth (founded by William of Wrotham c.1214); St Thomas' Southwark (refounded by des Roches and Amicius archdeacon of Surrey after 1212) and St Thomas' at Acre (refounded by des Roches c.1228). This is a quite remarkable total, all the more remarkable given that the early thirteenth-century is generally regarded as a period of decline in new monastic foundations. Des Roches founded more new abbeys and priories than were established by his three sovereigns, Richard I, John and Henry III, combined.[9] The favour he showed towards the Premonstratensian and Cistercian orders may well have been an inherited, family trait, shared with his kinsman, William des Roches, seneschal of Anjou, founder of the Cistercian abbey of Bonlieu and leading patron of the white canons of Perray Neuf.[10] The houses bishop Peter established in England share various other common features. Most relied heavily upon grants of land from the crown. Halesowen, Titchfield and Selborne were all sited on manors granted to des Roches by the king. Often their endowment included encumbered land; estates mortgaged to the Jews and hence bought out cheaply by des Roches; land which had long been the subject of complicated litigation, or which had escheated to the crown as the property of Normans who after 1204 defected to the French.[11] Such *Terra Normannorum* played a major role at Titchfield and Selborne.[12] Together with Netley, Titchfield and Selborne also benefited from the bishop's purchase of land or spiritualities from Norman religious houses, keen to divest themselves of their English estates and therefore willing to settle with the bishop at what may have

---

[9] A statement which excludes the large number of Dominican and Franciscan houses established during the reign of Henry III, in whose foundation the king played a leading role.
[10] *Gallia Christiana* xiv cols. 518,520,530,538–40; Vincent 'Des Roches' 15.
[11] For encumbered land, much of it mortgaged to the Jews, see *Selborne Charters* i 6, ii 46,51; *Monasticon* vi 933–4; *VCH Hampshire* iii 46–7,59–60; D. Le Faye, 'Selborne Priory 1233–1486', *Proceedings of the Hampshire Field Club* xxx (1973) 51; Meekings 'Netley Abbey' 12.
[12] Below no.67, appendix 4 nos.4,38.

been relatively low prices.[13] Most of the bishop's foundations were of middling wealth; none, so far as can be established, was under-endowed. However, the expense involved in proving title to encumbered property was to ensure that at least one of them, Selborne in Hampshire, experienced grave financial difficulties even before des Roches' death.

Vanity, a desire for posthumous recognition and a keen eye for a bargain, all played a role in des Roches' patronage of the religious. In general he seems to have spared less thought for the long-term well-being of his foundations than for the immediate kudos to be gained from their establishment. Pilloried by his enemies as an unregenerate curialist, the bishop may have regarded religious patronage as a means to improve his standing in the eyes of hostile contemporaries. His acta are remarkable for the extent of their quotation from scripture, borrowing passages from the gospels, the psalms and even the works of St Gregory the Great, almost as if the bishop were anxious to defy the worldly, militaristic reputation that clung to his name. By the same token, as founder of almost a dozen religious houses, des Roches might hope to go down in history as a pious benefactor rather than a hard-hearted statesman. None the less, we should be wary of treating his piety as mere window-dressing. Other examples spring to mind; Walter of Merton, Edmund Gonville, even Henry III, all of them patrons who could combine worldliness, a keen business sense and genuine piety in near-equal measure. Towards the end of his life, des Roches is known to have petitioned the Cistercian order for licence to have a Cistercian monk and two lay brothers living with him.[14] The care shown in his rule for Marwell (no.31) for the *minutiae* of liturgical observance, including the first known appearance outside Salisbury of the Sarum rite, suggests that the bishop's interest in the religious extended beyond a selfish desire for prestige. The Marwell rule is also eloquent upon the advantages of the common life. To live in common the chaplains and their prior were enjoined to follow strict instructions upon the distribution and accounting of their income. Here, as in all things, des Roches the man of piety is almost indistinguishable from des Roches the man of business.

THE BISHOP AND REFORM

Patronage of the established religious orders might well be viewed as a conservative impulse at a time when the church was turning away from the cloister to the needs of the world at large. We have seen already that des Roches can hardly be numbered amongst the vanguard of reforming bishops.

---

[13] Meekings 'Netley Abbey' 5–6,10,14–15,18–20; below nos.25,48, appendix 4 no.4.
[14] Meekings 'Netley Abbey' 1.

However, by the same token he should not be dismissed as an unregenerate curialist, oblivious to the initiatives undertaken by his reforming contemporaries. Several of the religious houses he helped to found – the hospitals at Portsmouth, Southwark and Acre, and the Winchester Blackfriars – answered to the needs of towns and cities with shifting, transient populations. The bishop appears to have responded to the church's urban mission as keenly as he sought to foster contemplative or secluded orders. His diocesan legislation is the earliest specifically to mention the new order of Dominicans, empowering them to serve as confessors throughout the diocese.[1]

In general, a distinction needs to be drawn between the intention and the practical application of much of the diocesan legislation of thirteenth-century England. We have seen already that des Roches' stated policy in respect to vicarages appears to have born little fruit in practice. By the same token, his diocesan statutes are unusual in omitting any reference to ordinations and in particular to the stipulation, increasingly prominent in reformist circles, that clergy, above all vicars, proceed to the orders of priest.[2] Admittedly des Roches' statutes imply that the parish clergy were discharging priestly functions, since they have much to say on the correct celebration of the mass.[3] They also require archdeacons or their officials to draw up lists of church ornaments. From 1232 inventories of service books, vestments and plate are to be found incorporated into the pipe roll accounts for the bishop's manors of Farnham and Calbourne, though more, one suspects, from a desire to catalogue the bishop's temporal possessions than through any reformist zeal.[4] In 1205 des Roches had obtained papal letters, enjoining all clergy, and especially archdeacons and deans, to take the orders appropriate to their offices within six months of monition.[5] Yet des Roches himself was only a deacon or subdeacon at the time of his election to Winchester, and it is a striking fact that not one of his clerical familiars is known to have advanced

---

[1] *C.& S.* 133 no.45.
[2] In general see J.R.H. Moorman, *Church Life in England in the Thirteenth Century* (Cambridge 1945) 223–5
[3] *C.& S.* 126–7 nos.3–7.
[4] *C.& S.* 128 no.13; Ms.28DR mm.8d, 16d. At Farnham the inventory specifies a missal, a temporal (that part of the breviary dealing with seasons and movable feasts), a gradual, an antiphoner, a troper, two (?) *text' de cupere*, a ?psalter (*salterium*), and a priest's vestments, all said to be kept "in the chapel". At Calbourne there were a silver chalice, two lectionary books (*libr' lectionar'*) comprising a missal and a breviary, and a set of painted or florid vestments. These details are repeated in the accounts for 1235–6 and 1236–7: Ms. 31DR m.7d,10d, which gives *sext'*, rather than *text'*, *de cupere*; Ms. 32DR mm.3,9d, which refers to *vestimentorum pictexico(rum)*, rather than *pudrefe'* or *pictrefact'*, which might otherwise be taken to mean that the vestments at Calbourne were damaged by rot.
[5] *Letters of Innocent III* no.645, in accordance with chapter 3 of the Third Lateran Council.

beyond minor orders.[6] Amongst the archdeacons this is only to be expected. As their title implies, the archdeacons were in theory the leaders of the deacons. Even so, master Hugh des Roches, archdeacon of Winchester, appears to have been only a subdeacon at the time of his death.[7] Elsewhere, although we may suspect that the bishop's chaplains were priests, it is surprising to find no specific reference to any ordination beyond the diaconate.

In other respects des Roches was more assiduous in his practical application of reformist principles. His legislation outlaws the holding of scotales, and we duly find that scotales vanish both from the episcopal estates and from the royal demesne during the years of Henry III's minority, years in which des Roches commanded great influence over the king.[8] By the same token, the elaborate nature of the bishop's one surviving indulgence (no.58), may reflect a clause of his diocesan legislation, prohibiting those seeking alms from preaching any sermon save to expound whatever indulgences they might possess. The intention here appears to have been to prevent fraudulent or exaggerated promises being made to those from whom alms were solicited; an intention well served by des Roches' indulgence, which could be read as a self-contained sermon, its message so clear as to rule out the need for extempore preaching.[9]

As royal councillor, in 1219 des Roches witnessed instructions to every justice in eyre, forbidding them to conduct ordeals by fire and water in accordance with the ban placed upon such ordeals by the Fourth Lateran Council.[10] Following the Council's decrees on penance, we find an officer known as the bishop's penitentiary introduced to des Roches' household.[11] Already, even before 1215, the bishop appears to have co-operated in an

---

[6] Des Roches was in legitimate orders at the time of his election to Winchester, but he was ordained priest by the pope: Migne *PL* ccxv col.672; *Letters of Innocent III* no.664.

[7] In general see C.N.L. Brooke, 'The Archdeacon and the Norman Conquest', in *Tradition and Change. Essays in Honour of Marjorie Chibnall*, ed. D. Greenway, C. Holdsworth and J. Sayers (Cambridge 1985) 3-4. For the orders of des Roches' clerks see below appendix 4 nos.3,7,10,27,29. Master Luke des Roches, archdeacon of Surrey, held the title of papal subdeacon, although at his death he willed a set of priests' vestments (*unum par vestimentorum sacerdotalium*) to Moses his clerk: PRO E210/11304 and see below, appendix 4 no.13.

[8] *C.& S.* 135-6 nos.63-4; *RLC* i 436b; Vincent 'Des Roches' 211-2.

[9] Below no.58, and see *C.& S.* 352-3,623,722,1043 for attempts to prohibit fraudulent or exaggerated claims being made to the faithful.

[10] *Pat.R.1216-25* 186.

[11] *WC* no.470; *Reg.Woodlock* 638; *C.& S.* 132-4 nos.42,46-7,49; Cheney *From Becket to Langton* 148n.

attempt to transfer royal markets to days other than the Sabbath.[12] On a darker note, whilst prepared to exploit Jewish money lending, both for personal profit and for the endowment of his various new foundations, des Roches did much as king's minister after 1231 to impose stricter regulation upon the Jews, supervising the imposition of heavy tallages and the first royal legislation specifically to incorporate the anti-Jewish decrees of the Lateran council.[13]

No-one could claim that Peter des Roches was a "reforming bishop" in the mould of archbishop Langton or bishop Richard Poer of Salisbury. Yet at the same time it would be foolish to speak of him, as commentators have spoken of him in the past, as an unregenerate curialist, a bishop in little more than name. With typical perception, Christopher Cheney long ago warned against any attempt to divide the bishops of the twelfth and thirteenth centuries into mutually opposed camps of reformers and curialists: "No clear-cut division produced two well-organized parties of churchmen: still less could one distinguish between a 'good' group of reformers and a 'bad' group of unregenerate maintainers of abuse".[14] No doubt Cheney was thinking of the distinction, so often yet so falsely drawn, between the "reformer" Stephen Langton and the "curialist" Hubert Walter. But, as always, where Cheney led we would be wise to follow. As he himself said of archbishop Hubert, so we could say of Peter des Roches: He knew his Angevin masters well. He had the sort of qualities they appreciated. While he lived he held *regnum* and *sacerdotium* in equal balance. Then, as now, the Church required the service of sinners as well as of saints.[15]

THE ACTA: EXTERNAL FEATURES

Of the more than 300 acta listed below, the majority consists of texts now lost. In all we have only 61 surviving texts of the 85 diocesan acta listed here, and only 25 of the 48 acta issued as royal official or confidant; making for a total of 86 texts in all. Twenty eight of these texts are represented by 31 originals, three of the texts for Selborne priory being preserved in duplicate originals (nos. 48, 52, 54).

---

[12] J.L. Cate, 'The Church and Market Reform in England during the reign of Henry III', in *Medieval and Historiographical Essays in Honor of James Westfall Thompson*, ed. J.L. Cate and E.N. Anderson (Chicago 1938) 27–65; Vincent 'Des Roches' 105, 210–11.
[13] Vincent 'Jews, Poitevins and the Bishop of Winchester' 119–32.
[14] Cheney *From Becket to Langton* 12.
[15] Adapted from Cheney's remarks in *From Becket to Langton* 41, and see *Ibid.* 32–41 for an extraordinarily perceptive sketch of Hubert Walter, much of which could be applied, word for word, to Peter des Roches.

FORGERY

In the entire collection there are only three items which are spurious or whose authenticity is seriously in doubt; no.55, now lost and which may never have existed; no.133, which may have been invented by the chronicler, Roger of Wendover, as propaganda against the bishop's political regime of the 1230s, and no.63 which is an outright forgery by the canons of St Mary Overy, Southwark. The witnesses to this last charter are invented, whilst historically and diplomatically it stands out as a late production, probably of the fourteenth century. Beyond these three distinctly dubious texts, some degree of suspicion must hover over the large number of originals from the archive of Selborne priory (nos.9, 48–54). The canons of Selborne possessed at least one forged charter of Henry III (no.48 B). The text of des Roches it recites appears to be genuine, although the fact that the two surviving originals of this text carry seals that are badly chipped around their tags, leaves room for suspicion. In general the witness lists, the language and the uncontroversial nature of the awards they recite, proclaim the Selborne charters as genuine. Certainly their authenticity has never before been questioned. Most of them carry genuine impressions of the bishop's seal, and on the whole pass muster; although a botched dating clause for one (no.54) and various oddities of phraseology, for example the appearance of the bishop's name in full, *Petrus* rather than the initial *P.*, in the *inscriptio* to no.52, and the peculiar sealing of no.54, all of them problems which are detailed below, strengthen the possibility that parts of the archive may have been 'retouched' at some time in the later thirteenth-century. It is unfortunate in this respect that nearly a third of our surviving originals are drawn from the Selborne archive. As will become apparent, the twelve Selborne charters are the work of only 3 or 4 scribes. It is suspicious that none of these scribes can be shown incontrovertibly to have worked on other, uncontested originals of des Roches, although this in itself is not conclusive evidence that the Selborne texts were forged. In terms of their subject matter, dating and witnesses, the Selborne charters can probably be accepted as genuine. Their diplomatic and palaeography, however, leave them open to a suspicion of later reworking.

THE BISHOP'S SEAL

Of the thirty one surviving originals, twelve preserve impressions of the bishop's seal and counterseal, of which six are in green and six in red or red-varnished wax. In addition, one charter retains a collection of wax crumbs, inaccessible within a seal bag (no.18), whilst another (no.74) bears traces of green wax on its tag. One original, described in the eighteeenth century but since lost, is said to have been sealed in white wax (no.107). A cast, taken from an unidentified original, is now BL Doubleday cast E69, and is included in W.de Gray Birch, *Catalogue of Seals in the Department of Manuscripts in the British Museum*, vol.1 (London 1887) no.2247.

 The seal itself is vesica-shaped, approx. 80 × 47 mm. Obverse: the bishop standing on a corbel, full length, dressed in cope and mitre, his right hand raised in blessing, a crozier in his left hand, the head of the crozier pointing inwards. Legend, embossed around the edge: +PETRUS DEI GRATIA WINTONIENSIS EPISCOPUS. Counterseal: two saints (Peter and Paul) with haloes, the saint on the left holding a key, the saint on the right holding a sword. Beneath them the figure of a bishop in mitre, presenting them with a gift. Legend, embossed around the edge: +SŪT M' SITQ BONI PETR' PAUL'QZ PATRONI (*Sunt Mihi Sitque Boni Petrus Paulusque Patroni*). The same legend appears on the counterseal of the second seal of Richard of Ilchester, bishop of Winchester (1173/4–1188), which in general appears to have provided the model for des Roches' seal, in preference to the seal of his immediate predecessor, bishop Godfrey de Lucy.

SEALING METHODS AND FORMAT

Two originals, both dating from before 1211, are sealed on coloured cords passing through two eyelets, punched through a fold at the foot of the document, the cords being knotted below the eyelets (nos.9,32). A third (no.73), issued jointly with other papal judges delegate in 1222, is sealed in a similar fashion on coloured cords. Another original, a polled cyrograph, has been trimmed at the foot and therefore defies a proper description. Of the rest, thirteen, all of them apparently sent as letters patent, are sealed on a parchment tag passing through a single set of slits cut through a turn-up at the foot of the document. The remaining fourteen, all but two of them addressed to the king or to individual correspondents, are sealed on a parchment tongue cut from the main part of the foot of the document. In most cases the tongue has been torn away, leaving no more than a step at the left hand corner of the foot. One (no.127) also carries traces of a step at the top right-hand corner, perhaps for a wrapping tie. Several of these, for the most part private letters, may have been sent as letters close. The one anomaly (no.54 A[1]), comes from the archive of Selborne priory: a licence for appropriation, with general address, apparently sealed on a tongue, the tongue and seal now missing.

 All but one of the originals are written on pieces of parchment broader than they are deep, the narrowest being 118 mm. (no.111) and the broadest 360 mm. (no.73). In depth, they vary between 23 mm. (no.118) and 300 mm. (no.73). One, exceptionally lengthy text (no.100), measures slightly more in depth than it does in width. In general, the impression is one of uniformity, both in sealing and measurement, implying that for the most part the bishop's charters were produced from within an episcopal chancery, rather than drawn up by outside beneficiaries, to be presented merely for the application of the bishop's seal. At the same time, various quirks of drafting, for example the similarities between the texts produced by des Roches for Mottisfont and Selborne priories (nos.34, 48–50), and the fact that various of the seal tags used in the Selborne charters appear to be cut from earlier drafts in Selborne's favour, suggest that des Roches' clerks had recourse to the archives of the bishop's beneficiaries both for forms and materials.

SCRIBES AND SCRIPTS

Whereas for most twelfth- or thirteenth-century episcopal chanceries the low survival rate of originals makes palaeographical analysis difficult if not impossible, for the chancery of bishop des Roches we have what amounts almost to an embarassment of riches: 31 original charters, besides 16 pipe rolls each of 10 or more membranes, drawn up in des Roches' treasury at Wolvesey. Beyond this, we can refer to the records of the the royal chancery and exchequer where des Roches was so active, besides the charter collections of the bishop's foundation at Selborne and the capitular archive at Winchester. In fact, so overwhelming is this wealth of evidence, that it is impractical to attempt more than a brief analysis of the acta and the bishop's pipe rolls, seeking in particular to identify the work of individual scribes. Even here we must tread warily. Hands may change or be deliberately disguised to accommodate a 'house style'. One scribe may write in several, superficially unrelated styles, whilst hands which to one observer may seem identical, to another may appear entirely distinct. That said, we may begin with a brief examination of the Winchester pipe rolls, the work of the bishop's treasury.[1]

As yet no serious palaeographical study has been made of the Winchester rolls, perhaps not surprisingly given that for des Roches' episcopate alone there survive nearly 200 membranes, each measuring two feet or more in length, covered both sides in a neat business hand. Over the thirty four years of des Roches' episcopate, at least four men served as *de facto* treasurers of

---

[1] In what follows I am extremely grateful to Professor Robert Patterson, who was kind enough to look through the collection on my behalf. In the process he saved me from numerous errors of attribution. However, as Professor Patterson would be the first to point out, the analysis below owes all too little to his meticulous eye for detail.

Wolvesey (appendix 4 nos.14-17). However, as with the royal exchequer, there seems little reason to suppose that it was the treasurer who actually wrote the pipe rolls. An admittedly superficial analysis suggests that although a "house-style" appears to have been adopted, and although individual rolls are generally written in a single hand, a large number of scribes was employed over the course of des Roches' years as bishop. Throughout, the accounts are heavily abbreviated. Manorial headings, and the names of manors supplied at the foot of each membrane, are written for the most part in capital letters which seem deliberately to mirror the style of the royal pipe rolls, a style which may derive ultimately from the lettering of the Domesday survey and which remains consistent throughout des Roches' years as bishop.[2] The margins supply titles for particular sections; *lib(eratio)*, *purch(asia)*, *expensa* and so on, which exhibit a marked uniformity both in form and script across the entire series of rolls between 1209 and 1237. Within the main body of the accounts, the capital letter *I* with tail looped to the left, and the capital *S* in the abbreviation *S(umma)* retain a characteristic form, as does the backward leaning *d*, especially noticeable in the abbreviation *Id'* for *idem*, where the upper stroke of the *d* trails backwards through the upright of the capital *I*. The superscript tittle *s*, generally regarded as a papal style, appears frequently in the very earliest roll, for 1208-9, although only seldom thereafter. Letters such as *x* representing the number 10, *vi* for 6, and the lower case *l* in *liberatio* with a bulge half way up the stem, remain consistent throughout. Yet, although there is no change of hand under des Roches comparable to that which occurs with the first of the rolls of his successor, William de Raleigh, and although all of des Roches' accounts appear to have been cast within a distinct "house style", they are the work overall of several, perhaps as many as a dozen individual scribes. It should also be noted that the hand of des Roches' accounts, save in the matter of capital headings, bears little relation to the script of the royal pipe rolls.

It would be interesting to discover the scribes of the Winchester pipe rolls at work on des Roches' charters, but as yet no evidence has emerged of any such cross-over between the functions of the bishop's treasury and chancery. The pipe roll hand is closest to that of nos.74 and 100 below, written within a few years of one another but in all likelihood by two distinct scribes. No.74, for Westminster abbey, employs the tittle *s* abbreviation. Like no.100 it displays a

---

[2] For suggestions as to the derivation of the royal pipe roll script see A.R. Rumble, 'A Domesday Postscript and the Earliest Surviving Pipe Roll', in *People and Places in Northern Europe 500-1600. Essays in honour of Peter Hayes Sawyer*, ed. I. Wood and N. Lund (Woodbridge 1991) 123-30.

tendency for the letters *p*, *r*, *s* and *x* to be looped and decorated below the line. Both scripts have a backwards leaning *d*, especially exaggerated in the case of no.74. As with the variations between these two charters, it is difficult elsewhere to distinguish changes brought about by the passage of time from real distinctions between hands. The best that we can do is to distinguish five or six scribes, with suggestions as to others of the charters which they may possibly have written.

SCRIBE 1 wrote at least two of the Selborne charters, nos.51 and 53, distinguished by the formation of the letters *f*, *l* and *s* and the tironian *et* decorated with four vertical strokes. Allowing for the passage of time, he may also have written no.9, a Selborne charter from twenty years earlier, which makes extravagant use of the tittle *s*, and yet another charter for Selborne, no.52 A². All four texts display the same splitting at the top of the vertical stem in the letters *b*, *d* and *l*. One of these charters (no.53) is dated by the hand of Peter Russinol, another (no.52 A²) by the hand of Peter de Chanceaux.

SCRIBE 2 wrote the three Selborne charters no. 54 A¹ and A², and no.48 A², all of which are in a rounded, rather untidy business hand. Again, these texts appear under the names of both the dataries, Russinol and Chanceaux.

SCRIBE 3 wrote the two book-hand or formal business-hand charters for Selborne, nos.48 A¹ and 50, which could be related to the work of scribe 1 above. He may also have written the other surviving book-hand charter no.49, alhough there are distinct differences here in the formation of certain letters, for example *d* and *p*. No.49 is the only one of des Roches' charters to make use of the letter 'thorn' which appears in the witness list to designate the *th* in *Swithuno*.

SCRIBE 4 wrote nos. 118 and 119, letters of 1220 to the legate Pandulph in a standard business hand.

SCRIBE 5 wrote the two significations of excommunication, nos.15 and 16, again in a hand reminiscent of that of the royal chancery.

SCRIBE 6 wrote the two late texts, nos.3 and 40, dated within a few months of one another and distinguished by the looped *d* of *divina* and the ligatures *st* and *tr*.

THE ACTA: INTERNAL FEATURES

The language of des Roches' charters follows the general principles established for the chanceries of all early thirteenth-century bishops, the only unusual features being the extent of scriptural quotation employed in the *arenga*, and the form of the dating clause. Given the bishop's close association with royal government, it is also intriguing to note the absence (one might

SCRIBES 1 AND 2

No. 51

Master and Fellows of Magdalen College, Oxford

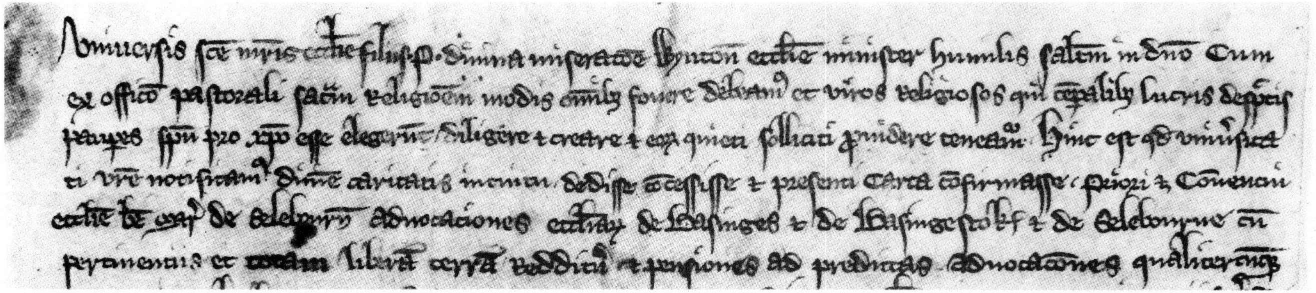

No. 48 (A²)

Master and Fellows of Magdalen College, Oxford

PLATE I

SCRIBE 3

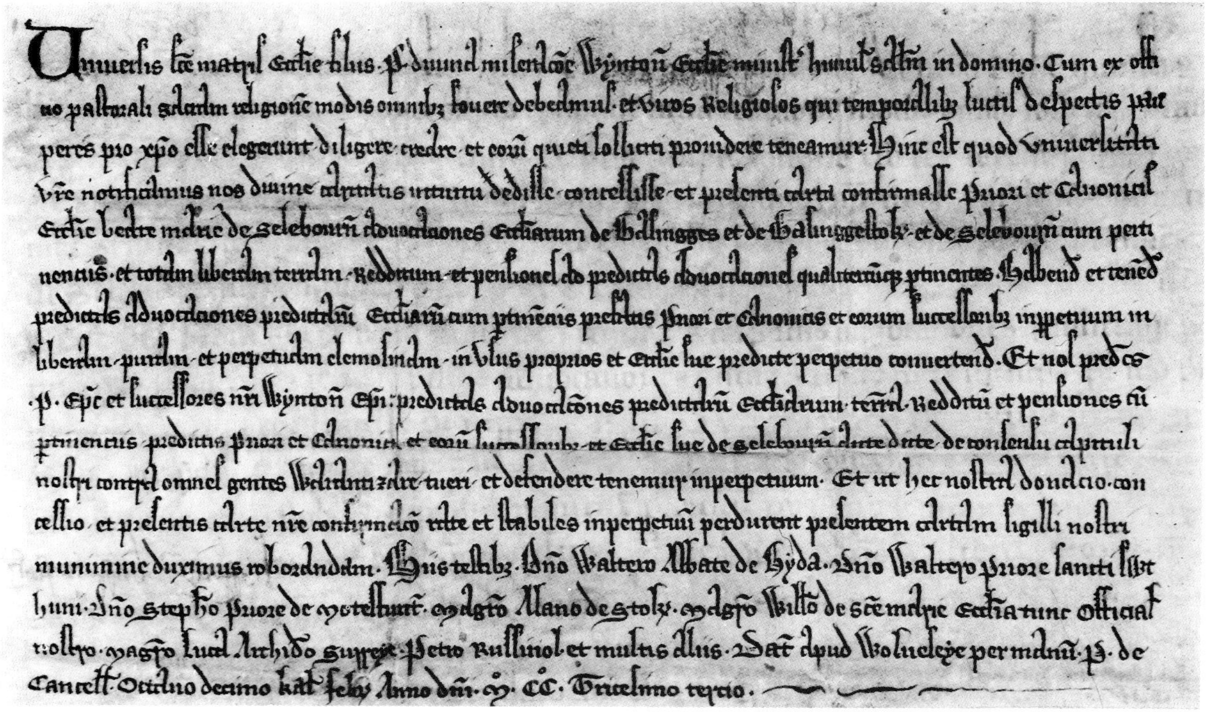

Omnibus sancte matris ecclesie filiis ad quos presens scriptum peruenerit humilis abbas et conuentus de pastura salutem in domino. Cum ex officio pastorali salutem religiosis mobis omnibus, fauore sedebimus, tanto amplius eorum temporalibus tuen debemus prouidere patronis pio zelo esse diligentur diligere tenentur. et eam quam salutare proponere tenemur. Hinc est quod uniuersitati uestre notum facimus nos diuine pietatis intuitu dedisse concessisse et presenti carta confirmasse priori et conuentui ecclesie beate Marie de Sellebourn donationes et terras de Billinges et de Ballingelok, et de Sellecompton pertinentiis. et totam illam terram Reddituum et pertinenciis ad prioris donationem quibusque pronomas Silbenis extentis predictis donationes predictam ecclesiam cum pertinentia predictas suos et Thomas et eorum heredibus imperium in liberam quietam et perpetuam elemosinam. in illius proprios et ecclesie sue predicte perpetuo connectens. Et nos predicti R. Episcopus et huiusmodi nisi Wyntoniensis Episcopi prouide donationes predictam ecclesiam terras Reddituum et pensiones in pertinentiis predictis prioris et conuentui et eorum successoribus in perpetuum. et ut uerum nostri donatio confirmatio nostra omnes gratias Williams anterius credendere tenemur in perpetuum. Et ut uerum nostra donatio confirmatio robur imperpetuum obtinere presentem cartam sigilli nostri munimine duximus roborandam. Huis testibus domino Waltero Abbate de Hyda. domino Waltero Priore sancti Siluestri domino Stephano Priore de Portchester capella. Magistro Johanne de Gatesden et Johanne de Bisselle, domino Willelmo de Sancta Anna tunc Officiali Wyntoniensis pacto hoc Ricardo, Suggesto Petro Russinell et multis aliis. Datum apud Wollebege per manus etc. Cancellari Octauo decimo kal' Maii Anno domini M. CC. Quadragesimo tertio.

No. 48 (A¹)   Master and Fellows of Magdalen College, Oxford

PLATE II

SCRIBES 4, 5 AND 6

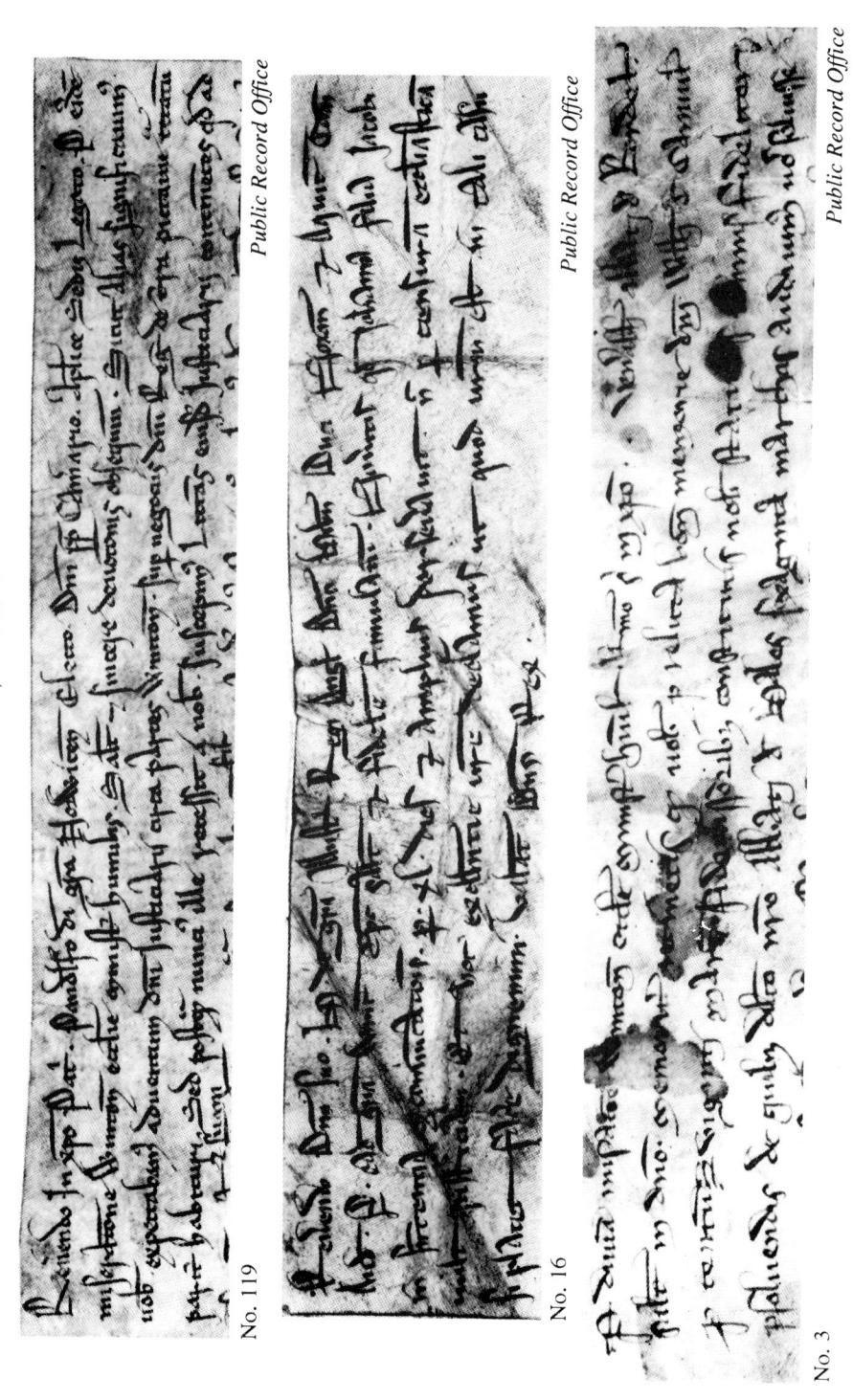

PLATE III

SEAL AND COUNTERSEAL OF BISHOP PETER

No. 9   *Master and Fellows of Magdalen College, Oxford*

PLATE IV

almost suggest the deliberate avoidance) of forms favoured by the chancery of the king. All of the texts are written in Latin, and as one would expect by this date, the vast majority, including all of the surviving originals, are given in the first person plural. Probably through errors in transcription, two cartulary copies, both from Southwick priory (nos.64,65), combine first person singular and plural, whilst a further two late copies from St Thomas' Southwark (nos.59,60) are phrased throughout in the first person singular. Cheney long ago drew attention to the use of the *cursus* in des Roches' charters.[1] Certainly, many of the bishop's texts, particularly the more solemn foundations and awards, contain rhythmical *clausulae* which follow the rules of the *cursus Romane curie*. However, to argue from this that they were directly modelled on Roman forms, would be to overstep the evidence. By the early thirteenth-century the rhythm of the *cursus* had become part and parcel of the language of the church. It required no special thought for a draftsman to write *Omnibus Cristi fidelibus* rather than *Omnibus fidelibus Cristi*, or on other occasions to dictate rhythmical rather than non-rhythmical phrases. In des Roches' case, the use of the *cursus* merely lends further support to the impression, substantiated by the present diplomatic analysis, that the bishop's charters were for the most part the work of professionally trained draftsmen and scribes.

INTITULATIO

The simplest form, *P(etrus) Dei gratia Wyntoniensis episcopus*, is the most common, appearing on at least twenty six occasions. Elsewhere we find a large number of variations on the more complicated theme *P(etrus) divina miseratione* (or *permissione*) *Wintoniensis ecclesie minister humilis*. Twenty six texts share various elements of this, the phrase *Wintoniensis ecclesie minister humilis* being their one common feature. The bishop's name is represented by the initial *P.* in all save two of the surviving originals, the exceptions being a duplicate pair of originals for Selborne priory (no.52) where *Petrus* is written out in full. Following a trend, observed in other English dioceses, letters addressed generally almost invariably place the bishop's title after the address. There are only five exceptions to this pattern, four of them to be dated before 1215 (nos.6,11,37,57,58).

---

[1] Cheney *English Bishops' Chanceries* 77–81, esp. 80n. which provides an analysis of no.32 below.

INSCRIPTIO

Letters addressed generally, to all Christ's faithful, or to all the sons of holy mother church, exhibit more than twenty variations on the basic forms familiar from countless documents of the period. The most common version, *Omnibus Cristi fidelibus* (or *Universis sancte matris ecclesie filiis*) *ad quos presens scriptum pervenerit*, occurs in nineteen instances, *scriptum* appearing far more often in the address than the word *cartam* favoured by the royal chancery.[2] In two cases a general address is preceded by such phrases as *reverendis amicis in Cristo* or *karissimis in Cristo* (nos. 33,37), normally reserved for private letters. Only one letter awarding gifts in perpetuity is addressed to its beneficiary rather than generally: *Religiosis viris priori et conventui ecclesie beate Marie de Seleborne nostre dyocesis* (no.50). Letters addressed to specific individuals follow the standard practices of the time. Archbishops, bishops, abbots, and in one instance a dean and chapter, are addressed as 'reverend' or 'venerable' (nos.3,47,81,100,114,119), with des Roches' name generally placed after the name of the addressee. Subordinates – the bishop's official, the keeper of des Roches' exchequer at Wolvesey, and a rural dean – are addressed as *dilecto*, with the bishop's title appearing first, in accordance with contemporary rules concerning precedence (nos. 20,39,46). Especially elaborate forms are reserved for kings and popes, invariably ranked above the bishop himself (nos. 15,16,127,132). In this context, it is interesting to note the uncertainty over precedence displayed in des Roches' two letters addressed to his arch-rival, Hubert de Burgh, one of which places Hubert, *amico in Cristo karissimo*, above the bishop; the other placing des Roches ahead of Hubert, *nobili viro et amico in Cristo karissimo*.

SALUTATIO

Of the twenty or more forms employed, by far the most common is the basic *salutem in domino*, which makes twenty five appearances. Next in frequency comes *eternam in domino salutem*, recorded at least fifteen times. Thereafter, we find numerous variations, ranging from the single word *salutem* (nos.6,15,116), to such elaborations as *cum omni subiectione et reverentia devota pedum oscula*, reserved for pope Gregory IX (no.132). As one would expect, private letters display an especially rich variety of forms. Greetings are mixed with pious references to the Godhead (nos.2,58,61,73); des Roches' affection for the addressee (nos. 47,111,114,119), or in one telling instance (no.100), a letter threatening its recipients, archbishop Langton and his

---

[2] For *cartam* in the *inscriptio* see nos. 9,21,52,67a,69,71, three of which are for Titchfield.

suffragans, with excommunication, to the ambiguous *salutem et debitum honorem*.

THE ARENGA

In general, the *arenga* or solemn preamble to a charter, was falling into disuse by the early thirteenth century. It is therefore significant that nearly a third of des Roches' diocesan acta display this archaic form, ranging from a mere fifteen words of pious introduction (no.33), to the extraordinary elaboration of the indulgence issued by the bishop for St Thomas' Hospital Southwark (no.58), a veritable sermon, comprising more than one hundred and sixty words, including lengthy quotations from the gospel of St Matthew and St Paul's first epistle to Timothy. All told, seventeen of des Roches' charters preserve some sort of *arenga*, for the most part attached to solemn confirmations or acts of foundation. Opinion is divided as to the significance of such clauses; some commentators dismissing them as mere platitude, others regarding them as a vital source of insight into the pre-occupations of the medieval church.[3] In the case of bishop des Roches, derided by his contemporaries for his lack of scriptural learning, there may be special significance to the frequency with which his acta resort to direct quotation. Nine texts include tags or echoes, mostly from the gospels and the epistles of St Paul (nos.13,31,48,49,50,58,80,81,100), introduced by such phrases as *unde propheta, sicut ait apostolus, sicut dicit beatus Gregorius* or *scriptura testante que dicit*. In England as opposed to France, the *arenga* tended to adapt rather than directly to recite phrases from scripture. It may be that the Frenchman, des Roches, should be seen as following a distinctively French tradition. As to the sentiments expressed, des Roches' *arengas* stress such common themes as the vanity of storing up treasure on earth (no.13), the obligation of those in episcopal office to encourage works of charity (nos.26,38,58,62,83–4), to bring succour to the poor and needy (no.80), and to foster the quiet and meditation of monastic communities (nos.34,48–50,54). Even the bishop's diocesan legislation includes direct quotation; "Whatever belongs to the clergy belongs to the poor", a tag derived ultimately from St Jerome.[4] Des Roches' earliest recorded charter, issued at Rome within a few weeks of his consecration, stresses the mutual obligations of bishop and monks, the head and members of the spiritual community (no.81). The language employed here, modelled on

---

[3] Compare the remarks of C.R. Cheney in *EEA* ii pp.lxi-iv, with those of J. Avril, 'Observance Monastique et Spiritualité dans les préambules des actes (x-xiiie s.)', *Revue d'Histoire Ecclésiastique* lxxxv (1990) 5–29.
[4] *C.& S.* 128 no.15.

St Paul's second letter to the Corinthians, may be borrowed from the papal chancery; certainly the body of the text, dealing with tithes, incorporates the papal catch phrases *decimis...tam de terris quam animalium nutrimentis* and *novalibus vestris que propriis manibus aut sumptibus colitis*, whilst the opening of the *arenga, cum simus secundum apostolum invicem alter alterius membrum*, mirrors the papal incipit *cum simus*.[5] Elsewhere, in charters for Mottisfont and Selborne priories (nos.34, 48–50 and see no.54), des Roches is known to have adapted a text first used by his predecessor, bishop Godfrey de Lucy. However, the opening words of the *arenga, cum ex officio pastorali tenemur*, appear in turn to mirror papal incipits such as *cum ex officio pastorali nobis incumbat* or *cum ex officio nobis iniuncto*.[6] By the early thirteenth century, the language of the papal chancery was so familiar in England, that it is hardly surprising to find echoes of it in the charters of des Roches. Some such echoes may reflect a quite deliberate adherence to papal models, others, as with the *arenga* adapted from bishop de Lucy, were picked up at second hand.[7]

NOTIFICATIO

As is standard by the early thirteenth-century, the *arenga* is most often linked to the *notificatio* or *dispositio* by an adverbial phrase or conjunction, the most common being *ea propter* (nos.26,32,38,50,80,83), *quapropter* (no.84), or *hinc est* (nos.48,49). As for the *notificatio* itself, by far and away the most common form involves the verb *noscere*, as in the phrase *noverit universitas vestra* which occurs fifteen times, and such variation as *noveritis nos* (nos. 18,40,45,52, 67a,69,71,75) or *noveritis quod* (no.16). The phrase *ad universitatis vestre volumus pervenire noticiam* occurs several times (nos.1,2,14,56,64), whilst papal commissions are prefaced by the standard formula: *litteras* (or *mandatum*) *domini pape suscepimus in hec verba* (or *recepimus in hac forma*) (nos. 5,47,100). Forms familiar from the royal chancery, involving the verbs

---

[5] For the clauses dealing with tithe compare *Letters of Innocent III* nos.162,202,296, and *Monasticon* i 211, a papal confirmation to St Swithun's Winchester, 3 March 1205 (*Letters of Innocent III* no.67). For the incipit compare Migne *PL* ccxvi 46 *Cum simus secundum apostolum et maioribus et minoribus* (to the diocese of Rheims, May 1209); *Ibid.* ccxiv 440 *Cum simus omnibus ex iniuncto nobis officio debitores iis specialiter et presertim providere tenemur* (addressed amongst others to the archdeacon of Poitiers, and hence possibly known to des Roches, December 1198).

[6] Migne *PL* ccxv 724 (*Letters of Innocent III* no.650, a papal mandate issued on des Roches' behalf in October 1205); *Epistolae Cant.* 336 (addressed to the suffragans of Canterbury, May 1191).

[7] See, for example, the similarity in sentiment, though not in wording, between the *arenga* to no.83 below and that printed by H. Fichtenau, *Arenga: Spätantike und Mittelalter im Spiegel von Urkundenformeln*, Mitteilungen des Instituts für Österreichische Geschichtsforschung xviii (Graz/Cologne 1957), p.52 no.69, from a charter of Philip Augustus (1183). Similar correspondences can be noted to others of Fichtenau's *exempla*.

mandare or *scire*, as in the injunction *mandamus vobis quatinus* (nos.20,39), or *sciatis nos* (nos.37,97 and see no.94), occur on only a handful of occasions, most of the them early on in the bishop's episcopate. This absence of royal forms is surprising, given des Roches' close association with royal government.

NARRATIO AND DISPOSITIO

Any number of devices and forms can be used for the main body of the text. As one would expect, letters issued as papal judge delegate may provide a brief resumé of the arguments of the opposing parties (nos.5,6 and see no.100). Foundation charters, besides listing the spiritual and temporal properties conferred, state the names of those for the sake of whose souls the bishop has made his award; for the souls of king John and his progenitors and of the bishops of Winchester past, present and future (no.13), or for the souls of Richard I, John and Henry III, the bishop's lords, and all his ancestors and successors (no.69). The early charters of Selborne are peculiar in this respect, naming only God and the Virgin Mary without any reference to mediation for specific individuals, even for the soul of the bishop, the priory's founder. On the whole, des Roches' charters avoid any lengthy narrative of past events, the exceptions being private letters reminding correspondents of payments or appointments to be made (nos.3,110,112,116), and an early charter for Mottisfont which rehearses the priory's efforts to obtain a cemetery (no.33). Grants and awards frequently end with a clause reserving the right of the bishop (nos.27,74,80), the right of the bishop's successors to make changes (no.31), or in a standard phrase, the dignity and/or authority of the church of Winchester (nos.26,28,34,35,37,49,50,83,85), a formula which already occurs in the charters of des Roches' predecessors, Richard of Ilchester and Godfrey de Lucy. The language here suggests an attempt to reserve the rights of the cathedral chapter, represented at Winchester by the monks of the cathedral priory of St Swithun. By the early thirteenth-century it was becoming accepted that all episcopal actions involving the transfer of property, spiritual or temporal, had to be ratified by the cathedral chapter. At Salisbury, Lincoln and the other secular cathedrals, this process of ratification resulted in the issue of large numbers of inspeximuses confirming episcopal awards in the name of the dean and chapter. At Winchester and the monastic cathedrals, the process is less well documented. Only in the fourteenth century did St Swithun's priory begin to maintain registers of confirmations in the name of the prior and convent. Even then, although the prior and convent might on occasion refuse their assent, the lapse of time recorded between episcopal award and capitular ratification suggests that the system was still somewhat haphazard.[8] In des Roches' day, when the bishop exercised far greater control over the prior and convent, St Swithun's could boast nothing to compare to the authority of the chapters of the secular cathedrals. Nonetheless, the reservation of the rights of the cathedral church as expressed in his charters, and the fact that a number of des Roches' awards are preserved in inspeximuses issued by the prior and convent, albeit inspeximuses issued for the most part long after the bishop's death, suggests that by the mid-thirteenth-century St Swithun's was asserting at least a limited degree of regulation over episcopal awards.[9]

THE INSPEXIMUS

Several of des Roches' charters involve the confirmation or ratification of grants by other people. Yet within the see of Winchester it appears that the full-blown inspeximus, reciting the text of an earlier award, was not perfected until relatively late in the episcopate of bishop Peter. As one would expect, charters issued by des Roches in company with other papal judges delegate may recite their papal mandate in full (nos.5,100), and in another case des Roches dispatched the full text of papal letters to the bishop of Salisbury (no.47). However, papal letters apart, only three of des Roches' charters provide a complete recital of any royal or episcopal act to be confirmed. Two of them, issued after 1230, are introduced by the infinitive *inspexisse* (nos.18,75), the third, from before 1218, employs the word *inspectione: sicut ex inspectione autentici intelleximus* (no.1). For the rest, the bishop's acta mention or paraphrase earlier grants without reciting them in full (e.g. nos.11,14,32,34,83). Twice in such circumstances, reference is made to the bishop 'following in the footsteps' of his predecessors (nos.11,34). The second of these texts (no.34), a confirmation issued on behalf of Mottisfont priory, is copied more or less word for word from an earlier confirmation by bishop Godfrey de Lucy, des Roches' immediate predecessor, yet at no point does des Roches' text acknowledge this borrowing save for the phrase *episcopi predecessoris nostri vestigiis inherentes*. In turn the Mottisfont text, first issued by bishop Godfrey, served as the basis of des Roches' foundation charter for

---

[8] *Reg.Common Seal* pp.xxi-iii.

[9] Nos.34, 66 and 75 appear to have been ratified by the prior and convent before 1238. Elsewhere, charters relating to St Thomas' Southwark, Selborne and Titchfield (nos.54,56,57,59,60-62,69,70) were ratified by the prior and convent between 1243 and 1266. In addition the survival of nos. 14 and 50 in the fourteenth-century St Swithun's cartulary suggests that the prior and convent obtained copies of certain episcopal awards, perhaps because they affected the rights of St Swithun's as a private corporation rather than as the cathedral chapter.

Selborne priory, providing further evidence that Selborne was colonized from Mottisfont, and suggesting that, for the new foundation, the canons of Mottisfont presented des Roches with his earlier text as a model to be used at Selborne.[10] The failure of des Roches' chancery to adopt the standard form of the inspeximus, common in the royal chancery, once again suggests that the bishop drew a conscious distinction between the forms appropriate to royal and episcopal office.

INJUNCTIO, CORROBORATIO AND SANCTIO

Following on from the main body of the charter, any one of several *clausulae* might be used to link the *dispositio* to the final clauses and the eschatocol. In isolated instances des Roches employs the injunctions *quare volumus et concedimus* (no.85) or *quare volumus et firmiter precipimus* (no.9) familiar from the royal chancery. Far more often his charters resort to a corroborative clause, most commonly introduced by the conjunctions *et* and *ut*: *et in huius rei testimonium*, yet another form used widely in royal documents (nos.20,70 and see nos.32,81,84); *et ut hec nostra confirmatio* (or *concessio, donatio*, or *ordinatio*) (nos.26,35, 38,45,48,51–4,61,64,69,71), or some variation on the theme *ut autem hec nostra concessio* (nos.6,13,18,20,31,65,83,85). Other words to be used in this context include *ne autem* or *ne igitur* (nos.1,2) and *quod ut* (nos.39,74). The *corroboratio* will then specify the authority by which the text is authenticated, most often the bishop's seal, a formula which occurs on more than twenty occasions, or by the present writing (or charter) and seal, which occurs a further eleven times. Three texts are said to be corroborated by the seals of the bishop and the others parties involved (nos.6,35,73). Only one, early, text employs a sanctions clause (no.33), although on other occasions the *corroboratio* may be elaborated; *ne autem venditio ista et concessio possit inposterum aliquatenus in dubium devocari* (no.2), or in the solemn foundation charter for Hales *ut autem factum istud plenius habeatur et firmum et nulla possit malitia vel fraude dissolvi* (no.13).

ESCHATOCOL

The acta may conclude with one or more of at least four elements: a witness list, the place of issue, the date, and a valediction. The selection and combination of these elements defies any simple ruling, although, as will

[10] It is far more likely that the model was provided by Mottisfont than that both texts were taken from a formulary. If they were from a formulary, one might expect to find the same text adapted on other occasions, which does not appear to have been the case.

become apparent, the combination of witness list and place/date clause is most commonly found in the first and the final decades of des Roches' episcopate. Several of the bishop's episcopal charters and the vast majority of his private correspondence end with a clause involving the words *valeat* or *valete*. Most often this form is reserved for private individuals; for example the bishop's official (no.46), the bishop of Salisbury (no.47), or the king (nos.15,16). However, two letters with a general address, both of them from St Thomas' Southwark, conclude either with the simple form *valete* (no.57), or *semper in domino valete* (no.58).

For the rest, thirty six of the surviving acta carry witness lists, whilst a further five, preserved in incomplete cartulary copies, retain some trace of a witness list now lost. Ignoring those copies where a witness list has clearly been heavily abbreviated (nos.64,65,72,81), the number of witnesses appended to a charter varies from five (nos.18,74, both of them originals) to 17 in the case of the foundation charter for Halesowen (no.13), several of the witnesses on this occasion being church dignitaries from outside des Roches' household. The average witness list names 9 individual witnesses. Virtually all of des Roches' witness lists conclude *et multis aliis*, a form not found in the royal chancery, although common in private charters. Churchmen appear as witnesses far more frequently than laymen, some lists consisting entirely of *magistri*, chaplains and clerks. In such cases, bishops and heads of religious houses are ranked above des Roches' official. In turn the official takes precedence over the remainder of the household including the archdeacons, save for a brief period during the 1230s when master Alan of Stokes, the outgoing official, is ranked above his successor, master William de Ste Mère-Église (nos.48–50). Men styled chaplain invariably appear without surname and in 9 out of 10 cases are ranked above other members of the household styled clerks (nos.11,13,34,38,45,49,56,59,83, the exception being no.80). However it should be noted that an individual styled clerk in one charter may appear without title in another. In general, titles are applied far more sparingly than in charters of the twelfth-century; terms such as *senescallus*, *camerarius*, *medicus*, or *pincerna*, which occur frequently in des Roches' pipe rolls, are almost unknown in the witness lists to his acta. Even when there is a reasonable degree of certainty that one or other of the witnesses was discharging an office such as seneschal or bishop's official, there is no guarantee that his office will be be specificied in the witness list.[11] Monks of St Swithun's priory appear only twice as witnesses. In both cases they seem to have been ranked above the bishop's household (nos.6,81). Two charters are

---

[11] See the case of master John of London, below appendix 4 no.1.

witnessed exclusively by laymen; a settlement over the episcopal estates (no.82), and a grant of a tenancy by knight service (no.9), both of them charters which relate to the bishop's temporal rather than his spiritual jurisdiction. The term *milites* is applied only once to knights serving as witnesses (no.83), although elsewhere the *milites episcopi* appear to have formed a recognized group, referred to as such in the Winchester pipe rolls and in the records of the royal chancery.[12]

THE PLACE/DATE CLAUSE

Des Roches' predecessor at Winchester, Godfrey de Lucy, was the first English bishop to introduce a regular dating clause to his charters, adopting a style modelled upon that of the charters of king Richard I. From 1192 Godfrey's acta are supplied with a clause specifying the place of issue, the name of a datary, the day and month in the Roman calendar, and the bishop's pontifical year. Something of this routine appears to have survived into the early years of bishop des Roches. By no means all of des Roches early charters are dated (see, for example, nos. 34,37), but from before 1210 we have three charters which supply the name of a datary, the place of issue and a date in the Roman calendar, in two cases adding the bishop's pontifical year (nos.21,34,81). All of these charters also carry witness lists, placed once before and on two occasions after the dating clause. Thereafter, from the next twenty years only a handful of dated charters survive. Two of them are closely associated with the royal court (nos.82,100). Two others are dated by the year of the incarnation with the day and month being supplied according to the Roman (no.6) or the church calendar (no.31), this latter specifiying the place of issue. Only with the bishop's return from crusade in 1231 does a full dating clause become a regular feature of his acta, and even then there is little uniformity to the styles adopted. Des Roches' charters of the 1230s almost invariably specify the place of issue, and occasionally the name of a datary. The date appears most often in the Roman style (nos.3,48,51–4,70,75,84) but sometimes according to the festivals of the church (no.18,40,49,50). The year is generally given according to the incarnation of Christ, but in one original no year is specified (no.52 A²), whilst a cartulary copy gives the year both according to the incarnation and the bishop's pontificate (no.75). Again, inconsistently, the dating clause may appear before (nos.75,84) or after (nos.18,48–54) a witness list, or without any witness list at all (nos.3,40,70). There seems to be no connection between these varying styles and any particular datary. Texts given by the hands of Peter de Chanceaux

---

[12] See, for example, *PR4DR* 15; Ms.7DR m.8d; *Rot.Lib.* 246.

(nos.48,49,52,75) or Peter Russinol (nos.34,53,54) exhibit most of the variations noted above. Altogether, throughout des Roches' episcopate, only 20 charters preserve any sort of dating clause. Sixteen of these are introduced by the formula *Dat'* familiar from both the royal and the papal chancery, although by giving the place of issue immediately afterwards in the form *Dat' apud*, des Roches diverges from both royal and papal practice. Four of his charters, including two originals, are introduced by the word *Act'* or *Acta* (nos.6,40,81,82), which might be accounted archaic.

The assumption with English episcopal charters of this period must be that we are dealing with dates according to the English church calendar, the year beginning on 25 March. This is an assumption substantiated in the case of at least one of des Roches' charters by the appearance of a date which makes sense according to the year calculated from 25 March but not according to the year beginning at Christmas or on 1 January (no.40). However, one text, surviving only in an imperfect copy, requires a year calculated from Christmas or 1 January rather than the Annunciation (no.70). In another case, an original, the scribe appears to have written 1237 in mistake for 1238 (no.54). Similar confusion occurs over the calculation of the bishop's pontifical year. Des Roches was elected to his see in January or February 1205, formally confirmed as elect by the pope around 21 June, consecrated on 25 September the same year, and installed at Winchester on 26 March 1206. A careful examination of the Winchester pipe rolls, which are dated by pontifical year, and whose individual manorial accounts are dated according to specific feast days from the year 16 des Roches onwards, suggests that the year 1 des Roches was presumed to begin on 25 September 1205, the date of the bishop's consecration.[13] Assuming this to be the case, the majority of dating clauses which include the bishop's pontifical year, appear to have been calculated

---

[13] W. Beveridge, 'The Winchester Rolls and their Dating', *Economic History Review*, ii part i (1929), 93–113. It was originally supposed that des Roches' episcopate was calculated from the restitution of his temporalities, acknowledged formally only in March 1206. Beveridge demonstrated that March 1206 could not be the beginning of the first pontifical year, suggesting instead the date of the bishop's consecration, 25 September. However, in practice, no issues from the bishopric were accounted to the king after 24 June 1205: *PR 7 John* 11–13. It may therefore be that the year 1 des Roches was calculated from June rather than September 1205. Either June or September would suit most of the calculations given below, although in one case June would fit better than September. In a nineteenth-century calendar appended to copies of the bpric customal, apparently drawn up to assist in the dating of material in the medieval pipe rolls, the year 1 des Roches is said to run from 1204 to 1205, a false calculation which may have led to yet further confusion in the dating of des Roches' rolls; Winchester, Hampshire Record Office ms. 68M74/R/E/A1 p.227.

correctly.[14] However, there are at least two exceptions, the first a charter of 1231 or possibly 1232 (no.75) said to have been issued in the year 26 des Roches (?*recte* year 27), and the other a Chertsey abbey charter dated 33 des Roches, apparently by mistake for year 32.[15] Since neither is an original, it seems reasonable to conclude that the copyist was at fault in both cases. We might bear in mind that des Roches' episcopate spanned much of the reign of John, the only English king whose regnal years are given according to a moveable feast, making them notoriously difficult to calculate.

EDITORIAL METHOD AND CRITERIA FOR DATING

In general the present collection is edited according to the principles of earlier volumes in this series. Originals are assigned the letter A, or in the case of duplicates A^1 and A^2. Copies are described as B,C and so on. All original episcopal acta and most of the bishop's original letters are printed in full, the exceptions being those few political letters for which a standard text already exists. Originals are punctuated as seen, retaining the *punctus elevatus*, though not the tyronian *et* which has been expanded in full. No stress marks have been found in any of the originals. Texts which survive only in cartulary or later copies are printed in full save those for which an accurate and readily accessible version has been published elsewhere. Copies are provided with modern punctuation intended to aid comprehension. In cartulary copies but not in originals, c/t and u/v spellings have, as far as possible, been standardized. The modern letters *i* and *j* are given as *i* throughout, according to the conventions of this series, although *j* is often written in later copies. The rubrics supply modern place and personal-names throughout. In the case of texts not printed in full here, I have attempted to make the rubrics as detailed as possible. With regard to the sealing of originals, the terms 'method 1' and 'method 2' are applied to documents sealed on a fold *sur double queue*, either through a single set of slits ('method 1'), or through two parallel sets of slits ('method 2'), in both cases cut through the fold at the foot of the document.[1]

---

[14] See *Chertsey Cart.* i nos.75,92; *WC* no.470, all of them deeds, preserved in late copies, provided with dates according to the incarnation and also according to des Roches' pontifical year. In the case of the Chertsey charters these details were added as rubrics, long after the deeds themselves were issued.

[15] *Chertsey Cart.* no.92. In this instance the calculation would work if the pontifical year ran from June rather than September 1205, in other words from the bishop's formal recognition as elect rather than from his consecration.

[1] For the distinction see *EEA* ii p.xlvii, where the two methods are designated 'd.q.1' and 'd.q.2'.

As noted above, the acta themselves are arranged in three basic sections. Those which relate to the bishop's affairs as diocesan or private land-holder are arranged alphabetically, according to beneficiary. Where two or more texts relate to the same beneficiary, these are given as far as possible in chronological sequence. The 50 or so texts which issue from des Roches' role as politician or courtier can for the most part be precisely dated, and are therefore arranged in a section of their own, in chronological order. The third section, drawn entirely from the Winchester pipe rolls, provides a calendar of some 200 lost writs mentioned as warranty for expenditure by the bishop's manorial bailiffs. These too are given in chronological sequence, being arranged thereafter according to the order of the membranes of the individual pipe rolls in which they appear.

Many of the bishop's charters are undated. However, external sources and in particular a detailed study of des Roches' household (below appendix 4) for the most part provide us with fairly narrow dating limits, supplied within brackets in the rubrics to each text and further explained in the notes. Where a text can be dated only from the beginning of the bishop's episcopate, I have assigned the outside limit March 1206, the date of des Roches' return to England. With the exception of one charter, clearly dated at Rome in October 1205 (no.81), it seems highly unlikely that des Roches was in a position to interfere in diocesan affairs before arriving in his see. Likewise it is improbable that any of the diocesan acta were issued between June 1227 and August 1231, the years of the bishop's absence in Palestine and Italy, a point of dating which is made clear in the notes whenever it arises.

PART 1: DIOCESAN AFFAIRS

1. **Alton Church**

Inspeximus of a composition between Hyde abbey and the church of Alton on the one hand, and Waverley abbey on the other, whereby Waverley agreed to pay Hyde forty shillings each year for the tithes of Neatham, with Hyde providing a chaplain for the community of Neatham, the settlement being drawn up in the synod of Winchester, in the presence of bishop Henry (de Blois). Bishop Peter confirms the settlement and ordains that Waverley should pay two marks of the forty shillings to Hyde, and one mark to the church of Alton, with the abbey of Hyde and the church of Alton making no further demand from Waverley, and Waverley selling no land in the parish of Alton to the loss of the parish church.

[March 1206 × July 1219, ?March 1206 × 1 August 1218]

    B = Winchester, Hampshire Record Office, ms. 21M65/A/1/11 (Register of bp William de Wykeham part ii) fo.224v. In an inspeximus by bp Wykeham dated 10 January 1386.
    Pd (calendar) from B in *Wykeham's Register* ed. T.F. Kirby, 2 vols., Hants. Rec. Soc. (1896–99) ii 386–7.

Universis sancte matris ecclesie filiis ad quos presens scriptum pervenerit Petrus Dei gratia Wyntoniensis episcopus salutem in domino. Ad universitatis vestre volumus pervenire notitiam antiquitus talem factam fuisse compositionem inter ecclesias de Hyda et Aulton' ex parte una et ecclesiam de Waverle ex parte altera super decimis de Netham sicut ex inspectione auctentici intelleximus: Sciant presentes et secuturi compositionem hanc factam fuisse in synodo Wyntoniensi in presentia domini Henrici episcopi inter ecclesiam sancti Petri de Hyda et ecclesiam beate Marie de Waverle super decima de Netham, quod ecclesia de Waverle persolvet ecclesie sancti Petri de Hyda singulis annis ad festivitatem sancti Michaelis quadraginta solidos pro supradicta decima, hac sane conditione quod ecclesia de Hyda providebit capellanum familie de Netham. Hiis testibus: Hugone abbate de Certes', Gaufrido priore Sancti Swithuni, Hugone archidiacono, Alardo priore Sancti Dionisii, Goscelino mag(istro). Set quoniam futuris temporibus posset ecclesie de Aulton' preiudicium generari ex eo quod non est in autentico evidenter distinctum quantum debeat de quadraginta solidis predictis solvi ecclesie de Aulton' et[a] quantum ecclesie de Hyda, cum tamen monachis de Hyda due marce solute sint tantum annuatim et una ecclesie de Aulton', unanimi consensu conventus de Hyda et conventus de Waverle et consilio prudentum, ordinamus ut ecclesia de Waverle persolvat singulis annis in festo sancti Michaelis pro omnimodis decimis de Netham ecclesie beati Petri de Hyda duas marcas et ecclesie de Aulton' unam marcam. Tenebitque ecclesia de Waverle predictas decimas de Netham quiete, pacifice et integre, ita quod decetero predicte ecclesie de Hyda et Aulton' amplius aliquid a predicta ecclesia de Waverle pro predictis decimis de Netham exigere non poterunt, et ecclesia de Waverle similiter occasione istius compositionis terras aliquas in parochia de Aulton' in dampnum ipsius ecclesie non poterit vendicare. Ne igitur imposterum aliquid possit venire in dubium predictam compositionem inter dictas ecclesias intervenisse illam sigilli nostri munimine confirmamus. Hiis testibus: Eustachio de Faukebergh, Hugone abbate Belli Loci Regis, magistro Alano de Stokes, Gregorio capellano, magistro Petro Roiusinol, magistro Roberto Basset, magistro Thoma de Eblesbourne, Stephano clerico et multis aliis.

[a] *erasure (illeg.) over line between* et *and* quantum

The settlement made in the time of bp Henry de Blois can be dated c.1149 × 1153, between the election of abbot Hugh of Chertsey and the promotion of Hugh du Puiset, archdn of Winchester, as bp of Durham. Bp des Roches' charter dates before the election of master Eustace de Fauconberg as bp of London, 25 February 1221, and the death of Peter Russinol I, d. before July 1219. The first two abbots of Beaulieu were both named Hugh (BL ms. Arundel 17 (Register of Newnham abbey) fos. 45r, 53r). Hugh I, the more likely witness to bp des Roches' charter, was promoted bp of Carlisle before 1 August 1218. Neatham was held by Waverley abbey from at least 1147 (*VCH Hampshire* ii 513). Hyde held the manor and advowson of Alton from an early date, and had been granted an annual pension of 40 shillings in the church there by bp de Lucy (*EEA* viii no.206). Des Roches' kinsman, Peter de Rivallis, was presented to Alton at some time before 1231, probably in succession to Roger archdn of Winchester (d. c.1208) (Appendix 4 no.41). For master Thomas of Ebbesbourne see below no.26.

2. Beaulieu Abbey

Notification that in the bishop's presence Walter Fortin, burgess of Southampton, acknowledged his sale to Beaulieu abbey for £100 of the houses in English Street, Southampton, which he built in the fee of Cormeilles abbey, which fee he acquired as shown in an authentic writing of the abbot and convent of Cormeilles. The property includes stone buildings partly lying in the fee of Rocelin Tyrel. Walter has granted all this property to Beaulieu in pure and perpetual alms, regardless of whether it may subsequently be valued at a higher price than that which the monks have paid, the present charter being drawn up at Walter's request. [March 1206 × 1 August 1218]

B = BL ms. Loans 29/330 (Beaulieu cartulary) fo.124r, witnesses omitted. s.xiii med.

Pd from B in *Beaulieu Cartulary* no. 242. The lost original is noticed in a thirteenth-century list of Beaulieu's charters for Southampton, arranged in a different order from those in the surviving cartulary: Southampton, City Archive Office ms. SC15/4 fo.1r no.3 (copy in BL ms. Facs. Suppl. X); *Confirmatio domini Petri Wynton' episcopi de donacione et concessione predicti Walteri facta conventui de domibus predictis.*

The sale recorded here was made in the time of abbot Hugh I of Beaulieu, promoted bp of Carlisle 1 August 1218. For the charter of Durandus abbot of Cormeilles mentioned above see *Beaulieu Cartulary* no.243. For Walter Fortin (fl.1198–1230), mayor of Southampton before 1217 and again c.1230, see C. Platt *Medieval Southampton, the port and trading community, A.D. 1000–1600* (London 1973) 241; *God's House Cart.* nos. 50, 159, pp.51n., 339; *St Denys Cartulary* no.109 and *passim*. In 1225–7 Walter received grain at bp des Roches' manor of Bishop's Waltham (Mss. 21DR m.12d; 22DR m.11). The present grant probably underlies a settlement reached before un-named papal judges delegate in 1221, whereby Beaulieu acknowledged various services owing to Philip abbot of Cormeilles for a house held of Cormeilles in Southampton, undertaking to provide lodging for the abbot, in the said house or elsewhere, whenever he should visit Southampton (BL ms. Add. 15669 (Newent cartulary) fo.71v).

3. Bordesley Abbey

Mandate to the abbot of Bordesley to pay sixty marks to Walter bishop of Worcester in settlement of an agreement on the abbot's part to stand pledge for the widow of William de Camville to the sum of one hundred and twenty marks. The bishop of Winchester has been told that the abbot has already paid sixty marks of this pledge to the abbot and convent of Hales, promising to pay the remainder at the coming Christmas and Easter. The money should be given to bishop Walter of Worcester who will issue a receipt.

Southwark, 3 December 1237

A = PRO E326/12579. No visible endorsement; approx. 145 × 50 mm.; originally sealed on tongue; tongue and seal torn away. Stained and creased. Scribe 6.

P. divina miseratione Winton(iensis) ecclesie minister humilis . karissimo sibi in Cristo .. venerabili abbati de Bordel' . salutem in domino . Memoriter retinetis quod vobis pro relicta bone memorie domini Willelmi de Camvil' pro centum et viginti marcis fideiussoribus constitutis nobis statutis terminis fideliter persolvendis de quibus dilecto nostro abbati de Hales sexaginta marchas audivimus vos solvisse . triginta marcas in presenti Natali domini et in Pascha proximo sequente residuum solvere promisistis . Volumus igitur et mandamus quod venerabili fratri domino Waltero Dei gratia[a] Wigorn' episcopo dictas sexaginta marcas nostro nomine numeretis . receptis eiusdem litteris de solucione nostro nomine facta eidem . Dat' apud Suwerk' .iii. non' Decembr' . anno domini .m°. ducentesimo tricesimo . septimo.

[a] Dei gratia *inserted over the line*

Demonstrates the bp's continued contacts with his foundation at Halesowen, and with the baronage of the West Midlands from amongst which Hales drew its principal patrons. Probably related to the bp's custody of the heir and barony of Robert de Marmiun after 1233 (*Cal. Chart. R. 1226–57* 186). The widow referred to must be Albreda de Marmiun (d. after 1233), wife of William de Camville the elder (d. 1207 × 1210), daughter and heiress of Geoffrey de Marmiun, lord of Llanstephan (*CP* iii 3, viii 508n.; Dugdale, *The Antiquities of Warwickshire*, ed. W. Thomas, 2 vols. (London 1730) 845; *CRR* v 54, xv no.394; *Warwickshire Feet of Fines*, ed. E. Stokes and F.C. Wellstood, Dugdale Soc. xi (1932), no.196). Albreda retained her interests in land at Arrow and Seckington, Warwicks., into the 1230s (*VCH Warwicks.*, iii 28, iv 106, 198; *CRR* xv no.394), but was succeeded at Llanstephan and elsewhere by her son Geoffrey de Camville (*Rot. Chart.* 53b–54). Geoffrey married firstly Felicia, sister of Simon of Frankley, a benefactor of bp des Roches' foundation at Halesowen (*CRR* xiii no.805; *VCH Worcs.* iii 120–1; *Lyttelton Charters* nos. 2, 4–6, 10, 15–17). Geoffrey's second wife was Leuca, grand-daughter of William de Braose. He died c.1219, when his lands passed to his son by Leuca, William de Camville the younger (d. c.1260), who was forced to recognize a claim by a step-brother, Richard, born to Geoffrey's first marriage (*CRR* xiii no.805). William de Camville the younger witnesses a number of Bordesley abbey deeds (*Cat. Anc. Deeds* i 286–7). He held land from his grandmother, Albreda de Marmiun, in Seckington, and elsewhere from Robert de Marmiun, head of the main branch of the Marmiun family (*CRR* xv no.394; *BF* 510). In 1228 Albreda quitclaimed to William the younger all her right in the barony of Llanstephan (*Cl.R. 1227–31* 114). It is unclear how Albreda came to owe bp des Roches the 120 marks mentioned in the present text; perhaps through the bp's custody of the barony of Robert Marmiun.

*4. Brian fitz Ralph

Indenture between P(eter) bishop of Winchester and Brian fitz Ralph concerning the latter's ransom to be paid to the bishop. [c.April 1218]

Mentioned in royal letters confirming the settlement, dated 18 April 1218 (*Pat.R.1216–25* 148). Compare nos. 44,79 below for similar conventions occasioned by the bp's ransoming of prisoners taken during the civil war, 1215–17. On 17 March 1218 the sheriff of Surrey and Sussex had been ordered to grant des Roches seisin of Brian's lands, Brian having failed to meet the terms of his ransom (*RLC* i 356, which includes similar orders for the lands of William Dacy and Robert le Hou).

5. Chertsey Abbey

Judgment by Peter bishop of Winchester, Thomas abbot of Stanley and R(ichard Grosseteste) archdeacon of Wiltshire, acting as papal judges delegate, reciting a mandate of pope Honorius (III) dated at the Lateran, 21 March 1217, concerning complaints by master Thomas rector of Chobham that Chertsey abbey prevents him from acquiring a cemetery for his church. The judges declare that the parties were summoned, and that master Thomas pleaded the need for a cemetery to save his parishioners the expensive and arduous journey they are at present forced to undertake. The abbot and convent of Chertsey feared the loss of tithes and offerings, but assented to the consecration of a cemetery, saving the right to Thomas' parishioners to opt for burial at Chertsey, their mother church; saving also the first mortuary legacies to Chertsey in accordance with the practice of the diocese, and saving all the tithes of the parish of Chobham, except all tithes of grain in 'Fletlande', a moiety of the tithe of grain in the abbot's demesne named 'Burierchas' and except all other tithes known to pertain to the chapel of St Lawrence, Chobham. In compensation, master Thomas and his successors, with the assistance of their parishioners, will pay twenty shillings and six pounds of wax each year to the sacrist of Chertsey, in Easter week and at Michaelmas in equal portions; the wax and half a mark of the money being supplied in offerings by the parish. This agreement has been sworn to by the parishioners of Chobham, binding their successors in perpetuity to the abbot and convent, their lords. Future clerks and chaplains of the chapel of Chobham will swear an oath to the abbot and convent for the observance of the same.

[c. April 1217 × 7 October 1221]

B = PRO E164/25 (Chertsey cartulary), fo. 44r-v. s.xv med.
Pd from B in *Chertsey Cart.*, part i, Surrey Record Soc v (1915), 62-4 no.74.

Dated between the arrival of the papal mandate and the death or retirement of Thomas abbot of Stanley, who had been succeeded by 7 October 1221. For Richard Grosseteste see *Fasti* iv 36. The settlement was revised 28 October × 24 March 1231 in a further judges delegate decision that mentions the present text (*Ibid.* 64-6 no.75). Both judgments played an important role in the establishment of Chobham as an independent parish, separate from Chertsey. Master Thomas of Chobham, the rector, is better known as subdean of Salisbury and author of an important work on confession. Before 1212 he had been in correspondence with master John of London, bp des Roches' official (Chobham *Summa* pp.xxvii-xxxviii). In 1234 master Thomas obtained letters from pope Gregory IX, requiring the precentor, chancellor and succentor of Salisbury to investigate his complaint that various parties, including Thomas, perpetual vicar of Chobham, had injured him in respect to certain legacies (Oxford, Magdalen College Muniments ms. Misc. Charter 222).

6. Chertsey Abbey

Settlement of a dispute between A(lan) abbot of Chertsey and Michael rector of Cobham, over a virgate of land which was held by Odo de la Hulle, and five acres in a certain island beyond the water, and a virgate called 'Hukelescroft', *and six acres at* 'Wolcroft', *and a purpresture called* 'La Rudynge', *and two hays in the vill of Cobham which Richard Telarius and Alswynus held. The bishop determines that all these lands belong to Chertsey abbey, except for the land called* 'Wolcroft' *which should remain to Michael the rector together with an ?alder grove* (alvetum) *called* 'Le Hezemore' *and a field called* 'Brembelacr'.

1 May 1224

B = PRO E164/25 (Chertsey cartulary), fo. 304r-v. s.xv med.

Pd (calendar) from B in *Chertsey Cart.*, II part i, Surrey Rec. Soc. (1958) 169-70 no.934.

Petrus Dei gratia Wyntoniensis episcopus omnibus ad quos presens scriptum pervenerit salutem. Noverit universitas vestra quod cum controversia coram nobis orta esset inter A(lanum) abbatem Certeseye et eiusdem loci conventum et Michaelem rectorem ecclesie de Coveham super[a] quibusdam terris quas idem abbas et conventus dicebant ad suum laycum feodum pertinere, prefatus vero Michael ad suam ecclesiam de Coveham spectare ut predia ecclesiastica asserebat, scilicet de una virgata terre quam Odo de la Hulle tenuit, et quinque acris in quadam insula ultra aquam, et una virgata terre que dicitur Hukelescroft, et sex acris apud Wolcroft, et una preprestura que dicitur La Rudynge, et duabus haiis in villa de Coveham quas Ricardus Telarius et Alswynus tenuerunt. Tandem post multas altercationes et allegationes hinc inde prepositas, cum nobis de iure singulorum liquido constaret tam per instrumenta tam[b] per testes fidedignos, et omni expeditione maioris deliberationis[c] prehabita et iuris ordine in omnibus observato, omnes predictas terras cum pertinentiis, preter illam terram que vocatur Wolcrofte, ad ius et proprietatem abbatis et monachorum ecclesie de Certeseye non obstante quacumque possessione sententialiter adiudicavimus, Michaeli rectori ecclesie de Coveham et successoribus suis super eisdem terris tam in petitorio quam in possessorio imperpetuum silentium inponentes. Cum etiam per inquisitionem diligentem nobis plenius constaret quod ecclesia de Coveham predictam terram que vocatur Wolcroft et unum alvetum quod dicitur Le Hezemore sicut antiquitus fossatum et clausum fuerat, et unam terram que dicitur Brembelacr' longo tempore preterito possidisset, ipsas ad ius et proprietatem predicti Michaelis et ecclesie sue de Coveham simili modo adiudicavimus, [et] ipsum Michaelem rectorem predicte ecclesie et successores suos ab impetitione predictorum abbatis et monachorum Certes' super predictis terris, scilicet Wolcroft, Le Hezemore et Le Brembulaker, imperpetuum declaravimus absolutos. Ut igitur hec nostra constitutio perpetue firmitatis robur optineat tam sigillo nostro quam sigillis partium eam duximus roborandam. Hiis testibus: domino A. capellano, monacho Sancti Swythuni, magistro Humfredo de Mulers, Roberto de Clincham', Iohanne persona de Colesdon', Willelmo de Batilly, Roberto decano de Guldeford, Roberto persona de Bocham, Roberto de Puntyngton', Gilberto de Haywode, Ricardo de Porta, Petro Portario et aliis. Acta anno ab incarnatione domini millesimo cc^o vicesimo quarto, kalend' Maii.

[a] ms. fo.304v [b] tamen per instrumenta tamen ms. [c] maiores deliberatione ms.

Unusual, though not unique (see below no. 81) amongst the bp's charters for being witnessed by a monk of St Swithun's. It is unclear whether the monk and A. the chaplain are one and the same person, or whether the copyist has omitted the monk's name or the gemipunctus.

*7. Chichester Cathedral

An assignment of fifteen marks a year from the church of Amport towards lights in the cathedral church of Chichester.

[27 June 1217 × 9 June 1238,? c.1217 × 1223]

Mentioned in instruments of 1279 and 1282 (*Chichester Cart.* nos. 648, 713 and see nos. 243, 658). The advowson of Amport had been held in the twelfth century by Adam de Port, lord of Basing, who in 1200 paid £6 to secure his rights in the church's subject chapels of Appleshaw and Cholderton (*CRR* i 111-12,191-2; PRO CP25(1) 203/2 nos. 15,16). Adam's son and heir, William de St John, a rebel during the civil war and thereafter subject to severe financial pressures, granted the church and its chapels to Chichester for lights in the cathedral, his charter being witnessed by bp R(ichard Poer) of Salisbury and bp S(imon) of Exeter, which dates it to the period June 1217 × 9 September 1223 (*Chichester Cart.* no. 371, although see no.903 part 31 which would suggest a slightly earlier date). Des Roches was asked by St John to induct the dean and chapter into possession, and his assignment of 15 marks probably occurred at much the same time (*Ibid.* no.370). By the summer of 1238 Amport church was held by master Hervey, the parson, who, in return for spiritual benefits, obtained pasture for his animals in Cholderton, a settlement made with the assent and at the wish of bp des Roches (PRO CP25(1) 203/7 no.7).

*8. Christchurch Twinham Priory

Confirmation to the priory of tithes of sheaves from the church of Milford-on-Sea and the chapel of Hordle.

[March 1206 × 16 April 1236, ?August 1231 × 16 April 1236]

Mentioned but not recited in a papal confirmation dated 16 April 1236 (BL ms. Cotton Tiberius D vi part i (Christchurch cartulary), fo. 71v, not in the papal register). Probably dates to the period after the bp's return from crusade in August 1231.

9. Eustace de Greinville

Confirmation of the grants made by Aimery earl of Gloucester in the manor of Mapledurham, to be held henceforth from the see of Winchester to which earl Aimery granted the service of one third of a knight's fee owed by Eustace and his successors.
[c.October × November 1210]

A = Oxford, Magdalen College, Muniments ms. Petersfield Charter 147. Endorsed: *Mapuldereham* (s.xv-xvi); approx. 212 × 141 + 16 mm.; seal and counterseal impressions in dark green wax, dark blue cords attached through 2 eyelets. ?Scribe 1.
Pd (calendar) from A in *Selborne Charters* ii 64.

Omnibus Cristi fidelibus . ad quos presens carta pervenerit . P. Dei gratia Wintoniens(is) episcopus salutem in domino . Noverit universitas vestra nos concessisse . et hac presenti carta nostra confirmasse . Eustachio de Greinvill' . et heredibus suis totam terram . et omnia tenementa . cum omnibus pertinentiis suis . que Amauricus comes Glocestr' filius Amauricii comitis de Evereus ei dedit . et carta sua confirmavit . scilicet in manerio suo de Mapeldreham . tenenda . et habenda . de se . et de heredibus suis .' ipsi Eustachio et heredibus suis per servitium tercie partis unius militis . scilicet quod servitium cum homagio predicti Eustachii . et heredum suorum .' predictus comes dedit ecclesie sanctorum Petri . et Pauli . et sancti Swithuni . Winton' . et nobis . et successoribus nostris episcopis Winton(iensibus) . tenendum . et habendum in perpetuum . Quare volumus . et firmiter precipimus . quod predictus Eustachius et heredes sui habeant et teneant omnia predicta tenementa cum pertinentiis de ecclesia Winton(iensi) et de successoribus nostris episcopis Winton(iensibus) . libere . et quiete . integre . et pacifice . in bosco . et plano . in viis et semitis . in pratis . et pascuis . in aquis . et molendinis . et in omnibus libertatibus .'quas predictus comes dedit . et carta sua confirmavit ipsi Eustachio et heredibus suis . per predictum servitium tercie partis unius militis inde faciendum .' cum evenerit . scilicet pro omni servitio et exactione . Et ut hec concessio nostra . et confirmatio rata permaneat et inconcussa .'hanc cartam impositione sigilli nostri corroboravimus . Hiis testibus . Waltero de Andeli . Henrico de Breibuf . Iohanne de la Charite . Mauricio de Turvill' . Roberto le Hot . Roggero de Ticheburne . Roberto de Hottot . Roggero Alys . Ad(a) de Cornhametone . Ricardo de Bera . Germano de Ranvill' . Nicolao de Westbur' . Thoma de Ho . Gilliberto de Esli . Petro de Hattinggeli . et multis aliis ibidem tunc presentibus.

> Follows a final concord made one month from Michaelmas 1210, by which the land and homage of Eustace de Greinville were transferred to the see of Winchester in return for a quitclaim from bp des Roches to earl Aimery of all right in a common pasture at Marlow (*Selborne Charters* ii 63). For Eustace de Greinville see below appendix 4 no.38. For Aimery's grant of the land to Eustace, to be dated 1200 × 1201, see *Gloucester Charters* no.94; *PR 3 John* 56; *Red Book* 154, although earl Aimery's charter itself must date after 1205 since it is witnessed by master Robert Basset as des Roches' steward. The present witness list is largely made up of the bp's secular tenantry, the charter being more appropriate to the bp's secular than his clerical household.

*10. Geoffrey de Lisle

Cyrograph between the bishop and Geoffrey de Insula, by which Geoffrey fined with the bishop for custody of the land and heirs of John de Andely, with the marriage of his heirs.   [29 September 1225 × 6 June 1227, ?June 1227]

> Confirmed but not recited by king Henry III, 6 June 1227, who, supposedly at the bidding of Geoffrey (more likely at that of des Roches), undertook to distrain Geoffrey and his heirs should they fail to make the payments due to des Roches (*Pat.R.1225-32* 127). John de Andely appears to have died c.1225, since in the years 1225-7 his lands in the bp's manor of

Ebbesbourne were sown and stocked, presumably because they were in the bp's hands (Mss.21DR m.2; 22DR m.1d, and see *CRR* xii nos. 1288,1918). The sale of custody to Geoffrey de Lisle was probably intended to boost the bp's finances on the eve of his departure for the crusade. For the Andely lands, 4 fees held of the see of Winchester in Candover, Ebbesbourne and elsewhere, see *Reg.Pont.* 387,593; *Red Book* 205-6; *BF* 735.

11. Godstow Abbey

Confirmation of an annual pension of twenty shillings in the church of Faringdon as granted by R(obert) bishop of Exeter and confirmed by R(ichard) and Godfrey bishops of Winchester. [March 1206 × 5 September 1211]

B = PRO E164/20 (Godstow cartulary) fos. 58v-59r (45v-46r). s.xiv in.

Middle English abstract from Bodl. ms. Rawlinson B 408 fo. 42r (52r) pd in *The English Register of Godstow Nunnery near Oxford*, ed. A. Clark, 2 vols., Early English Text Society (London 1905-11) i 167.

Petrus Dei gratia Wintoniensis episcopus universis ad quos presentes littere pervenerint salutem in domino. Sicut ex instrumento publico bone memorie Godefridi quondam Wintoniensis episcopi predecessoris nostri percepimus R(obertus) quondam Exon' episcopus monialibus de Godestowe .xx. solidos in ecclesia de Farendon' que sita est in Wintoniensi diocesi annuatim percipiendos, tanquam patronus caritatis intuitu assignavit, cuius assignationem bone memorie R(icardus) Wintoniensis episcopus tanquam diocesanus carta propria confirmavit, predictus autem Godefridus predecessor noster ipsius inherens vestigiis favorem suum predicte concessioni apposuit et donationem predictam per cartam suam autenticam approbavit. Nos igitur eis in hiis que pie gesserunt succedere cum Dei adiutorio cupientes, predictorum .xx. solidorum annuam pensionem, sicut in predecessorum nostrorum autenticis[a] continetur, predictis monialibus et ecclesie sue auctoritate pontificalis officii presentis carte patrocinio roboramus et sigilli nostri appositione munimus. Hiis testibus: magistro Iohanne de London', Gregorio capellano, magistro Roberto de Pavilly, magistro Roberto Basset, magistro Alano de Stokes, magistro Thoma de Ely, magistro Petro Russig(nol'), Galfrido de Calce et Dionisio clericis et multis aliis.

[a] *Foliation* auten/ticis ms fo.59r

Before the death of master John of London, below appendix 4 no.1. For the church of Faringdon near Alton (Hants.) see *VCH Hampshire* iii 20-22. For the grants by bps Richard and Godfrey of Winchester see *EEA* viii nos.150,203.

*12. Gunnora de Bendenges

Letters patent testifying that Gunnora de Bendenges was legitimately married to John fitz Hugh. [1220 × September 1223, ?c. September 1223]

> Produced by Gunnora in the course of litigation over her dower in September 1223 (*CRR* xi no.1037; *BNB* iii 503), a dispute arising from the bigamous, second marriage of her husband. For John fitz Hugh (d.1220 × 7 March 1222, on crusade), an important royal bailiff in Hampshire and elsewhere, see *Interdict Docs.* 12–14. The litigation between his two wives and their children was to last for the next twenty years.

13. Halesowen Abbey

Foundation charter of Halesowen Abbey. With the assent of (Walter) bishop of Worcester, the bishop grants the church and manor of Halesowen as given to him by king John, for the establishment of an abbey of Premonstratensian canons dedicated to the Blessed Virgin Mary, to pray for the souls of king John and his progenitors, the soul of bishop des Roches, his predecessors, and for the whole of Christendom. The bishops of Winchester are henceforth to act as the abbey's patrons. [25 July 1215 × September 1215, ?c. 8 August 1215]

> B = Untraced, formerly Birmingham Public Library, Hagley collection ms. 351147, an inspeximus by bp Adam de Orleton of Worcester, dated 4 January 1332, sold at Sotheby's on 12 December 1978 to a private collector and since inaccessible.
> Pd from B in Colvin *White Canons* 350–1 no.15, whence the version below. Pd (calendar) from B in *HMC 2nd Report* (1871) 38; *Lyttelton Charters* no.96; *The Lyttelton Papers* (Sotheby's sale catalogue, 12 December 1978) 22 no.13. Part only pd (inaccurately) from ?B in T. Nash, *Collections for the History of Worcestershire*, 2 vols. (London 1781–2) ii appendix p. xxviii no.11.

Omnibus sancte matris ecclesie filiis ad quos presens scriptum pervenerit Petrus Dei permissione Wintoniensis ecclesie minister humilis salutem in domino. Inproprie loquimur si domino Deo nostro cuius est terra et plenitudo eius[a] dare aliqua nos dicamus, qui etiam si qua bona fecerimus servos nos debemus inutiles[b] protestari. Nichil enim boni homo possidet nisi quod accepit de divine gratie largitate, unde propheta: Quid retribuam domino pro omnibus que retribuit michi[c]. Preterea si aliquibus bona facere videamur, non nobis proculdubio set divine sunt potius gratie ascribenda. Quia neque qui plantat neque qui rigat est aliquid secundum apostolum, set qui incrementum dat Deus[d]. Suum est ergo quod homo quandoque movetur ad bonum ad eum refertur si concepta voluntas perfectionis sortiatur effectum. Ideoque quia cooperatores suos esse voluit, qui tamen per se omnia fecit ad ingratitudinis vitium removendum satagere nos oportet ut eius auxilio suffragante ad ampliandum cultum eius specialiter operam demus, et aliis nichilominus pietatis operibus devotius insistamus, ne nos qui plus ceteris in hoc mundo recepisse aliquid cernimur sicut dicit beatus Gregorius[e] a mundi auctore gravius iudicemur. Inspirante igitur eo a quo omne datum optimum et omne donum perfectum[f], de assensu et voluntate venerabilis fratris Wygorn' episcopi diocesani concessimus omnipotenti Deo manerium et ecclesiam de Hales cum capellis et aliis pertinentiis, libertatibus et aliis consuetudinibus suis, ad fundandam ibi abbathiam sub Premonstratensis ordinis disciplina abbati et fratribus ibidem Deo servientibus qui pro tempore fuerint perpetuum profuturam. Quod quidem dominus Iohannes Anglie rex illustris nobis ad hoc contulerat et carta sua propria confirmarat, quam cartam ad maiorem rei evidentiam pariter cum nostra fratribus loci duximus concedendam. Volumus ergo ut iam(dicti) abbas et fratres predictum manerium et ecclesiam habeant et teneant libere et pacifice sicut in carta predicti domini plenius continetur. Volumus etiam ut ibi sit perpetuo abbas et fratrum conventus, sicut supranotavimus, qui Deo et gloriose virgini Marie devote et humiliter famulentur, et pro salute iamdicti domini regis Iohannis et animabus omnium progenitorum suorum et pro salute nostra et omnium episcoporum Wintoniensium specialiter tam predecessorum quam successorum nostrorum, et pro omni populo Christiano generaliter orationibus et tam hospitalitatis quam elemosinarum suffragiis interpellent et regulam Premonstratensis ordinis pariter et disciplinam observent. Nobis vero et omnibus episcopis Wintoniensibus successoribus nostris qui pro tempore fuerint in abbathia predicta ius patronatus perpetuo duximus retinendum. Ut autem factum istud ratum plenius habeatur et firmum et nulla possit malitia vel fraude dissolvi, concessionem istam presenti carta duximus confirmandam et sigilli nostri munimine roborandam. Hiis testibus: Waltero Wygorniensi episcopo, Simone Exoniensi, Willelmo Coventrensi episcopis, domino Pandulpho Norwycensi electo, Petro archidiacono Surr', Hugone Folioth[g] archidiacono Salepsbur', magistro Alano de Stok', Ricardo capellano nostro, Henrico de Leghia clerico, Dionisio clerico, Roberto de Clinchamp clerico, Willelmo de Batilli, Iohanne de Caritate, Rogero Alis, Hugone Maleth', Willelmo Cumyn, Waltero filio Radulphi.

[a] c.f. Ps. 23:1   [b] c.f. Mt.25:30   [c] Ps.115:12   [d] 1 Cor.3:7
[e] St Gregory *Homiliarum in Evangelia* 9.I.3 (Migne *PL* lxxvi col.1106)   [f] c.f. James 1:17
[g] ffolioth B

Dates after the election of Pandulph as bp of Norwich, 25 July 1215 (PRO E210/3480), and before the departure of the bps of Worcester, Coventry and the elect of Norwich for the Lateran Council, c. mid-September 1215 (Cheney *Innocent III and England* 388-9). On 8 August 1215 the king confirmed des Roches' intention to found a Premonstratensian house at Hales (*Rot. Chart.* 217). Des Roches, Pandulph, Simon bp of Exeter and master Peter archdn of Surrey all witness royal letters at Dover on 4 September 1215 (*RLP* 182). The present charter is remarkable for the extent of its scriptural and patristic quotation. For the foundation itself see Colvin *White Canons* 178-83. Halesowen had been amongst the churches held by des Roches prior to his election to Winchester (*RLP* 49). The grant of the manor and advowson by king John in October 1214 appears to have met a two-fold need; firstly as a reward for des Roches' service as regent-Justiciar during the previous six months, and secondly as compensation for his failure to achieve promotion to the archbpric of York. Halesowen lay within the diocese of Worcester whose incumbent, Walter de Gray, replaced des Roches as the royal candidate for York in the autumn of 1214 (Cheney *Innocent III and England* 162-5). De Gray's co-operation in the endowment of the new abbey might be regarded as a *quid pro quo* for des Roches' withdrawal from York. Unusual amongst foundations by a reigning bp, Halesowen lay far outside the diocese of its founder and its future patrons, the bps of Winchester. The abbey was colonized from Welbeck, a house under the patronage of the family of des Roches' clerk, master Eustace de Fauconberg (appendix 4 no.30). The peculiar circumstances of its foundation were to ensure that it later occupied an uneasy position, cut-off from the direct influence of its patrons. From the 1240s frequent disputes arose between the abbey and its tenants, who claimed exemption from any increase in services, on the grounds that the manor of Hales was of the crown's ancient demesne. In turn this dispute spread to another of des Roches' foundations, Titchfield, colonized from Hales in 1233, where an entirely spurious claim to ancient demesne status was lodged, almost certainly in emulation of the tenants of Hales (G.C. Homans, *English Villagers of the 13th Century* (Harvard 1942) 237,276-81; D.G. Watts, 'Peasant Discontent on the manors of Titchfield Abbey 1245-1405', *Proceedings of the Hampshire Field Club* xxxix (1983) 121-35). As late as the 1280s a team of artists responsible for tiling the floors of various of des Roches' foundations in Hampshire, appears to have extended its activities northwards to work on the re-tiling of Hales, whilst the literary productions of Hales and Titchfield proceeded in tandem into the fourteenth century (P.M. and A.R. Green, 'Medieval Tiles at Titchfield Abbey', *Proceedings of the Hampshire Field Club* xvii (1952) 6-30, esp. 27; B. Hill, 'Oxford Jesus College MS 29', *Notes and Queries* ccxx (1975) 98-105).

14. Hamble Priory

Confirmation of an annual pension of three marks granted by Richard bishop of Winchester (EEA viii no.151) payable from the church of Bishopstoke.

[March 1207 × July 1227, ?1217 × March 1223]

B = Winchester Cathedral Library, ms. Cartulary, part i, fo.36v no.73. s.xiv med. C = Winchester, Hampshire Record Office, ms. 21M65/A/1/2 (Register of Henry Woodlock) fo.196r. In an inspeximus by bp Woodlock dated 25 March 1315. D = Winchester College Muniments ms. 4222. A copy of C marked *extracta de regist'*. s.xiv med. E = Ibid. Muniments ms. 4227. An inspeximus of C by bp Wykeham dated 7 November 1400.

Pd from C in *Reg. Woodlock* 647, which misdates the inspeximus to 1308, confusing it with a similar confirmation of bp des Roches' charter which mentions but does not recite des Roches' text, for which see *WC* no.74. Pd (calendar) from B in *WC* no.73. Pd (calendar) from DE in *WCM* nos. 4222, 4227.

Universis sancte matris ecclesie filiis ad quos presens scriptum pervenerit[a] P(etrus) divina miseratione Wyntoniensis episcopus salutem in domino . Ad universitatis vestre notitiam volumus pervenire nos divine pietatis intuitu confirmasse .. priori et monachis ecclesie beati Andree de Hamele pensionem trium marcarum in ecclesia de Stokes[b] annuatim percipiendam quam ex concessione bone memorie Ricardi quondam Wyntoniensis episcopi predecessoris nostri predicti .. prior et monachi sunt adepti sicut in carta eiusdem episcopi plenius continetur. Ut autem hec nostra confirmatio^c futuris temporibus rata permaneat eam presentis carte et sigilli nostri appositione roboravimus. Hiis testibus: Philippo de Lucy^d, Philippo de Fakkenberge^e, magistro Hunfrido de Milleriis^f, magistro Alano de Stokes, Galfrido de Cauz, Willelmo capellano, Dionisio de Burgolio^g, Petro de Cancell', Willelmo filio Hunfridi^h, Hugone de Capell', Rogero de Rothomago et multis aliis.

^{a} ad quos presens scriptum pervenerit *only in* B filiis etc CDE ^{b} Stoke E
^{c} kmaco *cancelled* confirmatio C ^{d} Luci CDE
^{e} Philippo *not in* D; Fauc' D; Falkenberge CE
^{f} Humford *cancelled* Humfrido D. B *ends here* etc ^{g} Burgalio D ^{h} Humfr' D

Of the witnesses, Peter de Chanceaux only came to England in March 1207, making no appearance in the bp's household until after Michaelmas 1215. Master Humphrey de Millières makes no certain appearance there until 1217 × 1218. Master Philip de Fauconberg was appointed archdn of Huntingdon c. March × August 1223 and died 2 December 1228 (below appendix 4 nos.4,19,31). The bp set out for crusade in July 1227. The pension at Bishopstoke passed with the remainder of Hamble's property to Winchester College (*VCH Hampshire* iii 310).

15. Henry III, King of England

Notification addressed to king Henry (III), asking that he bring royal power to bear against fourteen named individuals, excommunicated for retention of tithes, who have remained unrepentant for more than forty days.

[28 October 1216 × 9 June 1238, ?August 1231 × 9 June 1238]

A = PRO C85/153/1. Dorse inaccessible; approx. 122 × 60 mm.; step at bottom left hand corner, tongue, seal and wrapping tie torn away. Originally sealed ?on tongue as letters close. Torn and rubbed, several words now illegible. Scribe 5.

Reverendo domino suo .H. Dei gratia illustri regi Anglie . domino Hibernie . duci Normannie et Aquitanie . comiti Andegavie .P. permissione divina Wint(oniensis) ecclesie minister humilis . salutem . et de hostibus triumphum . Vestra [no]verit^a excellentia Willelmum de Brademere . et Willelmum Hubersete . Ailwinum de Huberset' . Iohannem Nuittal . et Robertum filium Ailwini . Ricardum filium Osberti . et Reginaldum Molendinarium . Reinerum de Fonte . et Walterum Carrucarium . Reginaldum Carrucarium . Nicholaum Kynest . Hamonem de Berghes . Adam Stef . et Petrum filium Baldewini auctoritate ordinaria pro retencione decimarum feni . et molendin(i) et^b propter manifestam contumaciam eorum excomunicatos esse . et in excomunicatione illa per .xl. dies et amplius contumaciter . et incorrigibiliter perseverasse . Unde ....^{c} celsitudinem vestram devote rogamus quatinus quos brach(i)um spirituale correctioni subi..........^{c} prevalet vestre^d regie potestatis brachium ....at^e . Valete per tempora longa.

<blockquote>
^{a} the first two letters of noverit missing, supplied ^{b} et inserted over line
.^{c} approximately eight letters missing in ms. ^{d} vestre inserted over line
^{e} word illegible ?processat

To be dated after the accession of Henry III, and in all likelihood to the years following the bp's return from crusade in 1231, after the king's coming of age. The circumstances of the excommunication remain obscure.
</blockquote>

16. Henry III, King of England

Notification addressed to king Henry (III), asking the king to do his part following the excommunication of Joan daughter of James who has remained excommunicate for more than forty days.

[28 October 1216 × 9 June 1238, ?August 1231 × 9 June 1238]

<blockquote>
A = PRO C85/153/2. Dorse inaccessible; approx. 128 × 34 mm.; step at bottom left hand corner of foot, tongue, seal and wrapping tie torn away. ?Originally sealed as letters close. Scribe 5.
</blockquote>

Reverendo domino suo .H. Dei gratia illustri regi Anglie domino Hibernie duci Normannie et Aquitanie comiti Andegavie .P. eadem gratia Wint(oniensis) episcopus salutem et fidelem famulatum . Noveritis quod Iohanna filia Iacobi in sententia excommunicationis . per .xl. dies et amplius perseveravit . nec per censuram ecclesiasticam vult iusticiam . et hoc excellentie vestre declaramus ut quod vestrum est in tali casu si placet facere dignemini . Valeat dominus rex.

<blockquote>
Dating as no.15 above. The circumstances of the excommunication remain obscure.
</blockquote>

*17. Henry III, King of England

Notification addressed to king (Henry III), asking assistance from the secular arm against Edith de Gleddun who has remained excommunicate and unrepentant for forty days.   [August 1231 × February 1235]

<blockquote>
Mentioned but not recited in the course of litigation before the king's court. In February 1235 Edith sued the abbot of Titchfield (founded after August 1231), claiming that the abbot had obtained the above letters from the bp by false testimony, and demanding redress for the imprisonment which she had undergone as a result (CRR xv no.1360).
</blockquote>

18. Hurstbourne Church

Inspeximus and confirmation of a charter of king Henry (III). The king declares that, following the death of Simon de Peregore dean of Chichester, he presented Nicholas de Neville, king's clerk, to the vacant church of Hurstbourne, believing that the advowson lay with the crown. However, Robert bishop of Salisbury protested that Hurstbourne pertained to the Salisbury prebend of Burbage by reason of a pension that the prebendary of Burbage received from the dean (of Chichester). Nicholas de Neville having been admitted to Hurstbourne saving the rights of Salisbury, the king declares that he has now inspected charters of kings H(enry I), H(enry II), R(ichard I) and John, and a charter of bishop Godfrey of Winchester (EEA viii no. 238) who in the presence of king J(ohn) and at the presentation of bishop Herbert of Salisbury, with the assent of Richard Barre archdeacon of Ely, prebendary of Burbage, admitted (Simon) dean (of Chichester) as perpetual vicar of the church of Hurstbourne, paying an annual pension of five marks to Burbage. The king therefore acknowledges that Hurstbourne is annexed to the prebend of Burbage, reserving possession of Hurstbourne to Nicholas de Neville for life, rendering the aforesaid five mark annual pension. Given at Worcester, 21 May 1232. The bishop of Winchester hereby confirms the king's charter in as far as he is able, and declares that Hurstbourne church is annexed to the prebend of Burbage.

Woodstock, 8 January 1233

A = Salisbury Cathedral Library, D.& C. Muniments ms. Press III, Hurstbourne and Burbage box. Heavily stained on both dorse and recto. Endorsed: *Husseburn'* (s.xiii); *55. Ecclesia de Husseborn annex' praebende*........(illeg.) *An.1232* (s.xvii); approx. 280 × 173+30 mm.; parchment tag, sealing method 1; seal tied up inside linen bag, reduced to crumbs and inaccessible.

B = Ibid. ms. Liber Evidentiarum C pp.127-8 no.150. s.xiii ex. C = Trowbridge, Wiltshire County Record Office, Bp of Salisbury ms. D1/1/3 (Registrum Rubrum) fo.32v no.114. s.xiv. D = Ibid. Bp of Salisbury ms. D1/1/2 (Liber Evidentiarum B) fo.40r no.114. s.xiv. Pd (calendar) from A in *HMC Various Collections* i (1901) 359. The royal charter only, pd from ABCD in *Sarum Charters* 227-8 no.cciii. Further copies of the royal charter are to be found in PRO C53/26 (Charter Roll 16 Henry III) m.10 (calendared *Cal. Chart. R. 1226-57* 155); and in London, Inner Temple ms. Petyt 511/18 (Salisbury cartulary) pp.43-4.

.Universis Cristi fidelibus has litteras inspecturis .P. Dei gratia . Winton(iensis) . episcopus . salutem . in domino . Noveritis nos litteras domini .H. regis Anglie inspexisse in hec verba . H. Dei gratia . rex Anglie dominus Hybernie . dux Norman(nie) . Aquitanie . et comes Andegavie . archiepiscopis . episcopis . abbatibus . prioribus . comitibus . baronibus . iusticiariis . vicecomitibus . prepositis . ministris . et omnibus ballivis . et fidelibus suis . salutem . Ad omnium noticiam volumus pervenire quod cum per mortem magistri[a] . Symonis . de Peregore . quondam decani . Cicestren' . vaccaret ecclesia de

Hwusseburne . et ad ipsam presentassemus Nicholaum de Neovila clericum nostrum credentes ipsam esse de donacione nostra opponentibus autem se venerabili patre Roberto . Sar' . episcopo . et capitulo suo et asserentibus ipsam ecclesiam pertinere ad prebendam de Burebech' . que est prebenda ecclesie Sar' . ratione pensionis annue quam canonicus prebende predicte inde percipere consuevit per manum ipsius . decani . tandem ad dictam ecclesiam memoratus clericus noster de consensu partium taliter est admissus et in ea persona institutus *b*salvvo iure cuiuslibet quod nichil iuris nobis inde accresceret nec ecclesie Sar' . deperiret in predicta ecclesia de Hwseburne . Quoniam vero per inspectionem cartarum .H. regis avi .H. regis avi nostri . et ipsius .H. regis avi nostri . et .R. regis avunculi nostri . et domini .I. regis patris nostri *c*et etiam per inspectionem carte .Godefridi. quondam . Winton(iensis) episcopi . qui in presencia domini .I. regis patris nostri*c* de volumtate*d* eius et . assensu ad presentacionem .Hereb(erti). quondam Sar' . episcopi . sicut in carta ipsius Godefredi plenius continetur que est in ecclesia Sar' . et etiam de consensu et concessione .Ric(ardi). Bar' . archidiaconi . Eliens'. et canonici .Sar'. prebendati in dicta prebenda de Borbech' . sicut in carta predicti Ric(ardi) plenius continetur que similiter est in ecclesia .Sar'. in predicta ecclesia de Hwseburne . predictum . decanum . perpetuum vicarium constituit reddendo inde annuatim canonico predicte prebende quinque marchas . nobis plenius constitit de iure predicti .Sar'. et ecclesie sue quod habent in ecclesia predicta de Hwseburne . ratione donationis . concessionis . et confirmacionis predictorum progenitorum nostrorum nolentes quod iuri eorum aliquid depereat in dicta ecclesia de Hwseburne *e*presentibus litteris duximus protestandum pro nobis et heredibus nostris . quod dicta ecclesia de Hwseburne*e* libere et quiete pertinet ad ecclesiam predictam .Sar'. tanquam membrum predicte prebende de Borbech' . *f*Quare volumus et concedimus pro nobis et heredibus nostris quod ipsam in perpetuum teneat in puram et perpetuam helemosinam tanquam membrum predicte prebende de Borbech'*f*. sicut predictum est . salva predicto .N. clerico nostro possessione eiusdem ecclesie tota vita sua . cum solutione predicte pensionis annue quinque march(arum) . Hiis testibus . H. de Burgo . comite Cancie . iusticiario Anglie . Stephano . de Segrava . Godefrido de Craucumb'*g* . Iohanne filio Philippi . Galfrido . dispensatore . H. de Capella . et aliis . Dat' . per manum venerabilis patris .R. Cicestren' episcopi . cancellarii nostri . apud . Wigorniam . vicesimo primo die Maii . anno regni nostri . sextodecimo . Cum igitur ex ipsius carte inspectione nobis constiterit predictam ecclesiam de Hwseburne . ad prebendam de Borbech' . ac per hec ad Sar' ecclesiam pertinere . nos eiusdem domini regis concessionem ratam habentes eandem ecclesiam de Hwseburne . dicte . Sar' ecclesie tanquam membrum prenominate prebende de Bo[rb]ech'*h* .

pontificali auctoritate duximus concedendam . ipsius domini regis concessionem quantum in nobis est confirmantes . Volumus igitur et presenti carta precipimus ut eadem ecclesia dicte prebende in perpetuum sit annexa . Ut autem hec nostra concessio et confirmacio robur obtineat firmitatis .' presentem cartam sigilli nostri fecimus munimine roborari . Dat' . apud . Wodestoc . die sabbati . propxima^j . post Epiphaniam . domini . anno . domini . m°.cc°°. tricesimo secundo . Hiis testibus . magistris I. Lemovicen' . R. de Aplubi . P. de Chanceus . W. filio Hunfredi . Herberto clerico . et aliis.

minor variations in spelling and word order in copies not noticed
^{a} magistris A *with the final s erased* ^{b} B p.128 $^{c-c}$ et etiam........nostri *omitted* in CD
^{d} sic. A voluntate BCD $^{e-e}$ presentibus litteris...........Hwseburne *omitted in* C
$^{f-f}$ Quare volumus......Borbech' *omitted in* BCD ^{g} BCD *end here*
^{h} hole in A, *letters in brackets supplied* ^{j} sic A

The present text marks the closing stage in a long and complicated legal battle, stretching back to the 1150s, whereby the two churches of Hurstbourne and Burbage were first united under a single priest, then severed, and finally re-united as a prebend within Salisbury cathedral. In the meantime, despite acknowledging the dependence of Hurstbourne upon Salisbury, king John and his son, Henry III, were able to intrude their own candidates into the church. Between c.1213 and 1230 Burbage had been held by master Bartholomew des Roches, the bp's nephew, passing subsequently to master Luke des Roches, archdn of Surrey; hence the bp's particular interest in the case. Although the royal charter, inspected above, allows for life possession of Hurstbourne by Nicholas de Neville (d. 1245), the living was already annexed to master Luke's prebend of Burbage by 1236 at the latest (*BF* 1365, and for the date of Nicholas' death, see *CPR 1232-47* 448-9). For a full account of the dispute see N. Vincent. 'A Prebend in the Making: the churches of Hurstbourne and Burbage 1100-1250' (forthcoming).

19. Hyde Abbey

Assignment of sixty shillings each year in the mills of Durngate and Drayton, Winchester, in compensation for rents worth sixty shillings a year in the mill of Nunnaminster, the mill of Jocelin the Queen's brother and the mill of the Hospitallers, which rents bishop Godfrey de Lucy granted to Hyde (EEA viii no.207) in place of a similar sum which the monks once received from Roger de Molendinis.   [March 1206 × 29 September 1208]

B = BL ms. Cotton Domitian A xiv (Hyde abbey cartulary) fo.119r-v. s.xiii ex.

Universis Cristi fidelibus ad quos presens scriptum pervenerit P(etrus) Dei gratia Wintoniensis episcopus eternam in domino salutem. Cum bone memorie Godefridus predecessor noster concessisset dilectis filiis nostris abbati et conventui de Hyda sexaginta solidos annuos de molendino abbatisse sancte Marie et de molendino Hospitalariorum et de molendino Iocelini^a fratris regine singulis annis percipiendos in recompensatione .lx. solid(orum) quos Rogerus de Molend(inis) eis solvere consueverat, nos de consensu predictorum abbatis et conventus in recompensatione eorundem .lx. solid (orum) assignavimus eisdem abbati et conventui .lx. solidos, scilicet de molendino de Durngate .xxx. solidos, et de molendino de Draiton' .xxx. solidos, singulis annis ad festum sancti Michaelis percipiendos per manus eorum qui[b] eadem molendina de nobis vel de successoribus nostris tenuerint. Quos etiam .lx. solidos annuos presentium auctoritate ipsis confirmamus percipiendos donec in alio certo redditu eis pro voluntate sua competenter[c] satisfecerimus. Quod si forte de eisdem molendinis predicti .lx. solidi aliquando non potuerint eis reddi, nos vel successores nostri de aliis redditibus nostris ipsos .lx. solidos plene faciemus persolvi. Hiis testibus etc.

[a] Ioceline *corrected*   [b] *word erased (illeg.)*   [c] Ms fo.119v

The rents granted by bp de Lucy in the mills of Nunnaminster, Jocelin and the Hospitallers had already been restored to bp des Roches by the time of the earliest of the surviving episcopal account rolls, being paid for the whole year Michaelmas 1208-Michaelmas 1209 (*PR4DR* 77). Probably at about this time, abbot John Suthill of Hyde (*c*.1180- before June 1222) granted bp des Roches the whole course of the river Itchen within Hyde's lands, together with fishing rights and rights to grind corn as held by Roger de Molendinis, in consideration of the bp's grant recited above. The abbot had already confirmed to bp de Lucy part of a meadow called 'Inmede' and all rights on the Itchen granted to the bp's predecessors by Roger de Molendinis (*WC* nos.463-4; BL ms. Harley 1761 (Hyde cartulary) fo.32v). For Roger de Molendinis (fl.1148) see *Winchester in the Early Middle Ages*, ed. Barlow, Biddle *et al*, *Winchester Studies* i (Oxford 1976), no.874. For the mills named in des Roches' charter see below no.20.

20. Hyde Abbey

Mandate to the keeper and receiver of the bishop's house at Wolvesey to pay sixty shillings each year to the abbot of Hyde in compensation for the sixty shillings which the abbot used to receive from Roger de Molendinis, in accordance with charters on this subject issued by the bishop and his predecessor (no.19). The payment is to continue in perpetuity without any further instruction from the bishop beyond the present letters patent, which are to remain with the abbot and convent of Hyde. [March 1206 × 29 September 1208]

B = BL ms. Cotton Domitian A xiv (Hyde cartulary) fo.119v. s.xiii ex.

P(etrus) Dei gratia Wintoniensis episcopus dilecto et fideli suo custodi domus sue de Wulvesya et receptori suo ibidem quicumque pro tempore fuerit salutem[a] gratiam et benedictionem. Mandamus vobis quatinus sine occasione reddatis dilecto filio abbati de Hyda sexaginta solidos annuatim ad festum sancti Michaelis in quibus tenemur eidem pro recompensatione .lx. solidorum quos Rogerus de Molendinis ei reddere consueverat secundum tenorem cartarum dicto abbati super hoc a predecessore nostro et a nobis confectarum.

Volumus etiam quod dicta solutio sexaginta solidorum annuatim fiat predicto abbati vel attornatis suis non expectato a nobis aliquo alio mandato. Et in huius rei testimonium has litteras nostras patentes sigillo nostro signatas eidem fieri fecimus in testimonium predicte solutionis predicto abbati vel attornatis suis faciende, in perpetuum eidem abbati remansuras.

[a] *letter erased (illeg.)*

For the date see above no.19. For the mills of Winchester see Keene *Survey* pp. 59–63. Drayton mill is to be identified with the mill of St Cross and under bp des Roches was farmed by the bailiffs of St Cross Hospital at £4 10 shillings a year. Durngate mill was farmed during the same period for *c*. £7 a year (Mss. 7DR m.11d; 9DR m.10d; 13DR m.14d; 14DR m.12d). The present charter is significant in two respects. Firstly, it provides incontrovertible evidence that the bp's exchequer was situated at Wolvesey, and not, as has been mistakenly suggested, at Southwark (*PR6DR* pp.xxi-iii). Secondly, it demonstrates that the Winchester pipe rolls do not provide a full account of fixed outgoings at the episcopal exchequer. Despite the fact that the 60 shillings continued to be paid to Hyde from the Drayton and Durngate mills until at least the 1260s (see the recognisances issued in the time of bp J(ohn of Exeter or John de Pontoise) in BL ms. Harley 1761 (Hyde cartulary) fo.32v), no such payment is recorded as an expense on either the Winchester or the mills accounts of the episcopal pipe rolls under bp des Roches. A payment of £6 to cover two years of the 60 shillings, is first recorded in an account of the vacancy receipts after des Roches' death, in which context the money is said to be owing 'for the mill of Stanham which was demolished' (PRO SC6/1143/1). The form of the present letters, a patent delivered to Hyde presumably for annual submission at the bp's exchequer, provides an interesting indication of procedure at Wolvesey.

21. Hyde Abbey

Confirmation of grant made by Peter fitz Herbert of one hundred pounds of wax each year, to be paid from the church of Kingsclere towards lights at Hyde.

Meon, 28 May 1209

B = Winchester College Muniments ms. 12094 (copy). s.xiii med.
Pd (calendar) from B in *WCM* no. 12094. Confirmed but not recited in a charter of bp William de Raleigh, 19 Dec. 1249; *WC* no. 394. Cited but not recited in full, in a case before the Bench, Michaelmas term 20 Edward III; PRO CP40/348 m.370d.

Omnibus sancte matris ecclesie filiis ad quos presens carta pervenerit Petrus divina miseratione Wintoniensis ecclesie minister humilis eternam in domino salutem. Noverit universitas vestra nos divine pietatis intuitu concessisse et presenti carta confirmasse abbati et conventui sancti Petri de Hida elemosinam centum librarum cere quas Petrus filius Hereberti eis concessit in ecclesia de Kingesclera percipiendas annuatim ad festum sancti Michaelis ad luminaria eiusdem monasterii per manum persone predicte ecclesie de Kingesclere, sicut carta prenominati Petri rationabiliter protestatur. Ut autem hec nostra concessio stabilis et inconcussa permaneat eam presenti scripto sigillo nostro munito duximus roborandam. Hiis testibus: magistro Iohanne de Lond' officiali, domino Iohanne de Mara, Alano de Stokes, Roberto Basset, Roberto de Pavilli magistris, domino Iohanne de Briwes, Eustachio de Greinvill', Roberto de Hotot, Thoma de Huntifeld', Dionisio et Stephano clericis, Daniele Butt', Petro Iuvene, Rogero Wacelin', Roberto de Ponsout, Willelmo de Gimmeg', Rogero Aliz et multis aliis. Datum per manum domini Gregorii capellani nostri apud Menes, xxviii die Maii pontificatus nostri anno iiii°.

> For the background see *VCH Hampshire* iv 265. During the court case of 1346 Hyde produced the bp's charter, which the court interpreted as a confirmation of a charter of Peter fitz Herbert dated 24 June 1217 (PRO CP40/348 m.370d). However, the latter can have been only a confirmation by fitz Herbert of his earlier award. The month of June 1217 witnessed the reassertion of royal authority over Winchester following the French occupation during the civil war; hence perhaps the date of fitz Herbert's confirmation. The bp's charter was issued after the imposition of the papal interdict, and follows a dispute over the advowson of Kingsclere between fitz Herbert and Hyde, December 1208 (*CRR* v 312, 314).

*22. Isemania, widow of Philip le Arceveske

Letters testifying to the legitimate marriage of Isemania to Philip le Arceveske.

[?1227]

> Mentioned in the course of the Hampshire eyre of January-February 1236 (PRO JUST1/775 m.11: *litteris episcopi Wintoniensis*). Beginning in the Trinity term of 1227, Isemania had sought dower in various rents held by Philip in the suburbs of Southampton. However, Philip appears to have contracted a marriage at Nottingham to a woman named Dionisia in November 1214, before that to Isemania at Winchester on 1 May 1219 (*CRR* xiii nos. 365, 1195, 1324, 1609). At some time Isemania was able to obtain a decision by papal judges in support of her claim. She carried her letters from des Roches (presumably issued before the bp's departure for crusade) to the justice Martin Pattishull in London, but her case was remitted for hearing before the Hampshire eyre of 1228, because Isemania claimed her privileges as a burgess not to be impleaded outside the borough (of Southampton). The case dragged on into 1229, when Dionisia was still alive and was able to produce letters from archbp Walter of York, confirming her marriage. Isemenia's marriage was therefore judged invalid, and her claim to dower rejected without hope of appeal. The supposed letters of des Roches may actually have been issued by the bp's official rather than the bp himself, since in November 1228 the official was ordered to make an investigation into Isemania's claim (*CRR* xiii no.1195). Isemania's attempt to revive her claim in 1236, was remitted for hearing by the Bench in the summer of 1236, whereafter it disappears from view, presumably dismissed in light of the earlier decision.

*23. John the chaplain

Letters to pope Gregory (IX) testifying to the orders of John the chaplain.

[21 March 1227 × 9 June 1238]

> Mentioned (*litteras P(etri) Wintoniensis episcopi*) in undated letters of pope Gregory IX, notifying all recipients that the bearer, John the chaplain, has received papal licence to

celebrate (mass) wherever he may go (*ubicumque voluerit in itinere suo*) (PRO C270/36/21, a copy s.xiv. Not in the register of Gregory IX).

*24. Kingsmead Priory

Indulgence of fifteen days remission of enjoined penance for all those contributing to the building work of the priory. [*c*. 1218]

> Mentioned in BL ms. Wolley Charter XI.25, pd by Rose Graham, 'An appeal for the Church and buildings of Kingsmead Priory *c*.1218', *The Antiquaries Journal* xi (1931) 51–4, and see also *Acta Langton* 155 no.8. For the date see Graham, *loc.cit.*. Similar indulgences were issued by the archbp of Canterbury and the bps of London, Lincoln, Worcester, Salisbury, Bath, Chester and Exeter.

*25. Le Mans, Bishop, Dean and Chapter

A contract said to have been made between P(eter) bishop of Winchester on one side and the bishop, dean and the chapter of Le Mans on the other, by which the bishop, dean and chapter sold to the bishop of Winchester possession of their land at Kingston Deverill. [*c.* July 1233 × February 1237]

> Mentioned in letters from bp Geoffrey, R. the dean and the chapter of Le Mans directed to bp des Roches, dated 2 February 1236/7, complaining of des Roches' demand that Le Mans pay him the farm owing for the past year for the land. They complain that the land was deliberately sold to des Roches at a price £100 below its value in order that Le Mans might be spared the expense of litigation involved in buying out the previous farmer (PRO SC1/47/53). Des Roches' first recorded intervention in the neighbourhood came in July 1233 when the king confirmed a sale to him by Aimery Buche of all his lands in Little Kington for 100 marks (*Cal.Chart.R.1226–57* 183, misdated). On 13 April 1236 the pope confirmed the sale to des Roches by Le Mans for £40, of land at Kingston Deverill, through the agency of master Stephen de Arenis, canon of Le Mans (*Cal.Pap.Reg.* i 152–3). Confusion over the arrangement may well have arisen because of des Roches' absence abroad, 1235–6. This in turn may explain why the letters from Le Mans survive not in des Roches' archive but in the archive of the royal chancery, having perhaps been delivered to the king when Le Mans' proctors were unable to approach des Roches. The land itself was later put towards des Roches' posthumous foundation at Netley by the bp's executors (Meekings 'The Early Years of Netley Abbey' 5–6).

26. Lewes Priory

Appropriation of the church of St Olave Southwark saving a vicar's portion.
[5 September 1211 × 18 June 1224, ?1212 × 1219]

> B = BL ms. Cotton Vespasian F xv (Lewes cartulary) fo.189v (218v). s.xv med. C = PRO E40/14233. An inspeximus by Luke des Roches archdn of Surrey dated 11 August 1238. D = BL ms. Cotton Vespasian F xv (Lewes cartulary) fo.190r (219r). s.xv med., copy of C.

22   PART 1: DIOCESAN AFFAIRS

> Pd (calendar) from BC in 'Surrey Portion' 101–2 nos. 49, 50. Mentioned as a confirmation to Lewes from bp des Roches in letters of W(alter) prior of St Swithun's Winchester: PRO E40/15431.

Omnibus sancte matris ecclesie filiis ad quos presens scriptum pervenerit Petrus [divina]^a miseratione Wintoniensis ecclesie minister humilis eternam in domino salutem. Ecclesiasticis officiis^b desudantes dignum est ecclesiastica remuneratione gaudere et orationibus pro salute fidelium iugiter insistentes et hospitalitatis opera pia [devotione ex]cercentes^a pium est de bonis ecclesie locupletari. Ea propter attendentes quam pio affectu expansis caritatis^c visceribus domus sancti Pancratii de Lewes hospites passim admittat, dilectis filiis monachis ibidem Deo servientibus ecclesia[m sancti Olavi]^a in Suthewerk'^{d} cum omnibus pertinentiis suis in usus et refectionem hospitum convertendam duximus concedere et auctoritate episcopali^e confirmare, salva honesta et sufficienti^f sustentatione vicarii qui ad eorundem monachorum^g pre[sentationem et per nost]ram^a vel successorum nostrorum institutionem ministrabit in eadem, salva etiam in omnibus Wintoniensis ecclesie dignitate. Et ut hec^h nostra concessio perpetua gaudeat firmitate eam presenti scripto et sigilli nostri^j patrocin[io communivi]mus^a . Hiis testibus: magistro Alano de Stok' tunc officiali nostro, magistro Roberto Basset, magistro Roberto de Aren', magistro Thoma de Ebleburn^k, Umfrido capellano^l, magistro Henrico de Walton', magistro Raimundo de Winton'^{m}, Willelmo de Sancto Maxentio, Ricardo de Berking, Dionisio de Camera, Ricardo de Elmham^n et aliis multis^o.

^{a} *letters within brackets perished* C ^{b} oficiis C ^{c} karitatis CD ^{d} Suwerk' C Suthwerk' D ^{e} episcopali auctoritate CD ^{f} honesta et sufficiente CD honestate B
^{g} monacorum C ^{h} hec *not in* C ^{j} nostro C
^{k} Eppelburn CD ^{l} Humfrido cappellano C Humfrido capellano D
^{m} Rann' de Winterburn' D Rann' de ....a C ^{n} Helmam C ^{o} multis aliis CD

Dating problematical. Of the witnesses, master Robert de Airaines d. 18 June 1224. Master Alan of Stokes succeeded as bp's official after the death of master John of London (d. 5 September 1211). William de St Maixent last appears in 1219 (below appendix 4 nos.1,2,27,35). For master Thomas of Ebbesbourne, bailiff of des Roches' manor of Fonthill 1208–9, active in the bp's service to 1223–4, later subdean of Salisbury c.1237–1240, probably author of a medical receipt, implying expertise as a physician, see *Fasti* iv 39, 69; *PR 4 DR* 80; Mss. 7DR m.10; 9DR m.2d; 15DR m.7d; 19DR m.9. St Olave's church, or at least the church's advowson, had been granted to Lewes in the 1090s (*The Chartulary of the priory of St Pancras Lewes*, i, ed. L.F. Salzman, Sussex Rec. Soc., xxxviii (1932) 11, 22). Lewes farmed the church to the hospital of St Thomas Southwark, saving a pension of 6m p.a., in the time of R(obert?) master of St Thomas' c. 1215 × 1224, and it was confirmed as a possession of St Thomas' in a bull of pope Honorius III, 13 Jan. 1221 (BL ms. Stowe 942 (St Thomas' cartulary) fo.6v; 'Surrey Portion' 103–4). It may well be that St Thomas' had to be forced to relinquish possession by the bp's officers. This, together with the provision for a vicarage at St Olave's may explain the recital of the present text by master Luke, archdn of Surrey, who on 11 August 1238, following the bp's death, admitted Adam of Southwark as vicar, establishing a pension of 8 marks a year from Lewes ('Surrey Portion' nos. 50–53).

27. Lire Abbey

Institution of Robert fitz Robert to the church of Arreton at the presentation of the abbot and convent, reserving the customary pension to Lire.

[March 1206 × 9 June 1238]

B = BL ms. Egerton 3667 (Carisbrooke cartulary) fo.29v no.xxxvii, witnesses omitted. s.xiii med.

Pd from B in *Carisbrooke Cart.* no.38.

For Lire's annual pension of 40 shillings in Arreton church see *Ibid.* no. 23.

28. Lire Abbey

Institution of master Simon de Wautham to the church of St Michael, Niton, at the presentation of the abbot and convent, saving the entire income of the church to master P(eter) Russing(nol), perpetual vicar of the same, who is to bear all the church's expenses, paying an annual pension of five marks to master Simon, and saving the customary annual pension to Lire.

[March 1206 × July 1227, ?March 1206 × December 1219]

B = BL ms. Egerton 3667 (Carisbrooke cartulary) fo.29v no.xxxviii, witnesses omitted. s.xiii med.

Pd from B in *Carisbrooke Cart.* no.39 which gives Wantham in error for Wautham.

For Lire's annual pension of 5 shillings from Niton see *Ibid.* no.23. The present charter was cited in a court case of Hilary term 1229 when, following the death of master Simon, the advowson of Niton was retained by Lire in the face of a claim from the crown (*CRR* xiii no.1556). Bp des Roches was absent on crusade after July 1227. For master Simon de Wautham see also below no. 72. Des Roches employed two clerks named master Peter Russinol. Master Peter Russinol I, to whom this charter is more likely to refer, was dead by July 1219 (below appendix 4 nos. 18,20)

29. Lire Abbey

Letters to (the abbot and convent) thanking them for their presentation of the bishop's kinsman (nepotem) *to the church of Newchurch and assuring them of the bishop's goodwill in future.*

[March 1206 × 9 June 1238, ?August 1231 × 9 June 1238]

B = BL ms. Egerton 3667 (Carisbrooke cartulary) fos.29v-30r no.xxxix. s.xiii med.
Pd from B in *Carisbrooke Cart.* no.40, which omits one word.

P(etrus) permissione divina Wintoniensis ecclesie minister humilis amicis et cetera. Super eo quod nepotem nostrum ad ecclesiam de Niwechirche presentare voluistis, cuius ecclesie patronatus ad vos de iure noscitur pertinere, dilectioni vestre grates referimus copiosas, et noscat dilectio[a] vestra quod ad prove(he)nda vestra et ecclesie vestre negotia nos invenietis in quibuscumque poterimus preparatos. Valete.

a fo.30r

> The church of Newchurch was held from before 1204 by master Philip de Lucy (d. before Easter 1233) who last appears in 1230 (*Ibid.* no.37; below appendix 4 no.34). Des Roches only returned from crusade in August 1231. By 1244 the church was held by master Peter de Gattebrig' (BL mss. Egerton 2104a (Wherwell cartulary) fos. 99r-v,114v; Cotton Tiberius D vi (Christchurch cartulary) fo.103r). These dates would suggest that the kinsman presented to Newchurch was most likely the bp's nephew, Aimery des Roches, rector of Preston (below appendix 4 no.42).

*30. London, William bishop of

Assignment by P(eter) bishop of Winchester and B(enedict) bishop of Rochester, acting on papal authority, of the three manors of Clacton, Southminster and Wickham St Paul to William, formerly bishop of London, to support him in his retirement. [January 1221]

> Mentioned in royal letters of 27 January 1221 and in a bull of pope Honorius III dated 6 May 1221; *Pat.R.1216-25* 280; *Cal.Pap.Reg.* i 81. Followed by the election of des Roches' familiar, master Eustace de Fauconberg, as bp of London. The election itself appears to have been attended by the poet, master Henry of Avranches, a protégé of des Roches. Des Roches himself entertained a messenger at Farnham in 1221, coming from the elect of London, suggesting a keen personal interest in Eustace's promotion (below appendix 4 no.30).

31. Marwell College

Statutes for the chapel of Marwell founded by bishop Henry (de Blois), to be served by four chaplains, laying down rules of residence and attendance at the canonical offices, rules of behaviour, an order of services, and renewing the decree of bishop Henry (EEA viii no.80) concerning the chaplains' stipends, granting them £12 each year at the bishop's exchequer towards lights and vestments, £8 being retained to form a common fund. The bishop confirms the chaplains in possession of the land outside the gate, and grants them fifty quarters of grain each year from the church of Bishopstoke together with four carts of hay. The chaplain elected prior by the rest of the community, is to render annual accounts of income and expenditure, any residue being spent in the uses of the community, and not distributed amongst the chaplains individually.
Marwell, 7 March 1227

> B = Winchester, Hampshire Record Office ms. 21M65/A/1/2 (Register of bp Henry Woodlock) fo. 24r-v. *c.*1305. C = PRO C66/162 (Patent Roll 18 Edward II part ii) m.15. A royal confirmation dated 28 April 1325. D = Winchester, Hampshire Record Office ms. 21M65/A/1/5 (Register of bp John of Stratford) fos. 185v-186r. Copy of B *c.*1330.

Pd from B in *Reg. Woodlock* 74–6. Pd from C in *Monasticon* vi 1344. Pd (calendar) from C in *CPR 1324–27* 114.

Omnibus sancte matris ecclesie filiis Petrus Dei gratia Wyntoniensis episcopus eternam in domino salutem. Zelus domus Dei nos comedit[a] quo succensi cultum divinum in capella de Merewelle[b] a bone memorie domino Henrico episcopo predecessore nostro constitutum duximus renovare et pia adiectione ampliare. In primis igitur predecessoris nostri vestigiis inherentes statuimus quod quatuor capellani honesti per episcopum Wyntoniensem qui pro tempore fuerit eligantur et in dicta capella instituantur in ea perpetuo ac iugiter servituri. Adicimus quoque quod unum diaconum teneant[c] qui eis in ecclesia et domo deserviat. Isti autem capellani simul vivant, simul dormiant, simul comedant[d] in domibus quas ad opus eorum infra septa nostra eis fecimus edificari. Ipsi quoque singulis annis unum ex se eligant cui tam in[e] ecclesia quam extra tanquam priori obediant, et si ipsi in electione concordare nequiverint, unus eorum per officialem episcopi ad illum annum preficiatur. Nullus itaque sine prioris licencia speciali se ab horis canonicis absentet nec a communi refectione, et qui contra fecerit graviter arguatur nisi infirmitate excusetur. Item nullus extra iaceat de nocte sine speciali licencia prioris. Prior autem ad hoc licenciam non concedat sine iusta causa. Item prior nulli licenciam det se absentandi ut ultra octo dies extra maneat sed[f] si aliquem ex aliqua necessitate maiori tempore abesse oporteat, ab officiali episcopi licencia requiratur. Si quis autem sine licencia extra domum pernoctaverit, quot noctibus absens fuerit sine licencia tot denarii de suo vestitu subtrahuntur, preter penam a priore canonice infligendam. Item nulli licencia detur se absentandi per ebdomodam[g] ultra tres vices per annum, et si quis vel per licenciam prioris vel sine licencia per plures vices vel[h] diutius quam statutum est se absentaverit, pena gravior per officialem infligatur. Prior autem duobus simul se absentandi licenciam nulla[j] ratione concedat. Si autem aliquis ex consuetudine sine licencia se absentaverit nec per penas inflictas se emendaverit, tanquam criminosus per officialem a domo perpetuo expellatur et alter subrogetur. Ipse autem prior quod de aliis statuimus in se observet, scilicet ut sine iusta et rationabili causa [nec] ab horis canonicis nec a domo se absentet, et si frequentius vel diutius quam oportet absens fuerit officiali episcopi denuncietur ut per ipsum corrigatur. Cum autem ipse prior ex necessitate absens fuerit, vices suas interim alicui de sociis committat. Quicquid autem pene peccuniarie alicui infligitur per priorem communibus usibus applicetur. Preterea de hiis que communiter eis perveniunt undecumque absens ex quacumque[k] causa nichil recipiat sive cum licencia sive sine[l] licencia se absentaverit sed totum cedat in usus residencium. Item quicumque capellanus de incontinentia convictus fuerit vel per testes vel per facti evidenciam vel per prolis susceptionem sine spe restitutionis ammoveatur, similiter si de alio turpi crimine convictus fuerit. Item ad divina officia conveniant in cappis nigris et superpelliciis. Divina vero officia secundum Saresber' celebrentur, ita quod in aurora matutinum officium dicatur et prima continue. Post primam missa pro fidelibus celebretur. Postea vero fiat interpolatio ad privatas orationes et missas privatas. Deinde ad tertiam conveniant, qua finita missa canonica celebretur, et postea sexta dicatur, et si dies solempnis ieiunii fuerit nona ante prandium dicatur, aliis vero diebus post prandium. Vespere et completorium hora congrua dicantur et omnia cantando etiam ferialibus diebus. Post hec de stipendiis dictorum capellanorum, decretum domini Henrici episcopi renovantes, statuimus ut[m] duodecim libras et viginti solidos ad lumen ecclesie habeant de scac(c)ario Wynt(oniensis) episcopi qui pro tempore fuerit, quas prior ad opus eorum recipiat quatuor temporibus solitis, et unicuique viginti solidos tribuat ad vestitum. Octo vero libras residuas prior retineat et expendat in communes usus consilio aliorum. Preterea mansos extra portam ad usus proprios a predecessore nostro illis concessos eis ad usus communes confirmamus. Preter hec et quinquaginta quarteria grani ad usus eorum communes adicimus que rector ecclesie de Stokes eis annuatim ministrabit ad quinque terminos, quorum primus est festum sancti Michaelis in quo cariare[n] faciet tria quarteria frumenti[o] et tria ordei et quatuor parva quarteria avene apud Merewell', in festo sancti Nicholai totidem, in festo Purificationis totidem, in festo sanctorum Philippi et Iacobi totidem, in festo apostolorum Petri et Pauli totidem. Recipient etiam annuatim dicti capellani a predicto rectore ecclesie de Stokes quatuor rationabiles carectatas feni in tempore fenedii. Prior autem in fine anni de omnibus receptis et expensis rationem reddat, et si quid remanserit in communes usus communi consilio expendatur, ita quod inter eos minime dividatur. Hec ita ad presens statuimus, successoribus nostris potestate reservata secundum Deum et temporis conditionem ordinationem nostram in melius commutandi. Ut autem hec nostra constitutio seu collatio futuris temporibus rata et inconcussa permaneat, eam presentis carte et sigilli nostri munimine confirmavimus. Dat' apud Merewelle anno ab incarnatione domini m° cc°° xxvi, dominica secunda Quadragesime.

[a] c.f. Ps.68:10   [b] Merewell' C   [c] teneant *over the line in* C   [d] commedant C
[e] in *over the line in* C   [f] set C   [g] ebdomadam C
[h] vel *written over an erasure in* B *where* quam *has been erased*   [j] ullam C   [k] B fo.24v
[l] sine *over the line in* B   [m] ut *repeated in* B *erased*   [n] cariari C   [o] frumuenti B

Dated to 1227 on the understanding that the bp's charters follow the year beginning on 25 March. The statutes themselves are interesting in several respects. They provide one of the earliest references to the Sarum right being adopted outside Salisbury itself; an indication of the close links between des Roches and his one-time rival, bp Richard Poer of Salisbury. The concern shown for the financial management of the Marwell chaplains may well reflect des

Roches' own experience in matters financial. In addition, the role ordained for the bp's official in disciplining and overseeing the college provides yet further proof of the importance attached to the office during des Roches' episcopate. Between 1208 and 1211 the Winchester pipe rolls provide evidence of the refurbishment of the episcopal palace at Marwell and its decoration with paintings (*PR4DR* 53-4; *PR6DR* 13,16). A William chaplain of Marwell appears to have served as bp's bailiff for the manor of Meon Church before 1218, and in 1218/9 spent six weeks with his men at the bp's manor of Downton in Wiltshire (Ms. 14DR mm.2, 7). In 1226/7, the year of the present charter, the chaplains of Marwell received 6 shillings and 8 pence for hay purchased from them at Twyford (Ms.22DR m.11d). As with no.20 above, although the present grant allows for an annual pension from the bishop's exchequer, no such payment is recorded on the Winchester pipe rolls. Only after the bishop's death, do we read of the provision of fixed alms from the episcopal exchequer to Hyde (above nos.19-20), to the hospital of St Mary Magdalene near Winchester, and to Marwell; these latter payments apparently totalling more than £13 a year besides occasional gifts of cloth and money (*Cal.Lib.R. 1226-40* 359; PRO E372/85 mm.3,3d; E372/87 m.3d; E372/88 m.12).

32. Mont-Saint-Michel Abbey

Confirmation to the abbot and convent of the churches of Basing and Selborne together with various pensions specified in a charter of bishop Godfrey de Lucy (EEA viii no.224). [March 1206 × 5 September 1211]

A = Oxford, Magdalen College, Muniments ms. Basing Charter 34. Endorsed: *Exhit' .viii. Id'. Febr'.RH-M.*, followed by the notarial sign of Hugh Mussel, notary to archbp Robert Winchelsey (1294-1313) (for Mussel see Cheney, *Notaries Public* 35-6n, 74n; *Reg.Winchelsey* xix); *3d Basyngstoke olim pertinesse ad Michaelis Mount* (s.xvii); approx. 146 × 71 + 21 mm.; fine seal and counterseal impressions in green wax slightly chipped at top, coloured cords attached through two eyelets.

Pd from A in Cheney *English Bishops' Chanceries* 80 n.1. Pd (calendar) from A in Baigent and Millard *Basingstoke* 652; *Selborne Charters* ii 2.

Omnibus Cristi fidelibus ad quos presens scriptum pervenerit .' P. permissione divina Winton(iensis) ecclesie minister humilis eternam in domino salutem . Viris relligiosis in petitionibus suis honestis specialiter intendere nos oportet . ut tanto liberius suo serviant creatori .' quanto ecclesie Dei prelatos sibi senserint in neccessitatibus benignius suffragari . Eapropter nos relligionem et caritatem abbatis et conventus monasterii sancti Michaelis de Periculo Maris pia consideratione pensantes ecclesias de Basing' . et de Seleburn' . in diocesi nostra sitas cum omnibus pertinentiis suis eis auctoritate pontificalis officii confirmamus . et pensiones etiam quas in eisdem ecclesiis bone memorie .G. predecessor noster pietatis intuitu noscitur concessisse . sicut in eius scripto autentico plenius continentur . In huius autem confirmationis nostre testimonium presentem eis cartam duximus concedendam . Testibus hiis . magistro Iohanne de London' . magistro Alano de Stok' . magistro . Roberto de Pavill' . magistro .R(oberto). Bass(et) . R. decano Winton(iensi) . et multis aliis . per manum .P(etri). Russign' . sigilli nostri custodis.

28   PART 1: DIOCESAN AFFAIRS

Master John of London died 5 September 1211 (below appendix 4 no.1). For bp de Lucy's agreement to annual pensions of 3 marks in Selborne and 12 marks in Basing see *Selborne Charters* ii 2.

33. Mottisfont Priory

Notification that the convent of Mottisfont, newly founded by William Brewer, has petitioned for the consecration of a cemetery. A(mo) treasurer of York, to whose office the parish church of Mottisfont pertains, opposed the request, but withdrew his objections following a visit by William Brewer to York to discuss the matter, as Brewer has signified to the bishop in his letters patent. The bishop has therefore consecrated the cemetery.

[March 1206 × 1 March 1218, ?c.March 1206 × March 1208]

B = York Minster Library ms. M2(3)2 (Treasurers' cartulary) fo. 2r-v. s.xiv.
Pd from B in *Cart.Treas.York* 14-15 no.7.

Mottisfont priory was founded by Brewer in 1201 and colonized from Osney abbey (*Ibid.* no.10; *AM* iv (Worcester) 50). The present charter was clearly issued early in the priory's history, and before the promotion of Hamo as dean of York, *c.*1217 × 1 March 1218 (*York Minster Fasti* i 2,23). It seems unlikely that the bp's consecration of the cemetery can have taken place after the imposition of the papal interdict (March 1208), which elsewhere in the diocese of Winchester appears to have precluded burial within consecrated ground (*AM* ii (Waverley) 282). The advowson of Mottisfont parish church had lain with the church of York since before 1086 (*Cart.Treas.York* pp.viii-ix). By 1223 at the latest the living had been conferred upon the bp's kinsman, Peter de Rivallis (*RLC* i 543; Winchester, Hampshire Record Office ms. 13M63/2 (Mottisfont cartulary II) fos.148r-9r).

34. Mottisfont Priory

Confirmation to Mottisfont priory founded by William Brewer, of rents, churches and tithes, forbidding anyone to lay hands upon the canons, their servants, men or property, confirming that the canons should adhere to the rule of St Augustine, and reciting and confirming various grants made to the priory, including William Brewer's grant of land at King's Somborne, Oakley, and the churches of Longstock and Ashley; John Brewer's grant of the church of Little Somborne, and king John's grant of the church of Eling, with the appropriation of the said churches, saving vicars' portions in every case.

Southwark, 7 May 1208

B = Winchester, Hampshire Record Office ms. 1M63/2 (Mottisfont cartulary II) fo.105r. s.xv.
C = Ibid. fos. 104v-105r. s.xv. An extremely corrupt copy whose inadequacy appears to have led the same scribe to draw up the more accurate version B immediately after completing C.

Omnibus Cristi fidelibus ad quos presens scriptum pervenerit P(etrus) divina miseratione Wintoniensis ecclesie minister[a] humilis salutem in domino. Ex officio pastorali tenemur viros religiosos, qui pauperes spiritu esse pro Cristo neglectis[b] lucris temporalibus elegerunt, speciali affectu diligere et quieti sue sollicite providere ut tanto uberiores fructus de continua in lege Dei meditatione percipiant quanto a conturbationibus[c] malignorum amplius fuerint ex ecclesiastica defencione securi. Inde siquidem[d] est quod nos bone memorie G(odefridi) quondam Wintoniensis episcopi predecessoris nostri vestigiis[e] inherentes, dilectos in Cristo filios nostros canonicos regulares de Mottesfont locum et monasterium eorum quod dilectus in Cristo filius Willemus Briwer' in eadem villa in honore summe et individue[f] Trinitatis in solo suo fundavit, possessiones etiam et redditus, ecclesias[g] sive decimas quas in episcopatu nostro rationabiliter adepti sunt vel[h] in posterum Deo donante iustis modis poterunt adipisci, sub nostra et Wintoniensis ecclesie protectione suscepimus et specialiter[j] auctoritate officii confirmamus episcopali auctoritate, firmiter inhibentes ne quis locum in quo divino sunt mancipati officio seu alias possessiones eorum invadere vi vel fraude vel ingenio malo occupare audeat vel[k] retinere aut fratres, conversos, servientes vel homines eorum aliqua violentia[l] perturbare sive ad eos fugientes[m] causa salutis sue conservande[n] a septis curie sue violenter presumat extrahere. Precipimus autem ut in eodem monasterio de Mottesfont ordo canonicus et regularis conversatio secundum regulam magni patris Augustini quam primi inhabitatores professi sunt observetur et[o] ipsum monasterium a cuiuslibet alterius domus religiose subiectione liberum permaneat absolutum[p] . Inter possessiones autem ipsorum has specialiter duximus exprimendas[q] : ex dono predicti Willelmi Briwer' totam terram quam habuit apud Mottesfont, item decimas[r] de essartis de Sombourn et decimas de terra et[s] de essartis de Hackel', item decimam[t] acram terre de predictis essartis[v], item ecclesiam de Stok'[w] et ecclesiam de Essel'[x] et ex dono Iohannis Briwer'[y] ecclesiam de Parva Sombourn, et ecclesiam de Eling'[z] de dono domini regis[aa] Iohannis, cum omnibus libertatibus et[bb] earum pertinentiis, habendas et tenendas imperpetuum et in usus proprios[cc] ipsorum canonicorum, salvis competentibus vicariis perpetuis, convertendas, salvo in omnibus iure episcopali auctoritate et dignitate Wintoniensis ecclesie. Quod ut in posterum ratum permaneat et inconcussum presenti scripto et sigilli nostri patrocinio duximus confirmandum[dd] . Dat' apud Suthwerk per manum P(etri) Russinol'[ee] clerici nostri, non' Maii pontificatus nostri anno iii°. Testibus hiis[ff] : Willemo de Huntedon'[gg] et R. Wintoniensi archidiaconis[hh], Gregorio capellano, I. de Mara, magistro Alano de Stokes, magistro R. de Pavill', magistro I. de Ramesbir'[ii], G. de Caleston', Dionisio clericis et multis aliis.

30   PART 1: DIOCESAN AFFAIRS

^{a} magister C ^{b} negleg' C neglectis *cancelled beneath* neclectis B. c.f.Mt.5:3
^{c} conturdiccionibus *cancelled* B ^{d} signidi C ^{e} nostris C; vestigiis B
^{f} summe et dividue C ^{g} ecclesiarum B ^{h} C *inserts* rationabiliter *cancelled*
^{j} episcopalis C; specialiter B ^{k} vel *not in* C ^{l} violendicia C *with attempts at correction*
^{m} fugientes ad eos C ^{n} conservar' B ^{o} B *inserts* per ^{p} permaneat et absolicioni C
^{q} exprimendas *not in* C ^{r} C fo.105r ^{s} et *not in* C ^{t} decima C ^{v} essatis C
^{w} Stokes C ^{x} Eling *cancelled* Essel B ^{y} Briwerr' C, Briuer B ^{z} Elyng C ^{aa} rog' C
^{bb} et *not in* B ^{cc} propriorum C ^{dd} confirmandi C
^{ee} Russinok BC ^{ff} anno tertio testibus istis C ^{gg} W. de Huntendun C
^{hh} archidiaconus B ^{jj} Pamesbirg' B

On 12 April 1208, three weeks after the imposition of the papal interdict, the bp was granted custody of the religious houses of his see. A week or so later William Brewer received the restoration of abbeys and priories within his fee (*RLC* i 111, 112b). The present award may well be associated with the uncertainties aroused by the interdict. For the priory itself see above no.33. Bp des Roches' text is adapted from a charter of his predecessor, bp de Lucy, dated 30 August 1204 (*EEA* viii no.225). The present charter copies much of the wording of the earlier text. In turn des Roches' language is mirrored in confirmations by prior W(alter) of St Swithun's (c.1216 × 1239), and archbp Stephen Langton (not in *Acta Langton*) (Winchester, Hampshire Record Office mss. 13M63/2 (Mottisfont cartulary II) fo. 172r; 13M63/1 (Mottisfont cartulary I) fos. 43v-44r). The present text for Mottisfont also supplies the source for the wording of the bp's foundation charter for Selborne priory, suggesting that Selborne was colonized from Mottisfont (Below no.49).

35. Newark Priory

Notification that for the remission of his sins, with the advice of Pandulph bishop-elect of Norwich, papal chamberlain and legate, and with the assent of the chapter of Winchester, the bishop has appropriated the church of Woking, then vacant, to prior John and the convent of Newark priory, reserving an annual pension of £6 and fourteen shillings to the monks of the church of St John the Baptist, Stoke (-by-Clare), in two annual installments of sixty seven shillings, and saving reasonable maintenance for a vicar to be presented by the canons of Newark; copies of the present charter being given both to the canons of Newark and the monks of Stoke.   [September 1218 × 25 February 1221]

B = BL ms. Cotton Appendix xxi (Stoke by Clare cartulary) fo. 149r-v no.ccclix (sic.). s. xiii med.
Pd from B in *Stoke by Clare Cart.* ii 302-3 no.457.

Hiis testibus: magistro Alano officiali, magistro Hunfrido de Mille(r)iis, Galfrido de Lovers, Philippo de Luci et aliis.

Dates to the legation of Pandulph, and before the election of master Eustace de Fauconberg as bp of London, since Eustace witnesses a confirmation of the annual pension specified above, issued by John prior of Newark (*Ibid.* no.456).

*36. Newark Priory

Augmentation to five marks of an annual pension of half a mark granted by bishop Godfrey de Lucy (EEA viii no.228) to Newark priory in the church of Wield. [March 1206 × 9 June 1238, ? March 1206 × *c.* December 1225]

> Mentioned in a charter of bp William de Raleigh, 19 December 1249, augmenting bp des Roches' award to a pension of 12 marks (BL ms. Cotton Nero Ciii fo.218r: in an inspeximus of 3 June 1283). For bp de Lucy's award, dated September 1189, see *WC* nos. 90–1, and see *Ibid.* no.93 for letters of pope Honorius III, dated 11 January 1225, confirming the appropriation of Wield church to Newark. If such an appropriation took place it must presumably have postdated des Roches' award cited above. However, as late as 1291 Newark was regarded merely as a pensioner at Wield, where the advowson continued to be exercised by the bps of Winchester. The church was appropriated to Newark by bp John de Pontoise by a charter dated 23 December 1291: *Reg. Pont.* 245; *WC* 91 (misdated 13 December); *VCH Hampshire* iii 347 (which gives 1306 as the date of the appropriation).

37. Nuneaton Priory

Admission and institution of master R. de Grandon to the chapel of Blendworth at the presentation of the prioress and nuns, reserving the customary pension to Nuneaton. [March 1206 × 5 September 1211]

> B = BL ms. Add. Ch. 47398 (Nuneaton cartulary roll) m.1 no.10. s.xiv.

P(etrus) Dei gratia Wintoniensis episcopus karissimis in Cristo universis fidelibus has litteras inspecturis salutem et dilectionem in domino. Sciatis nos magistrum R. de Grandon' clericum ad presentationem dilectarum in Cristo priorisse et monialium de Etton' que habent ius patronatus in capella de Blaneword' intuitu Dei ad eandem capellam canonice admisisse et eum in ipsam personam instituisse, salva in omnibus nostra et Wintoniensis ecclesie dignitate et salva iamdictis monialibus antiqua et solita pensione, et ne contra factum nostrum de cetero super hoc indebite molestetur presentes litteras nostras eidem duximus in[a] institutionis sue testimonium concedendas. Hiis testibus: magistro I(ohanne) de London' officiali nostro, magistro I. de Rammesbiryg, magistro R(oberto) Basset et P(etro) Russig' clericis nostris et multis aliis.

> [a] in *inserted over line*

> Dates before the death of master J(ohn) of London. A Robert de Grendon, clerk, witnesses various Hampshire deeds *c.*1230 (*Selborne Charters* i 27, 30).

38. Nuneaton Priory

Confirmation of the church of Chalton granted by king Henry (II), and of annual pensions of nine marks from Chalton church with its chapel at Idsworth, seven marks from the chapel of Catherington, three marks from the chapel of Blendworth and three marks from the chapel of Clanfield.

[March 1206 × 25 February 1221]

B = BL ms. Add. Ch. 47398 (Nuneaton cartulary roll) m.1 no.13. s.xiv. C = Ibid. Add. Roll 47861* m.3 no.20, in a notarial inspeximus of the original, dated 1323.

Mentioned but not recited in a confirmation by bp John de Pontoise of Winchester, 20 April 1292 (BL ms. Add. Ch. 47398 m.2 no.16).

Omnibus Cristi fidelibus ad quos presens scriptum pervenerit P(etrus) Dei gratia Wintoniensis episcopus eternam in domino salutem. Que religiosis locis pia fidelium largitione sunt collata ut apud ea solidius permaneant auctoritate pontificali expedit confirmari. Ea propter ad notitiam omnium volumus pervenire nos divine pietatis intuitu confirmasse monialibus de Etton'[a] ecclesiam de Chauton' cum omnibus pertinentiis suis sicut a bone memorie Henrico rege Anglie eis legitime est concessa et a predecessoribus nostris confirmata. Confirmamus etiam eis antiquas et debitas pensiones, scilicet novem marcas de ecclesia de Chauton'[b] cum capella de Iddesworth'[c] et septem marcas de capella de Katerinton'[d] et tres marcas de capella de Blanewrth'[e] et tres marcas de capella de Clenefeld'[f]. Et ut hec nostra confirmatio perpetuum robur optineat firmitatis eam presentis sigilli nostri munimine roboravimus. Hiis testibus: magistro E(ustachio) de Fauceb'[g], magistro Roberto Basset, magistro Alano de Stok', magistro Philippo de Faukeberg', Ricardo capellano, Dionisio clerico, Ricardo de Berkyng'[h], Ricardo de Elmham clericis et aliis multis.

[a] Etone C [b] Chalgton' C [c] Idesworth' C [d] Katerington' C [e] Blaneworthe C [f] Clanefeld' C [g] Faukeb' C [h] Berkyngg' C

Before the election of master Eustace de Fauconberg as bp of London. For the churches and chapels see *VCH Hampshire* iii 210.

39. Nuneaton Priory

Mandate to T(homas), rural dean of Droxford, to install master John of Limoges, the bishop's clerk, to corporal possession of the church of Blendworth, to which he was admitted by the bishop at the presentation of the prioress and nuns. [March 1206 × 9 June 1238, ?1223 × November 1233]

B = BL ms. Add. Ch. 47398 (Nuneaton cartulary roll) m.1 no.11. s.xiv.

P(etrus) Dei gratia Wintoniensis episcopus dilecto in Cristo T. decano de Drokeneford' salutem, gratiam et benedictionem. Mandamus vobis quatinus magistrum Iohannem Lemovicen' clericum nostrum in corporalem possessionem ecclesie de Bleneword' nomine custodie sine dilatione mittatis qui ad presentationem priorisse et conventus sanctimonialium de Etton' ad predictam ecclesiam ipsum admisimus. Valete.

> Master John of Limoges is found in the bp's household from 1223 and last appears in England in November 1233 (below appendix 4 no.33). For Thomas, rural dean of Droxford (fl. c.1208 × 1230) see *Southwick Cart.* vol.2 iii no.111.

40. Nuneaton Priory

Admission and institution of Aimery of Limoges, clerk, as rector of the church of Blendworth at the presentation of the prioress and convent.

Wolvesey, 24 March 1238

> A = BL ms. Add. Ch. 47850. Endorsed: *Blendworth'* (s.xv); approx. 143 × 48mm.; wrapping tie and step for tongue, tongue and seal impression missing. Scribe 6.
> B = Ibid. Add. Ch. 47398 (Nuneaton cartulary roll) m.1 no.12. s.xiv.

.Omnibus Cristi fidelibus ad quos presentes littere pervenerint .P. divina miseratione Wint(oniensis) ecclesie minister humilis salutem in domino . Noveritis nos ad presentationem dilectarum filiarum .. priorisse et conventus de Eton' dilectum nostrum Emericum Lemovicen' clericum ad ecclesiam de Blenewurth' cum omnibus ad eam pertinentibus admisisse . ipsumque in eadem canonice rectorem instituisse . In cuius rei testimonium presentes litteras nostras sigillo nostro signatas eidem concessimus patentes . Act' apud Wolves' die mercurii in vigilia annunciacionis dominice . anno domini millesimo . ducentesimo . tricesimo . vii°.

> The fact that the vigil of Lady Day fell on a Wednesday in 1238, but not in 1237, helps to establish that the bp's charters are dated according to the English church calendar, the New Year beginning on 25 March.

*41. Peter Bydun

A charter granting Peter Bydun the office of keeper of the gate of the priory of St Swithun, Winchester, for life.

[March 1206 × 9 June 1238, ?August 1231 × 9 June 1238]

> Mentioned in letters of king Henry III, October 1261, and again April 1262, restoring the office to Peter, following inspection of the bp's charter; the king awarding compensation of £10 a year in wardships or escheats to Roger le Convers, king's serjeant, to whom he had previously granted Peter's office (*CPR 1258-66* 177, 209). Since Bydun was still alive in 1266, it

seems most likely that des Roches' charter was issued late in his episcopate, probably after his return from crusade in 1231. The office of gate-keeper was one of several domestic offices at St Swithun's whose appointment was claimed by the bishops. Bp des Roches' appointment of Peter Bydun and of various others, including the usher in the conventual kitchen, chief cook, chief serjeant in the almonry, and washer to the convent, is referred to in *Cal.Inq.Misc.* i no.261; an inquisition prompted by the king's intervention in such appointments following the banishment of Aymer de Valence, bishop elect of Winchester, and the consequent vacancy of the see. See below appendix 2 nos. 46-9.

*42. Peter des Roches, Bishop of Winchester

The bishop's will (confirmed by king Henry III on the eve of des Roches' departure on pilgrimage to Santiago di Compostela). [April 1221]

Mentioned in the king's confirmation (*Pat.R.1216-25* 286).

*42a. Peter des Roches, Bishop of Winchester

The bishop's will (confirmed by king Henry III on the eve of des Roches' departure for the crusade). [June 1227]

Mentioned in a royal charter forbidding any sheriff, bailiff or person to lay hands upon the bp's movables or to impede the proper implementation of his will. Similar awards were made on the same occasion to the bps of Bath and Lincoln (*Cal.Chart.R.1226-57* 42).

*43. Peter des Roches, Bishop of Winchester

The bishop's will. [1238]

Probably to be identified with the will drawn up by des Roches' at Rochester, 31 October 1236, and confirmed by the king on 4 November, whose executors were to include Ralph de Neville, bp of Chichester, master P. archdn of Winchester (probably a scribal error for H[ugh des Roches]), L(uke des Roches), archdn of Surrey and master Elias of Dereham (*CPR 1232-47* 166). Later evidence suggests that the abbot of Beaulieu and Peter de Rivallis also served as executors (*Ibid.* 423). By the time of his death des Roches was an incredibly wealthy man. It is likely, therefore, that his will involved the disposal of considerable sums in cash. Paris (*CM* iii 489-90) states that the bp left 500 marks to the hospital of St Thomas at Acre, described as the least of his gifts. Elsewhere, we catch glimpses of individual benefactions. To the cathedral of Tours, he left money to endow an obit worth 40 shillings p.a of the money of Tours; a considerable sum, more indeed than that of the obit celebrated at Tours for king Richard I (*Martyrologe de Tours* 44, and for Richard see p.34). At least 123 marks was spent by des Roches' executors in purchasing rents at Easton and at Awbridge near Michelmersh, for the endowment of obit celebrations in St Swithun's priory (*WC* no.310; *Reg.Pont.* 389; BL ms. Add. 29436 fo.42r; Winchester Cathedral Library, D.& C. muniments provisional listing W52/1 m.1d). The executors also delivered 50 marks to Christchurch priory, so that des Roches' obit could be celebrated each year in the priory church, with annual payments of five shillings to the refectory and five shillings to be given in bread to the poor at the priory gate

(BL ms. Cotton Tiberius D vi part ii (Christchurch cartulary) fos. 32r,34r). The sacrist of Chertsey abbey held rents worth thirty shillings a year at *Elsteham* from which to celebrate des Roches' anniversary (BL ms. Cotton Vitelius A xiii (Chertsey cartulary) fo.80v). In addition, as founder of Halesowen and Titchfield, des Roches had already obtained participation in the prayers of the Premonstratensian order (Colvin *White Canons* 180, 258-9). His appearance in the kalendar of Wintney nunnery probably reflects the grants he made to the nuns during his lifetime (below nos.83-4). To his successors as bps of Winchester, he bequeathed a considerable quantity of stock, stipulating that it pass in perpetuity from one bp to the next as a basic endowment for the bpric estates (*CPR 1330-34* 230; and see *Cl.R.1237-42* 62-3,79; *Cl.R.1247-51* 327,329). Ranulph of Warham, bp of Chichester, had made similar provision for the see of Chichester before his death in 1222, whilst the executors of Hugh Foliot, bp of Hereford (d.1234), granted £26 in perpetuity to sow the bpric's lands. A similar pattern was later to be followed at Canterbury and elsewhere (*Chichester Cart.* no. 903 part 32; *Reg.Cantilupo* 38-9; M. Howell, *Regalian Right in Medieval England* (London 1952) 76-7). As might be expected, des Roches' grant to Winchester was on a lavish scale, reserving 1556 oxen, compared to the 252 bequeathed by bp Ranulph to Chichester, and more than 12000 sheep, compared to Ranulph's 3150. Finally, and most importantly, des Roches left money for the foundation of two Cistercian houses, at Clarté Dieu, near his birthplace in the Touraine, and at a site known as *Locus Sancti Edwardi*, subsequently transferred to Netley on Southampton water. Clarté Dieu is said to have been endowed with lands purchased with a sum of money entrusted by the bp's executors to the abbot of Cîteaux (*Gallia Christiana* xiv 327-30, instr. col.91-2). Netley was endowed with various manors purchased before the bp's death from the religious corporations of northern France. Credit for its foundation was later claimed by king Henry III, although in fact the king bestowed very little upon Netley in terms of land or money (Meekings 'The Early Years of Netley Abbey' *passim*, and see *Cl.R.1242-7* 22; *CPR 1232-47* 423; *AM* ii (Waverley) 323).

*44. Ralph Monachus

Convention between P(eter) bishop of Winchester and Ralph Monachus, concerning the latter's debt to the bishop of seventy two and a half marks, half of which is to be paid by 1 September 1218 and the remainder at the following Christmas. Ralph pledges all his lands as security. The bishop may transfer either the lands or the debt to whomsoever he choses, should Ralph fail to meet the terms for repayment. [*c.*August 1218]

Confirmed by royal letters dated 13 August 1218 (*Pat.R.1216-25* 175). Compare *infra* nos. 4,79. For Ralph le Moine, perhaps an adherent of the leading rebel Robert fitz Walter, see *RLC* i 271. For his lands, including those in Hampshire, see *BF* 74,77,257.

45. Salisbury, Bishop and Canons

Confirmation of the possessions of bishop Herbert, the canons and church of Salisbury, within the diocese of Winchester.

[March 1206 × 7 January 1217]

B = Trowbridge, Wiltshire County Record Office, ms. Bp of Salisbury D1/1/1 (Register of St

Osmund) fo. 27r. s.xiii in. C = Salisbury Cathedral Library, Dean and Chapter ms. Liber Evidentiarum C p.418 no.549. s.xiv ex.
Pd from B in *Reg. St Osmund* i 223-4.

Omnibus sancte matris ecclesie filiis ad quos presens scriptum pervenerit P(etrus) Dei gratia Wintoniensis episcopus salutem in domino. Noveritis nos divine caritatis intuitu auctoritate pontificali confirmasse Deo et beate Marie et venerabili fratri nostro Herberto Sarr'*a* episcopo et eius successoribus *b*et eiusdem ecclesie canonicis omnes possessiones et libertates*b* omniaque iura sibi in diocesi nostra canonice collata tam in ecclesiis et decimis quam omnibus aliis rebus sicut carte donatorum rationabiliter testantur. Et ut hec nostra confirmatio perpetue firmitatis robur optineat eam presentis scripti testimonio et sigilli nostri patrocinio communimus*c* . Hiis testibus: magistro E(ustachio) de Faucunberg', magistro Roberto Basset, magistro Alano de Stok'*d* , magistro Philippo de Faucunberg', Ricardo capellano, Dionisio clerico, Ricardo de Berching*e*, Ricardo de Elmham, Roberto de Cerne, Willelmo filio Humfridi*f* clericis et multis aliis.

a Saresbur' C   *bb* et.....libertates *underlined* in B   *c* communivimus C   *d* Stokes C
e Berking' C   *f* Umfr' C

Bp Herbert Poer d. 7 January 1217.

46. Salisbury, Dean and Chapter

Mandate to master Alan (of Stokes), the bishop's official, to permit the dean and chapter to collect the (tax of a) sixteenth owing from the prebends of Salisbury within the see of Winchester, in accordance with instructions from king (Henry III). [March 1227]

B = Trowbridge, Wiltshire County Record Office, ms. Bp of Salisbury D1/1/1 (Register of St Osmund) fo.74v (p.150). s.xiii in.
Pd from B in *Reg. St Osmund* ii 75-6.

P(etrus) Dei gratia Wintoniensis episcopus dilecto sibi magistro Alano offic(iali) suo salutem in domino. Noveritis dominum regem nobis per litteras suas mandasse quod permittamus decanum Sar' colligere per manum suam sextamdecimam de prebendis ecclesie Sar' que sunt in episcopatu nostro Wintoniensi, et ideo vobis mandamus quod ipsum decanum predictam sextamdecimam in episcopatu nostro Wintoniensi de predictis prebendis colligere permittatis, et si quid de predicta sextadecima prebendarum dicte ecclesie cepistis, id eidem decano restitui faciatis. Val(e)t(e) in domino.

The king's letters were issued on 5 March 1227 (*RLC* ii 174), and were carried to bp des Roches by Nicholas of Potterne 'about the beginning of Lent' (28 February) (*Reg. St Osmund* ii 75).

The next dated entry in the register after the present mandate is for 22 March 1227 (below no.47). For the churches concerned, Odiham, Hurstbourne Tarrant, King's Somborne and Godalming, see *Reg. St Osmund* ii 76.

47. Salisbury, Richard Bishop of

Letters to bishop R(ichard) of Salisbury, reciting letters of pope Honorius (III) notifying bishop (Peter) of Winchester that the pope has made peace between the emperor F(rederick II) and the Lombards. The Emperor therefore intends to cross (to the Holy Land) in the August passage, a large fleet having been prepared. The pope exhorts the bishop of Winchester to renew his preaching of the crusade, to induce both great and small to join the general passage of August and to compel those who delay, to fulfil their vows as crusaders. Given at the Lateran, 13 January 1227. The bishop of Winchester instructs the bishop of Salisbury to fulfil this commission within the diocese of Salisbury. [March 1227]

B = Trowbridge, Wiltshire County Record Office, ms. Bp of Salisbury D1/1/1 (Register of St Osmund) fo.75r (p.151). s.xiii in.
Pd from B in Wilkins *Concilia* i 559b-60; *Reg. St Osmund* ii 77-8.

Venerabili fratri in Cristo domino R(icardo) Dei gratia Sar' episcopo P(etrus) divina permissione Wintoniensis ecclesie minister humilis salutem et fraterne dilectionis augmentum. Litteras domini pape suscepimus in hec verba: Honorius episcopus servus servorum Dei venerabili fratri episcopo Wintoniensi salutem et apostolicam benedictionem. Benedictus Deus qui ex his que mala videntur frequenter bona dignatur elicere et que timentur cedere ad dispendium, ad compendium revocare. Cum igitur super discordia que inter karissimum in Cristo filium nostrum F(redericum) illustrem Rom' imperatorem semper augustum et regem Sicilie, et Lombardos exorta fuerat, per studium nostrum auctore pacis Deo cooperante salubriter sit provisum ut idem imperator ad transfretandum in instanti Augusto magnifice prout imperialem decet excellentiam se attingat, multitudinem navium et alia que tanto negotio congruunt studiosissime preparando sicut eo nobis per sollempnes nuntios et litteras insinuante letantes audivimus et ipsa rei evidentia prout ab his qui viderunt asseritur clarius manifestat. Sperantes quod Deus exurgens iudicare disposuit causam suam et Ier[usa]lm' sicut diebus pristinis restaurare mandatum quod de predicanda cruce dudum a sede apostolica accepisti duximus innovandum, fraternitatem tuam sollicitantes et hortantes attente ac per apostolica scripta tibi mandantes quatinus iniuncte tibi predicationis officium exerceas studiose, magnos et parvos secundum datam sibi a Deo prudentiam sedulis exortationibus inducendo ut in instanti

38   PART 1: DIOCESAN AFFAIRS

Augusto quo generale passagium est indictum impendant suum dicte Terre Sancte subsidium modis omnibus quibus possunt. Crucesignatos autem ut occasione cessante in ipso passagio transeant efficaciter moneas et si necesse fuerit per censuram ecclesiasticam, nisi quos evidens necessitas excusaverit, sublato appellationis impedimento compellas. Dat' Lat' id(ibus) Ian(uarii), pontificatus nostri anno xi°. Harum igitur auctoritate litterarum vobis mandamus quatinus negotium secundum *a* formam prescriptam in vestra diocesi exequamini diligenter. Valeat in domino fraternitas vestra.

a single letter erased, now illegible B

Bp Richard Poer transmitted des Roches' letters to the dean of Salisbury on 22 March 1227 (*Reg. St Osmund* ii 77 [misdated]). The letters of pope Honorius III, sent generally to archbps and bps engaged in preaching the crusade, are recorded in the papal register (*Reg. Hon. III* no.6157, dated 11–13 January 1227).

48. Selborne Priory

Grant of the advowsons, pensions, rents and lands of the churches of Selborne, Basing and Basingstoke.   Wolvesey, 15 January 1234

A[1] = Oxford, Magdalen College Muniments ms. Basingstoke Charter 5. Endorsed: *ii[a] exn'* (s.xiv); *G-Hsecheford* +, the notarial mark of Gilbert, called Hammergold, of Seckford (fl.1310) (Cheney, *Notaries Public* 114–5, 129n.); *Basyng' Basyngstok'* (s.xvii); approx. 250 × 146 + 42mm.; book hand; parchment tag, perished above seal, sealing method 1; parchment slip wrapped around the tag, bearing the words *tercia.carta* (s.xiii/xiv); fine impressions of the bp's seal and counterseal in red coloured wax, chipped at top and bottom. Scribe 3.

A[2] = Ibid. Basing Charter 35. Endorsed: *donacio ecclesiarum de Basyng Basyngstok' et Selborn'* (s.xv); *ii[u]* (s.xiv); approx. 160 × 75 + 18mm.; parchment tag, sealing method 1; parchment slip tied to top of tag, bearing the words *tercia vel quarta carta* (s.xiv); right hand strand of tag below seal bears vertical arrangement of letters *si p...t in........confir......t perpe....a de......in pu....eat......llirs....de D.......enuz......et ma....*, presumably because cut from a draft; fine impressions of bp's seal and counterseal in red wax, top broken away. Scribe 2.

B = Ibid. Selborne Charter 100, a version of des Roches' text apparently copied from A[2] but here dated 24 rather than 15 January 1234. Purports to be an inspeximus by king Henry III, dated at Windsor, *nono kl' Novembr' anno regni nostri decimo octavo* (24 October 1234). Marked in a slightly later hand on the outside of the turn-up, *ad instanciam Imberti de Montferrant li. xvi.* Imbert was a Savoyard knight active at Henry III's court in the 1250s. There is no known connection between him and Selborne priory. The inspeximus is suspect in many respects. It employs the Roman calendar, more or less unknown in dating royal documents, and it purports to be witnessed by Stephen of Seagrave as Justiciar and by Peter de Rivallis, both of whom were disgraced in May 1234 at which time Seagrave was forced to resign the Justiciarship. These witnesses apparently 'borrowed' from Ibid. Basing Charter 18 (*Selborne Charters* ii 4), a genuine charter of Henry III, dated 9 July 1233, enrolled in the charter roll for 1232–3 (PRO C53/27 m.3, whence *Cal.Chart.R.1226–57* 182–3, where the date 9 June is given by mistake). The great seal attached to B (green wax on coloured cords) is the second seal of Henry III, used 1259–1272. There is no obvious explanation for the forgery, particularly since the text of des Roches inspected appears to be genuine.

Pd (calendar) from A¹/A² in Baigent and Millard *Basingstoke* 653, 655; *Selborne Charters* ii 4, 15–16 (which states incorrectly that Oxford, Magdalen College Basing Charter 15 is a further copy of the present text); *English Historical Documents 1189–1327*, ed. H. Rothwell (London 1975) 742–3 (from A²). Pd (calendar) from B in *Selborne Charters* i 14, which misdates the text of des Roches' charter, and states incorrectly that it is identical with that of Oxford, Magdalen College, Selborne Charter 97 (for which see no.50 below).

Universis sancte matris ecclesie filiis[a] P. divina miseratione Wynton(iensis) ecclesie minister humilis salutem in domino . Cum ex officio pastorali sacram religionem modis omnibus fovere debeamus . et . viros religiosos qui temporalibus lucris despectis pauperes spiritu[b] pro Cristo esse elegerunt diligere et[c] creare et eorum quieti solliciti providere teneamur . hinc est quod universitati vestre notificamus nos[d] divine caritatis intuitu dedisse . concessisse . et[e] presenti carta confirmasse priori et canonicis[f] ecclesie beate Marie de Selebourn'[g] advocationes[h] ecclesiarum de Basingges et de Basinggestok'[j] . et de Selebourn'[g] cum pertinentiis . et totam liberam terram . redditum . et pensiones ad predictas advocationes qualitercumque pertinentes . habendas et tenendas predictas advocationes predictarum ecclesiarum cum pertinentiis prefatis priori et canonicis[f] et eorum successoribus inperpetuum in liberam . puram . et perpetuam elemosinam . in usos proprios et ecclesie sue predicte perpetuo convertendas . [k]et nos predictus .P. episcopus et successores nostri Wynton(ienses) episcopi .' predictas advocationes predictarum ecclesiarum . terra(m) . redditum et pensiones cum pertinentiis . predictis priori et canonicis et eorum successoribus . et ecclesie sue de Selebourn' antedicte . de consensu capituli nostri contra omnes gentes warantizare . tueri . et defendere tenemur in perpetuum[k] . Et ut hec nostra donacio . concessio . et presentis carte nostre confirmacio rate et stabiles in perpetuum perdurent presentem cartam sigilli nostri munimine duximus roborandam . Hiis testibus . domino Waltero abbate de Hyda . domino Waltero[l] priore Sancti Swthuni . domino Stephano[m] priore de Motesfunt'[n] . magistro Alano de Stok' . magistro[o] Willelmo de Sancte Marie Ecclesia tunc officiali nostro . magistro[p] Luca archidiacono Surreye[q] . Petro Russinol . et multis aliis . Dat' apud Wolveseye[s] per manum .P. de Cancell' . octavo decimo kal'[t] Febr' anno domini .mº. ccº. tricesimo tercio.

[a] filiis *not in* B   [b] cf. Mt. 5:3 spiritu *not in* A¹   [c] et *not in* A¹   [d] nos *not in* B
[e] B *inserts* hac   [f] canonicis A¹ conventui A² B   [g] Seleburn' B
[h] advocationes *given in the singular throughout in* B *and other words altered to agree*
[j] Basing' et de Basingestok' B   [k-k] et nos predictos ...... in perpetuum *omitted in* A² B
[l] Waltero *not in* B   [m] Stephano *not in* A² B; gemipunctus *in* A²   [n] Motesfont' A²
[o] magistro *not in* B'   [p] magistro *not in* A² B   [q] Surr' A²B
[s] Wlves' A² Wulves' B   [t] xviiiº. kal' A² nono kl' Febr' B (i.e. 24 January 1234)

The *arenga* to this and the following two texts is adapted from no.34 above, a confirmation to Mottisfont priory, from which des Roches' new foundation appears to have been colonized (below no.49). Selborne priory was sited on land purchased by des Roches *c*.1231–2 from local tenants, several of whom appear to have been in debt to the Jews. A poor house from the start, the greater part of its endowment consisted of the appropriated churches of Selborne, Basing and Basingstoke, whose advowsons the bp had purchased from the Norman abbey of Mont-Saint-Michel, following the death of their former rector, master Philip de Lucy (d. 27 September 1231/2, probably 1232), a sale confirmed by the king on 9 July 1233 (*Cal. Chart.R.1226–57* 182–3, misdated, and see below appendix 4 no.34). In addition, due no doubt to his political influence at court, des Roches obtained a grant of part of the royal manor of Selborne, previously farmed at £4 a year to master Stephen de Lucy, master Philip's brother (d. *c*.March 1232) (for the date of his death compare *Rot.Hugh Welles* ii 303,319; *Pat.R.1225–32* 26,30,51,292,468). This land was managed directly by the crown until the spring of 1233 and only awarded to des Roches by royal charter on 4 May 1233, being confirmed together with other privileges by royal charters of 9 March and 10 April 1234 (*Selborne Charters* i 7,10,13–14; PRO E372/76 mm.3,3d). Taken together, this evidence suggests beyond reasonable doubt that the present charter and nos. 49–50 below are dated according to the English church calendar and should be assigned to 1234 rather than to 1233.

49. Selborne Priory

Foundation charter, granting the Augustinian canons established at Selborne all the land which the bishop had by gift of James of Oakhanger, James of Norton, and king H(enry III) to assist the canons in their reception of travellers and the poor; granting them also the appropriation of the churches of Selborne, Basing and Basingstoke, saving sufficient vicarages in each church. The priory is to be dedicated to the Blessed Virgin Mary, and is to remain under the patronage of the bishop and see of Winchester, free from subjection to any other house.

Wolvesey, 20 January 1234

A = Oxford, Magdalen College Muniments ms. Selborne Charter 6. Endorsed: *carta Petri episcopi de situ prioratus* (s.xiii); approx. 267 × 172 + 31 mm.; fine book hand; parchment tag sealing method 1; parchment slip wrapped around tag, bearing the words *sexta carta* (s.xiv); chipped remnants of seal and counterseal impressions in green wax; traces of wool tied to tag and seal with thread; text pierced by 30 or more tiny holes made by ?insects. ?Scribe 3.

Pd from A in *Monasticon* vi 511; White *Selborne* i 375–7; *Selborne Charters* i 8–9.

Omnibus Cristi fidelibus ad quos presens scriptum pervenerit .P. divina miseratione Winton(iensis) ecclesie minister humilis salutem in domino . Ex officio pastorali tenemur viros religiosos qui pauperes spiritu^a esse pro Cristo neglectis lucris temporalibus elegerunt .'speciali affectu diligere . fovere pariter et creare . eorumque quieti sollicite providere . ut tanto uberiores fructus de continua in lege Dei meditatione percipiant .' quanto a conturbationibus malignorum amplius fuerint ex patroni provisione et ecclesiastica defensione securi . Hinc est quod universitati vestre notificamus nos divine caritatis instinctu de assensu conventus ecclesie nostre Winton(iensis) fundasse domum religiosam ordinis magni patris Augustini in honore Dei et gloriose semper virginis eiusdem Dei genitricis Marie apud Seleburne ibidemque canonicos regulares instituisse . ad quorum sustentationem et hospitum et pauperum susceptionem .ᵃ dedimus . concessimus . et presenti carta nostra confirmavimus eisdem canonicis totam terram quam habuimus de dono Iacobi de Acangre . et totam terram . cursum aque . boscum et pratum .ᵃ que habuimus de dono Iacobi de Nortone . et totam terram boscum et redditum .ᵃ que habuimus de dono domini .H. regis Anglie . cum omnibus predictarum possessionum pertinentiis . Dedimus etiam et concessimus in proprios usuus[b] eisdem canonicis ecclesiam predicte ville de Seleburne et ecclesias de Basing' et de Basingestok' cum omnibus earundem ecclesiarum capellis libertatibus . et aliis pertinentiis . salva honesta et sufficienti sustentatione vicariorum in predictis ecclesiis ministrantium . quorum presentatio ad priorem predicte domus religiose de Seleburne et canonicos eiusdem loci in perpetuum pertinebit . Preterea possessiones et redditus . ecclesias sive decimas quas in episcopatu nostro adempti sunt vel in posterum Deo dante iustis modis poterunt adipisci .ᵃ sub nostra et Winton(iensis) ecclesie protectione suscepimus et episcopalis auctoritate officii . confirmavimus eadem auctoritate firmiter inhibentes .ᵃ ne quis locum in quo divino sunt officio mancipati seu alias eorum possessiones invadere vi vel fraude vel ingenio malo occupare audeat . vel etiam retinere . aut fratres conversos . servientes . vel homines eorum aliqua violentia perturbare . sive fugientes ad eos causa salutis sue conservande a septis domus sue violenter presumat extraere . Precipimus autem ut in eadem domo religiosa[c] de Seleburne ordo canonicus et regularis conversatio[d] secundum regulam magni patris Augustini quam primi inhabitatores professi sunt .ᵃ in perpetuum observetur[e] . et ipsa domus religiosa a cuiuslibet alterius domus religiose subiectione libera permaneat . et in omnibus absoluta . salva in omnibus episcopali auctoritate . et Winton(iensis) ecclesie dignitate . Quod ut in posterum ratum permaneat . et inconcussum presenti scripto et sigilli nostri patrocinio duximus confirmandum . Hiis testibus . domino Waltero abbate de Hyda . domino Waltero priore de Sancto Swipuno . domino Stephano priore de Motesfonte . magistro Alano de Stok' . magistro Willelmo de Sancte Marie Ecclesia . tunc officiali nostro . Luca archidiacono de Surr' . magistro Hunfrid' de Millers . Henrico et Hugone capellanis . Roberto de Clinchamp' . et Petro Rossinol clericis . et multis aliis Dat' apud Wlves' per manum .P. de Cancellis . in die sanctorum martirum Fabiani et Sebastiani . anno domini . millesimo ducentesimo tricesimo tercio.

[a] cf. Mt. 5:3   [b] sic A   [c] letter erased beneath the o of religiosa now illeg.
[d] letter m erased after conversatio   [e] observerur A

42   PART I: DIOCESAN AFFAIRS

A comparison between the present text and no.34 above, a confirmation of 1208 issued in favour of Mottisfont priory, shows that much of the present charter, including the *arenga*, the inhibition against attacks upon the convent and the clause specifying the rule to be adopted, are copied more or less verbatim from the earlier Mottisfont text, in itself little more than an adaptation of a charter of bp Godfrey de Lucy in favour of Mottisfont. This does much to substantiate the argument that Selborne priory was first colonized from Mottisfont. The first prior of Selborne, John de Wich, may well be identified with a namesake, attorney of the prior of Mottisfont in 1230. He may have been a kinsman of Richard Wich, future bp of Chichester (D. Le Faye, 'Selborne Priory 1233–1486', *Proceedings of the Hampshire Field Club*, xxx (1973) 51–2; *CRR* xiii no.2409). For the date of the present award see no.48 above.

50. Selborne Priory

Licence to appropriate the churches of Selborne, Basing and Basingstoke, saving vicars' portions. Wolvesey, 22 January 1234

A = Oxford, Magdalen College Muniments ms. Selborne Charter 97. Endorsed: *exhit' .viii. id' Febr' .H-M* (the notarial mark of Hugh Mussel, notary (fl.1290) for whom see *infra* nos.32,54); *G-Hsecheford*+ (the notarial mark of Gilbert of Seckford, notary (fl.1310) for whom see *infra* nos.48,54); approx. 290 × 79 + 34 mm.; book hand; parchment tag, sealing method 1; impressions of the bp's seal and counterseal in green wax, wrapped in wool, the lower half missing. Jagged tear in left hand side of the recto, measuring 40 mm. from top to bottom and extending 35 mm. into the text, part of which has been lost. Scribe 3.

B = Winchester Cathedral Library, ms. Cartulary part iii fo.70r no.398. s.xiv med.

Pd from A in *Selborne Charters* i 9–10; Baigent and Millard *Basingstoke* 658 (calendar). Pd from B (calendar) in *WC* no.398. *Selborne Charters* i p.14 states incorrectly that Magdalen College Selborne Charter 100 is a copy of A (see above no.48).

P. divina miseratione*ᵃ* Wynton(iensis) episcopus religiosis viris priori et conventui ecclesie*ᵇ* beate Marie de Seleborne*ᶜ* nostre dyocesis salutem . in omnium salvatore . Cum ex officio pastorali sacram religionem plantare . et plantatam modis omnibus fovere debeamus et viros religiosos qui temporalibus lucris neglectis*ᵈ* .' pauperes pro Cristo esse*ᵉ* elegerunt . diligere creare . et eorum quieti sollicite providere teneamur . Eapropter dilecti in domino filii ecclesias de Selebourne*ᶜ* . de Basinges et Basingestok'*ᵍ* in dyocesi nostra sitas que de advocatione vestra propria esse conprobantur .' cum capellis ab eisdem dependentibus . ac*ʰ* aliis pertinentiis suis ad uberiorem sustentationem ves[tr]a[m]*ʲ* et ad suscepucionem hospitum et pauperum ad domum vestram de Seleborn'*ᶜ* confluencium vobis et domui vestre predicte deputamus . concedendo vobis*ᵏ* per presen[tes ut easd]em*ʲ* ecclesias nunc de iure et de facto vacantes auctoritate vestra propria ingredi . et in proprios usus*ˡ* perpetuo tenere valeatis . salva honesta et com[petenti susten]tacione*ʲ* vicariorum in predictis ecclesiis ministrancium . quorum presentatio ad priorem predicte domus religiose de Seleborne*ᶜ* et canonicos*ᵐ* eius[dem loci perpetu]o*ʲ* pertinebit . iure et dignitate*ⁿ* ecclesie nostre Wynton(iensis) semper salvis . Quod ut in posterum ratum permaneat et inconcussum presentes litteras sigilli [nostri muniminje^j roboratas^o vobis fieri fecimus patentes . Hiis testibus . domino Waltero abbate de Hyda . domino Waltero priore Sancti Swthuni . domino Stephano priore [de Mo]tesfunt'^{p} . magistro Alano de Stok' . magistro Willemo de Sancte Marie Ecclesia tunc officiali nostro . magistro Luca archidiacono Surreye . magistro H(umfrido) de Millers^q . Henrico et Hugone capellanis . Roberto de Clincha(mp) . Petro Russinol^r . et multis aliis . Dat' apud Wlvesey^s in die sancti Vincentii . anno domini .m°.cc°. tricesimo tercio^t .

^{a} Petrus miseratione divina B ^{b} ecclesie *not in* B ^{c} Seleburne B ^{d} necglectis B
^{e} c.f. Mt.5:3; esse *not in* B ^{g} Basynges et Basyngstok' B ^{h} et B
^{j} *letters in brackets perished in* A ^{k} vobis *not in* B
^{l} usus proprios B ^{m} canonicis B ^{n} indignitate B ^{o} roboratis B
^{p} domino priore Stephano de Motesfonte B *letters in brackets perished in* A ^{q} Milliers B
^{r} Russynol B ^{s} Wolvesye B ^{t} m°.cc°. xxxiii° B

For the date see no.48 above.

51. Selborne Priory

Confirmation to the prior and canons of land and a mill at Sheet in Mapledurham granted by Eustace de Greinville.

Kingston (upon Thames), 16 February 1238

A = Oxford, Magdalen College Muniments ms. Petersfield Charter 152. Endorsed: *pro terris et molendinis in Mapulderham* (s.xvi); approx. 167 × 108 + 16 mm.; parchment tag, sealing method 1; fine impressions of the bp's seal and counterseal in red coloured wax, chipped around the edges. Scribe 1.
Pd (calendar) from A in *Selborne Charters* ii 65.

Omnibus Cristi fidelibus ad quos presens scriptum pervenerit .P. divina miseratione Winton(iensis) ecclesie minister humilis salutem in domino . Noverit universitas vestra nos divine caritatis intuitu concessisse et presenti carta nostra confirmasse Deo et ecclesie beate Marie de Seleburn' et priori et canonicis regularibus ibidem Deo servientibus et eorum successoribus in perpetuum . totam terram de Sithe cum molendino et cum omnibus aliis pertinentiis suis quam habent de dono Eustachii de Greinvile in manerio de Mapelderham . habendam et tenendam eisdem priori et canonicis et eorum successoribus inperpetuum . libere . pacifice . integre et quiete ab omni servitio et seculari exactione . salvo servitio capitalium dominorum . Et ut hec nostra concessio et confirmatio firma et stabilis in perpetuum permaneat .' presens scriptum sigillo nostro roboravimus . Hiis testibus . magistro Hunfrido de Milliers tunc officiali nostro . Hug(one) archidiacono Wint(oniensi) . Luca archidiacono de Surr' . magistro Elya de Derham . Petro de Rivallis . Roberto de Clynchamp . Iohanne de Colesdon' . Rogero Alis . Galfrido de Rupibus . et multis aliis . Dat' apud Kingeston' . xiiii°. kal' . Marcii . anno gratie . m°.cc°.xxxvii°.

> For Sheet, now in Petersfield, see *VCH Hampshire* iii 116. For Eustace's land there and his grant of it to Selborne see above no.9; *Selborne Charters* ii 65; *Reg. Pont.* 389–90. The present text is mentioned and confirmed in a charter of Walter prior of St Swithun's before November 1239 (*Selborne Charters* ii 5; *WC* no.400).

52. Selborne Priory

Grant in perpetual alms of land at West Tisted purchased from Henry le Savage. Faversham, 3 March 1238

> A¹ = Oxford, Magdalen College Muniments ms. Selborne Charter 250. Endorsed: vi^{ta} (s.xv); *I virgat' West Tysted* (s.xvi); approx. 208 × 153 + 27 mm.; parchment tag, sealing method 1; the opening words *Universis Cristi* repeated in the same hand as the rest of the text, on the left hand side at the bottom of the recto, now under the fold; the letters *am'* written on the right hand strand of the tag drawn out below the seal; impressions of the bp's seal and counterseal in red wax, chipped at the top.
> A² = Ibid. Selborne Charter 233. Endorsed: v^{ta} (s.xiv); *Westestede* (s.xv); approx. 201 × 121 + 19 mm.; parchment tag, sealing method 1; fragments of red wax with identifiable portions of the bp's seal and counterseal, tied up in wool padding. The lines of the text heavily underscored. Final line written in a very cramped version of the ?same hand. ?Scribe 1.
> Pd from A¹/A² (calendar) in *Selborne Charters* i 16–17, 23.

Universis Cristi fidelibus presentem cartam inspecturis vel audituris Petrus Dei gratia Wynton(iensis) episcopus eternam in domino salutem . Noveritis nos dedisse concessisse et presenti carta nostra confirmasse in perpetuam elemosinam Deo et ecclesie beate Marie de Seleburn' et priori et canonicis ibidem Deo servientibus et eorum successoribus imperpetuum[a] unam virgatam terre quam emimus de Henrico le Savage[b] in Westtysted'[c] cum mesuagio . bosco . pastura . et omnibus aliis pertinentiis et libertatibus suis . habendam et tenendam predicte ecclesie et predictis priori et [d]canonicis et successoribus suis de dominis feodi[d] libere . solute . et quiete . reddendo inde predicto Henrico le Savage[b] et heredibus suis . annuatim unam libram cimini ad Nativitatem beate Marie pro omni servitio demanda et exactione seculari . Et ut[e] nostra donacio concessio et carte nostre confirmacio rate et stabiles imperpetuum[a] permaneant ;' presens scriptum sigilli nostri appositione roboravimus . Hiis testibus . domino Ricardo abbate de Tychefelde[f] . magistro Alano de Stok . magistro Willelmo de Sancte Marie Ecclesia . magistro Humfrido[g] de Milers . Roberto de Wautham . Rogero Alis . Galfrido de Roches . Thoma Alis . Rogero Wascelin[h] . Iohanne de Teford . Iohanne de Suwerk et aliis . Dat' apud Faverisham per manum Petri de Chaunceles quinto non'[j] Marcii[k] anno domini m°.cc°. tricesimo septimo.

a in perpetuum A²   *b* Sauvage A²   *c* Westistude A²
d-d canonicis et eorum successoribus inperpetuum libere etc. A²   *e* A² *inserts* hec
f Tychefeld' A²   *g* Umfrido A²   *h* Wacelin A²   *j* nonas A²   *k* A² *ends here*

For the lands purchased and given by Henry le Savage and his widow see *Selborne Charters* i 3-4, 21-2, 33, 48. The present grant is mentioned and confirmed in a charter of Walter prior of St Swithun's Winchester, before November 1239 (*Ibid.* ii 5; *WC* no.400).

53. Selborne Priory

Confirmation to the prior and canons of grants made by Joan, widow of Robert le Hod, in West Tisted.   Farnham, 4 June 1238

> A = Oxford, Magdalen College Muniments ms. Selborne Charter 257. Endorsed: *Westystede* (s.xv); approx. 168 × 91 + 19 mm.; parchment tag, sealing method 1; fine impressions of the bp's seal and counterseal in green wax; right hand strand of tag cut from a draft, with letters arranged vertically; *Osbi...sal'....nosta' ....succ....in vill....in libera....perman....de m....Elya.....Rob' de.* Scribe 1.
> Pd from A in *Selborne Charters* i 23.

Omnibus Cristi fidelibus ad quos presens scriptum pervenerit .P. divina miseratione Winton(iensis) ecclesie minister humilis salutem in domino . Noverit universitas vestra nos divine caritatis intuitu concessisse et presenti carta nostra confirmasse ecclesie beate Marie de Seleburn' . priori et canonicis ibidem Deo servientibus et eorum successoribus in perpetuum . totam terram quam habent de dono Iohanne que fuit uxor Roberti Lohod in villa de Westistede . habendam et tenendam eisdem ecclesie priori et canonicis et eorum successoribus in liberam puram et perpetuam elemosinam . Et ut hec nostra concessio et confirmatio firma et stabilis in perpetuum permaneat .' presens scriptum sigilli nostri inpressione roboravimus . Hiis testibus . magistro Hunfrido de Milliers tunc officiali nostro . Hugone archidiacono Wint(oniensi) . Luca archidiacono Sureye . magistro Elya de Derham . Petro de Rivallis . Roberto de Clynchamp . Iohanne de Colesdon' . Iohanne de Venuz . Roberto Marmiun . Willelmo de Ho . et multis aliis . Dat' apud Farnham per manum Petri Rusinol pridie nonas . Iunii . anno gratie . m°.cc°.xxxviii°.

> For Joan's grant see *Selborne Charters* i 23, 55, 71-2. In 1233 Robert and Joan le Hod had given the bp 300 marks to hold the manor of West Tisted from the bp and his successors in perpetuity, save for 1 virgate there held by Henry le Savage, all of which lands the bp had recovered by fine in the king's court against Richard bp of Salisbury (PRO CP25(1) 203/6 no.8). Joan was still active in litigation there, against Ralph Camoys, in 1241 (PRO CP25(1) 203/7 no.17).

46   PART 1: DIOCESAN AFFAIRS

54. Selborne Priory

Licence to appropriate the church of West Tisted as granted by to the prior and canons by Joan the widow of Robert le Hod, saving a vicar's portion.

Farnham, 4 June 1237 [?1238]

A¹ = Oxford, Magdalen College Muniments ms. Selborne Charter 273. Endorsed: *Westystede rectorur* (s.xv); approx. 174 × 65 mm. Step on left of foot for tongue, tongue and seal torn away; step at the left hand top corner for a wrapping tie; a more accurate copy than A². Scribe 2.

A² = Ibid. Selborne Charter 269. Endorsed: *Exhit' viii. Id' Febr' R-H-M*, followed by the notarial mark of Hugh Mussel (fl.1290) (see above nos.32,50); *.ii.ᵃ* (s.xiv); *GHseheford* +, the notarial mark of Gilbert of Seckford (fl.1310) (see above nos.48,50); *Whitestede* (s.xv); approx. 172 × 66+21 mm.; parchment tag, sealing method 1; impression of the bp's seal and counterseal in red wax wrapped in wool padding, the lower half missing. Scribe 2.

B = Ibid. Selborne charter 245. An inspeximus by Valentine prior of St Swithun's, 4th Sunday in Lent 1265 (17 March 1265/9 March 1266), apparently taken from A². C = Winchester Cathedral Library, ms. Cartulary part iii fo.71r no.403. s.xiv med. D = Ibid. fo.71v no.404. s.xiv med.. A copy of the inspeximus B, merely reciting the opening few words of the bp's charter.

Pd (calendar) from A¹/A²/B in *Selborne Charters* i 21, 58. Pd (calendar) from C/D in *WC* nos. 403-4.

Universis sancte matris ecclesie filiis*ᵃ* [presens scriptum inspecturis] P. divina miseratione*ᵇ* Wynton(iensis) ecclesie minister humilis salutem in domino . Cum ex officio pastorali sacram religionem modis omnibus fovere debeamus et viros religiosos qui temporalibus lucris necglectis*ᶜ* pauperes spiritu pro Cristo esse elegerunt diligere . creare . et eorum quieti sollicite providere teneamur .'pium propositum*ᵈ* Iohanne que fuit uxor*ᵉ* Roberti le Hod*ᶠ* paterna affectione sequentes . ecclesiam de Westysted'*ᵍ* quam dicta Iohanna ecclesie de Seleburn'*ʰ* . priori et canonicis ibidem Deo servientibus pro animabus nostris et*ʲ* antecessorum et successorum suorum contulerat .' in proprios usus convertendam . cum tota terra quam idem religiosi*ᵏ* habent de dono eiusdem Iohanne divine caritatis intuitu ecclesie beate*ˡ* Marie de Seleburn'*ʰ* priori et canonicis ibidem Deo servientibus et eorum successoribus in perpetuum*ᵐ* concedimus et presenti carta*ⁿ* confirmamus . habendam*ᵒ* et tenendam in puram et perpetuam elemosinam . salva honesta et sufficienti sustentatione vicarii in predicta ecclesia ministrantis . cuius presentatio ad priorem et conventum*ᵖ* dicte domus in perpetuum*ᵠ* pertinebit*ʳ* . Et ut hec nostra concessio et confirmacio firma et stabilis in perpetuum permaneat .' presens scriptum sigilli nostri inpressione*ˢ* roboravimus . Dat' apud*ᵗ* Farnham per manum .P. Russinol pridie nonas Iunii anno domini*ᵘ* .mº.ccº.xxxviiº.

ᵃ filiis *not in* A² C
ᵇ *words in square brackets supplied from* BCD, *not in* A¹A². *In* C *they are inserted over the line.*
  BCD *read* Petrus miseratione divina Winton' ecclesie. D *ends* miseratione divina etc etc
ᶜ neglectis C   *ᵈ* prepocitum C   *ᵉ* uxor *not in* B   *ᶠ* Lood B

g Westistede A²B Westystede C   *h* Selleburne BC   *j* nostris *not in* B; et *not in* A²BC
k reliosi A²   *l* sancte B   *m* imperpetuum C   *n* B *inserts* nostra
o habentdam A²   *p* canonicos A²B; conventum A¹C
q inperpetuum *not in* B; imperpetuum C   *r* pertinebat C, B *ends here*
s impressione C   *t* aput A¹   *v* domini *not in* A²C

For the church of West Tisted see *Selborne Charters* i 23–4, 54–5, 71–2. The similarity between the date, place and subject matter of this grant and no. 53 above, combined with the fact that none of the texts of the present award is particularly accurate, suggests that 4 June 1238 was intended as the date rather than 4 June 1237.

*+55. Southampton, God's House (?*spurious*)

A writing between the bishop and the burgesses of Southampton, touching upon the appointment of Warin, a canon of the priory of St Denys' Southampton, as master of God's House.

Cited in 1290, in a dispute between bp John de Pontoise and the king over the advowson of God's House. The writing is said to have been delivered to bp John, but when it was demanded in court, the bp was unable or unwilling to produce it (*God's House Cart.* pp. xl–xlv; *Reg. Pont.* 707–10).

56. Southwark, Hospital of St Thomas

Notification that A(micius) archdeacon of Surrey has bought out all the tenements and lands held from him and his predecessors in Southwark from the fee of the church of Winchester for an annual rent of five shillings and fourpence, and with the bishop's consent has used these lands to provide a new site for the hospital of St Thomas, for the salvation of the bishop's soul. The bishop hereby confirms the hospital in possession of all these lands, buildings, gardens, apple trees and water courses. The annual rent of five shillings and fourpence is to be paid by the hospital to the archdeacon of Surrey.

[11 July 1212 × August 1215, ?c. 1213 × 1214]

B = Winchester, Hampshire Record Office ms. 21M65/A/1/5 (Register of bp John of Stratford) fo. 170r–v. In an inspeximus of 1324. C = Bodl. ms. Rawlinson D763 (St Thomas' cartulary) p. 5. s.xv in.

The first few words of the charter are further recited in Bodl. ms. Rawlinson D763 p. 11 and in BL ms. Stowe 942 (St Thomas' cartulary) fo. 1v (s.xvi med.); both of them heavily abbreviated copies of an inspeximus of bp des Roches' charter issued by prior John of St Swithun's 1243 × 1244, or 1247 × 1250. For the second of these copies see *St Thomas' Cartulary* no. 2.

Universis Cristi fidelibus ad quos presens scriptum pervenerit P(etrus) Dei gratia Wyntoniensis episcopus salutem in domino*a*. Ad universitatis vestre notitiam volumus pervenire quod cum dilectus in Cristo A. archidiaconus Surr' preeunte assensu et auctoritate nostra*b* de proprio emisset omnia tenementa et omnes terras eorum[c] qui de ipso et archidiaconis[d] predecessoribus suis in Suwerk[e] iure hereditario per annuum servicium quinque solidorum et quatuor denariorum de feodo Wyntoniensis ecclesie tenuerant, statim[f] de consilio nostro in remedium salutis anime mee[g] fundavit et edificavit quoddam hospitale in eisdem terris in honore Dei et beati martiris Thome ad perpetuam pauperum susceptionem, unde quoniam ea que in usus pauperum pia fidelium devotione conferuntur perpetua debent pace ac securitate gaudere, laudabile factum eiusdem archidiaconi pio favore prosequentes, prescriptas terras cum omnibus ad eas pertinentibus in edificiis, in ortis, pomariis et aquis, Deo et prefato hospitali in perpetuam elemosinam concessimus et presenti carta nostra confirmavimus, salvo archidiacono Surr' annuo et debito censu quinque solidorum et quatuor denariorum. Hiis testibus[h]: domino Eustachio de Fakenbergh, magistro Alano de Stokes, magistro (Roberto)[j] Basset, domino Gregorio, domino Ricardo capellano, Dionisio clerico, Amico clerico, Symone capellano, Thoma clerico et aliis.

[a] in domino *not in* C   [b] B fo.170v   [c] eorum *not in* B   [d] C *inserts* Surr'
[e] Suthewerk C   [f] statum C   [g] sue C   [h] C *ends here* Hiis testibus etc.
[j] *name omitted* B

On 11 July 1212 Southwark was swept by a fire, perhaps started during the rowdiness of a scotale. The fire destroyed part of the conventual buildings of St Mary Overy, including the priory's infirmary dedicated to St Thomas of Canterbury, a visitor to the priory in the month before his martyrdom in 1170 (Paris *CM* ii 536; *Flores Hist.* ii 141–2; M. Bateson, 'A London Municipal Collection of the Reign of John', *EHR* xvii (1902) 729–30; *Diceto* i 342; *Becket Materials* iii 122–3). The present text must date after the fire of 11 July 1212, and before the retirement or death of master Amicius as archdeacon of Surrey, which had taken place by August/September 1215 (below appendix 4 no.11). It is unlikely that des Roches would have been in a position to influence developments at Southwark after the rebel seizure of London, 17 May 1215. Sufficient time must have elapsed since the fire to allow for some degree of new building. In October 1213 master Amicius secured quitclaims before the Bench of various properties in Southwark which he had put towards the new hospital (PRO CP25(1) 225/3 nos. 46–7, 50). In each case the dorse of the final concord notes that Stephen archbp of Canterbury staked his claim; quite why, is uncertain. Possibly, as successor to St Thomas, the archbp claimed rights as the hospital's patron, a claim which was to lead to violent dispute between Canterbury and Winchester in the 1250s (Paris *CM* v 348–52). Various of the lands held by master Amicius appear to have passed to his successor as archdn, master Luke des Roches, providing an interesting but rare example of the temporal endowment of an archdeaconry (below appendix 4 no.13).

57. Southwark, Hospital of St Thomas

Notification that following the destruction by fire of the hospital of St Thomas, Southwark, with the counsel of prior Martin and the canons of St Mary (Overy), Southwark, and in consideration of the inadequate nature of its former site, the bishop has decreed that the hospital be refounded in a more ample location where the air is cleaner and the water supply better. The hospital has therefore been removed to a site within the bishop's fee, and in consideration of the efforts made on its behalf by A(micius) archdeacon of Surrey, the bishop hereby appoints him the hospital's keeper.

[11 July 1212 × August 1215, ?c.1213 × 1214]

B = Winchester, Hampshire Record Office ms. 21M65/A/1/5 (Register of bp John of Stratford) fo.170r. In an inspeximus of 1324. C = Bodl. ms. Rawlinson D763 (St Thomas' cartulary) pp.4-5. s.xv in. Marked *an 1207* (s.xvii). This and various dates in the margin apparently derived from the inaccurate account in *AM* iii (Bermondsey) 451, 457. D = Ibid. pp.10-11. s.xv in. In an inspeximus by prior John of St Swithun's 1243 × 1244, or 1247 × 1250.

A heavily abbreviated copy of the inspeximus marked D above, reciting only the first few words, is given in BL ms. Stowe 942 (St Thomas' cartulary) fo.1r: *P(etrus) Dei gratia etc ut supra in prima carta huius libri*; whence, *St Thomas' Cartulary* no.2. The ms. lacks at least one and possibly more fos. at the beginning.

P(etrus) Dei gratia Wyntoniensis episcopus^a universis Cristi fidelibus ad quos presens scriptum pervenerit salutem in domino. Cum^b hospitale sancti Thome martiris de Suthewerk^c incendio miserabili totum esset in cineres et favillas redactum, tractatum habuimus cum Martino priore et canonicis sancte Marie de Suthewerk super statu hospitalis reformando et de assensu eorundem canonicorum propter angustiam loci^d in quo prius hospitale fuit fundatum^e et propter alias multas et varias incommoditates, communicato bonorum et prudentum virorum consilio, decrevimus et ordinavimus quod hospitale transferetur in locum maioris amplitudinis ubi aer est serenior et aquarum decursus uberior. Translatum est itaque pro parte et de novo fundatum in fundo Wyntoniensis ecclesie ubi, domino cooperante, iam ad usus pauperum quedam sunt edificia constructa. Attendentes autem pium^f devotionis affectum quem dilectus in Cristo A. archidiaconus Surr' gerit circa^g fratres, sorores et pauperes^h eiusdem hospitalis, auctoritate pontificali concessimus et commisimus^j ei tam in exterioribus quam in interioribus curam et custodiam dicti hospitalis et fratrum ac sororum et omnium bonorum ad eos pertinentium. Valet(e).

^a D starts here   ^b D p.11   ^c Suwerk D   ^d locim C   ^e fundatum in fr..do propter D
^f p'me B   ^g circa *not in* C   ^h et pauperes *not in* C   ^j C p.5

For the date see no. 56 above. In the final concords of October 1213 master Amicius is described merely as archdn of Surrey, with no title relating to his custody of the hospital. By a settlement dated 1215 (i.e. probably after 25 March 1215), Martin prior of Southwark agreed to the refoundation and transfer of the hospital and established a division of property with master Amicius whereby no other hospital was to be built on the abandoned site (Bodl. ms. Rawlinson D763 pp.12-13). The present text may well have been known to the compiler of the late Bermondsey annals, who in his garbled and misdated account of the refoundation of the hospital speaks of bp des Roches transferring St Thomas' to a site *ubi aqua est uberior et aer est sanior* (*AM* iii 457). There may also be an echo of the language of des Roches' charter in the mid-thirteenth-century Dunstable annals, which record the bp's translation, whilst on crusade in 1227–9, of the site of the hospital of St Thomas at Acre: *quia locus infirmus erat, translata est habitatio versus mare in aerem puriorem* (*Ibid.* 126; below appendix 2 no. 30).

58. Southwark, Hospital of St Thomas

Indulgence of twenty days remission of enjoined penance for all contributors to the fabric of the new hospital of St Thomas, Southwark.

[11 July 1212 × May 1215]

B = Bodl. ms. Rawlinson D763 (St Thomas' cartulary) pp.9–10. s.xv in. C = BL ms. Stowe 942 (St Thomas' cartulary) fos. 3v–4r. s.xvi med.
Pd (calendar) from C in *St Thomas' Cartulary* no.10.

Petrus Dei gratia Wyntoniensis episcopus^a universis Cristi fidelibus per episcopatum Wyntoniensem constitutis salutem in eo qui salus est fidelium^b. Sicut ait apostolus: Exercitatio corporis que consistit in ieiuniis, vigiliis et aliis carnis macerationibus ad modicum utilis est^c; pietas vero ad omnia valet promissionem habens vite que^d nunc est et future. Dominus noster Ihesus Cristus inter opera pietatis sex quasi pre ceteris laudabiliora et magis meritoria nobis enumerat, commendat et docet implenda dicens^e: Esurivi et dedistis michi manducare, sitivi et dedistis michi bibere, hospes fui et collegistis me, nudus et cooperuistis me, infirmus et visitastis me, in carcere et venistis ad me^f. Executoribus horum pietatis operum benedictionem et celestis regni^g gloriam repromittit^h, dicens: Venite benedicti patris mei, percipite regnum quod vobis paratum est ab origine mundi^j. Neggligentibus autem et omittentibus opera misericordie maledictionem intentat et penam ignis eterni, dicens: Ite maledicti in ignem eternum qui preparatus est diabolo et angelis eius^k. Attendendum est igitur filii karissimi et altius in cordibus vestris reponendum quam necessarium et quam sit meritorium saluti animarum nostrarum pietatis operibus perpetuis^l indulgere, per que nobis benedictio repromittitur et eterne vite beatitudo comparatur. Ecce apud Suthewerk^m hospitale vetus ad susceptionem pauperum pridem constructum miserabili quodam incendio totum in cineres et favillas est redactum, et quia locus in quo illud vetus hospitale^n erat fundatum ad susceptionem et inhabitationem pauperum^o minus idoneus et minus erat neccesarius tum propter angustiam loci, tum propter aque defectum et alias multas incommoditates, de nostro et prudentum virorum consilio transferetur et transplantatur in locum alium maioris^p amplitudinis ubi aer purior est et serenior et aquarum uberior affluentia. Cum autem huius novi hospitalis edificium multas et multiplices exigat expensas, nec sine suffragio fidelium fine debito valeat consummari, universitatem vestram rogamus, monemus et exhortamur attentius et in remissionem peccatorum vestrorum vobis iniungimus quatinus secundum facultates vestras de bonis a Deo vobis^q collatis, in edificium huius novi hospitalis manum misericordie porrigatis et nuntios eiusdem hospitalis ad vos pro necessitatibus pauperum ibidem suscipiendorum et alendorum^r venientes pie caritatis affectu recipiatis, ut per hec et alia que feceritis pietatis opera, ab ipso qui bonorum omnium^s retributor est pius et misericors Deus, post huius vite decursum eterne beatitudinis mercedem reportetis. Nos autem de misericordia Dei et meritis gloriose virginis Marie et apostolorum Petri et Pauli et sancti Thome martiris et sancti Swithuni^t confidentes, omnibus in Cristo pie credentibus qui predictam domum et pauperes ibidem cohabitantes largitionibus elemosinarum suarum pietatis oculo^v respexerint, confessis videlicet, corde contritis et vere penitentibus, de iniuncta sibi penitentia viginti dies relaxamus et orationum ac beneficiorum que fiunt in ecclesia Wyntoniensi et aliis ecclesiis per episcopatum Wyntoniensem constitutis auctore domino participes esse concedimus. Semper in domino valete.

^{a} episcopus *not in* C ^{b} qui est salus fidelium C ^{c} *1 Tim. 4:8* ^{d} quod C
^{e} docens B ^{f} *Matt. 25: 35-6* ^{g} C fo.4r ^{h} reperun' B ^{j} *Matt. 25:34*
^{k} Matt. 25:41 ^{l} perpentuis B perpensuis C ^{m} Suthwerk' C ^{n} illud *not in* C illud verus hospitale B ^{o} pauperum *not in* C ^{p} moris B ^{q} vobis *not in* C
^{r} et alendorum *not in* C suscipiendorum B *corrected* from sussipiendorum ^{s} B p.10
^{t} Swithini C ^{v} occlo B

Probably of much the same date as no. 57 above, whose language is mirrored here. As an indulgence, the present text is remarkable for the extent of its quotation from scripture, mostly taken from St Matthew, but referring at length to 1 Timothy. Indulgences were often intended as miniature sermons, in this case to be proclaimed by the representatives of the hospital as they toured the diocese. See, for example, an indulgence of Hugh (of Balsham) bp of Ely on behalf of the Domus Dei at Ospringe, containing a similar miniature sermon with scriptural quotation, and ending with a specific injunction: *Nolumus autem quod per has litteras nostras predicator aliquis admittetur, sed illa sufficiat predicatio que presentibus litteris continetur* (Cambridge, St John's College Muniments ms. D9/9). A similar injunction is found in des Roches' diocesan legislation, composed after 1222, prohibiting those licenced to seek alms within the diocese from preaching any sermon save to expound whatever indulgences they might have and to make plain their need for assistance, a prohibition passed 'lest the word of God be sullied' (*C.& S.* 129 no.19).

59. Southwark, Hospital of St Thomas

Grant in free alms of a tithe of the bishop's mills at Southwark.
[11 July 1212 × February 1216, ?c.1213 × 1214]

B = Winchester, Hampshire Record Office ms. 21M65/A/1/5 (Register of bp John of Stratford) fo.170r. In an inspeximus of 1324. C = Bodl. ms. Rawlinson D763 (St Thomas' cartulary) p.3. s.xv in. D = Ibid. p.3. s.xv in. In an inspeximus by bp William de Raleigh, 4

52   PART 1: DIOCESAN AFFAIRS

January 1246. E = Ibid. p.11. s.xv in. In an inspeximus by prior John of St Swithun's 1243 × 1244, or 1247 × 1250. Incomplete.

There are heavily abbreviated copies of the texts marked D and E above in BL ms. Stowe 942 (St Thomas' cartulary) fos. 1v, 2v, neither of which recites more than the first few words, followed by notes *ut supra in carta de decimis molendinorum* (see *St Thomas' Cartulary* nos. 2,4). At least one and possibly more folios are missing from the beginning of BL ms. Stowe 942.

Universis sancte matris ecclesie filiis^a ad quos presens^b scriptum pervenerit Petrus Dei gratia Wyntoniensis episcopus salutem in domino. Noverit^c universitas vestra me pro salute anime mee^d et pro salute animarum successorum meorum episcoporum Wyntoniensium concessisse, dedisse et hac presenti carta mea confirmasse Deo et novo hospitali sancti^e Thome de Suwerk^f et fratribus ac pauperibus eiusdem hospitalis totam decimam molendinorum meorum de Suthewerk in liberam et perpetuam elemosinam singulis annis^g in usus pauperum plene^h et integre per manus custodum de ipsis molendinis recipiendam^j. Hiis testibus: domino Eustachio de Faukenbergh, magistro Henrico, magistro Thoma, Roberto capellano, Henrico capellano, Thoma clerico, Willelmo Alemanno et aliis.

^{a} filiis BE salutem C ^{b} presens *not in* C ^{c} Noverint CDE; D *begins here* Noverint
^{d} E *ends here* ut supra in iiii folio, *now fo.2 (p.3) implying that 2 folios are missing at the start of the ms.* ^{e} beati CD ^{f} Suthewerk CD ^{g} annis *not in* C
^{h} C *ends here* plene etc. Hiis testibus etc. ^{j} D *ends here* recipiendam. Hiis testibus etc.

After the fire of 11 July 1212 and before a papal confirmation of the present grant, given by pope Innocent III, 6 March 1216 (*St Thomas' Cartulary* no.19; *Letters of Innocent III* no.1059). Probably of much the same date as nos.56–8 above.

60. Southwark, Hospital of St Thomas

Grant in free alms of all tithes from the bishop's meadows in Southwark.
[11 July 1212 × 13 January 1221, ?c.1213 × 1214]

B = Winchester, Hampshire Record Office ms. 21M65/A/1/5 (Register of bp John of Stratford) fo.170r. In an inspeximus of 1324. C = Bodl. ms. Rawlinson D763 (St Thomas' cartulary) p.3. s.xv in. D = Ibid. p.3. s.xv in.. In an inspeximus by bp William de Raleigh, 4 January 1246. E = Ibid. p.11. s.xv in. In an inspeximus by prior John of St Swithun's 1243 × 1244, or 1247 × 1250. Incomplete.

There are heavily abbreviated copies of the texts marked D and E above in BL ms. Stowe 942 (St Thomas' cartulary) fos. 1v, 2v, neither of which recites more than the first few words, followed by notes *ut supra in carta de decimis prati* (see *St Thomas' Cartulary* nos. 2,4). At least one and possibly more fos. are missing from the beginning of BL ms. Stowe 942.

Universis sancte matris ecclesie filiis ad quos presens scriptum pervenerit Petrus Dei gratia Wyntoniensis episcopus salutem in domino. Noverit^a universitas vestra me pro salute anime mee et pro salute animarum^b

successorum meorum^c episcoporum Wyntoniensium concessisse, dedisse et hac presenti carta mea confirmasse Deo et novo hospitali sancti Thome de Suwerk^d et fratribus ac pauperibus eiusdem^e hospitalis omnes decimas de pratis meis de Suwerk^f in liberam et perpetuam elemosinam singulis annis in usus pauperum plene et integre per manus custodum de ipsis pratis recipiendas^g. Hiis testibus: domino Eustachio de Faukenbergh, magistro Henrico, magistro Thoma, Roberto capellano, Henrico capellano, Symone capellano, Thoma clerico, Willelmo Alemanno et multis aliis.

^{a} Noverint CDE ^{b} animatim C
^{c} E ends here ut supra in iiii° folio now fo.2 (p.3) implying that 2 folios are missing at the start of the ms. ^{d} Suthewerk CD ^{e} eisdem D ^{f} Suthewerk C Southewerk D
^{g} recipiendam BCD, CD end here Hiis testibus etc.

After the fire of 11 July 1212 and before a papal confirmation by pope Honorius III, 13 January 1221 (*St Thomas' Cartulary* no.22). Probably of the same date as no. 59 above. Beginning in 1223/4 (Ms.19DR m.12) the Southwark account of the Winchester pipe rolls records grain given to the hospital in lieu of tithe.

61. Southwark, Hospital of St Thomas

Confirmation to the new hospital of St Thomas, Southwark, of the tithe of a mill in Beddington granted by Robert of Beckenham, and of the tithe of hay in all his meadows granted by Reginald of Briddinghurst.

[11 July 1212 × July 1227, ? c.1224 × July 1227]

B = Winchester, Hampshire Record Office ms. 21M65/A/1/5 (Register of bp John of Stratford) fo.170v. In an inspeximus of 1324. C = Bodl. ms. Rawlinson D763 (St Thomas' cartulary) pp.11-12. s.xv in. In an inspeximus by prior John of St Swithun's 1243 × 1244, or 1247 × 1250. D = BL ms. Stowe 942 (St Thomas' cartulary) fo.1r. s.xvi med. Incomplete. E = Ibid. fo.1v. s.xvi med. A heavily abbreviated copy of the inspeximus C above: *Universis Cristi fidelibus etc ut supra in carta confirmationis de donis Roberti Bekeham et Reginaldi Brettyngherst.*
Pd (calendar) from DE in *St Thomas' Cartulary* nos. 1,2.

Universis Cristi fidelibus ad quos presens scriptum pervenerit P(etrus) divina miseratione Wyntoniensis ecclesie minister humilis salutem in eo qui est salus hominum^a. Noverit^b universitas vestra nos episcopali auctoritate concessisse et confirmasse fratribus et sororibus novi hospitalis sancti Thome martiris de Suwerk'^{c} rationabile^d donum quod Robertus de Begeham^e fecit et carta sua confirmavit eis de decima proveniente de molendino suo^f de Bedynton'^{g} cum pertinentiis secundum tenorem carte sue, eadem etiam^h episcopali auctoritate dedimus^j et presenti scripto confirmavimus^k eisdem^l fratribus predicte domus et sororibus rationabile donum quod Reginaldus de Bretingehirst^m fecit fratribus predictis et sororibus et carta sua confirmavit eis de decima^n feni de omnibus pratis suis secundum tenorem*°* carte sue. Et ut hec*°* nostra confirmatio perpetue firmitatis robur optineat eam presenti sigillo nostro signatam*°* roboravimus. Hiis testibus*°*: magistro Humfrido de Millers, Dionisio clerico, Roberto de Climchamp, Willelmo de Batill' senescallo, Galfrido de Loveriis, Willelmo capellano, Radulpho de Navar', Willelmo de Alem', Willelmo le Hirm' et aliis.

<blockquote>
a omnium C   *b* Noverint C   *c* Suthewerk C   *d* ratonabile C   *e* Regeham B
f D *starts here, the first folio of Stowe ms. 942 being lost*   *g* Bedyngton' C; Bedinton' D
h eadem et C; eadem in D   *j* concedimus CD   *k* confirmamus D
l C p.12 eisdem fratribus et sororibus novi hospitalis sancti Thome martiris de Suthewerk predicte domus et sororibus rationabile C   *m* Brettynghurst C; Brettyngherst D
n proveniente de molendino *cancelled* D   *o* tenorem *not in* C
p et ut hec *repeated, cancelled* D   *q* presenti *and* signatam *not in* B
r C *ends here* etc B *ends here* Hiis testibus: magistro Humfrido de Millers, Dionisio clerico, Willelmo de Batill' senescallo, Willelmo capellano et aliis
</blockquote>

After the fire of 11 July 1212. William de Batilly died during the bp's absence on crusade after July 1227. He appears with the title seneschal on the Winchester pipe rolls between 1224 and 1227 (below appendix 4 no.22). For Reginald of Briddinghurst, a landholder in Peckham and Camberwell (fl.1218–30), see *St Thomas' Cartulary* no.7; *Surrey Fines* 12,13,17. For Robert of Beckenham see *Surrey Fines* 11.

62. Southwark, Hospital of St Thomas

Licence to the new hospital of St Thomas, Southwark, granting that services may be held in the hospital's chapel and bells be rung to summon the hospital's inhabitants to prayer.   [11 July 1212 × 1234, ?c.1213 × July 1227]

<blockquote>
B = Winchester, Hampshire Record Office ms. 21M65/A/1/5 (Register of bp John of Stratford) fo.170r. In an inspeximus of 1324. C = Bodl. ms. Rawlinson D763 (St Thomas' cartulary) p.4. s.xv in. D = Ibid. p.5. s.xv in. E = Ibid. p.11. s.xv in. In an inspeximus by prior John of St Swithun's 1243 × 1244 or 1247 × 1250. Incomplete.
</blockquote>

Universis sancte matris ecclesie filiis presentes litteras inspecturis P(etrus) miseratione divina*°* Wyntoniensis episcopus salutem eternam in domino*°*. Cum inter opera caritatis illa potissime*°* debeant computari que et corporum honestatem et salutem respiciunt animarum, nos volentes dilectis filiis fratribus domus hospitalis sancti Thome martiris de Suthewerk et pauperibus infirmis in eadem domo pro tempore existentibus in hiis*°* que ad salutem pertinent animarum quantum secundum Deum*°* possumus affectione providere paterna, dicte domus fratribus et infirmis concessimus intuitu caritatis ut in capella eiusdem domus divina officia nocturna pariter et diurna sibi et servientibus suis in domo ipsa commorantibus sollempniter*°* sibi faciant celebrari, similiter*°* concedentes eisdem ut campani sui ad missam et ad horas*°* canonicas de die et nocte utantur ad fratres suos et servientes ad divina officia convocandos. In cuius rei[j] testimonium presentes litteras nostras sigillo nostro signatas eisdem fratribus duximus concedendas.

[a] divina *not in* E; divina miseratione episcopus Wintoniensis CD
[b] salutem in domino sempiternam E   [c] E *ends here*   [d] his C   [e] dominum CD
[f] C *inserts* et cum nota cum neccesse fuerit   [g] similiter *not in* D   [h] set ad oras D
[j] D *ends here* In cuius rei etc. C *ends here* In cuius rei etc. sigillo nostro signat' etc.

Presumably to be dated after some degree of rebuilding had taken place. The licence to celebrate mass must nevertheless have been an early priority for the new hospital, and undoubtedly predates a settlement by H(umphrey) prior of St Mary's Southwark (c.1225 × 1234), sealed by bp des Roches, granting that the hospital might have two bells weighing no more than 100 lbs (below appendix 2 no.29). The present licence therefore probably predates the bp's departure for crusade in July 1227. It is interesting that the text refers only to the brothers and the servants of the hospital. Earlier awards (above nos. 57,61) speak of the brothers and sisters.

+63. Southwark, Priory of St Mary Overy (*spurious*)

Notification that it was proved in synod that the chapel or altar of St Mary Magdalene within the aisles of the conventual church of St Mary (Overy), Southwark, has pertained since its foundation to the conventual church. Having regard to the poverty of the conventual church, the bishop confirms it in possession of the said altar or chapel.

B = PRO C53/131 (Charter Roll 18 Edward III) m.3. In a royal confirmation of 12 January 1344. C = PRO C66/328 (Patent Roll 13 Richard II) m.14. In a royal confirmation of B, 20 August 1389. D = BL ms. Add. Charter 44694. A confirmation of C by king Henry VI, 13 February 1441. E = PRO C66/448 (Patent Roll 19 Henry VI) m.6. The enrolment of D. F = PRO C66/507 (Patent Roll 3 Edward IV) m.21. Royal confirmation of C, 4 November 1463.

Pd (calendar) in *Cal. Chart.R. 1341–1417* 34–5 (from B); *CPR 1388–92* 110 (from C); *CPR 1436–44* 499 (from E); *CPR 1461–67* 307 (from F).

P(etrus) Dei gratia Wintoniensis episcopus omnibus Cristi fidelibus salutem et benedictionem Dei et suam. Notum vobis omnibus facio quod capella sive altare beate Marie Magdalen' constructa infra alas ecclesie prioratus ecclesie beate Marie de Suthwerka disrationata est coram me et omni clero ecclesie nostre in sinodo nostra ad pertinentiam ecclesie sancte Marie predicte ab initio constructionis capelle sive altaris predicti. Quare ego intuitu oneris et paupertatis ecclesie predicte eandem disrationem[a] concessam et ratam habeo et presenti cartula auctoritate Dei et beati Petri apostoli dictam capellam sive altare imperpetuum stabilem esse et in usus proprios priori et conventui ecclesie predicte et successoribus suis perpetuis temporibus pertinere et debere confirmo. Hiis testibus: magistro Galfrido archidiacono Surr' et magistro Oseberto officiali nostro et aliis. Valete.

a disrationationem F

The charter is noticed in James Howell, *Londinopolis or an historicall discourse or perlustration of the City of London* (London 1657) 338, whence Manning and Bray, *The History and Antiquities of Surrey*, 3 vols. (London 1804-14) iii 560 and the *VCH Surrey* ii 108 (*VCH London* i 481), all of which cite Howell's reference to construct a wholly fictitious account, whereby the bp built a 'spacious' chapel of St Mary Magdalene at Southwark following the fire which destroyed the priory buildings in 1212. In fact, it is far from clear how much of the conventual building at Southwark was destroyed in 1212, beyond the hospital of St Thomas. The present text is a gross forgery. The witnesses are invented (which requires emendation to *Fasti* ii 94 where Geoffrey archdn of Surrey is accepted as genuine). Its wording, including the inconsistent use of the first person singular and plural, accords with no other charter of des Roches. A supposed charter of bp William (Giffard or de Raleigh) recited in the same set of royal confirmations is also spurious. The sheer number of confirmations obtained for these charters is in itself an indication of their controversial nature, although not all of the charters recited on these occasions are forged (see *Monasticon* vi 172; *Regesta* iii no.829-30). The supposed charter of Giffard/Raleigh purports to grant the parish church of St Margaret to Southwark priory, and it may well be that a claim to the churches of St Margaret and St Mary Magdalene provided the motive for the forgeries. There is evidence from des Roches' years as bp that the church of St Mary Magdalene was attached in some way to the priory (*St Thomas' Cartulary* no.2). Its building adjoined the walls of the conventual church (*VCH Surrey* iv between pp.156-7). There is no evidence in any of the later Winchester bishops' registers of a rector being admitted to the church, although until at least 1309 St Mary Magdalene appears to have functioned to some extent as an independent entity (*Reg. Woodlock* 395, 894, and see *Taxatio 1291* 207b, 209). It may be that the spurious charters of des Roches and Giffard/Raleigh were drawn up in the early fourteenth century to compensate for the lack of title deeds to churches which legitimately belonged to the priory, or, alternatively, to legitimize the complete annexation of what had previously been no more than the priory's pensioners.

64. Southwick Priory

Confirmation to the prior and canons of the advowson of the church of St Andrew Farlington as granted by Roger de Merlay.

[1206 × 9 June 1238, ? c. 1215 × July 1227]

B = Winchester, Hampshire Record Office ms. 1M54/3 (Southwick cartulary III) fos. 53v-54r no. 226. s.xiv ex. Stored in the priory archive amongst the contents of *cophinus* 4 in 1366 (Ibid. mss. 1M54/3 fo.258r; 1M54/2 fo.34v), and earlier (temp. king Henry III) amongst the contents of *cophinus* 15 (Ibid. ms. 1M54/2 fo.42v).
Pd (calendar) from B in *Southwick Cart.* vol.2 part iii no.226.

Omnibus sancte matris ecclesie filiis presentes litteras inspecturis P(etrus) miseratione divina Wyntoniensis ecclesie minister humilis salutem in domino. Ad universitatis vestre notitiam volumus pervenire nos divine pietatis intuitu confirmasse priori et canonicis beate Marie de Suthewyk ius patronatus ecclesie beati Andree de Farlyngton prout Rogerus de Merlay illud eis canonicis*a* concessit et contulit. Et ut hec nostra confirmatio rata et stabilis permaneat*b* eam presentis scripti testimonio et sigilli mei*c* munimine roboravimus. Hiis testibus: magistro Alano de Stoke, magistro Roberto Basset et multis aliis.

a canonice B   *b* Ms. fo.54r   *c* sic ms.

Roger de Merlay held Farlington from Matthew fitz Herbert and later, from Matthew's son, Herbert fitz Matthew, from at least 1201, granting it before his death, c.1250, to his daughter Agnes, wife of Nicholas de Gymminges (*CIPM* i no.221; *BF* 708; *CRR* i 438). In 1201 he was sued over the advowson there by Robert de Curci (*CRR* i 438). His award to Southwick is to be dated 1206 × 1215 (*Southwick Cart.* vol.2 iii no.220; *Daventry Cartulary* pp.xliii-iv, xlviii). The present confirmation predates no.65 below, since like Merlay's grant it fails to mention any pension in the church payable to Southwick. By August 1231 the church itself was held by Ernisius, chaplain of Philip de Aubigné (*Cl.R.1227-31* 551).

65. Southwick Priory

Award of an annual pension of twenty shillings to the infirmary of the priory, to be paid by the rector of the church of Farlington.

[1218 × 9 June 1238, ? 1218 × July 1227]

B = Winchester, Hampshire Record Office ms. 1M54/3 (Southwick cartulary III) fo.49v no.222. s.xiv ex. In an inspeximus by bp William de Raleigh, 21 June 1246, issued because the seal to des Roches' original charter was damaged. Raleigh's inspeximus stored amongst the contents of *cophinus* 4 of the priory archive in 1366 (Ibid. mss. 1M54/3 fo.258r; 1M54/2 fo.34v), and earlier (temp. king Henry III) amongst the contents of *cophinus* 15 (Ibid. ms. 1M54/2 fo.42v).
Pd (calendar) in *Southwick Cart.* vol.2 part iii no.222, where Raleigh's inspeximus is misdated.

Universis sancte matris ecclesie filiis ad quos presens scriptum pervenerit Petrus divina miseratione Wyntoniensis episcopus eternam in domino salutem. Noverit universitas vestra nos intuitu caritatis concesisse priori et conventui de Suthewyk viginti solidos*a* nomine beneficii ad opus infirmarie sue de ecclesia de Farlington' annuatim persolvendos, quos videlicet rector predicte ecclesie de Farlington' eidem infirmarie singulis annis ad duos terminos persolvet, scilicet decem solidos ad Pascha et decem solidos ad festum sancti Michaelis. Ut autem hec nostra concessio stabilis et firma permaneat eam presentis carte et sigilli mei*b* munimine confirmavimus. Testibus: magistro Hunfrido de Millers, magistro Nicholao de Viene et multis aliis.

a solidi B   *b* sic ms.

Master Nicholas de Vienne makes his first recorded appearance in the bp's service in 1218/19, and is unrecorded after the bp's departure on crusade in July 1227 (appendix 4 no.36). The pension from Farlington is further recorded in *Southwick Cart.* vol.1 i no.184.

*66. Taunton Priory

Confirmation of the possessions of the prior of St Andrew, Taunton, held since the time of bishops William Giffard and Henry (of Blois).

[March 1206 × 9 June 1238]

> Mentioned in letters of Walter prior of St Swithun's (*c*.1216 × *c*.1239), confirming confraternity between St Swithun's and Taunton priory (*WC* no.460).

*67. Titchfield Abbey

Foundation Charter. [22 August 1231 × 20 September 1232, ?*c*.3 May 1232]

> B = BL ms. Loans 29/55 (Rememoratorium de Tychefeld) fo.1r. s.xiv ex. Not the foundation charter but an abstract headed *Carta de Fundacione abbathie de Tichefeld'*. C = Winchester, Hampshire Record Office ms. 5M53/998 m.1. Inspected in royal letters patent of 3 February 1560, probably copied from B.
> Pd (translation) from B in Colvin *The White Canons* 185. The opening few words are further copied from B in Winchester, Hampshire Record Office ms.46M48/109 fo.80(a)r, a ms. which includes copies of numerous Titchfield charters, not previously noticed.s.xvi.

Venerabilis pater dominus Petrus de Roches Wyntoniensis episcopus fundavit monasterium de Tichefeld' ordinis Premonstratensis super solum ecclesie parochialis tanquam super firmam petram anno domini millesimo ducentesimo tricesimo tertio et dedit in puram et perpetuam elemosinam absque ullo retenemento sui vel successorum suorum ecclesiam parochialem beati Petri apostoli de Tichefeld' cum capellis de Crofton' et de Chark' et pensione viginti solidorum percipienda de ecclesia de Wykeham singulis annis, et manerio de Berton' cum terris et tenentibus eidem ecclesie annexis et pertinentibus, videlicet in villis de Tichefeld', Uptone, Sarebury et Southbroke, qui vocantur tenentes ecclesie. Dedit etiam in puram et perpetuam elemosinam manerium de Tichefeld' cum manerio de Leghe et villatis de Mune, Chullyng, Pleystowe, Abbechute, Felde, Broke, Weresassch', Schitehagge[a], Uptonne, Curebrigge et Quabbe, eidem manerio pertinentibus. Concessit etiam in puram et perpetuam elemosinam maneria de Swanewyke, Walesworthe et duas partes de Porcestre et tertiam partem manerii de Cornhampton' cum bosco de Chirlewode et terra de Flexlond' cum omnibus dominicis et aliis ubique suis pertinentiis[b].

> [a] Schitebagge C   [b] pertinentibus C

> The manor of Titchfield escheated to the crown in the 1190s from John de Gisors, an early defector to the French. Thereafter, it was held as Norman land from the crown by a series of keepers, the last being Geoffrey de Lucy, kinsman of master Stephen and master Philip de Lucy, whose lands played an important role in the foundation of Selborne priory (*Cal.Chart.R.1226–57* 71, 82; *Pat.R.1225–32* 242; *Cl.R.1227–31* 179, 572; above nos.48–9). Des Roches is said to have founded his abbey at Titchfield on 15 August 1231 (*Collectanea*

Anglo Premonstratensia, ed. F.A. Gasquet, 3 vols., Camden Society 3rd series (1904–6) iii no.578), perhaps as a thanks offering on his safe return from crusade; the abbey lying close to the bp's harbour at Fareham, where des Roches may well have landed in August 1231. He was granted the advowson of the church of Titchfield by the king on 22 August 1231 and on the same day received a royal charter confirming the church's appropriation to the Augustinian (*sic.*) canons whom he had brought to refound there from Bristol (the words Bristol being erased in the chancery enrolment of this text) (PRO C53/25 m.4; *Cal. Chart. R. 1226–57* 139; *WC* nos. 94(1), 321 (misdated)). On 20 October 1231 the manor itself was transferred to des Roches for his foundation, still described as Augustinian, the king's award being made in compensation for a punitive fine of £500 which des Roches had been forced to agree to, following his political downfall in 1224 (PRO C60/30 m.1). A year or so later, on 20 September 1232, the king issued a further charter confirming des Roches' grants to the abbey, including the manor of Titchfield, the canons now being described as belonging to the Premonstratensian order (*Cal. Chart.R. 1226–57* 168; *Monasticon* vi 931–2; *WC* no.94(3)). The canons themselves were drawn from the bp's earlier foundation at Halesowen. In the fifteenth century they claimed to have been granted seisin of Titchfield on 3 May 1232 (Gasquet, *op.cit.* no.578). For the place names above, see Colvin, *The White Canons* 185. The bp's grants mentioned here are unlikely all to have been made at the time of the foundation. Various of the constituent properties are awarded in nos.67a–71 below.

67a. Titchfield Abbey

Grant to the abbot and canons in free, pure and perpetual alms, of the manor of Swanwick and two parts of the manor of Portchester, for the souls of king Henry (III), the bishop himself, kings Richard (I) and John, the bishop's lords, and the souls of all the bishop's predecessors and successors, the bishop having bought these lands from master Humphrey de Millières for four hundred marks and having obtained confirmation from the bishop and chapter of Avranches from whom master Humphrey bought the land. [3 May × 20 September 1232]

> B = Winchester, Hampshire Record Office ms. 5M53/998 m.1. Original letters patent of Elizabeth I, dated 3 February 1560, inspecting *quasdam legendas sive libros quosdam in quibus continentur veri ut dicitur tenores diversarum concessionum, confirmationum cartarum et aliorum scriptorum ac munimentorum*, apparently from cartulary or other copies rather than from the originals. Not in the patent roll of Elizabeth I.

Universis Cristi fidelibus presentem cartam inspecturis Petrus divina miseratione Wintoniensis ecclesie minister humilis eternam in domino salutem. Noveritis nos divine pietatis intuitu et pro salute domini regis Henrici filii regis Iohannis et nostra nostrorumque salute et pro animabus dominorum nostrorum Ricardi et Iohannis regum Anglie et omnium antecessorum et successorum nostrorum dedisse, concessisse et presenti carta nostra confirmasse Deo et ecclesie beate Marie de Tych' et abbati et canonicis ordinis Premonstratensis ibidem Deo servientibus et eorum successoribus totum manerium de Swanewyk cum omnibus suis pertinentiis, et duas partes manerii de Porcestre cum omnibus suis pertinentiis et libertatibus infra villam et extra, quod etiam manerium cum predictis duabus partibus de Porcestre cum omnibus suis pertinentiis emimus a magistro Humfrido de Milers pro quadringentis marcis argenti, et de quo manerio similiter cum predictis duabus partibus de Porcestr' cum pertinentiis confirmationem domini Abricensis episcopi et ecclesie sue capituli impetravimus, a quo episcopo et ecclesie sue capitulo magister Humfridus predictam terram cum pertinentiis emit, tenendum et habendum predicte ecclesie et abbati et canonicis predictis et eorum successoribus in liberam, puram et perpetuam elemosinam. Hiis testibus etc.

> To be dated after the foundation, and included amongst the endowment confirmed by royal charter on 20 September 1232 (*Cal.Chart.R.1226–57* 168; *Monasticon* vi 931). For master Humphrey de Millières see below appendix 4 no.4. Des Roches' purchases from him of Swanwick and Portchester were confirmed to the bp by the king in August 1231 (*Cal.Chart.R.1226–57* 140).

*68. Titchfield Abbey

Grant of the manor of Wellsworth and its appurtenances in the manor of Cosham.   [3 May × 20 September 1232, ?September 1232]

> Cited in a royal confirmation to the abbot and convent by king Edward II, 15 May 1318 (PRO C66/149 m.15, calendared in *CPR 1317–21* 143: *donationem, concessionem et confirmationem quas Petrus quondam Wyntoniensis episcopus fecit*). Dated after the foundation, for which see above no.67, and included amongst the properties confirmed to the abbey on 20 September 1232 (*Monasticon* vi 931). Around 25 September, the abbot of Titchfield fined 15 marks with the king to have the same liberties in his manors of Portchester, Wellsworth, Swanwick and Cosham as he had in the manor of Titchfield itself (PRO C60/31 m.3; E372/77 m.2d).

69. Titchfield Abbey

Appropriation to the abbot and convent of the church of Titchfield, for the souls of king Henry (III), the bishop himself, kings Richard (I) and John, the bishop's lords, and the souls of all the bishop's predecessors and successors, saving a vicar's portion.   [29 September 1232 × 22 February 1235, ?c.October 1232]

> B = Winchester Cathedral Library ms. Cartulary part i fos.46v–47r no.94(ii). s.xiv med. In an inspeximus by prior Andrew of St Swithun's, 27 September 1258. C = Ibid. ms. Book of Endowments fo. 48r–v no.28 (ii). s.xvi ex. Copy of B.
> Pd (calendar) from B in *WC* no.94(2).

Universis sancte matris ecclesie filiis presentem cartam inspecturis vel audituris Petrus divina miseratione Wyntoniensis ecclesie minister humilis eternam in domino salutem. Noveritis nos[a] divine pietatis intuitu et pro salute domini regis Henrici filii regis Iohannis et nostra nostrorumque salute et pro animabus dominorum nostrorum Ricardi et Iohannis regum Anglie et omnium antecessorum et successorum nostrorum dedisse, concessisse et presenti carta confirmasse Deo et ecclesie beate Marie de Tychefeld' et *b*abbati et canonicis ordinis*b* Premonstr(atensis) ibidem Deo servientibus et eorum successoribus *c*ecclesiam parochialem de Tychefeld'*c* cum capell(is) et omnibus aliis pertinentiis et libertatibus ad predictam ecclesiam pertinentibus in puram et perpetuam elemosinam et in proprios usus suos, habendam et possidendam imperpetuum*d*, salva in predicta ecclesia honesta sustentatione vicarii, ad quam quidem predicti abbas et canonici et eorum successores nobis et successoribus nostris ydoneum presentent vicarium cui committi possit cura animarum. Et ut hec nostra donatio, concessio et presentis carte confirmatio inconcusse et stabiles permaneant, presens scriptum sigilli nostri patrocinio duximus confirmandum. Testibus etc.

a C fo.48v   *b-b* abbati...ordinis *underlined in* B
c-c ecclesiam......Tychefeld' *underlined in* B   *d* B fo.47r

The present text must date after the bp's decision to install Premonstratensian rather than Augustinian canons at Titchfield, and before no.70 below. The church is unmentioned in the royal confirmation of 20 September 1232. The Winchester pipe roll for Michaelmas 1232–1233 records that over 100 quarters of grain, presumably the proceeds of the harvest of 1232, were delivered from Titchfield church to the bp's manor of Twyford, suggesting that, as yet, the proceeds of the church were not in the hands of the new abbey. In the same year a pair of bells was carried to Titchfield from the Isle of Wight (Ms.28DR mm.4,16d). The church had undoubtedly been appropriated by the time of the next surviving pipe roll, when the abbot owed arrears of just over £6 for the cultivation of grain sown after the church came into his hands (Mss.31DR m.10d; 32DR m.2). In all likelihood the present award is to be dated shortly after Michaelmas 1232. Since at least 1224–5 the bp had been in receipt of large quantities of grain, apparently from the manor rather than the church of Titchfield, in 1225–6 as much as 267 quarters, probably by purchase from the manor's keeper (Mss. 20DR mm.7d,8; 21DR mm.11d,12d,13; 22DR mm.11,11d).

70. Titchfield Abbey

Taxation of a vicarage for the church of Titchfield. Simon the chaplain, hereby appointed vicar, is to have all obventions and oblations of the altar, tithes of geese, tithes of hay in Crofton and Chark, and an acre of meadow in Crofton, tithes of gardens and two quarters of salt. All other tithes, Easter offerings of eggs, and the first legacies of the dead are to remain to Titchfield abbey.

Wolvesey, 22 February 1235

B = Winchester Cathedral Library ms. Cartulary part i fo.47r no. 94 (iv). s.xiv med. In an inspeximus by Andrew prior of St Swithun's, 27 September 1258. Marginal note: *taxatio vicarie. originale non est exhibitum.* C = Ibid. ms. Book of Endowments fo. 86v no.66 (formerly no.64). s.xvi ex. Apparently copied independently from the original or another copy. D = Ibid. fo.48v no.28(iv). s.xvi ex. Copy of B.

Pd (calendar) from B in *WC* no.94 (4).

Omnibus Cristi fidelibus presens scriptum visuris vel audituris P(etrus) Dei gratia Wyntoniensis episcopus eternam in domino salutem. Noverit universitas vestra vicariam ecclesie de Tychefeld'[a] a nobis coram clericis nostris magistro[b] Alano de Sancta Cruce, magistro Roberto de Forde, et coram Radulpho et Henrico capellanis nostris et coram Simone capellano tunc facto vicario[c] eiusdem ecclesie et aliis clericis nostris quamplurimis, sic fore taxatam, scilicet quod vicarius[d] habebit omnes oblationes et obventiones altaris exceptis decimis, de quibus habebit decimam aucarum et decimam feni de Crofton' et de Chark'[e] et acram prati de Crofton' et decimas de hortis; habebit etiam duo quarderia salis. Omnes autem cetere decime maiores et minores et omnia ova Pasche que colliguntur et offeruntur, et prima[f] legata omnium morientium remanebunt abbati et conventui de Tychefeld'[a], et in huius rei testimonium huic scripto sigillum nostrum apposuimus[g]. Dat' apud Wulvesey viii kal' Martii anno gratie m° cc° xxxv° [h].

[a] Tichefelde C   [b] C *inserts* scilicet   [c] BD *insert* et   [d] vicaria C   [e] Charke C
[f] premia C   [g] BD *end here* Dat' etc.
[h] mccxxxv° *repeated* C *with the first version crossed out*

The dating clause appears to have given the copyist of C some trouble. Des Roches was away from Winchester in February 1233 and 1234, and probably in 1232 and 1238. His whereabouts in February 1237 are unknown. In February 1236 he was overseas. In this isolated instance it appears that the scribe was reckoning according to the year beginning 1 January or Christmas, and that the text should indeed be dated to 1235. Alternatively, the copyist of C may have misread iv for v, or deliberately corrected a date from the outmoded English church calendar. In 1267/8 the vicar's portion was augmented by master Thomas, official of bp Nicholas of Winchester, whose settlement mentions but does not recite that by des Roches (Winchester Cathedral Library ms. Book of Endowments fos. 87r-88r no.67). Yet another augmentation was decreed in 1533, on the eve of the dissolution (Ibid. fos. 88r-90r no.68).

71. Titchfield Abbey

Grant to the abbot and canons in pure and perpetual alms of all the land in Corhampton which the bishop purchased from Adam of Corhampton, and the wood of 'Cherlewode' *bought from Jordan de Wakervile.*

[29 September 1233 × 25 August 1234]

B = PRO C66/148 (Patent Roll 11 Edward II part i) m.4. A royal inspeximus dated 10 January 1318. C = Winchester, Hampshire Record Office ms. 5M53/998 m.2, inspected in royal letters patent of 3 February 1560.
Pd from B in *Monasticon* vi 934. Pd (calendar) from B in *CPR 1317-21* 73.

Universis Cristi fidelibus ad quorum notitiam presens carta pervenerit Petrus divina miseratione Wyntoniensis ecclesie minister humilis eternam in domino salutem. Noveritis nos divine caritatis intuitu dedisse, concessisse et presenti carta nostra confirmasse Deo et ecclesie beate Marie de Tychefeld'[a] et abbati et canonicis Premonstratensis ordinis ibidem Deo servientibus et eorum successoribus totam terram nostram quam habuimus in Cornhampton' cum omnibus pertinentiis suis sine aliquo retinemento, in puram et perpetuam elemosinam, quam quidem terram emimus de Ada de Cornhampton'[b]. Dedimus etiam predicte ecclesie abbati et canonicis et eorum successoribus totum boscum nostrum cum omnibus[c] suis pertinentiis qui vocatur Cherlewode, in puram et perpetuam elemosinam quem emimus de Iordano de Wakervile[d]. Et ut hec nostra donatio, concessio et carte huius confirmatio inposterum perseveret, presenti scripto sigillum nostrum apponere dignum duximus. Hiis testibus[e]: magistro Alano de Sancta Cruce, Willelmo atte Henne, Ada de la Berre, Emerico de Sace, Reginaldo de Cunde, Galfrido de Rupibus, Iacobo de Northton', Henrico de Wodecote, Ricardo de Cardevile, Thoma de Gumming', Iohanne de Shifford, Thoma Thable et multis aliis.

[a] Tich' C   [b] Cornehampton' C   [c] omnibus *over the line* B   [d] Warkerwyll' C
[e] C *ends here* Hiis testibus etc.

After the royal confirmation of 20 September 1232 (above nos. 67–8). Corhampton was in the bp's hands for the full year Michaelmas 1232–1233, appearing as an item of account on the Winchester pipe roll (Ms. 28DR m.3d). It had apparently been transferred to Titchfield by the time of the next surviving account, taken at Michaelmas 1236. Des Roches had gone into exile abroad in February 1235. Adam of Corhampton may have been forced into his sale by Jewish debt (Colvin *White Canons* 186–7; Cole *Documents* 302). The wood of *Cherlewode* is probably to be identified as the wood belonging to Jordan de Wakervile at Hambledon, purchased by des Roches as long ago as 1223–4, when it was enclosed with a ditch at a cost of 16 shillings (Ms.19DR m.4). Jordan's charter transferring the land to des Roches is preserved in Winchester, Hampshire Record Office ms. 5M53/998 m.2, in an inspeximus of 1560. It states that the bp had paid 65 marks for the wood, described as being bounded by the bp's wood of *Havleherod*, the wood of the prior of Southwick and the woods that once belonged to the son of Roger de Scures and John of Wallop, together with all the clearings (*landa*) extending from John's wood as far as the ride (*chiminuus*) of *Hameleweye* extends to the ride of *Huntborn'*; the bishop and his successors to pay a quitrent of a wax candle weighing 1lb each year to the church of St Swithun's Winchester, for the soul of Jordan, his ancestors and successors. The same inspeximus of 1560 preserves quitclaims dated 1272 to Titchfield abbey from Philip de Hoyville, Agnes de Cobham and Agnes de Port her sister, daughters of Thomas de Venuz, of all right in *Cherlewoode* in return for 10 marks. On 25 August 1234 the constable of Portchester was instructed to allow the abbot of Titchfield to have the timber he had felled in *Chorlewud* for the fabric of his church, notwithstanding that the wood lay within the royal forest of Bere (*Cl.R.1231–4* 506). The present text is therefore to be dated before February 1235 and probably before August 1234.

72. Waltham Abbey

Institution of master Simon of Waltham to the church of Caterham at the presentation of the abbot and convent, saving the vicarage held by Thomas de

Gatton who is to pay two marks each year to master Simon.

[25 September 1205 × July 1227]

B = BL ms. Cotton Tiberius C ix (Waltham cartulary) fo.158r. s.xiii med.
Pd from B in *Waltham Charters* no.629.

Hiis testibus: domino Gregorio capellano et cetera.

Master Simon of Waltham (?Essex) is probably to be identified with a namesake, clerk of king John, rector of the churches of Hastings, Slapton and Coldred, and canon of Chichester, active 1205–1217 (*RLP* 61b, 68, 75; *Pat.R.1216–25* 83; *Rot. Chart.*, 141b, 158b, 190, 190b, 198b); probably the same as the master Simon de Wautham, rector of Niton, who died before February 1229 (Above no.28). Bp des Roches set out for crusade in July 1227. The advowson of Caterham had been granted to Waltham Abbey, 1205 × 1209, by Roger son of Everard de Geiste; a gift made for the sake of the souls of various kings, kinsmen and bps Godfrey de Lucy and P(eter) des Roches (*Waltham Charters* no.627). Thomas de Gatton was already vicar by 1198 (*Ibid.* no.628).

73. Westminster Abbey

Judgement delivered by Stephen archbishop of Canterbury, Peter bishop of Winchester, Richard bishop of Salisbury, Thomas prior of Merton and Richard prior of Dunstable in a suit between Eustace bishop of London and William abbot of Westminster. Acting as papal judges delegate, the judges determine that Westminster abbey and the church and parish of St Margaret are to be exempt from the jurisdiction of the bishop of London, being directly subject to Rome, and the clergy within the parish of Westminster being placed under the authority of the abbots of Westminster. The blessings of abbots, dedications of churches and chapels, consecration of altars, ordination of monks and clerks, confirmations and the blessing of chrism are all to be administered by whatever bishop the monks may chose. The judges recite the bounds of the parish (given in detail). Beyond these bounds the vills of Knightsbridge, Westbourne and Paddington, together with their chapels, are to be considered as part of the abbot's jurisdiction. The church of Staines is to remain appropriated to the abbey of Westminster for the charitable works of the monks, saving a vicarage of twenty shillings each year. The parson of Staines is to be presented by the abbot and instituted by the bishop of London. In compensation, the bishopric of London is to have the whole manor and church of Sunbury, saving a perpetual vicarage, the remaining profits of Sunbury church being spent on masses and candles within St Paul's cathedral. The present judgement is drawn up in the form of a cyrograph under the seals of the judges and the parties. Dated 1222.

[25 March × 20 April 1222]

A = Westminster Abbey Muniments ms. no.12753. Described in full in *Acta Langton* no.54.

Six seals attached on cream and light brown silk cords through eyelets. Fine impression of the bishop of Winchester's seal and counterseal in green wax, tied up in a silk bag.
B = Ibid. ms. Domesday fos. 634v-635v. s.xiii/xiv. C = BL ms. Cotton Faustina A iii (Westminster abbey cartulary) fo.250v. s.xiii ex.
Pd from ABC with full references in *Acta Langton* no.54. Pd from A in H. Wharton, *Historia de episcopis et decanis Londoniensibus necnon de episcopis et decanis Assavensibus* (London 1675) 247-54. Pd from Wharton in Wilkins *Concilia* i 598.

Abbot William of Westminster d. 20 April 1222. On 18 September 1222 it was des Roches who, in accordance with the present settlement, was nominated by the Westminster monks to bestow blessing on William's successor, abbot Richard of Barking (Paris *CM* iii 74–5; *Flores Hist.* ii 176; below appendix 2 no.23). Such connivance in the new found independence of Westminster may well have contributed to a cooling of relations between des Roches and his former protégé, Eustace de Fauconberg bp of London. After 1223 bp Eustace was to side with Langton and the English bps in the movement which brought about des Roches' downfall at court (appendix 4 no.30).

74. Westminster Abbey

Institution of master Thomas of Shenfield to the church of Morden at the presentation of the abbot and convent, reserving a customary annual pension of half a mark to the church of Westminster.

[March 1206 × 9 June 1238, ?c.1217 × July 1227]

A = Westminster Abbey Muniments ms. no.1846. Endorsed: *de pensione de Mordone videlicet de dimidia marca..........ecclesie Westm' de .... persolvenda* (s.xiv); approx. 161 × 71 + 20 mm.; parchment tag method 1, seal impression missing, fragments of green wax on tag.
B = Ibid. ms. Domesday fo.336r. s.xiv in.

Universis sancte matris ecclesie filiis ad quos presens scriptum pervenerit .P. Dei gratia Wintoniensis ecclesie minister humilis eternam in domino salutem . Noverit universitas vestra nos ad presentationem dilectorum filiorum abbatis et conventus Westm' magistrum Thomam de Senesfeld'[a] clericum ad ecclesiam sancti Laurentii de Mordon' divine miserationis intuitu admisisse et ipsum in eadem personam canonice instituisse . salva antiqua et debita pensione dimidie marce annuatim ecclesie Westm' persolvenda . salvo etiam nobis et successoribus nostris iure episcopali . Quod ut ratum perseveret in posterum .' presentis scripti testimonio et sigilli nostri inpressione duximus roborandum . Hiis testibus . magistro Humfrido de Miller' . magistro Nicholao de Fernham . magistro Nicholao de Humeto . Uliano Chesneduit . Humfrido clerico et multis aliis.

[a] Menesfeld' B

Master Humphrey de Millières makes no appearance in the bp's household before 1217 (appendix 4 no.4). A master Nicholas de Hommet witnesses Westminster charters in the time of king John, being in all likelihood a kinsman of abbot William de Hommet (1214-1222) (*WAC* no.431). Master Nicholas of Farnham is probably to be identified with a namesake,

66   PART 1: DIOCESAN AFFAIRS

royal physician and later bp of Durham (1241-1249), who makes his first recorded appearance before December 1218 when he was presented by Westminster abbey to the church of Aldenham (Hunts.) (*Rot.Hugh Welles* i 137). By 1221 he also held the living of Charlton-on-Otmoor (*RLC* i 468b). Master Thomas of Shenfield (Essex) also appears to have been a physician, being described as *medicus* in his only other recorded appearance, in a suit against Thomas de Camville over land in Shenfield, conducted between Michaelmas 1225 and Easter 1226 (*CRR* xii nos. 1267, 2397, 2403). Julian Chenduit appears as a regular witness to Westminster abbey charters after 1200, and as the abbey's steward, serving as Westminster's attorney at the exchequer in 1218-19 (PRO E159/1 m.2; E368/1 m.6; *WAC* nos. 328, 333, 424, 426, 439, 461, 484, 486). In January 1221 he served as ordainer in a dispute between the abbey and bp Hugh of Lincoln, in which context he is described merely as clerk (Westminster Abbey Muniments ms. Domesday fos. 448r-449r; Oxford, Brasenose College Muniments ms. Wheathampstead deed 1).

75. Wherwell Abbey

Inspeximus and confirmation of a charter of Godfrey bishop of Winchester (EEA viii no.251) granting and confirming to abbess M(athilda) and the nuns of Wherwell the tithes of Matthew de Poteria's demesne in Wallop, the tithes of Anketil de Brayboef's demesne in Drayton Cannes, the tithes of three virgates of land in Cholderton; all of which tithes the nuns were granted by former patrons; together with the nuns' own tithes of Wherwell and the tithes of their demesne in the Isle (of Wight), reserving an annual pension of half a mark to the parish church of Newchurch; confirming also the tithes of the nuns' new assart in Tufton, reserving an annual pension of two shillings to the parish church of Tufton. Given at Winchester, 21 March 1201.

Marwell, 20 December 1232 (?1231)

B = BL ms. Egerton 2104a (Wherwell cartulary) fos. 98v-99r no.191. s.xiv ex.

Further copies of bp Godfrey's charter are to be found in Ibid. fos.98r-v, 100r-v (in an inspeximus by Walter prior of St Swithun's, 4 January 1232/3). Ibid. fo. 200r-v gives a copy of the same inspeximus by prior Walter, witnessed by master Alan, *dominus* Robert de Beauchamp, master Walter de Schirborn' and *dom*. Simon parson of Goodworth, followed by a note (s.xiv): *originale scriptum dicti domini Godefridi episcopi ex vetustate fere consumptum sub eiusdem episcopi sigillo tenor' quo supra continetur et est in custodia sacriste*.

Omnibus sancte matris ecclesie filiis ad quos presens scriptum pervenerit Petrus divina miseratione Wyntoniensis ecclesie minister humilis salutem in domino . Noveritis nos inspexisse cartam bone memorie Godefridi predecessoris nostri in hec verba: Omnibus Cristi fidelibus ad quos presens[a] scriptum pervenerit Godefridus Dei gratia Wyntoniensis ecclesie minister eternam in domino salutem. Ad iniunctum nobis spectat officium ea que religiosis pia fidelium largitione conferuntur, scriptis autenticis communire et ne qua possent tractu temporis calliditate malignantium inquietari, pastorali solicitudine providere. Eapropter universitati vestre volumus innotessere nos paci et tranquillitati dilectarum in domino M(athilde) abbatisse et monialium de Wherewell' prospicere volentes, concessisse et pontificali auctoritate confirmasse Deo et monasterio de Wher' et monialibus ibidem Deo servientibus et servituris in perpetuum decimas dominii Mathei de Poteria in Wellop, et decimas dominii Anketilli de Brayboef in Drayton', et decimas trium virgatarum terre in Chelewarton' quas scilicet decimas de donatione patronorum ab antiquo possiderunt, decimas etiam dominii sui de Wher' et decimas dominii sui de Insula per pencionem annuam dimidie marce argenti solvendam parochiali ecclesie de Nywecherch', et decimas novi asarti sui de Tokynton' per pensionem annuam duorum solidorum solvendam parochiali ecclesie de Tokynton', in liberam et perpetuam elemosinam possidendas et[b] in usus proprios ipsarum convertendas. Ut autem hec nostra confirmatio perpetuam opptineat firmitatem eam presentis scripti serie et sigilli nostri inpresione roboravimus, salvo in omnibus iure episcopali et auctoritate et dignitate Wyntoniensis ecclesie. Dat' apud Wynton' per manum Philippi clerici nostri, xii° kal' April' pontificatus nostri anno duodecimo. Testibus: Rogero archidiacono Wyntoniensi, magistro Amico archidiacono Surr', Gregorio, Humfrido, Terric' capellanis, Alex(andro), Willemo de Turr', Thoma de Chaliton' magistris, Roberto de Cornevill', Godefrido, Regny', Simone, Thoma, Petro clericis, Gervasio, Iordano de Camera et multis aliis. Nos igitur hiis que per bone memorie predictum predecessorem nostrum legitime facta noscuntur prebentes favorem, gratiam et assensum, prefatam cartam ipsius presentis scripti patrocinio communimus cum nostri appositione sigilli. Dat' apud Merewell' per manum Petri de Cancell' xii° kal' Ianuar' anno ab incarnatione m°cc°xxxii pontificatus nostri anno xxvi°.[c] Testibus hiis: magistro Alano officiali nostro, magistro Roberto Basset, domino Henrico capellano, Roberto de Clinchamp, Petro Russinol, Willemo filio Humfridi et multis aliis.

[a] presens *repeated ms. underlined for deletion*   [b] fo.99r
[c] *sic ms. ?for xxviii. The 26th year of des Roches' consecration ran from 25 September 1230–24 September 1231. The bp was still abroad in December 1230. December 1232, the 28th year, would tie in with the date of the inspeximus by prior Walter (January 1233) mentioned above, although December 1231 is also possible.*

For the date see note *c* above. For Wherwell's lands on the Isle of Wight, at Ashey near Ryde, see Hockey *Insula Vecta* 66–70.

*76. Wherwell Abbey

Confirmation of a settlement over tithes made between the nuns of Wherwell and the rector of Barton Stacey as drawn up by the prior of St Augustine's Bristol and other papal judges delegate. [23 February 1232 × 22 June 1235]

Mentioned but not recited in a papal confirmation dated 22 June 1235 (BL ms. Egerton 2104a (Wherwell cartulary) fo.23r-v; not in the register of Gregory IX). The settlement confirmed on this occasion had been drawn up at St Augustine's Bristol on 23 February 1232 by the prior of St Augustine's and the dean of Bristol, the prior of St James' Bristol excusing his attendance; between G. Teutonicus rector of Barton acting via master Giles of Bridport his attorney, and the nuns of Wherwell via Nicholas their chaplain, over the tithes of Herbert de Caune, knight, in Drayton and tithes of hay in Barton. Letters of the rector of Barton to the bp survive, asking him to confirm this settlement (Ibid. fo.201r-v).

*77. **Wherwell Abbey**

Confirmation of a settlement over tithes made between the nuns of Wherwell and the rector of Over Wallop drawn up in the presence of the precentor of Salisbury and other papal judges delegate.

[March 1206 × 27 June 1235, ?August 1231 × 27 June 1235]

Mentioned but not recited in a papal confirmation dated 27 June 1235 (BL ms. Egerton 2104a (Wherwell cartulary) fos. 200v-201r; not in the register of Gregory IX). Probably to be dated to the period after the bp's return from crusade.

*78. **William de Beauchamp**

Convention between P(eter) bishop of Winchester and William de Beauchamp, whereby William is to pay the bishop five hundred marks in return for custody of the lands of Robert de Aubigny and the marriage of Robert to William's daughter, saving to the bishop all wardships, escheats and marriages falling vacant within Robert's lands from the time when the lands were in the bishop's custody until 6 November 1221. William is to pay three hundred marks of the said money to the bishop or his representative at the octave of Easter (10 April) 1222 at the New Temple, London, on pain of paying six hundred marks to the bishop or his representative at the same place at the following Pentecost (22 May 1222). The residue of two hundred marks is to be paid within two years of Easter 1222, William finding sound pledges from the religious for its repayment. William pledges all his lands to the king as security, to be distrained by the king should he fail to meet his terms with the bishop. [c. 5 November 1221]

For Robert de Aubigny, heir to the barony of Cainhoe (d.s.p. 1233), see Sanders *Baronies* 26. The present agreement is confirmed but not recited in royal letters of 5 November 1221 (*Pat.R.1216–25* 317–8). The terms were rescheduled around 7 December 1221 when it was agreed that the residue of 200 marks might be paid at 100 marks a year within two years of Michaelmas 1222 (*Ibid.* 321–2). The sale of Robert, one of des Roches' more valuable wards, was almost certainly intended to help finance the bp's proposed participation in the Fifth Crusade, cancelled after the collapse of the expedition at Damietta late in 1221. For a similar convention by which des Roches sold the lands and marriage of Robert son and heir of Robert Marmiun to the earl of Chester for £400, the earl delivering a charter pledging repayment, see *Ibid.* 319 (13 November 1221).

*79. William de Pontearche

Convention between P(eter) bishop of Winchester and William de Pontearche concerning the latter's fine for his ransom which is to be paid at set terms (not recited). William meanwhile pledges all his lands as security in similar terms to no.44 above. W(illiam) Marshal earl of Pembroke, William de St John and W(illiam) Marshal junior stand as guarantors for Pontearche's payments.

[c. August 1218]

Confirmed by royal letters dated 18 August 1218 (*Pat.R.1216-25* 175-6). The ransom payment was still being disputed as late as 1227, when Pontearche was sued for £11 owing to des Roches and William de St John, the latter having presumably been forced to pay part of his pledge to the bp. In the suit of 1227 the present convention is referred to as a charter which Pontearche gave the bp (*CRR* xiii no.357; *RLC* ii 196,199b). For Pontearche, tenant of the St John lords of Basing, and nephew by marriage of William Marshal the elder, see *VCH Hampshire* iii 23, iv 387; *BF* 94, 418; *CP* ii 126; *Hist.Maréchal* line 7265. In March 1217 he and John fitz Hugh (for whom see above no.12) had suddenly deserted the rebel cause, an action for which, according to the *Histoire de Guillaume le Maréchal*, 'they are despised even to this day' (*Hist. Maréchal* lines 15828-33, and see *Gervase* ii 111-12). On 10 August 1218, a few days before the present agreement, Pontearche acted as witness at Winchester to a settlement involving des Roches (below no.82).

80. Winchester, St Cross Hospital

Appointment of master Alan of Stokes as keeper for life of the house of St Cross, Winchester, in order that the relief of the poor ordained by bishop Henry de Blois may be maintained. [1212 × July 1227, ?1218 × July 1227]

B = Winchester, St Cross Hospital ms. Liber Secundus fo.3r. s.xv in.
Pd (translation) from ?B in L.M. Humbert, *Memorials of the Hospital of St Cross and Alms House of Noble Poverty* (Winchester 1868) 18. Pd (English translation, without witnesses) from ?Humbert, in W.T. Warren, *St Cross Hospital near Winchester: its History and Buildings* (Winchester/London 1899) 73-4 (misdated).

Universis Cristi fidelibus Petrus Dei gratia Wintoniensis episcopus eternam in domino salutem. Inter opera pietatis non modicum indicatur cum pauperum utilitati prospicitur et eorum sustentationi ut stet et maneat provida discretione providetur, scriptura testante que dicit: Beatus qui intelligit super egenum et pauperem, in die mala liberabit eum dominus[a]. Eapropter nos precavere cupientes ne erogatio in Cristi pauperes facienda per constitutionem domini Henrici episcopi in domo sancte Crucis Wynton(ie) defraudetur vel pereat, magistrum Alanum de Stokes quem virum novimus providum et fidelem elegimus et ei dicte domus curam cum pertinentiis commisimus, habendam et tenendam libere et quiete et pacifice toto tempore vite sue, salva nobis et successoribus nostris auctoritate et dignitate in eadem. Et in huius ordinationis testimonium ei presentem cartam fecimus sigillo nostro roborata(m). Hiis testibus: magistro Roberto Basseth, magistro Nicholao de Vien', Dionisio de Burgoil clerico, Roberto de Clinchamp, Petro Russingnol, magistro Hugone capellano, Rogero Wascelin' et multis aliis.

[a] Ps.40:1

Master Alan of Stokes appears to have succeeded master John of London (d.5 September 1211), both as master of St Cross and as bp's official (below appendix 4 nos.1,2). As early as Michaelmas 1214 master Alan is found owing pannage for the pigs of St Cross (Ms.9DR m.4). However, it is likely that the present text was drawn up some time after his actual appointment to St Cross. Of the witnesses, master Nicholas de Vienne is first recorded in the bp's household in 1218/19. Denis de Bourgueil retired or died at about the time of the bp's departure for crusade in 1227 (appendix 4 nos.15,36).

81. Winchester, St Swithun's Priory

Confirmation to prior S(tephen de Lucy) and the convent, of tithes of land, of animal feed stuffs, of hay and of mills arising from the priory's demesne within the manor of Barton Priors; granting the monks exemption from the payment of tithes from newly cultivated land which they manage directly.

Rome, 15 October [1205]

A = Lost, see below C.
B = BL ms. Add. 29436 (Winchester cartulary) fo.33v (formerly 23v). s.xiii in. C = Winchester Cathedral Library ms. The Book of John Chase fo.18v: *Sine dat' Carta Petri episcopi de decimis dominicorum nostrorum de Barton' et aliis et de iunctalibus (sic.) excolendis. Actum Rom' ides Octobr. Testibus Iohanne bartonario et Germano et Waltero monachis Wintoniensibus etc.* Taken from a list of the cathedral muniments c.1643. The text itself is not recited.

P(etrus) dignatione divina Wintoniensis ecclesie minister humilis S(tephano) venerabili priori totique conventui Wintoniensi salutem et sincere vinculum caritatis. Cum simus secundum apostolum invicem alter alterius membrum[a], mutuis nos debemus caritatis operibus prevenire et ita sollicite providere quod partes singule suo toti conveniant, et ad honorem corporis cedat membrorum proportio singulorum. Nos autem quibus in hoc spirituali corpore gratia divina, cooperantibus vobis, capitis attribuit dignitatem, sicut loco sic volumus officio premunire et taliter nos gerere domino suffragante quod totius corporis forma de capite capiat decoris pariter et honoris augmentum[b]. Vestris igitur postulationibus annuentes, decimas de dominico Berthone vestre

Wint(onie) tam de terris quam animalium nutrimentis, decimas etiam feni et molendinorum de dominicis vestris sicut eas hactenus predecessorum nostrorum temporibus pacifice possedistis, vobis et per vos ecclesie Wintoniensi episcopalis auctoritate officii confirmamus. Concedimus insuper vobis ne de novalibus vestris que propriis manibus aut sumptibus colitis cuiquam decimas exolvere debeatis, sed usibus conventus cedant penitus profuture. In huius autem rei testimonium presentem cartam annotari fecimus et sigilli nostri munimine roborari. Act' Rom' id' Octobris per manum P. Russinol sigilli nostri custodis, sub hiis testibus: Iohanne[c] bartonario et Germano et Waltero monachis Wintoniensibus etc.

[a] c.f. Eph.4:25   [b] c.f. I Cor.12: 12–31   [c] B *ends here* Iohanne etc.

Given at Rome and addressed to prior S(tephen de Lucy) (*c.*1199–1214), the charter can only date to the bp's time at the papal curia in 1205–6, immediately after his promotion to the see of Winchester. This supposition is strengthened by the text itself, with its stress upon the obligations of one raised to a higher dignity through the help of the Winchester monks. The witnesses were presumably amongst the election committee sent by St Swithun's to Rome to support des Roches. The language of the charter may well be based on some as yet unidentified model from the papal chancery. Certainly the phrases used here to describe tithes of newly cultivated land are common to papal letters of the period, including a papal privilege in favour of the prior and convent of St Swithun's: *Monasticon* ii 211 (*Letters of Innocent III* no.1095), and see *Letters of Innocent III* nos.162, 296, 428.

82. Winchester, St Swithun's Priory

Composition between Peter bishop of Winchester and Walter prior and the convent of St Swithun's on one hand, and on the other William Brewer, made in the presence of William bishop of London, Richard bishop of Salisbury, and Eustace de Fauconberg the king's treasurer, appointed arbiters by the opposing parties in the presence of the papal legate Guala to settle a dispute over customs and demands claimed by William Brewer from the men of the bishop and prior of Winchester, in the forest of William's bailiwick and in the hundred of Somborne. For the sake of his soul, the soul of his wife, heirs and ancestors, William quitclaims to the bishop, prior and their men all customs and demands which he received or claimed from them in the aforesaid forest, save those granted in the Charter of the Forests issued by king Henry III to the barons of the realm, saving the king's right. In the same way Brewer quitclaims all suit to the hundred of Somborne and all exactions, customs and demands which he claimed to have received there in the time of the king's ancestors, provided that the legate, the king and his council will assent to this

St Swithun's Winchester, 10 August 1218

A = Winchester Cathedral Library, Alchin Scrapbook II (Henry III–Richard II) p.1 no.2.

Dorse inaccessible; approx. 184 × 91 mm.; foot cut away, no indication of sealing. Perished.

B = Ibid. Baigent Papers, Cathedral Records fo.266r-v. s. xix transcript of A, recording parts of the text now perished.

CYROGRAPHUM[a]

Hec est compositio facta inter dominum Petrum Wintoniensem episcopum et Walterum priorem et conventum ecclesie sancti Swithuni Wint(oniensis) ex una parte et dominum Willelmum Briewerr' ex alia coram domino Willelmo Lond' et domino Ricardo Sar' episcopis et Eustachio de Faucoberg domini regis thesaurario arbitris de voluntate partium coram domino Gal' sedis apostolice legato constitutis super querelis et exactionibus inter eos diu habitis de consuetudinibus et demandis quas idem Willelmus habuit et habere clamavit de hominibus predictorum episcopi Wintoniensis et prioris et conventus tam de foresta ballie predicti Willelmi quam de hundredo de Sumburn' . Scilicet quod idem Willelmus pro salute anime sue et uxoris sue et antecessorum et heredum suorum remisit et quietas clamavit de se et heredibus suis predictis episcopo Wintoniensi et priori et conventui et eorum successoribus et hominibus eorum imperpetuum omnes demandas consuetudines et exactiones quas percipere consuevit et percipere clamavit de eis de foresta predicta preter illas que continentur in carta regis Henrici tercii facta baronibus regni de foresta salvo iure domini regis . Idem etiam Willelmus pro se et heredibus suis concessit predictis episcopo Wint(oniensi) et p[riori et][b] conventui et eorum successoribus et hominibus eorum in perpetuum quod quieti sunt de secta hundredi de Sumburn' de [omnibus exac]tionibus[b] consuetudinibus et demandis quas idem Willelmus dicebat ad se ratione predicti hundredi pertinere tempore antecessorum [domini regis si][b] predictus dominus legatus et dominus rex et eius consilium assensum prebuerint . Acta sunt autem hec apud Sanctum Swithunum [Win]t'[b] decima die Augusti anno regni domini regis Henrici tercii secundo . Hiis testibus . Willelmo de P[ondelarche][b] . Rogero filio Henrici . Waltero de [Rumy]es'[b] . Ricardo de Anesy . Mauricio de Turevill' . Rogero de Baalun' . Willelmo de War[belto]n[b] . Galfrido de Sumborn' . Radulpho filio Ricardi . Gilleberto de Esseleg' . Philippo de Totteford . Nicholao de Kil[ham][c] . [Walter]o de [Scoteney][b] . Petro Rufo . Rogero de Vilers et multis aliis.

[a] polled   [b] letters in square brackets supplied from B perished in A
[c] Kil[ham] supplied from WC no.329, for which see note below.

Significant as an indication of the way magnates such as Brewer were forced to extend the liberties extracted from the crown under the Forest Charter of 1217, within their own bailiwicks. Brewer's award of the privileges mentioned in the composition survives as Winchester Cathedral Library ms. Alchin Scrapbook II p.1 no.3, copied into WC no.329. It adds little to the present agreement, save to state that attachments are to be included amongst the forest demands exempted from Brewer's quitclaim, as rights pertaining to the crown. Brewer's charter is undated, makes no mention of the legate Guala, and has only a few witnesses in common with the agreement above. The forest bailiwick mentioned in both charters is almost certainly the Hampshire forest of Bere Ashley, held by Brewer's family from the crown in hereditary right (*Rot. Chart.* 39b; PRO E32/161 m.1d; Turner *Men Raised from the Dust* 71–2). According to a list of foresters compiled c. 1217 × 1220, Brewer held Dartmoor in Devon and the forest *iuxta Wincestr' hereditarie* (PRO E32/253). Bere Ashley passed to Brewer's son and namesake (*Cl.R. 1227–31* 463). In an early thirteenth-century inquest, it was said to lie between Winchester and Southampton, and between the rivers Test (*Gurst*) and Itchen (*RLC* ii 156b; and see *Southwick Cart.* vol.1 i pp.192–3 for a later and more detailed extent). In 1228 it was judged to be of ancient creation and therefore immune from the disafforestation provided for in the 1217 Forest Charter (*Cl.R. 1227–31* 102–3). Brewer's hundred of Somborne lay within the forest bounds. By contrast to bp des Roches', none of whose manors lay in either the forest of Bere or the hundred of Somborne, the prior and convent possessed several manors in Brewer's bailiwick, including Sparsholt, Compton, St James outside the Westgate and Fulflood, all of them members of the great estate known as Barton Priors. In the 1230s Brewer's heirs quitclaimed to the convents of Nunnaminster and Hyde all rights exacted from their men within the hundred of Somborne, in return for payments of 12 marks and 18 marks respectively, the rights being described as services of grain, cheeses, chickens, eggs, and of ploughing, harvesting and carrying grain (PRO DL42/2 (Coucher Book of the Duchy of Lancaster II) fos. 189r, 191v). Brewer himself had founded Mottisfont priory within the forest bounds, endowing it with assarts and newly tilled lands (above no.34).

83. Wintney Priory

Confirmation to the nuns that the church of Sparsholt is to be appropriated to them after the death of William of London, the present rector, as granted by bishop Godfrey (de Lucy) (EEA viii no.256), saving the establishment of an adequate vicar's portion.

[5 September 1211 × 25 February 1221, ?5 September 1211 × c.1217]

B = PRO C53/124 (Charter Roll 11 Edward III) m.30. In a royal inspeximus of 18 March 1337, confirming an inspeximus by prior William and the convent of St Swithun's, dated 7 June 1287.
Pd (calendar) from B in *Cal. Chart. R. 1327–41* 394 (given wrongly as ms. m.31).

Omnibus Cristi fidelibus ad quos presens scriptum pervenerit P(etrus) Dei gratia Wyntoniensis episcopus eternam in domino salutem. Que a predecessoribus pie et rationabiliter sunt statuta rati successorum habitatione convenit confirmari ut que bono inchoata sunt principio pluribus munita suffragiis procedant firmius ad effectum. Ea propter ad notitiam omnium volumus pervenire nos divine pietatis intuitu et pro salute anime nostre et animarum predecessorum nostrorum Wyntoniensium episcoporum confirmasse ecclesie[a] beate Marie de Wynteneia et monialibus ibidem Deo servientibus ecclesiam sancti Stephani de Westpersolte cum omnibus ad eam pertinentibus, post decessum Willelmi de London' eiusdem ecclesie rectoris, in usus proprios convertendam sicut bone memorie Godefridus predecessor noster eis noscitur concessisse. Statuimus etiam quod ad earundem moniali(um) presentationem per nos vel per successores nostros Wyntonienses episcopos perpetuus vicarius in eadem ecclesia instituatur, ad sui sustentationem rationabilem vicariam habitturus, salvis in omnibus iure episcopali auctoritate et dignitate Wyntoniensis ecclesie. Ut autem hec nostra concessio et confirmatio perpetuum robur optineant scriptum istud sigilli nostri testimonio duximus roborandum. Hiis testibus: magistro Alano de Stok' officiali nostro, magistro Eustacio de Faucunberge, magistro Roberto Basset, Gregorio, Ricardo capellanis, Dionisio clerico, Ricardo de Elmaham, Ricardo de Berking', Roberto de Cerne clericis, Willelmo de Sorwell, Iohanne de Karitate, Iohanne de Colevill' militibus et multis aliis.

[a] ecclesia B

Master Alan of Stokes was appointed bp's official following the death of master John of London (d. 5 September 1211). William of Shorwell became sheriff of Hampshire in 1217. Eustace de Fauconberg was promoted royal treasurer in the same year, and was elected bp of London before 25 February 1221 (below appendix 4 nos.1,2,23,30). Bp des Roches was commemorated as a benefactor of Wintney in the nuns' obit list (BL ms. Cotton Claudius D iii fo.151r: *obiit Petrus episcopus Wintoniensis benefactor noster*).

84. Wintney Priory

Institution of Richard de la Hull, priest, as vicar of the church of Sparsholt at the presentation of the nuns of Wintney to whom the church was appropriated by bishop Peter in accordance with a charter of bishop Godfrey (de Lucy) awarding the church to the nuns after the death of the then rector, William of London; an award later confirmed by bishop Peter (above no.83). The bishop hereby endows a vicarage in the church with the tithes of lambs, wool, milk, cheeses and all small tithes of the parish, all obventions of the altar and all tithes of the land of Geoffrey Porcar', Segar, Warin and Richard fitz Hernicus in Westley. The messuage of the church is to remain to the nuns who are to provide a suitable house for the vicar elsewhere. The vicar and his successors are to bear the customary burdens of the church. Farnham, 4 June 1238

B = PRO C53/124 (Charter Roll 11 Edward III) m.30. In a royal inspeximus of 18 March 1337, confirming an earlier inspeximus by prior William and the convent of St Swithun's, dated 7 June 1287.

Pd (calendar) from B in *Cal.Chart.R. 1327–41* 394 (given wrongly as ms. m.31)

Universis Cristi fidelibus P(etrus) divina miseratione Wyntoniensis ecclesie minister humilis salutem in domino. Ea que auctoritate predecessorum nostrorum locis religiosis data sunt[a] rationabiliter et concessa per nos nolumus diminui set potius augmentari. Quapropter ecclesiam de Westpersolte quam bone memorie Godefridus Wyntoniensis episcopus divine pietatis intuitu dedit et concessit Deo et ecclesie beate Marie Magdalen' de Wynteneia et monialibus ibidem Deo imperpetuum famulantibus, ac pontificali auctoritate confirmavit post decessum Willelmi de London' rectoris eiusdem in usus proprios convertendam, salvo eo quod ad earundem monialium presentationem perpetuus vicarius institueretur in ecclesia memorata, nos eandem ecclesiam predictis monialibus quamvis alias confirmaverimus iterum confirmavimus et eas post decessum Willelmi de London' supradicti in possessionem[b] eiusdem ecclesie cum pertinentiis induci fecimus corporalem, salva nobis et successoribus nostris episcopali auctoritate et Wyntoniensis ecclesie dignitate. Ad presentationem etiam monialium earundem Ricardum de la Hull' presbiterum vicarium instituimus in ecclesia sepedicta, cuius vicaria arbitrio officialis[c] nostri taxata in subscriptis consistit, videlicet in decimis agnorum, lane, lactis et caseorum, et in aliis minutis decimis totius parochie, et in omnibus obventionibus altaris, et in omnimodis decimis de terris Galfridi Porcar', Segar', Warini, Ricardi filii Hernic', quas habent in Westeleia. Mesuagium etiam prefate ecclesie predictis monialibus totaliter remanebit, ita quod dicto vicario domum invenient alibi competentem. Sustinebit etiam idem vicarius et successores eiusdem omnia onera ad dictam ecclesiam pertinentia, debita et consueta. In huius autem rei testimonium presens scriptum fecimus sigilli nostri munimine roborari. Dat' apud Farnham, ii non' Iunii, anno domini m°cc° xxx° octavo. Hiis testibus: magistro H. de Miller' tunc officiali nostro, Hugone Wyntoniensi et Luca de Surr' archidiaconis, magistro Reginaldo de Wautham, Petro Russinol, Roberto de Clinchamp[d] ' Simone persona de Crundal, Rogero Wacelin', magistro Drocon', Iohanne vicario sancte Trinitatis de Guldeford, Willelmo clerico presentium scriptore persona de Bagehurst et multis aliis.

[a] *lacuna in* B   [b] in possessionem *over the line* B   [c] offic' B   [d] Clinclamp ms.

85. Witley Church

Institution of Robert de Ferles to the church of Witley at the presentation of Gilbert de Aigle.

    [c. March 1227 × 20 December 1231, ? August × 20 December 1231]

    A = PRO E135/5/35. Endorsed: *P.Wynton' de ecclesia de Witle* (s.xiii/xiv); *ad presentationem domini de Aquila ix* (s.xiv); approx. 169 × 73 mm. (originally 169 × 35 + 19 mm. before turn-up flattened out); slits for tag, sealing method 1; tag and seal missing.

Omnibus Cristi fidelibus ad quos presens scriptum pervenerit P. divina miseratione Winton(iensis) episcopus . eternam in domino salutem . Noverit universitas vestra nos divine pietatis intuitu ad presentacionem Gileberti de Aquila dedisse et concessisse*a* Roberto de Ferles clerico .'ecclesiam de Witle et eum in eadem canonice instituisse . Quare volumus et concedimus quod dictus Robertus dictam ecclesiam habeat et possideat libere integre et quiete cum omnibus pertinentiis suis . salvis in omnibus nobis et successoribus nostris iure episcopali et Wint(oniensis) ecclesie dignitate . Ut autem hec nostra concessio seu institucio perpetuam optineat firmitatem . eam presentis carte et sigilli nostri munimine .'confirmavimus . Hiis testibus . domino Luca archidiacono Surr' . magistro Willelmo de Dammartin . magistro I. Lemovic' . magistro Simone de Sabilio . Roberto de Wisle . Rogero de Depa . Ricardo persona de Chidingesaud' . Luca et Stephano capellanis et multis aliis.

a corrected from concessesse

For Witley see *VCH Surrey* iii 63, 69. Gilbert de Aigle was only restored to his lands in England in the spring of 1227 (PRO C60/25 m.12), and then only by proxy. Bp des Roches was absent on crusade from July 1227 to August 1231. De Aigle died before 20 December 1231 when his lands escheated to the crown (PRO C60/31 m.7; *Pat.R.1225-32* 458). Of the witnesses, master John of Limoges appears to have been exiled from England between 1224 and 1229 (appendix 4 no.33). Robert de Ferles was active in 1241 in the management of the former Aigle lands (*CPR 1232-47* 259, and see p.473). He is almost certainly the R. rector of Witley, party to a lost settlement with Waverley abbey made before bp des Roches in 1233 (*WC* 244; below appendix 2 no.38). At Easter 1229 a Robert de Ferles served as attorney to Bartholomew des Roches, the bp's nephew, archdn of Winchester (*CRR* xiii no.2044). In 1239 a namesake received money from the sheriff of Surrey on behalf of Hubert de Burgh (*CRR* xvi no.869). By 1248 the church of Witley was held by a Simon de Vercers, possibly a Savoyard, presented by Peter of Savoy to whom the Aigle lands had been awarded (*Cal.Pap.Reg.* i 253). The Aigle lands had previously passed via des Roches' kinsman Peter de Rivallis, to Gilbert Marshal, earl of Pembroke, who in 1238 made an unsuccessful attempt to grant Witley church to the convent of St Mary de Gloria, Anagni, the pope requesting that des Roches confirm the award (*Reg.Greg.IX*.nos.3817-19). For Robert de Ferles as attorney of Gilbert de Aigle, 1222, see *Sussex Fines* no. 175.

PART 2: THE BISHOP IN POLITICS

86. *Letters testimonial of W(illiam) bishop of London, G(ilbert) bishop of Rochester, H(enry) bishop of Exeter, H(erbert) bishop of Salisbury, E(ustace) bishop of Ely, G(eoffrey) bishop of Coventry, M(auger) bishop of Worcester, J(ohn) bishop of Norwich, W(illiam) bishop of Lincoln, S(imon) bishop of Chichester and P(eter) bishop of Winchester to pope I(nnocent III), informing him of the canonical election of master Jocelin, canon of Wells, as bishop of Bath following the death of bishop Savaric, and of the assent granted to the election by king John, petitioning the pope to confirm the election, the see of Canterbury being vacant.* [March × May 1206, ?c.23 April 1206]

> B = Wells Dean and Chapter Library ms. charter 39. An inspeximus by William Brewer bp of Exeter, April 1242. C = Ibid. charter 40. An inspeximus by Jocelin bp of Bath and Wells, William Brewer bp of Exeter and William de Raleigh bp of Norwich, July 1242. D = Ibid. ms. Liber Albus I fo.56r no.ccxi. s.xiii med.
>
> Pd from B in *Archaelogia* lii (1890) 106; C.M. Church, *Chapters in the Early History of the Church of Wells* (1894) appendix S pp.404-5. Pd (calendar) in *HMC Wells* i 64 (from D); ii 554 (from BC). Listed with full references in *EEA* iv no.221.

Jocelin was elected to the see March × April, and was consecrated 28 May 1206 (Cheney *Innocent III and England* 154-5). King John's letters in similar terms to the above are dated 23 April (*Archaeologia* lii (1890) 106). It is interesting to note that the bps are ranked in the address by seniority of appointment rather than the traditional hierarchy of the province of Canterbury in which Winchester ranked above all other suffragans save London.

87. *Similar letters from the same bishops, addressed to J(ohn) cardinal deacon of Santa Maria in Via Lata, legate of the pope.* [March × May 1206, ?c.23 April 1206]

> B = Wells Dean and Chapter Library ms. charter 39. An inspeximus by William Brewer bp of Exeter, April 1242. C = Ibid. ms. Liber Albus I fo.56r-v no.ccxii. s.xiii med.
>
> Pd from B in *Archaeologia* lii (1890) 106-7; C.M. Church, *Chapters in the Early History of the Church of Wells* (1894) appendix S pp.405-6. Pd (calendar) in *HMC Wells* i 64 (from C); ii 554 (from B). Listed with full references in *EEA* iv no.222.

*88. *Letters to pope (Innocent III) from (William) bishop of London, (Peter) bishop of Winchester, (Gilbert) bishop of Rochester, (Eustace) bishop of Ely and J(ocelin) bishop elect of Bath, on the preservation of the dignity of king John and his realm.* [c.May 1206]

Mentioned in letters of king John, 8 May 1208, requesting the abbots and the remaining eleven suffragan bps of the province of Canterbury to seal similar letters to the pope, presumably concerned with the election dispute at Canterbury, the letters to be issued in duplicate to forestall the perils of the journey to Rome (*RLP* 64–64b; *Foedera* 92).

*89. *Writ respiting Hugh de Neville from rendering account at the royal exchequer until 25 June 1208*   [29 September 1207 × 25 June 1208]

Mentioned in *Memoranda Roll 10 John* 35 and see p.64. Issued in the exchequer year 10 John and so dated after 29 September 1208.

*90. *Letters of Peter bishop of Winchester, (Geoffrey fitz Peter) the justiciar and William Brewer, instructing the keepers of English ports on how to proceed in respect to chattels belonging to the men of Flanders.*   [c. July 1208]

Mentioned in royal letters of 26 July 1208 (*RLP* 85b).

*91. *Letters to W(illiam) bishop of London, E(ustace) bishop of Ely and M(auger) bishop of Worcester, notifying them that their safe-conduct to 15 August (1209) has expired.*   [c.15 August 1209]

Mentioned in letters addressed to des Roches by the three bishops, the commissioners of the papal interdict whose vain negotiations at Dover in the summer of 1209 preceded the implementation of full papal sanctions (*Gervase* ii p.civ no.28 and see p.cv no.29).

*92. *Letters patent guaranteeing safe conduct to W(illiam) bishop of London, E(ustace) bishop of Ely and M(auger) bishop of Worcester, to come to England.*   [c.23 August 1209]

Mentioned in letters addressed to des Roches by the three bishops (*Gervase* ii p.civ no.28), complaining that the king's letters of safe-conduct (issued on 23 August, *Ibid.* p.ciii no.27) are inadequate as they exclude master Simon (of Langton) and the monks of Canterbury, whom the bishops consider essential to negotiations. The king has done nothing to restore lands seized from the church, despite des Roches' assurance that such restitution would be made (*Gervase* ii p. civ no.28). The present letters probably to be identified with letters of (Peter) bp of Winchester, (Jocelin) bp of Bath, H(ugh) bp of Lincoln and an un-named bp-elect of Exeter offering safe-conducts, mentioned in a letter to the king (*Ibid.* ii p.civ no.29).

*93. *Charter of Peter bishop of Winchester deposited at the royal treasury, cited by the king in ordering that a prest of £100 be paid to Walter de Gray, the royal chancellor.*   [c. 26 May 1212]

Mentioned in royal letters alongside a similar charter issued by Walter de Gray himself; *RLC* i 118. Des Roches' charter was presumably a guarantee of repayment on de Gray's behalf.

PART 2: THE BISHOP IN POLITICS 79

94. *Letters of P(eter) bishop of Winchester, J(ohn) bishop of Norwich, William Longespée earl of Salisbury, G(eofrey) fitz Peter earl of Essex, R(anulph) earl of Chester, W(illiam) earl of Arundel, Aubrey earl (of Oxford), H(ugh) de Neville, master R(ichard) de Marsh archdeacon of Northumberland, H(enry) archdeacon of Stafford, W(illiam de Cornhill) archdeacon of Huntingdon, William Brewer, Peter fitz Herbert, Thomas of Sampford, Robert of Burgate, Robert de Ropell', John of Bassingbourn, Engelard de Cigogné, Henry fitz Count, master E(rnold) of Auckland, master Robert of Gloucester, John fitz Hugh, Philip of Oldcotes, Brian de Lisle and Walter de St Ouen, who have obtained from king John the restoration to Peter de Maulay of his custodies. The above-said guarantors undertake to deliver up de Maulay to the king, for the king to do his will with him, should he in any way contradict royal orders in future or knowingly offend against the king; nor will they think badly of whatever vengeance the king may see fit to inflict. Earlier written conventions between de Maulay and the king are to remain in force as are any charters that the king may demand from de Maulay in future. Sealed by the said guarantors, with W(illiam) earl of Arundel sealing at the petition of bishop P(eter) of Winchester on his own behalf and on behalf of the bishop. The king undertakes that bishop (John) of Norwich will append his seal.* 'Meves', 27 May 1212

B = PRO C53/10 (Charter roll 14 John) m.7d.
Pd from B in *Rot. Chart.* 191–191b.

Des Roches offered 20 palfreys as pledge should de Maulay offend the king after 8 May 1212, with Walter de Gray and master Arnold of Auckland standing as the bp's guarantors. William de Cantiloupe offered two palfreys in the same manner (*Ibid.* 191b). Painter (*King John* 230–1) regards this entire transaction, which included an undertaking by Henry fitz Count, the king's cousin, to be beaten, or perhaps more correctly 'to be chastised', should de Maulay offend, as demonstrating the king's capricious sense of humour. More likely it referred to some temporary set-back in de Maulay's career, now obscured by the loss of the chancery rolls for the years before 1212. Nonetheless it provides a significant demonstration of the ties binding des Roches and his fellow alien, de Maulay. In 1211–12 de Maulay's men are recorded at the bp's manor of Farnham for three nights with nine horses (Ms. 7DR m.7d). The place-name *Meves* is unidentified. On the same day the court was at Bishop's Sutton, passing on to Winchester (*Misae Roll 14 John* 232). Duffus Hardy (*RLP* intro. Itinerary *sub* 27 May), identifies *Meves* as Woolmer. It might just as easily be the bp's manor of Meon, south of Bishop's Sutton.

*****95.** *Letters patent of H(enry) archbishop of Dublin, P(eter) bishop of Winchester, J(ohn) bishop of Norwich and of twelve named barons addressed to (Stephen) archbishop of Canterbury, with similar letters addressed to the prior of Canterbury and the bishops of London, Hereford, Ely, Bath and Lincoln, undertaking to uphold the form of peace established by the pope between king John and the church in England. Should the king contravene this settlement its* *guarantors will abide by papal mandates and the king will for ever forfeit his right to the custody of vacant churches.* [c. 24 May 1213]

> Mentioned in letters of the king (*RLP* 98b-99; *Foedera* 112; *Reg.Antiq. Lincs.* i 133). The baronial letters to Langton are enrolled in *RLP* 114b; *Foedera* 112. Those to the prior and convent of Christ Church Canterbury survive in the original; Canterbury Cathedral Library ms D. & C. Cart. Antiq. S236.

*96. *Letters of (Peter) bishop of Winchester, (John) bishop of Norwich, the earls of Chester and Winchester, William Brewer and William Marshal earl of Pembroke, or the earls of Ferrers and Arundel acting as William Marshal's proxies, undertaking to ensure that king John makes full satisfaction for any damages suffered by the church during the period of papal interdict. The king is to pay forty thousand marks before the interdict can be relaxed, and the remainder in annual instalments of twelve thousand marks.* [c. June 1214]

> These guarantees were solicited by the pope and mentioned in letters issued by the king on 17 June 1214 (*Letters of Innocent III* no.976; *SLI* 161 n.1; *Rot. Chart.* 199; *Foedera* 122). Those issued by William earl Ferrars, dated 16 June, pledging all his goods for the terms of the payment, are printed in *RLP* 139; *Foedera* 123. In the event the interdict was lifted in July 1214 after only twenty seven thousand marks had been paid in damages, des Roches and his fellow guarantors standing pledge for the thirteen thousand marks needed to make up the intended initial instalment of forty thousand (Cheney, *Innocent III and England* 348-55).

*96a. *Letters to Gilbert fitz Reinfrey, J(ohn de Lacy) constable of Chester, R(obert) de Vieuxpont, G(eoffrey) de Neville the chamberlain, Philip of Oldcotes and Brian de Lisle, containing instructions (?on the defence of the king's castles).* [c.23 April 1215]

> Mentioned in royal letters of 23 April 1215 to the same addresses (*RLP* 134), presumably intended to co-ordinate resistance to the baronial rebellion in the north. Both Painter (*King John* 305-6) and West (*The Justiciarship* 211) misinterpret the king's letters in stating that des Roches made a personal tour of inspection amongst the castles of the north.

*96b. *Letters to effect the release of Winchester castle to Savaric de Mauléon so that Savaric may receive the king's Poitevin mercenaries there.* [May 1215]

> Requested by the king in letters to des Roches of 11 May 1215 (*RLP* 135). On 14 May the bp's knights, William de Falaise and Maurice de Turville, were ordered to follow the instructions of Richard Marsh and Hugh de Neville on the disposition of Winchester castle and the person of Henry, the king's son. However, on the same day the king restored to them custody of the castle under the bp's authority, ordering that the queen and the future Henry III be taken to Marlborough (*RLP* 136). On 16 May John de Caritate was told to release Winchester castle to Savaric de Mauléon and his Poitevins (*RLP* 136b). Here and elsewhere the contradictory royal instructions appear to result from confusion between the king's castle at Winchester and the bp's castle at Wolvesey.

PART 2: THE BISHOP IN POLITICS   81

97. *Inspeximus of Magna Carta 1215 issued by Stephen archbishop of Canterbury, Henry archbishop of Dublin, William bishop of London, Peter bishop of Winchester, Jocelin bishop of Bath and Glastonbury, Hugh bishop of Lincoln, Walter bishop of Worcester, William bishop of Coventry, Benedict bishop of Rochester and master Pandulph papal subdeacon.* [June 1215]

> B = PRO E164/2 (Red Book of the Exchequer) fos. 234r-6v. s.xiii med.. The original was still amongst the treasury archives at the time of Stapleton's array in the 1320s: *Antient Kalendars* i 103.
>
> Address and corroboration pd from B in C. Bémont, *Chartes des Libertés Anglaises (1100–1305)* (Paris 1892) 39n; *Acta Langton* no.16; McKechnie *Magna Carta* 478–9n. Same portions pd, with variant readings noted from the remainder of the text, in *Statutes* i pp.xciii, 9–13.
>
> For the background see C.R. Cheney, 'The Church and Magna Carta', *Theology* lxviii (1965) 266–72; A.J. Collins, 'The Documents of the Great Charter of 1215', *Proceedings of the British Academy* xxxiv (1948) 244.

98. *Letters testimonial of S(tephen) archbishop of Canterbury, H(enry) archbishop of Dublin, W(illiam) bishop of London, P(eter) bishop of Winchester, J(ocelin) bishop of Bath and Glastonbury, H(ugh) bishop of Lincoln, W(alter) bishop of Worcester, W(illiam) bishop of Coventry, acknowledging king John's intention to reform the law of the forests by means of an enquiry by twelve knights in each county, and declaring that both sides (king and barons) are agreed that this should not be done in such a way as to make the forests unmanageable.* [15 June 1215 × September 1215, ? July 1215]

> B = PRO C54/12 (Close roll 17 John) m.27d. C = PRO C54/13 (Ibid.) m.21d.
> Pd from BC *Foedera* 134; McKechnie *Magna Carta* 496 no.6; Holt *Magna Carta* 498–9 no.14.
> For the background see Holt *Magna Carta* 352, 486–7.

99. *Letters testimonial of Stephen archbishop of Canterbury, Henry archbishop of Dublin, William bishop of London, Peter bishop of Winchester, Jocelin bishop of Bath and Glastonbury, Hugh bishop of Lincoln, Walter bishop of Worcester, William bishop of Coventry, Richard bishop of Chichester and Pandulph papal subdeacon, noting that when peace was made, the barons promised to issue whatever securities the king might require, save in castles or hostages. However, when the king required them to issue charters (recited) promising fealty to the king, his custom and right, and that of his heirs, in life, limb and land, the barons refused.* [15 June 1215 × September 1215, ? July 1215]

> B = PRO C66/14 (Patent roll 17 John) m.21d.
> Pd from B *RLP* 181; *Foedera* 134; McKechnie *Magna Carta* 497 no.7; *Acta Langton* no.17; Holt, *Magna Carta* 498 no.13. Perhaps to be dated to the same meeting as no.98 above, although see Holt *Magna Carta* 486–7.

100. *Letters of P(eter) bishop of Winchester, (Simon) abbot of Reading and Pandulph papal subdeacon to S(tephen) archbishop of Canterbury and his suffragans, reciting letters of pope Innocent (III) informing the said (Peter, Simon and Pandulph) that king John deserves the protection customarily afforded to a crusader. The archbishop and his suffragans having ignored previous papal mandates, ordering them to discipline those who conspire against the king, the pope hereby empowers (Peter, Simon and Pandulph) to excommunicate such conspirators; empowering them also to suspend any bishop who refuses to enforce this sentence. Given at Ferentino, 7 July 1215. The three commissioners accordingly inform the archbishop of Canterbury that nine named individuals and others have defied the king, refused to enforce a threefold form of peace* (triplex forma pacis), *and occupied and armed the city of London with the assistance of various of its citizens, to the great prejudice of the king and in contempt of mother church. The commissioners therefore pronounce excommunication and interdict against the conspirators, and quash any new customs or grants of land unlicenced by the king; denouncing in particular the city and citizens of London, G(iles) bishop of Hereford, W(illiam) his archdeacon, Alexander the clerk, Osbert de Samara chaplain, J(ohn) de Fereby clerk, R. chaplain of Robert fitz Walter and all who disturb the realm or impede the king's privileges as crusader.* Dover, 5 September 1215

    A = Canterbury Cathedral, Dean and Chapter Library ms. Cart. Antiq. M247. Approx. 275 × 285 + 15 mm.; slits for three tags, sealing method 1; tags and seals missing.
    Pd from A in F.M. Powicke, 'The Bull 'Miramur Plurimum' and a Letter to Archbishop Stephen Langton, 5 September 1215', *EHR* xliv (1929) 90-93; *HMC 5th Report* 454 (calendar). The papal bull only pd from A in *SLI* no.80; *Letters of Innocent III* no. 1016 (calendar).

Venerabilibus in Cristo patribus . domino S. Dei gratia Cant(uariensi) archiepiscopo . totius Anglie primati . et sancte Rom(ane) ecclesie cardinali . et suffraganeis eius . P. eadem gratia Winton(iensis) episcopus .. abbas de Rading' . et Pand(ulfus) . domini pape subdiaconus et familiaris . salutem . et debitum honorem . Litteras domini pape recepimus in hac forma . Innocentius episcopus servus servorum Dei . venerabili fratri .. Winton(iensi) episcopo . et dilectis filiis .. abbati de Rading' . et Pand(ulfo) . subdiacono et familiari nostro . salutem . et apostolicam benedictionem . Mirari cogimur et moveri . quod cum karissimus in Cristo filius noster .I. rex Anglie illustris . supra spem Deo et ecclesie satisfecerit . et presertim venerabili fratri nostro S. Cant(uariensi) archiepiscopo . sancte Rom(ane) ecclesie cardinali . et coepiscopis eius . quidam eorum minus quam oportuerit et decuerit ad sancte crucis negotium . apostolice sedis mandatum et fidelitatis prestite iuramentum debitum habentes respectum .'nullum ei contra perturbatores regni quod ad Romanam ecclesiam ratione dominii pertinere dinoscitur prestiterunt auxilium vel favorem . quasi conscii ne dicamus socii coniurationis inique . quia non caret scrupulo societatis occulte . qui manifesto facinori desinit obviare . Ecce qualiter patrimonium ecclesie Romane pontifices prefati defendunt . qualiter crucesignatos tuentur . immo qualiter se opponunt hiis qui destruere moliuntur negotium crucifixi . peiores proculdubio Saracenis . cum illum nitantur a regno depellere . de quo precipue sperabatur . quod deberet succurrere Terre Sancte . Sed forte non sine causa ille rex regum cuius ipse rex obsequio se devovit . permisit huiusmodi scandalum ante quam iter peregrinationis arriperet suscitari . ut fideles ab infidelibus discernantur . et cogitationes occulte de multorum cordibus revelentur[a] . singulis secundum sua merita recepturis .'nisi perversi prevenerint humili satisfactione vindictam . Nam etsi dictus rex remissus esset aut tepidus in hac parte .' nos non dimitteremus tantam nequitiam incorrectam . cum sciamus per Dei gratiam et possimus huiusmodi presumptionis audaciam castigare . Unde ne talium insolentia non solum in periculum regni Anglie verum etiam in pernitiem aliorum regnorum . et maxime in subversionem totius negotii crucifixi . valeat prevalere .' nos ex parte omnipotentis Dei patris et filii et spiritus sancti . auctoritate quoque beatorum apostolorum Petri et Pauli ac nostra . omnes huiusmodi perturbatores regis ac regni Anglie cum complicibus et fautoribus suis excommunicationis vinculo innodamus . et terras eorum ecclesiastico subicimus interdicto . prefatis archiepiscopo et coepiscopis suis . in virtute obedientie districtissime iniungentes .'quatinus utramque sententiam singulis diebus dominicis et festivis . pulsatis campanis et extinctis candelis per totam Angliam sollempniter publicare procurent . donec satisfecerint dicto regi de dampnis et iniuriis irrogatis . et ad eius obsequium humiliter revertantur . universis insuper eiusdem regis vassallis in remissionem peccaminum ex parte nostra firmiter iniungentes . ut contra perversores huiusmodi prefato regi oportunum tribuant consilium et iuvamen . Siquis autem ipsorum hoc nostrum preceptum neglexerit adimplere .' sciat se ab officio pontificali suspensum . et subiectorum sibi obedientiam esse subtractam . quia iustum est . ut inferiores ei nequaquam obediant . qui suo superiori obedire contempnit . Ne igitur mandatum nostrum tergiversatione cuiusquam valeat impediri .' executionem omnium predictorum cum ceteris que ad hoc negotium pertinuerint . vobis duximus committendam . per apostolica vobis scripta precipiendo mandantes . quatinus omni appellatione postposita procedatis sicut videritis expedire . Quod si non omnes hiis exequendis potueritis interesse .'duo vestrum ea nichilominus exequantur. Dat' Ferentin' non' Iulii. pontificatus nostri anno octavodecimo. Certi quoque sumus quod dominus papa vobis plures litteras monitorias et preceptorias destinavit pro domino .I.

Anglie rege illustri . contra perturbatores regie dignitatis . et utinam eorum efficaciam rei evidentia loqueretur . Notorium enim est . quod nobiles viri Robertus filius Walteri . qui exercitus Dei se nominat marescallum . comes Winton' . comes de Clara . comes Gloucestr' . Eustac(hius) de Vescy . Ricc(ardus) de Percy . I. constabularius Cestr' . W. de Abbeny . W. de Mombray . et multi alii complices et fautores ipsorum coniurationibus et conspirationibus illicitis contra dignitatem regiam colligati . contra ipsum regem et pacem regni quod est patrimonium beati Petri . arma movere nequiter presumpserunt . et contra triplicem formam pacis quarum quelibet honesta et rationabilis erat . et a viris dominum timentibus merito acceptanda . ipsum dominum suum contemptibiliter diffidarunt . fidelitatis sibi prestite vinculo dissoluto . Civitatem quoque London' . corone pariter et regni sui caput . fraudulentis machinationibus occuparunt . contempta forma pacis quam dominus papa presentibus et consentientibus eorum nuntiis providerat observandam . quibus ipsi cives non sine reatu periurii consenserunt . civitatem ipsam contra dominum suum regem et suos nequiter munientes . et quos fideles ipsi regi noverant .'penis corporalibus et pecuniariis prout gravius poterant punientes . unde quidam ipsorum licet crucesignati petentes iustitiam .' optinere nullatenus potuerunt . Ideoque predictis civibus cum aliis regis hostibus coniuratis . domino regi dignitates regie sunt subtracte . cum ipsi quod inauditum est . terras donent . consuetudines regni approbatas evacuent . nova iura constituant . et que a rege domino suo de consilio magnatum qui tunc erant eius familiares provide ordinata fuerant .'dissipent et immutent . et ita cum nemo tam nefariis eorum ausibus contradicat . nec ad sancte crucis negotium nec ad apostolice sedis mandatum . sive ad fidelitatis prestite iuramentum respectus debitus adhibetur . prout dominus papa conqueritur in suorum continentia mandatorum . Propter quod rex catholicus et tam Ihesu Cristo quam ecclesie illius sanguine rubricate devotus . in quantum ipsi possunt regia spoliatus est dignitate . et contemptibilis clericis factis et laicis quia sedis apostolice super se et terram suam presidium invocavit . ut possit dominus papa vere dicere cum propheta . filios enutrivi . et exaltavi . ipsi autem spreverunt me[b] . Cum plerique simplices causam predictorum vel ut dicamus verius inimicorum eo favorabiliorem existiment quo prelatos Angl(ie) conspiciunt aut manifeste sibi assistere aut contemptis mandatis apostolicis in eis ausus nefarios non punire . Cum igitur faciente prodolor malitia subditorum inde pericula fortius invalescant . unde ad sedanda illa dominus papa interponit sollicitius partes suas .' ne nos quibus super hiis dominus papa executionem mandatorum suorum specialiter demandavit . qui dudum expectavimus pacem et non venit . quesivimus bona et ecce turbatio . si ultra tacuerimus in periculum animarum et ordinum incidamus .' predictos nobiles domini regis et regni turbatores et hostes . et omnes eorum consiliarios complices et fautores denuntiamus auctoritate apostolica excommunicationis vinculo innodatos . et terras eorum interdicto subiectas . vobis eadem auctoritate firmiter precipiendo mandantes . quatinus utramque sententiam singulis dominicis et festivis diebus . extinctis candelis et pulsatis campanis denuntietis publice et faciatis singuli vestrum per proprias dioceses in singulis parrochialibus ecclesiis secundum formam prescriptam firmiter observari . Coniurationes quoque et conspirationes seu confederationes omnes contra predictum regem factas .'auctoritate apostolica denuntiamus omnino cassatas . constitutiones etiam vel assisas seu infeudationes vel donationes terrarum per eos vel per eorum aliquos factas . seu iudicia si qua sunt ab eis vel eorum aliquibus sine regis auctoritate presumpta . vel si qua faciant in futurum .' auctoritate sedis apostolice in irritum revocamus . omnes illos excommunicationis sententie specialiter supponentes .'qui aliquibus predictorum usi fuerint vel utentur . vel in aliquibus possessionibus libertatibus aut liberis consuetudinibus occasione seu auctoritate tali ius sibi presumpserint vendicare .'quibus utique sententiis civitatem . et cives London' involuimus tanto expressius et seorsum ab aliis .'quanto excessas ipsorum magis detestabilis reperitur . Qui cum essent peculiaris eius populus et in corona sua quasi lapides pretiosi fulgerent .'conversi sunt in arcum pravum[c] . et de fidelibus persecutores effecti . Interdictum vero sub hac forma precipimus observari . ut nullum omnino sacramentum ecclesiasticum celebretur in terris omnium predictorum vel ubi presentes extiterint . preter baptismum parvulorum et viaticum in extremis . Specialiter autem .E. dictum Hereforden(sem) episcopum . et .W. archidiaconum suum . Alexandrum clericum . Osbertum de Samara capellanum .I. de Fereby clericum . et R. capellanum dicti Roberti filii Walteri . qui sine superioris mandato ausus fuit Cristum domini pollutis labiis diffidare .' auctoritate apostolica propter manifestos eorum excessus quos in prefatum regem crucesignatum et beati Petri patrimonium commiserunt .' simili sententie duximus supponendos . et tamdiu hanc et alias predictas sententias firmiter observari precipimus .' donec de dampnis et iniuriis irrogatis et de tanta contumacia tam ecclesie Rom(ane) quam ipsi regi satisfecerint competenter . et sic promovere studuerint negotium crucifixi .'sicut illos constat idem negotium in illata guerra predicto regi et eius sociis crucesignatis contra concessum sibi privilegium disturbasse . Simili quoque sententia innodamus omnes qui dicto regi in regno suo denegant subtrahunt . et minuunt iura sua . Ne igitur que premissa sunt tergiversatione aliqua remaneant inexpleta .'vobis ex parte domini pape districte precipimus . quatinus circa observantiam predictorum talem diligentiam vigilantiam et sollicitudinem apponatis .' ne propter absentiam seu negligentiam alicuius quod absit aliqua de premissis suo defraudentur effectu . Dat' apud Dover' . non' . Sept' . pontificatus domini Innocentii .iii. pape . anno octavodecimo . regni vero domini .I. regis Anglie .anno . xvii°.

^a Luc.2:35   ^b Is.1:2   ^c c.f. Ps.(Gr.) 20:4, 77:57

The present letters mark a crucial turning point in relations between king John and the rebels, signalling the outbreak of the second stage of the civil war and an end to the peace obtained at Runnymede in June 1215. The letters are remarkable in that they nowhere refer to Magna Carta, a settlement of which the pope was still ignorant when his letters of 7 July were dispatched, and one which des Roches and his fellow commissioners clearly hoped to consign to oblivion. Instead des Roches and his companions refer to a *triplex forma pacis*, almost certainly a lost papal proposal communicated to England in the spring of 1215, which the barons are said to have rejected. The commissioners appear to have been particularly anxious to prey upon Innocent's concern for the crusade, making repeated reference to John's privileges as crusader and the damage done to the enterprise by his enemies in England. For commentary see Holt *Magna Carta* 370–75, 413–7; Cheney *Innocent III and England* 379–80, and for the *triplex forma pacis* see also C.R. Cheney, 'Gervase, Abbot of Prémontré: a Medieval Letter Writer', in Cheney, *Medieval Texts and Studies* (Oxford 1973) 253. The present letters were followed swiftly by the suspension of archbp Langton and a further sentence of excommunication against rebel barons (below appendix 2 nos. 6–7). Of the clerks excommunicated in the present letters, William archdn of Hereford was a brother and close adherant of the leading rebel, Robert fitz Walter; John de Fereby a clerk of the rebel Eustace de Vescy (*RLC* i 146,165b; *RLP* 101b). Osbert de *Samara* (?Seamer, Yorks. N.R.) is almost certainly the Osbert, chaplain of Richard de Percy, who in March 1215, together with John de Fereby, acted as representative of the rebel barons in Rome (*DD* no.19, and for Osbert as witness to a charter of Richard de Percy, see Bodl. ms. Rawlinson B455 (Cartulary of St Leonard's York) fo.221v). In 1223–4 a Richard de *Sammar'*, monk of Whitby, was admitted as master of the Benedictine nunnery at Stainfield, Lincs., with the assent of Richard de Percy, the house's patron (*Rot.Hugh Welles* iii 126–7). For various suggestions as to the identity of Alexander the clerk, see Powicke 'The Bull 'Miramur Plurimum'' 90 and references. The present excommunication of bp Giles of Hereford is further referred to in *Cal.Pap.Reg.* i 40–41; *Reg.Hon.III* no.28; Vincent *Guala* appendix 2 no. 165.

*101. *Letters patent to the burghers of Ypres authorising an attorney to collect money owed to the crown on behalf of bishop des Roches.* [c. 12 September 1215]

Mentioned in royal letters of 12 September 1215 (*RLP* 155). In 1214, acting as regent, des Roches had paid various unspecified subsidies to Ypres from his own purse. It appears that, rather than reimburse him directly, the king merely offered to help him recover the money from the men of Ypres who had not put it to the use for which it was intended, presumably to strengthen John's alliances in the Low Countries in the lead up to the Bouvines campaign; *RLP* 122b, 123b, 124, 177, 184b; *RLC* i 175, 176. The money was still owing to des Roches in 1222 (below no.124). The mission of 1215 for the money's recovery marks the earliest known appearance of Robert Passelewe, later a prominent figure in des Roches' regime of the 1230s.

*102. *Letters of the legate (Guala) and of (Peter) bishop of Winchester, testifying to expenses of £52 made on the king's behalf by the burgesses of*

Gloucester, to be remitted from the burgesses' liability for tallage.
[November 1217 × 1218]

Cited as warranty for a writ of *allocate* during an audit of expenses following the civil war: PRO E159/2 (King's Remembrancer memoranda roll 3 Henry III) m.18d: *litteras domini legati et domini Wyntoniensis,* also Vincent *Guala* no. 31. For similar quittances to the men of Gloucester, to cover the king's expenses there during the war, see *RLC* i 360,370. For the tallage of the royal demesne taken after 9 November 1217, see *Pat.R.1216-25* 170-1; Carpenter *The Minority* 67. The present quittance of £52 is mentioned in the pipe roll drawn up after Michaelmas 1218, but as a writ of the king (*PR 3 Henry III* 41, and see below appendix 1 for the way in which courtiers' letters are often disguised as royal writs in the pipe rolls).

*103. *£55 and ten shillings spent by Geoffrey de Martigny, constable of Northampton, by writ of (Peter) bishop of Winchester.* [1216 × 1217]

Mentioned during an audit of war-time expenses to be deducted from the arrears owing by the burgesses of Northampton (PRO E368/1 (Lord Treasurer's Remembrancer memoranda roll 2 Henry III) m.6: *Galfrido de Martino .... per breve domini Wintoniensis*).

*104. *Letters of king H(enry III), H(enry) archbishop of Dublin, (Walter) archbishop of York, (William) bishop of London, (Peter) bishop of Winchester, (Jocelin) bishop of Bath, and (Silvester) bishop of Worcester addressed to pope Honorius (III), complaining that the canons of Carlisle have ignored papal mandates and the orders of the legate Guala, associating themselves with the king's excommunicate enemies, celebrating services in locations placed under interdict, and electing an excommunicate clerk at the behest of the king of Scotland.* [26 April × June 1217]

Mentioned in papal letters to the legate Guala, dated 13 July 1217 (PRO SC7/18/3; pd *Foedera* 147. Calendared from the version in the papal register, dated 6 July 1217 in *Cal.Pap.Reg.* i 48; *Reg. Hon. III* no.650, and see Sayers *Honorius III* 214 no.3). The papal letters are in turn a reply to royal letters of 26 April 1217 (*Pat.R.1216-25* 111; *Foedera* 147). The present letters are further referred to in papal letters of 24 August 1218, ordering the legate Guala to remove the Augustinian canons from Carlisle, to institute seculars in their place, quash the disputed election and divide the revenues of the see between the bishop and his canons (*Reg.Hon.III* no.1596; *Cal.Pap.Reg.* i 57). In the outcome, the bpric passed to Hugh, former abbot of Beaulieu, a close associate of des Roches (*infra* nos. 1,2,110; Sayers *Honorius III* 175-6; Vincent *Guala* nos. 12-14).

105. *Mandate of the papal legate Guala cardinal priest of St Martin (in Montibus), P(eter) bishop of Winchester and W(illiam) Marshal earl of Pembroke, addressed to Peter de Maulay, informing him that William de Aubigné has delivered up William son of Henry de Neville of Hale and William son of Osbert of Boothby as hostages to obtain the release of Alice Trussebut his*

wife, and has promised to deliver up his son Nicholas de Aubigné, a clerk, to Philip Mark by 9 April 1217 to the same end, to serve as Agatha's pledge until he can deliver his son Robert to the king. De Maulay is instructed to release Agatha to her husband. [16 × 22 March 1217]

B = PRO C54/16 (Close roll 1 Henry III) m.21d. C = PRO C54/17 (Ibid.) m.13d.
Pd from BC *RLC* i 335b; Vincent *Guala* no. 82.

Enrolled on the dorse of a membrane of the close roll whose recto carries writs dated 16 × 22 March 1217. For Peter de Maulay and his detention of prisoners during and after the civil war see Carpenter *The Minority* 46, 197, 338. For William de Aubigné and his wife see *Ibid.* 25 n.3.

106. *Notification by the papal legate Guala cardinal priest of St Martin (in Montibus), H(enry) archbishop of Dublin, W(alter) archbishop of York, P(eter) bishop of Winchester, J(ocelin) bishop of Bath and Glastonbury, S(ilvester) bishop of Worcester, W(alter) de Lacy, J(ohn) of Monmouth and H(ugh) de Mortimer addressed to Peter de Maulay, informing him that because of his retention of William of Lancaster, whom the king ordered released to R(anulph) earl of Chester, the earl has threatened to set out immediately on crusade, leaving the realm desolated by his absence. At the petition of the said legate, bishops and magnates, the earl has delayed his departure in order to take up the cross against the enemies of God and the king. Since the earl has served so well in this capacity, de Maulay is ordered to send William of Lancaster without delay under safe-conduct to Gloucester where he is to be released into the custody of the sheriff of Gloucestershire, Ralph Musard. Unless this is done as quickly as possible, great harm may befall the king and his realm.* [16 × 22 March 1217]

B = PRO C54/16 (Close roll 1 Henry III) m.21d. C = PRO C54/17 (Ibid.) m.13d.
Pd from BC *Foedera* 146; Vincent, *Guala* no. 83.

Enrolled on the dorse of a membrane of the close roll covering the period 16 × 22 March 1217. For the circumstances see Carpenter *The Minority* 26–7, 197

107. *Notification by Walter archbishop of York, William bishop of London, Peter bishop of Winchester, Richard bishop of Durham, Richard bishop of Salisbury, Hugh bishop of Lincoln, Jocelin bishop of Bath and Glastonbury, Simon bishop of Exeter, William bishop of Coventry, William Marshal (I) earl of Pembroke, Hubert de Burgh justiciar of England, Saher earl of Winchester, John Marshal and Thomas of Erdington, that at the request of the legate Guala, king Henry (III) has granted the church of Chesterton to the canons of S. Andrea, Vercelli.* [*c.*8 November 1217]

A = Untraced. In 1767 amongst the charters of the abbey of S.Andrea at Vercelli, a collection dispersed between 1802 and 1830. Said to have been sealed with 15 (*sic*) seals, all of them in white wax save for the Marshal's which was in green wax.

B = Vatican, Archivio Segreto ms. Reg. Vat. 12, part 2 (Register of Honorius III, year 8) fo.188v no.428. In an inspeximus by pope Honorius III, 2 May 1224. C = Cambridge, Trinity College Muniments Box 22 Chesterton no.6. In a notarial inspeximus, 23 March 1405, of letters of pope Urban IV, 11 May 1262 (not in pope Urban's register) reciting various early charters associated with the grant of Chesterton to S. Andrea Vercelli.

Pd from A in 'Philadelfo Libico' (G.Frova) *Gualae Bicherii presbyteri cardinalis S. Martini in Montibus Vita et Gesta* (Milan 1767) 100–1 note s. Pd (calendar) from Frova's version in J. Foster, 'The Connection of the Church of Chesterton with the Abbey of Vercelli', *Proc. Cambridge Antiq. Soc.* xiii new series vii (1908-9) 187-9. Pd (calendar) from B in *Reg.Hon.III* no.4955; *Cal.Pap.Reg.* i 97. The papal confirmations, B and C, are noticed from originals, now lost, once at S.Andrea Vercelli, in Frova *Op.Cit.* 101-3n.

Walterus Dei gratia Eboracensis[a] archiepiscopus, Willemus Londoniensis[b], Petrus Wintoniensis, Ricardus Dunelmensis[c], Ricardus Sarresburiensis[d], Hugo Lincolniensis, Ioscelinus Bathoniensis et Glastoniensis, Simon Exoniensis, Willelmus Coventrensis[e] eadem gratia episcopi, Willelmus Marescallus comes Pembrocii, Hubertus[f] de Burgo iusticiarius[g] Anglie, Saerus comes Wintoniensis, Iohannes Marescallus, Thomas de Erdington, universis Christi fidelibus litteras presentes visuris salutem. Universitati vestre notum facimus quod dominus noster karissimus Henricus rex Anglie illustris, nobis presentibus et consencientibus ac consulentibus, intuitu Dei et pro salute sua propria et animabus predecessorum suorum dedit et concessit ad preces domini Guale[h] tituli sancti Martini presbiteri cardinalis tunc apostolice sedis legati in Anglia, qui pro pace sua et regni diu et multum laboravit, ecclesiam de Cestretune[j] que de sua erat advocatione cum omnibus pertinentibus ad illam Deo et ecclesie beati Andree Vercellensis et canonicis ibidem Deo servientibus ad sustentacionem eorundem, quam quidem ecclesiam idem dominus Guala in honore Dei et beati Andree construxit ibidem, habendam et possidendam in liberam et puram et perpetuam elemosinam, consentiente et confirmante domino Roberto tunc Elyensi electo in cuius episcopatu consistit ecclesia memorata. In cuius rei testimonium presentibus litteris sigilla nostra fecimus apponi.

[a] *Here and elsewhere the episcopal titles are expanded in* A *but not in* BC
[b] Londonensis A   [c] Riccardus Dunelmen' B Dunhelmensis A
[d] Riccardus Sarresburien' B Saresburien' C   [e] Covventren' C
[f] Humbertus A   [g] iusticiariis C   [h] Gual' B Gualis C   [j] Cestretun' B Cestreton' C

Henry III's letters granting Chesterton to S. Andrea, sealed by the Marshal, are dated 8 November 1217 (Frova *Op. Cit.* 100; *Cal.Chart.R. 1226–57* 234; *Cal.Pap.Reg.* i 97; Cambridge, Trinity College Muniments Box 22 Chesterton nos. 6,7). The original survives as Vercelli, State Archive ms. Pergamene 149. Amongst the other instruments printed by Frova *Op.Cit.* 100–104, letters of Guala granting Adam of Wisbech a perpetual vicarage at Chesterton survive in a notarial copy of 1293 (Turin, Biblioteca Reale ms. Pergamene sec. xiii no.88). The original letters of Robert, bp elect of Ely, dated at London, 13 November 1217, confirming the grant of the church, are untraced although copies of a confirmation by Roger prior of Ely survive in *Reg.Hon.III* no.4955 (pd from the lost papal original in Frova *Op.Cit.* 101–2); Cambridge Trinity College, Muniments Box 22 Chesterton no.6. In general see Vincent *Guala* no. 16.

108. *Letters of credit issued by W(illiam) bishop of London, P(eter) bishop of Winchester, R(anulph) bishop of Chichester, W(illiam) bishop of Worcester, W(illiam) Marshal earl of Pembroke, H(ubert) de Burgh justiciar of England, W(illiam) Brewer and Fawkes de Bréauté, promising repayment to all merchants who may lend up to six thousand marks to R(anulph) bishop of Chichester, P(eter) Saracen citizen of Rome, Geoffrey de Caux and brother Richard, a monk of Abingdon, or to two or three of them, towards the expedition of the king's affairs in the papal curia.* [November × December 1218]

> B = PRO C66/20 (Patent roll 3 Henry III) m.6d (cancelled). C = PRO C66/21(Ibid.) m.6d (cancelled).
> Pd from BC *Pat.R.1216–25* 208–9 and see *Ibid.* 181.

> Geoffrey de Caux was a member of des Roches' household (below appendix 4 no. 29). Similar letters appear to have been drawn up in August 1218, covering loans of 5000 marks (*Pat.R.1216–25* 167). The delay in the mission's departure probably reflects the crown's financial predicament in the aftermath of civil war. The mission itself was charged with obtaining respite in payments of the annual census owed to Rome since 1213 (*DD* no.25).

*109. *Letters of (Peter) bishop of Winchester cited as warranty for royal letters close, ordering the sheriff of London to make payments said to total £30 four shillings and sixpence (actually sevenpence) towards cloth and garments for the king's household knights for the next feast of Pentecost.* [c.16 May 1219]

> Cited as warranty (*RLC* i 391b).

110. *Letters of R(ichard Marsh) bishop of Durham and P(eter) bishop of Winchester to R(alph) de Neville, dean of Lichfield, notifying him of the circumstances surrounding the recent election to the bishopric of Lismore. The bishop of Waterford has shown the legate (Pandulph) that when Guala cardinal priest of St Martin (in Montibus) was legate he visited the north for the consecration of the bishop of Carlisle. There he and the king's council were approached by master R(obert) of Bedford, accompanied by master Macrobius and master David, canons of Lismore, who bore letters from their chapter empowering them to elect to the vacant see of Lismore. However, the bishop of Waterford now claims that he and his predecessors have held the see of Lismore as part of the see of Waterford, and although in the past, 'in the time of the Irish',* *the sees were separate, he has shown the legate and the bishops of Durham and Winchester letters of J(ohn) cardinal priest of St Stephen in Celio Monte, then legate in Ireland, testifying that the sees have been united. In this way the letters which master R(obert) of Bedford obtained from the king, ordering the justiciar of Ireland to grant him the possessions of the see of Lismore, were obtained fraudulently. The legate (Pandulph), anxious to remedy this mistake, has authorised the bishops of Durham and Winchester to instruct Ralph de Neville to write to the justiciar of Ireland, ordering him to restore the possessions of Lismore to the bishop of Waterford, despite the letters obtained by master R(obert). In the same way, Neville should write to the archbishop of Cashel, to whom master R(obert) obtained similarly fraudulent letters, notifying him of the situation on behalf of the bishop of Waterford, especially since Odo Wegan, who is known to archbishop D(onnchad) of Cashel, should not be given possession of the aforementioned (see of Lismore)* (maxime cum Odo Wegan quem D. archiepiscopus Cassellensis ....perinnoscens fuit, rerum predictarum possessionem nonquam habuerit). [June 1219]

A = PRO SC1/6/24. No visible endorsements; approx. 164 × 67 mm.; step at left hand corner of foot; tongue and seal torn away. Heavily stained and now illegible in parts.

Pd from A in J. Boussard, 'Ralph Neville Éveque de Chichester d'après sa correspondance', *Revue Historique* clxxvi (1935) 220–1n.

The election of master Robert had taken place by 13 December 1218 when the king wrote to Geoffrey Marsh, justiciar of Ireland, informing him that the canons of Lismore, having elected master Robert, had at first been forced to renounce their election, before the council, because they had not sought royal licence to proceed. They had since gone on to re-elect Robert and obtained the king's assent (*Pat.R.1216–25* 183). By 28 April 1219 des Roches and de Burgh were clearly concerned by the circumstances of master Robert's promotion, ordering the Irish justiciar to delay the release of Lismore's temporalities (*RLC* i 391). The present letters from des Roches and the chancellor, Richard Marsh, elicited a response on 7 June when royal letters patent and close were sent to the Irish justiciar and the archbp of Cashel, *Teste* Hubert de Burgh, *per litteras suas et litteras dominorum Winton' et Dunolm' episcoporum factas auctoritate domini legati*. The king's letters are virtually identical to those of the bps above, save for their omission of the final clause concerning Odo Wegan, whose meaning remains obscure (*RLC* i 392). Both the king's and the bps' letters contain an obvious inaccuracy: the visit by the legate Guala to the north cannot have been for the consecration of bp Hugh of Carlisle, which did not take place until February 1219. More likely the visit was concerned with Hugh's election (*Fasti* ii 20). In the aftermath an enquiry was held by papal authority, in which the legate Pandulph, archbp Langton and the bp of Rochester judged that master Robert should be restored to Lismore, which was to remain independent of Waterford (*RLC* i 425b, 475b–6; *Cal. Pap. Reg.* i 69–70).

III. *Letters to H(ubert) de Burgh, justiciar, forwarding a letter from the legate (?Pandulph), and asking that de Burgh appoint a suitable cleric by whom the business proposed by the legate can be prosecuted. The bishop himself is ready to do what he can for the affairs of the king and the justiciar.*

[May 1216 × 19 July 1221, ?January 1220]

A = PRO SC1/1/200. No visible endorsement; approx. 118 × 23 mm.; step at bottom left hand corner; tongue and seal torn away; two slits for the insertion of tongue or wrapping tie. Badly rubbed. Letters close.
Pd from A in *DD* no.68.

The mention of a legate dates the document between the arrival of Guala in England and the resignation of Pandulph. The text is fairly certain to be linked to *DD* nos. 62-4, letters of Pandulph to de Burgh, asking that a suitable clerk be chosen to negotiate a prolongation of the truce with France, and to *DD* no.61, a letter in similar terms addressed to des Roches in January 1220. The letters enclosed, mentioned above, might possibly be *DD* no.59. For the truce see Carpenter *The Minority* 172-4,176-8.

112. *Letters to H(ubert) de Burgh, justiciar of England, informing him of the wishes of king (Henry III) and the legate (Pandulph) in respect to the lands of Walter de Hauville.* [January 1220]

A = PRO SC1/1/198. No visible endorsment; approx. 151 × 35 mm.; step on left hand side, tongue and seal torn away. Stained.
Pd (English abstract) from A in West *The Justiciarship* 245.

P. Dei gratia Wintonien(sis) episcopus . nobili viro et amico in Cristo karissimo . domino H. de Burg' iusticiario Anglie . salutem in domino . Super hoc quod significastis nobis per litteras vestras de terra que fuit Walteri de Hauvill' committenda Galfrido et Gilleberto de Hauvill' . qui custodiunt gilfalcones domini regis . sciatis quod multum placeret nobis eorum promotio . et inde honorem plurimum domino regi credimus provenire . Volumus tamen vos scire quod dominus rex super eadem terra scripsit domino legato in vigilia circumcisionis cum apud Saresb' fuimus ut custodiam eiusdem terre committerit de consilio vestro Henrico de Hauvill' . et credimus quod dominus legatus super eodem negocio scripserit vobis . Unde consulimus vobis quatinus huic negocio supersedeatis donec inde tractatum habueritis cum domino legato . Nos quidem predictorum Galfridi et Gilleberti . et aliorum qui domino regi servicium faciunt gratam haberemus promotionem . Super hiis et aliis voluntatem vestram . nobis rescribatis . scientes quod multum desideramus loqui vobiscum . Valet(e) semper in domino.

Des Roches, the king and the legate Pandulph were at Downton (Wilts.) on 28 December 1219, where they were entertained by the bp's bailiffs (Ms. 15DR mm.7d,8). They reached Salisbury on the following day, two days before the dispatch of the letters of the king referred to above (*CACW* 2). Walter de Hauville d. after 21 September 1219 (*RLC* i 400b). Despite the king's instructions, presumably prompted by des Roches, de Burgh authorised an award on 12 February 1220 whereby Walter's custodies in Bladon (Oxon.) passed to Walter's nephew, Geoffrey de Hauville, and his lands in Essex to Gilbert de Hauville (*RLC* i 411). In general see Carpenter *The Minority* 172.

PART 2: THE BISHOP IN POLITICS   93

113. *Letters to Hubert de Burgh, justiciar of England. The bishop has heard from G(eoffrey) de Neville, chamberlain, that the archbishop of Bordeaux has died. He therefore advises de Burgh to write to the dean and chapter of Bordeaux under the king's seal, that they chose a pastor who has the confidence of the king and his council, and who will work for the good of the king, currently under the protection of the church of Rome. De Burgh should also write to Adam de Monte Morelli, that he keep faithfully the king's castles he holds by order of the late archbishop. De Burgh may see fit to send the king's letters (to the dean and chapter) together with the letters the bishop himself has written on this subject.*
[*c*.April 1219 × 9 March 1220]

A = PRO SC1/1/199. No visible endorsement; approx. 141 × 47 mm.; tongue and seal torn away. ?Letters close.
Pd from A in Prynne *Records* iii 45; *Foedera* 164.

PRO Lists and Indexes 15 (Revised edn., New York 1968) nos. 1/20, 1/199 dates this letter and no.114 below to 1227, presumably because William, archbp of Bordeaux since 1207, died in September 1227 (*Gallia Christiana* ii cols. 820-2; Gams *Series Episcoporum* 520). But, by September 1227 des Roches had already set out on crusade. The whole tone of the letter requires a date April 1219 × January 1221, the period of Henry III's minority when des Roches and de Burgh worked in tandem as the king's chief ministers. It and its companion, no.114 below, appear to have been inspired by a false rumour of the death of archbp William, due no doubt to the archbp's prolonged absence on the Fifth Crusade, in which enterprise he remained abroad from 1218 until after October 1221 (*Pat.R.1216-25* 152; R.Röhricht, *Studien zur Geschichte des fünften Kreuzzuges* (Innsbruck 1891) 43 no.6, 74-5 no.54). For the concerns to which his absence gave rise, see *DD* no.103. The rumour must be assumed to have been proved false by 9 March 1220, when a nuncio from archbp William was received at court in England (*RLC* i 414). Geoffrey de Neville, who communicated the rumour to des Roches, left Poitou *c*. June 1220 and was not in the province again until December 1222, by which time archbp William had returned safely from crusade (*Pat.R.1216-25* 152, 235, 275, 357).

114. *Letters to the dean and chapter of Bordeaux admonishing them to elect a worthy archbishop.*   [*c*.April 1219 × 9 March 1220]

A = PRO SC1/1/20. No visible endorsement; approx. 147 × 45 mm.; step, tongue and seal torn away. Stained and rubbed.
Pd from A in Prynne *Records* iii 46.

Viris venerabilibus et in Cristo dilectis . decano et capitulo . Burdegalen' .P. divina miseratione Wintonien(sis) ecclesie minister humilis . salutem et sincere dilectionis affectum . Quia bone memorie .. archipresulem vestrum audivimus decessisse . discretionem vestram rogamus atque monem[us in domino]^a diligenter quatinus circumspectis status terre vestre et regis Angl(ie) circumstantiis . qui in defensione et custodia Rom[an]e^a ecclesie consistit . in provisione pastoris taliter vos habeatis . quod et dignitas ecclesie vestre illibata servetur et vos et ecclesia vestra quantum ad regem naturalem dominum vestrum gaudere debeat . temporalibus incrementis . Et quoniam huiusmodi negotium aliquantam deliberationem capit . consulimus in domino . quatinus pensatis omnibus sicut prediximus honore videlicet ecclesie vestre et utilitate . atque patrie tranquillitat[ate]_a_ in hoc facto provide et consulte vos studeatis habere . Nos vero quantumcumque cum honestate poterimus . ad executionem boni propositi vestri et alios [consiliaros]_a_ domini regis . pro nos habebitis et benignos . Valeat universitas vestra semper in domino.

a letters in square brackets now illegible, supplied from Prynne

Inspired by a false rumour of the death of archbp William, and therefore probably never sent. For the dating and circumstances see above no.113.

*115. *Letters stating that (Richard fitz Regis) sheriff of Berkshire was in the king's service on the morrow of the Sunday after Easter (6 April 1220) and should therefore not be penalised for his failure to attend at the Exchequer.*
[c.April 1220]

Cited in the memoranda roll PRO E159/3 (King's Remembrancer memoranda roll 4 Henry III) m.6: *litteras domini Wintoniensis*, and see West *The Justiciarship* 240.

116. *Letters of P(eter) bishop of Winchester and (Hubert) de Burgh, justiciar of England, to R(alph) de Neville dean of Lichfield, instructing him to listen to Roger fitz John's account of the proceedings in a law suit involving Henry fitz Count (sheriff of Cornwall) and the sheriff of Devon, and to issue instructions to the sheriff of Cornwall as seems expedient, following the advice of Martin of Pattishall.*
[c.April 1219 × 27 April 1220]

A = PRO SC1/6/26. No visible endorsement; approx. 145 × 45 mm.; step at left hand corner, tongue, tie and seal missing. Irregular tear at the top of the document, which is badly rubbed and illegible in parts.

P. Dei gratia Wint(oniensis) episcopus [et Hubertus] de Burg' iusticiarius Anglie dilecto sibi R. de Nevill' decano Lichef' salutem . Mittimus ad vos Rogerum filium Ioh[annis] ut querelam et narrationem suam quam nobis exposuit . audiatis quia nobis videtur quod faciendum sit ei breve domini regis [direc]tum H. filio Comitis . ut venire faciat per .iiii.^{or} milites de comitatu Cornub' coram iusticiis in banco iudicium et recordum ad diem certum . et ut interim ponat in respectum duellum quod contra eum invadiatum est . Nobis etiam videtur quod mandandum sit vic(ec)omiti Devon' ut securos plegios ab eo capiat de clamato suo prosequendo quia vicomes Cornub' non vult ut dicitur mandatis domini regis prout debet parere et accepta securitate . id(em) vicomitem Cornub' scire faciat demonstratum_a_ quicquid nobis inde visum

PART 2: THE BISHOP IN POLITICS 95

fuerit .ᵃ habito consilio Martini de Patishill' et aliorum virorum discretorum prout iustius et melius secundum narrationem suam videritis expedire .ᵃ faciatis . Valet(e).

ᵃ *word badly rubbed:* ?d'.mitum.

Henry fitz Count was removed from the sheriffdom of Cornwall on 27 April 1220, for just the sort of high-handed treatment of royal orders complained of above (*Pat.R.1216–25* 231). No trace of a case involving Roger fitz John can be found in either the Devon eyre of 1218–9 or the records of the Bench (although see possibly *PR 2 Henry III* 74, for a namesake purchasing a writ in pursuit of legal action at Barking, Essex). The nature of his case and of Henry fitz Count's determination that it should proceed by duel, against Roger's wishes, remain obscure.

*117. *Letters of W(alter) archbishop of York, P(eter) bishop of Winchester, R(ichard) bishop of Durham the king's chancellor, H(ugh) bishop of Carlisle and S(imon) abbot of Reading given to A(lexander II) king of Scotland, promising to support with ecclesiastical censure a settlement whereby king H(enry III) will give Joan his eldest sister in marriage to king A(lexander) at Michaelmas 1220 should she be available, or otherwise grant Isabella his younger sister within a fortnight thereafter. King H(enry) undertakes to provide husbands for Margaret and Isabella, the sisters of king A(lexander) within a year of 9 October 1220 or to return them to their brother's keeping.*

[*c.*15 June 1220]

Mentioned in royal letters of 15 June 1220 reciting the marriage settlement with Scotland, referring also to similar undertakings by fifteen lay magnates (*Pat.R.1216–25* 235; *Foedera* 160). For the circumstances see Carpenter *The Minority* 194–7, 245–6, 252–3.

118. *Letters to Pandulph, bishop-elect of Norwich, papal chamberlain and legate. In answer to his order to attend to the affairs of Poitou, the bishop of Winchester replies that he hopes to arrive in Winchester on the coming Sunday after the decollation of St John the Baptist (30 August 1220), where he will discuss the matter fully with the justiciar; the legate's own attendance at these discussions is advisable. Concerning the dean of Poitiers, who has now come to the bishop, the bishop is keeping him with him, considering him useful to the negotiation of a truce or in other matters. The bishop will keep the legate informed of developments and meanwhile detain the dean until he hears the legate's wishes. Concerning those who participate in tournaments, the legate should know that the bishop has never knowingly received those who were at the latest tournament in his household* (domo), *nor were any of the bishop's knights there. Had they been, they would have been expelled forever from his following* (familiarite) *and only find the bishop's forgiveness if the legate willed it. As the*

legate wishes, the bishop has pronounced an interdict on the lands of all those he knows to have been present, throughout his diocese. For the same reason, he has sent back to his homeland (ad partes nostras) *certain of his kinsmen* (nepotes) *and knights so that they may not bear arms in England. Concerning Robert de Cardin(an)* (pro negocio domini Roberti de Cardin') *who has sustained great injury in the king's service, the bishop asks that the legate act as is fitting to the honour of God, the church and the king, and so that others may be dissuaded from causing offence to the legate.* [24 × 29 August 1220]

A = PRO SC1/2/13. No visible endorsement; approx. 139 × 60 mm.; step at left of foot; tongue and seal torn away; slit for insertion of tongue. Scribe 4.

Pd from A in Prynne *Records* iii 48; and inaccurately in *Foedera* 162-3, which gives *Rob' diac' cardinalis* for *Rob' de Cardin'*.

In the summer of 1220 Poitou was disturbed by the appointment of Philip of Oldcotes as seneschal. Philip dean of Poitiers had been presented to the church of Wearmouth, Durham, during des Roches' period as justiciar in 1214 (*RLP* 120, 130b). Arriving in England in the summer of 1220, he returned to France after 17 September, where he was employed to seek the release of Joan, the sister of Henry III, from the custody of Hugh de Lusignan (*RLC* i 430; *Pat.R.1216-25* 250; *DD* nos. 95-6). No letter from Pandulph to des Roches survives concerning tournaments; however, on 25 August the legate issued a general prohibition directed to Hubert de Burgh, ordering him to seize the lands of any who participated in tournaments (*Foedera* 162; *AM* iii (Dunstable) 60). There is no direct evidence of des Roches reducing his household in line with the promise contained in the present letters, nor is it possible to determine which of the bp's kinsmen were required to return to France. Pandulph's anxiety may have been prompted by the tense situation in the west country, where the court was about to attempt to bring Henry fitz Count to heel. Henry was deposed as sheriff of Cornwall in April 1220, and on 10 July Robert de Cardinan was promoted in his place (*Pat.R.1216-25* 241), a local knight who had been prominent in the royalist defence of Devon and Cornwall during the civil war 1215-17 (*Pat.R.1216-25* 207; *RLC* i 2b, 246b, 265, 275, 283, 298b, 340b). Cardinan's appointment as sheriff had been more or less disregarded by the disobedient fitz Count; no doubt one of the subjects for discussion at the Winchester council planned for 30 August.

119. *Letters to Pandulph, bishop-elect of Norwich, papal chamberlain and legate, reiterating the bishop of Winchester's earlier message that he was waiting near Winchester for the arrival of the justiciar, to discuss the king's affairs in Poitou. However, after this earlier message had been sent, the bishop received letters from the justiciar, asking him to come to the justiciar and the legate to discuss Poitou and how to proceed in respect to Henry fitz Count. Since the bishop does not wish the king's affairs to suffer by his absence, despite the fact that it is advisable for the peace of the realm and the liberty of the bishop's markets that he remain near Winchester, he is now hurrying to meet the justiciar and legate, bringing with him the dean of Poitiers who, as he signified in other letters, he believes to be useful to the prosecution of affairs in Poitou.*

[24 August × 1 September 1220]

PART 2· THE BISHOP IN POLITICS 97

A =PRO SC1/2/14. No visible endorsements; approx. 132 × 43 mm.; step at left of foot; tongue and seal torn away. Scribe 4.
Pd from A in *RL* i no.cxxxiii.

The earlier message referred to here is clearly no.118 above, which had envisaged a council at Winchester on 30 August 1220. The great Winchester fair was held in the week of St Giles (1 September), hence the bp's anxiety to remain near Winchester. By 25 August both Pandulph and de Burgh were already on their way into the west country (*Foedera* 162; PRO SC1/1/182). De Burgh appears to have reached Exeter on 1 September, where des Roches joined him on the following day (*RLC* i 428b-429). The court remained there until 7 September. Meanwhile Henry fitz Count came in to make his peace (*Pat.R.1216-25* 266-7).

120. *Mandate addressed to (Hugh de Bolebec) sheriff of Northumberland, declaring that the bishop has inspected a charter which Robert fitz Roger has from the king over the manor of Newburn and the service of Robert de Trokelawe, who holds in the appurtenances of the said manor by service of forty shillings. Since the bishop has been shown that the said service arises from the knight's fee which the said Robert (?fitz Roger) holds in the manor, he forbids the sheriff to levy tallage against the said Robert.* [October 1220]

B=PRO E159/4 (King's Remembrancer memoranda roll 5 Henry III) m.2d.
Pd (without address) West *The Justiciarship* 240n. The address, presumably abbreviated, reads *Dominus Vintoniensis vicecomiti Norhumberlandie*.

For the circumstances see Carpenter *The Minority* 225-6.

*121. *Petitions from (Peter) bishop of Winchester (*Vintoniensis*), P(andulph) bishop-elect of Norwich, king Philip Augustus of France, king Henry (III) of England, and queen Berengaria, to pope Honorius (III) protesting against infringements of the liberties of the canons of St Pierre de la Cour, Le Mans, by (John) archbishop of Tours and (Maurice) bishop of Le Mans.*
[c.1220 × June 1221]

Cited in a bull of pope Honorius III dated 24 June 1221 (*Cartulaire du Chapitre-Royal de Saint Pierre de la Cour du Mans*, ed. M. d'Elbenne & L.J. Denis, Archives Historiques du Maine iv (1903-7) pp.59-61 no.48 which notes the altenative ms. readings of *Vincervensis/Vinconiensis*; whence Potthast *Regesta* i no. 6694). Given the involvement of so many figures associated with England, and bearing in mind des Roches' many links to Maine and the Touraine, it seems more likely that the mysterious bp of *Vinconiensis/Vincervensis* should be identified as the bp of Winchester (*Wintoniensis*) rather than as the bp of Vendôme (*Vincestrensis*). The pope appointed (Hugh de Rupibus) abbot of Marmoutier, the dean of St Martin's Tours and (Geoffrey de Thouars) treasurer of St Hilaire at Poitiers to ensure the canons' liberties. Hugh de Rupibus was almost certainly a kinsman of bp Peter; Geoffrey de Thouars had succeeded Peter as treasurer of St Hilaire.

*122. *Letters of P(eter) bishop of Winchester and R(anulph) earl of Chester* *concerning a safe-conduct to England for Hugh de Lacy so that he may speak with the king.* [c.17 September 1221]

<small>Cited as the warranty for royal letters patent to this effect (*Pat.R.1216–25* 301). For the circumstances see Carpenter *The Minority* 22,271,306–7.</small>

*123. *Letters informing the mayor and commune of La Rochelle that five hundred marks have been deposited at the New Temple in London.*
[c.5 October 1221]

<small>Mentioned in royal letters and in letters from the mayor and commune of London asking that the money serve as security for credits to be paid to Engelard de Cigogné and Aimery de Sacy sent on the king's service into Poitou (*Pat.R.1216–25* 303, 313–4).</small>

*124. *Letters of attorney, appointing representatives to collect money owed by the burghers of Ypres to king John, which the said king ordered paid to the bishop of Winchester.* [c.9 February 1222]

<small>Mentioned in *Pat.R.1216–25* 326. For the circumstances see above no.101.</small>

*125. *Letters of (Peter) bishop of Winchester, the bishop of Killaloe (*Laon'*), and the priors of Lanthony and St Oswald's, addressed to the bishops of Kildare, Meath and Ossory, notifying them that letters of pope (Honorius III) were shown to the archbishop of Cashel at Gloucester in the presence of the said bishops of Winchester and Killaloe, the said priors and many others, on 30 July 1222.* [c.30 July 1222]

<small>Mentioned following a recital of the letters of Honorius III, dated 19 May 1222, notifying the archbp of Cashel of complaints by king Henry III that in contravention of the statutes of the Lateran Council, the archbp had placed the king's lands and men under interdict without reasonable cause and after an appeal had been launched by the king to Rome. The archbp is ordered to relax his sentence within 15 days of the receipt of the present letters, the pope having committed the hearing of the original dispute to the bps of Kildare, Meath and Ossory (*RLC* i 517). For the background to the dispute, see A. J. Otway-Ruthven, *A History of Medieval Ireland*, 2nd ed. (London 1980) 131–2.</small>

*126. *Letters of (Stephen) archbishop of Canterbury, (Eustace) bishop of London, (Peter) bishop of Winchester, (Richard) bishop of Salisbury, and (Ranulph) bishop of Chichester guaranteeing safe-conduct to England for Hugh de Lusignan.* [c.August 1222]

<small>Mentioned in royal letters (*Foedera* 167–8). For the background see Carpenter *The Minority* 291–2, 294–5.</small>

PART 2: THE BISHOP IN POLITICS 99

127. *Letters to king Henry (III), informing him that on 6 August (1225) the king's treasure set out from Portsmouth in good order. The bishop did his best to ensure that the treasure ship was given good and trusty escort, and personally administered an oath to the captains of each of the twelve vessels that they would not leave the treasure ship until it had safely made land. The bishop trusts that the treasure ship will be safe, since he has received a reliable report that nearly three hundred ships from the king's realm are in Brittany, gathered in the port of St-Matthieu, where the king's ship and treasure may put in if necessary, without danger. Philip de Aubigné was with the bishop at Portsmouth on 6 August, ready to sail with his men to the place where the king has sent him. The bishop intends to remain on the coast, prosecuting the king's orders and taking care to transmit whatever news he hears. He asks the king to reply with instructions. In regard to Normandy, the bishop has heard no news* (rumores) *beyond that which he lately passed on; it is now eight days since the last ship arrived from those parts.*

[c.6 August 1225]

A = PRO SC1/4/167. No visible endorsement; approx. 162 × 57 mm.; step at top right hand corner, step at left of foot; tongue and seal torn away.
Pd from A in *RL* i no.ccxix; *DD* no.180.

For the equipment and arming of the king's great ship at Portsmouth, 14–20 July 1225, see *RLC* ii 50b-51. It was commanded by Reginald de Bernevall and Thomas de Haye (alias Thomas of the Temple), and reached its destination in Gascony, rather than in Brittany, before 25 August when its treasure was delivered to Richard, the king's brother, William earl of Salisbury, Philip de Aubigné, Geoffrey de Neville and the other leaders of the expeditionary force sent against Poitou (*RL* i no.ccxx; *DD* no.181). Both Bernevall and Thomas of the Temple enjoyed connections to the see of Winchester. In 1225-6 Bernevall is found delivering wine to the bp's manor of Fareham near Portsmouth (Ms. 21DR m.12).

128. *Undertaking by S(tephen) archbishop of Canterbury, E(ustace) bishop of London, P(eter) bishop of Winchester, J(ocelin) bishop of Bath, H(ugh) bishop of Lincoln, R(ichard) bishop of Salisbury, R(alph) bishop of Chichester the king's chancellor, W(alter) bishop of Carlisle, G(eoffrey) bishop of Ely and Thomas bishop of Norwich, that at the petition of king H(enry) III they will work steadfastly to ensure that the settlement arrived at between him and H(ugh) de Lusignan count of La Marche and Angoulême, Queen I(sabella) H(ugh)'s wife, H(ugh) viscount of Thouars and William le Archevêque be maintained.* London, 20 December 1226

B = PRO C66/35 (Patent roll 11 Henry III) m.11d.
Pd from BC *Foedera* 184; *Acta Langton* no.95; *Pat.R.1225-32* 152-3.

129. *Newsletter of G(erold) patriarch of Jerusalem and papal legate, P(eter)* *archbishop of Caesarea, N(icholas) archbishop of Nazareth, (Peter) archbishop of Narbonne, P(eter) bishop of Winchester, W(illiam) bishop of Exeter and the masters of the orders of the Hospitallers, the Templars and the Teutonic knights, giving general notification that the Emperor (Frederick II) has failed to join the August passage to Syria, as a result of which more than 40,000 pilgrims have abandoned the venture, returning in the same ships by which they came. 800 of the knights who remained demanded that the truce (with the Sultan of Damascus) be broken or they would return home. The duke of Limburg has been appointed to command the army on behalf of the Emperor and, following a solemn council, has declared that he intends to break the truce (with Damascus). It was put to him that such a course of action would be unwise, and against the wishes of the pope, but he and his councillors replied that the pope himself must have intended that the truce be broken. Others argued that the Saracens themselves were more than likely to break the truce of their own accord; others, that this was a good time to act, as the Sultan was pre-occupied with wars of his own against the rulers of Hama, Edessa and Aleppo, and would be likely to seek terms if the crusaders attacked. Eventually it was agreed to march on Jerusalem, but first to fortify Caesarea and Joppa, which it is hoped may be accomplished before the next August passage. This determination was made public outside the city of Acre on 28 October (1227), when the crusaders were told to be ready to set out for Caesarea on 2 November; an announcement which did much to boost the army's morale.* [c.28 October 1227]

    B = Wendover *Flores* iv 145-7, in a copy of papal letters, addressed to all Christ's faithful, demanding assistance for the crusade, dated 23 December 1227, not in the register of Gregory IX. C = Paris *CM* iii 128-9, copied from B. D = BL ms. Cotton Vespasian A xvi fo.20r. s.xiii ex. Bound up with the annals of Waverley abbey.

    Further pd (calendar) ?from Wendover/Paris in J.D. Mansi, *Sacrorum Conciliorum nova et amplissima collectio*, vol.xxiii (1225-68) (Venice 1779) cols.40-1. Pd (calendar) from the published versions noted above, in R. Röhricht, *Regesta Regni Hierosolymitani (MXCVII-MCCXCI)*, 2 vols. (1893-1904) no.984; Potthast *Regesta* no.8090.

    It is interesting to note the preservation of the newsletter together with the annals of Waverley. This in turn adds weight to the possibility that a further newsletter, apparently preserved only in the Waverley annals, but without its opening or address (*AM* ii pp. 305-7) may have been sent by des Roches. Dated at Acre, 20 April (1228), it recounts the Emperor's entry into Jerusalem in terms flattering to Frederick II; details the truce with the Sultan of Cairo, the Emperor's coronation, and the infighting amongst the Saracens. The letters are written in the first person plural, and speak of king Richard I as *perpetue memorie dominum nostrum*, both features that might support an ascription to des Roches. Their preservation at Waverley, within des Roches' diocese, lends further credence to this conjecture.

130. *Letters of (Hugh) archbishop of Arles, (Peter) bishop of Winchester and (Miles) bishop of Beauvais, issued at the request of John bishop of Sabina and*

PART 2: THE BISHOP IN POLITICS   101

Thomas cardinal priest of St Sabina, legates of the apostolic see, testifying to the conduct of the said John and Thomas in respect to the emperor (Frederick II); reciting instruments of John and Thomas, demanding that the emperor permit free elections to churches and monasteries and grant satisfaction to the counts of Celano, the sons of Rainald of Aversa, the Templars, the Hospitallers and other churchmen, providing pledges within eight months of his absolution, these pledges to be nominated by the church from amongst the princes, counts and barons of Germany, the communes of Lombardy, Tuscany, the March and the Romagna, and the marquises, counts and barons of those provinces; the emperor's obligations with regard to the Holy Land are to remain in force. The said John and Thomas further declared on the pope's behalf that the emperor is to reimburse the pope for all expenses incurred outside the kingdom for the defence of ecclesiastical liberties and the patrimony of St Peter. They further pronounced a sentence of excommunication against the emperor, to be enforced should he break any of the conditions agreed above. Done at St Justa near Ceprano, the feast of St Augustine 1230.

St Justa near Ceprano, 28 August 1230

Pd from the papal registers and other sources in Italy in *Constitutiones et Acta Publica Imperatorum et Regum, vol. II MCXCVIII-MCCLXXII*, ed. L.Weiland, *MGH Leges* iv part ii (Hanover 1896), 177–8 no.141. Pd (calendar) from the papal registers only, in *Epistolae saeculi xiii e registris pontificum Romanorum selectae per G.H. Pertz*, ed. C. Rodenberg, 3 vols. *MGH Epistolae saec. xiii*, i-iii (Berlin 1883) no.415; *HDF* iii 218–9; *Reg. Greg.IX* no.421. Pd (English translation) in *Select Documents of European History, Vol.I 800–1492*, ed. R.G.D. Laffan (London 1930) 126–8.

Des Roches' activities as peace maker between pope and emperor are noticed in *AM* i (Tewkesbury) 76, iii (Dunstable) 126; *San Germano* 163.

*131. *Letters of (Peter) bishop of Winchester, (William Brewer) bishop of Exeter and (Gerold) patriarch of Jerusalem testifying to the authenticity of a cross brought to St Albans from the valley of Josophat.* [c.1231]

Mentioned by Matthew Paris (*Gesta Abbatum* i 291) who refers to a more detailed account in the *Gesta Abbatum* 'at the end of the book', no longer to be found.

132. *Inspeximus by P(eter) bishop of Winchester and H(ugh) bishop of Ely addressed to pope G(regory IX), reciting letters of pope Honorius (III) addressed to R(alph) bishop-elect of Chichester, vice-chancellor of king Henry (III), declaring that the king, although lacking in years, being of full age in wisdom, should have disposal of the affairs of his realm. (Peter) bishop of Winchester, (Hubert de Burgh) the justiciar, and W(illiam) Brewer have been instructed in other letters to give the king full authority over his realm. Neville is*

therefore to grant the king disposal of the royal seal, sealing no letters without the king's permission; assisting the king to do good by faithful counsel and sound advice as he is known to have done for the king's father. Given at the Lateran, 13 April 1223. The bishops of Winchester and Ely have issued their inspeximus lest doubt arise over the aforesaid matters.
[August 1231 × 9 June 1238, c.July × December 1232]

A = PRO SC1/1/21. No visible endorsement; approx. 180 × 130 mm.; a draft, probably never sealed or dispatched in its present form.
Pd from A in *RL* i no.ccclviii; *DD* no.243.

The present text is undoubtedly a draft, as shown by the repetition (?through scribal error) of its opening few words; a fact which may explain its survival in the chancery archive. It must date after the election of Hugh of Northwold as bp of Ely, and hence after des Roches' return from crusade in August 1231. The bull of Honorius III recited here is amongst those originally solicited by des Roches in 1222-3, but subsequently used by Hubert de Burgh to secure des Roches' dismissal from government (Carpenter *The Minority* 301-6). The papal letters to Neville are preserved alongside similar letters to des Roches, de Burgh, William Brewer, the earl of Chester and the English baronage, in the Red Book of the Exchequer, whose scribe wrongly copied them out as if they were letters of Gregory IX (PRO E164/2, fo. 274r-v; Norgate *The Minority* 286-90). A similar letter of the same date, addressed to Walter de Clifford, and copied correctly as a bull of Honorius III, is preserved in PRO E36/274 (Exchequer Liber A) fo.104v. The present inspeximus by des Roches and the bp of Ely is almost certainly related to events in the summer and autumn of 1232, when following the dismissal of Hubert de Burgh, the king sought papal backing for the resumption of custodies and titles formerly bestowed upon de Burgh and his satellites. Its outcome may well have been the papal letters of 10 January 1233, empowering the king to resume all liberties, powers, possessions and bailiwicks granted out to the king's prejudice (*Cal. Pap. Reg.* i 131).

133. *Letters said to have been composed by Peter bishop of Winchester and Peter de Rivallis, to which the king was persuaded to put his seal; sealed also with the seals of eleven of the king's councillors, directed to the magnates of Ireland, announcing that Richard Marshal has been proscribed by judgment of the king's court because of his manifest treason, and condemned to the loss of his inheritance and the destruction of his property. He nonetheless persists in his attacks against the king. The magnates of Ireland are therefore exhorted to capture him, alive or dead, should he come to Ireland, for which service they will be rewarded with hereditary possession of all the Marshal's lands and belongings in Ireland, currently in the king's mercy, to which promise by the king the said (eleven councillors) stand guarantor.* [January × February 1234, ?Fictitious]

B = Wendover *Flores* iv 292-3. C = Paris *CM* iii 265-6, copied from B.

The story of the letters as told by Wendover and repeated by Paris, and the letters themselves as recited in the chronicles (without their address), are highly suspect. The chroniclers report that the Irish wrote back to the king, in secret, requesting reassurance that the promise to divide the Marshal lands would be honoured. The king's councillors then drew up royal charters repeating the earlier undertaking, having first removed the seal from the chancellor, Ralph de Neville, who was unwilling to connive in their iniquity. Later in Wendover's account, the letters are portrayed as a factor in the Marshal's supposed murder. Wendover claims that archbp Edmund revealed their contents at a council held after the fall of des Roches, to refute the king's protestation that he bore no reponsibility for the death of the Marshal. However, Wendover's account may be fictitious, designed to blacken the reputation of des Roches and his associates. Whilst it seems highly probable that the Irish magnates hoped to gain the Marshal lands in reward for putting down Richard Marshal's rebellion, there is no reason to suppose that des Roches or the Irish were involved in the sort of conspiracy portrayed by Wendover. For further discussion see B. Wilkinson, 'The Council and the Crisis of 1233-4', *Bulletin of the John Rylands Library* xxvii (1942-3) 387-91.

PART 3: WRITS CITED IN THE WINCHESTER PIPE ROLLS

The present calendar provides details of some 200 writs cited as warranty for expenditure or activities noted in the individual manorial accounts of the bishop's pipe rolls. A typical entry will note that a specified sum of money has been spent on a particular project 'by the bishop's writ' (*per breve episcopi*) or 'by the bishop's letters' (*per litteras episcopi*), the formula *per breve (domini) episcopi* being abbreviated below to *pb(d)e*. The name of the manor where the expenditure took place is supplied here from the headings in the individual pipe rolls. Although the writs themselves are all of them now lost, it is possible to reconstruct something of their basic form by analogy with other, surviving collections of thirteenth-century estate correspondence. Thus, the in-coming correspondence of bishop Ralph de Neville of Chichester (1222–1244) preserves numerous letters from the bishop's stewards and bailiffs, replying to or requesting instruction by the bishop's writ on such matters as the harvest, the purchase of grain and livestock, payments to the bishop's pensioners and the entertainment of guests on the episcopal estates[1]. Even closer in form, the archive at Westminster abbey preserves a large collection of the writs of abbot Walter of Wenlock (1282–1307). As with the lost Winchester writs, those of abbot Walter were filed away by their recipients as warranty for individual payments. When accounts were taken, the abbey's bailiffs returned their writs as vouchers to the abbot's exchequer, and a note was entered on the account roll that such and such a payment had been made *per breve*. In many ways the letters of abbot Wenlock provide us with a model for the lost writs of des Roches. Despite their relatively late date, many of them lack a formal dating clause. A large number appear to have been issued as letters close, sealed with the abbot's secret seal. They are composed in French as well as in Latin, even though at Westminster, as at Winchester, the account rolls appear to have been written exclusively in Latin[2]. Only one important distinction needs to be drawn between the collection at Westminster and the references assembled below. By comparing the surviving writs with the abbey accounts it is possible

---

[1] J. and L. Stones, 'Bishop Ralph de Neville, Chancellor to King Henry III, and his Correspondence: A Reappraisal', *Archives* xvi (1984), 227–57.
[2] *Documents illustrating the rule of Walter de Wenlok, Abbot of Westminster, 1283–1307*, ed. B.F. Harvey, Camden Society 4th series ii (1965), esp. pp.1–3, 10–17.

to show that at Westminster expenses warranted by writ could be entered on the account roll either as expenses *per breve* or *per preceptum*. This last formula, *per preceptum*, occurs on innumerable occasions within the Winchester pipe rolls, indeed far more frequently than the formula *per breve* or *per litteras*. Yet, by contrast to Westminster in the 1280s, there seems no guarantee that expenses entered *per preceptum* at Winchester in the early thirteenth century necessarily involved the issue of a writ. Our most important evidence here is an entry on the pipe roll for 1218/19 (no.207 below), where the scribe originally entered a payment by the bailiff of Ebbesbourne as having been made *per preceptum episcopi*, but then crossed out the words *per preceptum* and replaced them with *per breve*. In other words, a conscious distinction appears to have been drawn between writ and precept, the simplest explanation being that the formula *per breve* was applied to letters in writing whilst *per preceptum* might involve no more than verbal instruction. This interpretation is as yet unproved. Nonetheless, in what follows I have assembled references only to payments where there is unequivocal evidence for the issue of writs or letters. Even within these limited parameters, it will be apparent to anyone familiar with the sheer bulk of the Winchester pipe rolls that the present collection is unlikely to be comprehensive.

1208/9

134. Seven shillings and ten pence spent at Witney, 28 January 1209, in the expenses of Baldwin de Cumin' *per breve domini episcopi* (Ms.4DR m.3, pd *PR4DR* 17)

135. Twenty five shillings spent at Witney around 2 February 1209 in the expenses of master A. and master Richard de Pavilli, staying there 15 days *pbde* (Ms.4DR m.3, pd *PR4DR* 17)

136. £33, six shillings and eight pence released by master J(ohn) of Lond(on) *per dominum episcopum* to the lady Iseut de Laci at Farnham *pbde* (Ms.4DR m.5d, pd *PR4DR* 38)

For master John see appendix 4 no.1.

137. Three pence spent at Farnham in carrying the bishop's writ (*breve*) to Merdon (*Meredon'*) (Ms.4DR m.5d, pd *PR4DR* 38)

138. Sixpence spent at Farnham in carrying the bishop's writ (*breve*) to Nicholas Haring (Ms.4DR m.5d, pd *PR4DR* 38)

139. Six shillings, three and a half pence spent at Farnham in the expenses of clerks of Savar(ic) de Mauléon (*Mauliun*) and the viscount of Thouars (*vic' de Thuars*) *pbde* (Ms.4DR m.5d, pd *PR4DR* 39)

1210/11

140. A colt (*pull'*) released to Robert de Lurdun at Twyford *per breve episcopi* (Ms.6DR m.1d, pd *PR6DR* 15)

141. £5, fourteen shillings and nine pence released to the men of *Wik'* at Downton *pro terra sua gwarata pbde* (Ms.6DR m.2d, pd *PR6DR* 34)

142. Eightpence in the expenses of Brito at Farnham carrying the bishop's writ to ?Nornis (*In exp' Briton' quando portavit breve episcopi Nornis*) (Ms.6DR m.3, pd *PR6DR* 42)

Possibly to be read as 'the writ of the bp of *Nornis*' (?Norwich).

143. Thirty two pence spent at Witney on birds purchased for the queen *pbe* (Ms.6DR m.4, pd *PR6DR* 65)

144. Fifty shillings released to master Will(iam) at Wargrave *pbde* (Ms.6DR m.5d, pd *PR6DR* 85)

145. Five quarters of corn released from Fareham to clerks imprisoned at Portchester (*clericis incarceratis apud Porcestr'*) between 29 September (1210) and 2 February (1211) *pbe* (Ms.6DR m.7, pd *PR6DR* 110)

146. Twenty quarters of oats released at Fareham to Charles de Ceresia (*Karol' de Ceresia*) *pbde* (Ms.6DR m.7, pd *PR6DR* 110)

147. Twenty pence spent at (Bishop's) Waltham on two of the bishop's men with greyhounds (*leporar'*) for ten days *pbe* (Ms.6DR m.7, pd *PR6DR* 116)

148. The treasurer (*hordarius*) at (Bishop's) Waltham quit from further account for nine shillings of arrears *pbde* (Ms. 6DR m.7, pd *PR6DR* 116)

The treasurer in question being the hordarian of St Swithun's priory.

149. The bailiffs of Alresford quit from arrears from the time of Henry la Mart(re) and Osbert the reeve *pbe*, having fined with the bishop over arrears of £14 three shillings and ten pence (Ms.6DR m.9, pd *PR6DR* 146)

150. Eleven pence spent at Southwark in carrying letters from the bishop (*litteras ex parte episcopi*) to earl Aubr(ey de Vere) and J(ohn) fitz Hugh (Ms.6DR m.9d, pd *PR6DR* 155)

151. Eighteen shillings and eightpence spent at Taunton on a fur cape (*capa forrata*) for G(eoffrey) de Caux *pbe* (Ms.6DR m.10d, pd *PR6DR* 172)

For Geoffrey see appendix 4 no.29.

152. Daniel (bailiff of Alresford Forum) continues to owe fifty seven shillings, two and a half pence *pbe*; Full', Reimund, and Walter Niwel continue to owe thirty seven shillings *pbe* (Ms.6DR m.11d, pd *PR6DR* 186)

1211/12

153. Four quarters of oats used at Clere in entertaining verderers (*viridar'*) *pbe* (Ms.7DR m.2)

154. Fifty four shillings and five pence spent at Fareham in the expense of carpenters, sawn planks and the wages of master Henry the crossbowman, for carpentry at the pond of Southwick *pbe* (*In carpent' locand' et planchis sicand' et liberat' magistri Henrici arbal' ad carpent' meiremium stagni de Suwerk' per breve episcopi*) (Ms.7DR m.4d)

155. Ten quarters of oats given to Thomas Painel at Fareham *pbe* (Ms.7DR m.4d)

156. Six shillings, nine and a half pence spent at Farnham in making shingle(s) (*cendula facta*) *pbe*, sent to Winchester to cover the cloister (*ad claustr' cooperiend'*) (Ms.7DR m.7d)

157. Twenty four shillings and two pence spent at Farnham in entertaining the bishop for one night and in the expenses of Ernold (of Auckland) the king's treasurer *pbe* (Ms.7DR m.8)

158. Twenty seven pence spent at Southwark in the expenses of Richard de Grai with two horses for three days *pbe* (Ms.7DR m.10)

159. Four shillings and four pence spent at Southwark in the wages (*vad'*) of T. de Huntinges' *pbe* (Ms.7DR m.10)

160. Thirty two pence spent at Southwark in the expenses of master Henry de Walton' for three days *pbe* (Ms.7DR m.10)

161. Seventy shillings, four and a half pence spent at Southwark in the expenses of the bishop's knights going to Wales (*milit' episcopi eunt' apud Wall'*) *pbe* (Ms.7DR m.10)

162. £9, two shillings and six pence owed in arrears at Taunton by W(illiam) the constable, pardoned *pbe* (Ms.7DR m.10).

For William of Shorwell, constable of Taunton, see appendix 4 no.23.

163. Forty shillings spent in work (*expens' in operat'*) at Ivinghoe *pbe* (Ms.7DR m.11d)

1213/14

164. Twenty five shillings and eight pence spent at Fareham in buying twenty two quarters of oats *pbe* (Ms.9DR m.2d)

165. Thirteen shillings and four pence spent at Alresford in buying one dozen barrels (*barill'*) *pbe* (Ms.9DR m.3d)

166. One and a half quarters of barley used at Clere in feeding geese *pbe* (Ms.9DR m.5d)

167. Four shillings and four pence spent at Farnham in the wages of ?assistants for fishing (*mercede iuvat..eum ad piscand'*) *pbe* (Ms.9DR m.6)

168. Forty shillings and two pence spent at Farnham in entertaining (Eustace) bishop of Ely *pbe* (Ms.9DR m.6)

> The bp also received three kids, priced at tenpence halfpenny.

169. Three shillings and ten pence spent at Farnham buying ridge-tiles (*crest'*) sent to Winchester *pbe* (Ms.9DR m.6)

170. Seventeen and a half pence spent at Farnham in the expenses of R. de Hasting' and the archdeacon (?of Winchester) *pbe* (Ms.9DR m.6)

171. Three shillings and nine pence spent at Farnham in the expenses of the mayor of London (*maioris Lond'*) *pbe* (Ms.9DR m.6)

172. Eight pence spent at Brightwell in the expenses of Roger de Andeli with the bishop's hounds (*canibus*) *per breve eius* (Ms.9DR m.8)

173. Seventeen shillings, three and a half pence arrears owed by the bailiffs of Witney, quit *pbe* (Ms.9DR m.8d)

174. Forty two pigs given in alms at Taunton *pbe* (Ms.9DR m.10)

1215/16

175. Twelve pence spent at Brightwell on J. monk of Winchester who brought the bishop's letters patent (Ms.11DR m.2d)

176. £8 released at Downton to master William, the man of Laur(ence) de Duniun' *pbe* (*In lib' Will' magistro homini Laur' de Duniun'*) (Ms.11DR m.4)

177. £42 released at Wycombe to the Templars (*Templar'*) by one tally, by the bishop's letters (*litteras*) (Ms.11DR m.5)

> During the civil war, when the bp's manors were over-run by rebel forces, the Templars served frequently to collect and transport the bp's rents.

110   PART 3: WRITS CITED IN THE WINCHESTER PIPE ROLLS

178. £12 released at Adderbury to brother Ralph de Templo by one talley *pbe* (Ms.11DR m.5d)

179. Twenty four shillings and one penny spent at Adderbury in the purchase of seventeen ells of russet cloth (*uln' russestt'*) for the bishop *pbe*, at seventeen pence the ell (Ms.11DR m.5d)

180. Seven shillings and sixpence spent at Wargrave in threshing twenty one quarters of corn with haste *pbe*, and an estimated twenty three and a half quarters of mancorn (Ms.11DR m.5d)

181. Three shillings spent at Wargrave in the expenses of Alured the clerk for three days before 23 February (1216) *pbe* (Ms.11DR m.5d)

182. £17 and two shillings spent from Witney in buying robes and furs (*robis et forrur'*) at Oxford *pbe* (Ms.11DR m.6)

183. Twenty shillings given to Richard de Elman' at Witney by the bishop's gift *pbe* (Ms.11DR m.6)

184. Sixpence spent at Meon in carrying the bishop's writ (*breve*) to the earl of Arund(el) (Ms.11DR m.6)

185. Forty shillings spent at Alresford in buying three dozen barrels (*barill'*) *pbe* (Ms.11DR m.6d)

186. Seven shillings and nine pence spent at Alresford in making three hundred hurdles (*clat'*) *pbe* and in sending them to Wolvesey (*Wlves'*) (Ms.11DR m.6d)

1217/18

187. Thirty two shillings spent at Fawley (*Falelia*) in the expenses of Gervase the huntsman (*venator*) *pbe* (Ms.13DR m.3)

188. Twenty two shillings and eight pence spent at (Bishop's) Waltham in the expenses of Herbert de Foxcote and two of his men *pbe* (Ms.13DR m.3d)
   Herbert is recorded with 4 greyhounds for 22 days.

189. Three quarters of grain given to William de Iussai at Fareham *pbe* when he crossed overseas (Ms.13DR m.8)

190. Eightpence given at Taunton to a certain man (*cuidam*) to carry the bishop's letters (*litteris*) to W(illiam) de Sor(well) in Cornwall (*Cornub'*) (Ms.13DR m.12)
   For William of Shorwell see appendix 4 no.23.

PART 3: WRITS CITED IN THE WINCHESTER PIPE ROLLS 111

191. Three shilling and four pence spent at Taunton on W. Mautravers and John his companion who came with the bishop's hounds to hunt pigs in the park (*cum canibus ad currend' ad porcos*) for 10 days *pbe* (Ms.13DR m.12)

192. Thirteen shillings and four pence released to master J(ohn) of Dartford and to master Winemerius at Taunton, to cover their expenses on the road when they carried money (*denarios*) to the bishop, and fifteen pence in master Winemerius' expenses for five days when he came for the money with the bishop's writ (*breve*) (Ms.13DR m.12)

> For master John of Dartford, a clerk in des Roches' service 1218–37, particularly associated with building work, see below no.296; *Cl.R.1227–31* 206; *Cal.Lib.R.1226–40* 204,209; *Selborne Charters* i 21, ii 6.

193. Forty two shillings, three and a half pence spent at Taunton in entertaining (*hospiscio*) P(eter) de Maulay (*Malo Lacu*) for one night *pbe* (Ms.13DR m.12)

> Peter de Maulay, a native of the border region between the Touraine and Poitou, was a close political associate of des Roches. See above no.94.

1218/19

194. Three shillings and three pence spent at Meon church on Britonis and his companions for 3 days *pbe* (Ms.14DR m.2)

195. Nine shillings and ten pence spent at (Bishop's) Sutton on Robert Lurd' and Walter the fisherman (*piscat'*) and their helpers at *Wlvem'* for nine days *pbe* (Ms.14DR m.3)

196. Fourteen shillings and ten pence released at Burghclere to the wife of Thomas Notehat *pbe* (ms.14DR m.4d)

197. Ten shillings spent at Witney on master Humphrey *pbe* (Ms.14DR m.5d)

> Probably master Humphrey de Millières, for whom see appendix 4 no.4.

198. Three shillings and six pence spent at Witney on brother Azo going to Winchcombe (*Winchec'*) to buy furs (*forrur'*) and returning *pbe* (Ms.14DR m.5d)

> Azo bought just over seventy five shillings worth of furs.

199. Seven shillings and six pence spent at Witney on J(ohn) de Chinun, ill there, *pbe* (Ms.14DR m.5d)

> For John de Chinon see appendix 4 no.14.

200. Twenty five shillings and a halfpenny spent at Witney on two carters, two carriers (*caret' et summar'*) and nine other servants (*servient'*) for seven days after 15 August (1219) *pbe* (Ms.14DR m.5d)

201. Six shillings spent at Witney on William the huntsman (*venator*) with the king's hound (*cane*) for thirty six days *pbe* (Ms.14DR m.5d)

202. £7 released at Adderbury to Walter de Verdun *pbe* (Ms.14DR m.6)

For Walter, sheriff of Essex and Hertfordshire, see Carpenter *The Minority* 115-16, 319n.

203. Twelve pence spent at Adderbury on the abbot of Hales *pbe* (Ms.14DR m.6)

In the same year the abbot received sixty four shillings of the profits of the manor.

204. £6 and fourteen pence released from Ivinghoe to master Humphrey *pbe* (Ms.14DR m.6)

For master Humphrey see above no.197.

205. One hundred and fifteen and a half quarters of corn sent from Wycombe to Southwark (*Suwerk*) by the bishop's writ (*pbe*) and by tally against Peter Bacun; and one hundred and forty nine quarters of oats sent to the same *pbe* (Ms.14DR m.7)

For Peter, a former clerk of the sheriff of London, later bailiff of des Roches' manor of Southwark 1217-1219, see *RLP* 134; *Memoranda Roll 14 Henry III* 31; *Cal.Lib.R.1226-40* 369; *Cal.Lib.R.1240-45* 74; Mss.13DR, 14DR, 15DR *passim*.

206. Ten quarters of corn released from Downton to Aimery de St Amand (*Amaur' de S. Amand*) *pbe* (Ms.14DR m.7)

Aimery de St Amand was a Norman, subsequently promoted steward of the royal household during des Roches' ascendancy at court in 1233.

207. Twenty two shillings and six pence spent at Ebbesbourne in buying one hundred ells of wool *(? lane col')* *pbe* (Ms.14DR m.7d)

The words *per breve episcopi* are written over *per preceptum episcopi*, implying that here and elsewhere *per preceptum* is a formula intended to cover unwritten instructions.

208. Two shillings and three pence spent at Fareham in the purchase of birds sent to Wolvesey (*Wlves'*) against Christmas *pbe* (Ms.14DR m.8d)

209. Twelve shillings and ten pence spent at Taunton on A(ndrew) de Chanceaux (*Cancellis*) for three days *pbe* (Ms.14DR m.10d)

For Andrew see above p xl.

210. Two shillings spent at Taunton on Nicholas de Brun for eight days *pbe* (Ms.14DR m.10d)

1219/20

211. £6 thirteen shillings and four pence released to Henry the crossbowman (*balistarius*) at ? (unidentified) *pbe* (Ms.15DR m.2d)

212. Fifteen shillings spent at ? (unidentified) in carrying one hundred and twenty six salted hogs (*baconis*) to London *pbe* (Ms.15DR m.2d)

213. Twenty quarters of oats released from Fareham to the constable of Portchester *pbe* (Ms.15DR m.3)

214. Forty shillings released from Meon to master Walter de Hida *pbe* (Ms.15DR m.3)

215. Eight shillings and six pence spent at (?Wycombe) in sowing and tending three hundred and fifty ?pear trees and two hundred ?apple trees (*pilis et . . . apuall'*) and sending them to London *pbe*; with seventeen shillings and six pence spent on carrying them to Marlow (*Merlave*) by the same writ (Ms.15DR m.5)

216. Twenty six shillings spent at (?Wycombe) in carrying 5350 pieces of firewood (*busc'*) to Marlow (*Merlave*) *pbe* (Ms.15DR m.5)

217. One mark released to Herbert the clerk at Wycombe Ecclesia *pbe* (Ms.15DR m.5)

218. Eight shillings and two pence spent at Wycombe Ecclesia in entertaining (*in hospicio*) the archdeacon (?of Buckingham) *pbe* (Ms.15DR m.5)

219. £25 and ?(illegible) released to William fitz Humphrey at Adderbury *pbe* (Ms.15DR m.5)

For William fitz Humphrey see appendix 4 no.32.

220. Twelve quarters of grain released to the abbot of Hales at Adderbury *pbe* (Ms.15DR m.5)

221. Twenty two shillings, six and a half pence spent at Brightwell on William fitz Humphrey overseeing the threshing for nine weeks (*qui prefuit tric'*) *pbe* (Ms.15DR m.5d)

For William fitz Humphrey see above no.219.

222. Eight shillings spent at Highclere on N. de Fluri for three weeks *pro securitate meridin'* (sic) *pbe* (Ms.15DR m.6)

223. Three shillings and eight pence spent at Overton on John de Valt' staying in the autumn for the harvest (*ad collect' blad'*) *pbe* (Ms.15DR m.7)

224. Five shillings spent at Taunton on J. the seneschal coming for three days with the bishop's letters (*litteris*) (Ms.15DR m.10)

114   PART 3: WRITS CITED IN THE WINCHESTER PIPE ROLLS

225. Twenty three shillings and three pence spent at Taunton on P(eter) de Rupibus, W. de Vilers, Giles the crossbowman (*balist'*), N. de Fluri and R. Funten' *pbe*, and on W(illiam) de Swapham and A. the clerk who were at table (*ad mensam*) for seventeen days (Ms.15DR m.10)

For Peter de Rupibus see appendix 4 no.44.

226. Eight shillings spent at Taunton on P(eter) de Rupibus who held the bishop's writ (*cum breve episcopi*) and on W(illiam) de Swapham and A. the clerk who were with him at table for fifteen days in the autumn (Ms.15DR m.10)

For Peter see above no.225.

227. Three shillings given at Taunton in the entertainment of master W. the physician (*medicus*) who undertook his cure (*qui curam eius egit*) for twelve days, and in his wages of half a mark *pbe* (Ms.15DR m.10)

The 'cure' being that of Peter de Rupibus.

228. Thirty two cheeses weighing two weys (*pond'*) released from Witney to the abbot of Hales *pbe* (Ms.15DR m.11)

1220/21

229. £44 eighteen shillings and six pence owed by John de la Buluse at Merdon for corn (*blado*) sold to him *pbe*, of which he says that £10, twenty and a half pence were used in sowing (Ms.16DR m.1)

230. Fifty four shillings and two pence spent at Highclere in the bishop's expenses *per breve ipsius* (Ms.16DR m.2)

231. Twenty pence spent at Brightwell on Geldewin de Duay for 1 night *pbe* (Ms.16DR m.2d)

Joldewin de Douhet was a Poitevin or Gascon knight of the royal household, closely associated with des Roches.

232. Ten shillings of arrears released at Witney to master Laurence *pbe* (Ms.16DR m.3)

233. Fourteen shillings and two pence spent on the bishop's household (*familie episcopi*) staying at Witney *pbe* (Ms.16DR m.3)

234. Thirty nine shillings paid from Witney to Germanus the baker (*pistori*) of Oxford *pbe* (Ms.16DR m.3)

The same G., the bp's baker, received thirty one shillings and six pence halfpenny of corn, whilst a further twenty seven shillings and eight pence was spent on making an oven (*rogo*) for him.

PART 3: WRITS CITED IN THE WINCHESTER PIPE ROLLS 115

235. Sixty six shillings and eight pence released from Witney Burgh to master William Scot *pbde* (Ms.16DR m.3)

For master William, a scholar, see Emden *Biographical Register of Oxford* 1657-8. He had received a similar payment from des Roches at Witney in 1215-16; Ms.11DR m.6.

236. Six pence spent at Adderbury in summoning a ?driver (*aur' locat'*) to carry sixteen cheeses to London *pbe* (Ms.16DR m.3)

237. The same sixteen winter cheeses (*cas' yem'*) weighing three and a half stones (*petras*) sent to London from Adderbury *pbe* (Ms.16DR m.3d)

238. Twenty pence spent at Wycombe on two serjeants (*servient'*) going to and from Berkhamsted (*Burhamstede*) and on three boys and one carter (*caret'*) for two nights, and fifteen pence on their commons (*prebendo*) *pbe* (Ms.16DR m.3d)

239. Five sheep given at North Waltham to N. the chaplain of *Hart'* after shearing *pbe* (Ms.16DR m.4d)

240. Forty two shillings, eight and a half pence spent at Farnham on hay bought for the bishop *per breve eiusdem* (Ms.16DR m.5)

241. Five shillings and two pence spent at Farnham in carrying twenty hauberks (*loric'*), 2000 hurdles (*hurrdl'*) and 8000 herrings (*allecis*) to (Bishop)stoke *pbe* (Ms.16DR m.5)

242. Six shillings, five and a half pence spent at Farnham on G. the huntsman (*venator*) at Christmas *pbe* (Ms.16DR m.5)

243. Three quarters of mancorn released at (Bishop's) Sutton to Richard the clerk *pbe* (Ms.16DR m.5d)

244. Sixty six shillings and eight pence pardoned Robert the reeve (*preposito*) at Meon *pbe* (Ms.16DR m.6)

245. Eleven shillings and three pence spent at Meon in buying fish at Southampton (*Suhamton'*) for the legate (Pandulph) *pbe* (Ms.16DR m.6)

246. Six shillings and eight pence spent at Fareham on P(eter) de Rupibus staying at Portchester castle (*castr' Porcestr'*) *pbe* (Ms.16DR m.6d)

For Peter see above no.225.

247. Eight and a half pence spent at Downton on R. the succentor (*succentoris*) (?of Wells) *pbe* (Ms.16DR m.9)

The succentorship of Salisbury was held by a man named Anastasius, c.1215-1227 (*Fasti* iv 41), so that the present order is more likely to refer to Wells.

248. Ten sheep given at Downton to Iakel' as the bishop's gift after their birth and before shearing (*post part' et ante tons'*) *pbe* (Ms.16DR m.9)

249. Thirteen shillings and four pence given at Taunton as the bishop's gift to the subdean of Wells (*subdecano Wellens'*) for the work of the church of Wells *pbe* (Ms.16DR m.10)

250. Two shillings and eleven pence spent at Taunton on R(obert) de Lurdon who came for ten days with the bishop's letters (*litteras*) for R. de Bernevall and J. de Pavill(y) (Ms.16DR m.10)

251. Two shillings and three pence spent at Taunton on William de Vad' who came for nine days with the bishop's letters (*litteris*) that he might hunt in the park (*ut curreret in parco*) (Ms.16DR m.10)

1223/24

252. Three quarters of oats given at (Bishop's) Waltham in the commons of W. de Chinun and Henry Bacun *pbe* (Ms.19DR m.2d)

253. Four shillings and six pence spent at Farnham in carrying 12,000 herrings (*allec'*) to Winchester *pbe* (Ms.19DR m.4d)

254. Twelve pence spent at Farnham in carrying kitchen equipment (*utensilibus coquine*) to Winchester *pbe* (Ms.19DR m.4d)

255. Eight pence spent at Farnham on Ralph Singe and Sayr who came with the bishop's letters (*litteris*) on the munitioning of the castle (*ad muniend' castrum*) (Ms.19DR m.4d)

> The same year saw the installation of five keepers in the castle for sixteen days by the bp's orders at a cost of thirteen shillings and fourpence. An indication of the tensions that surrounded the bp's dismissal from court.

256. Ten pence spent at Farnham on J. Sarace(n) and W(illiam) de Batilli who came with the bishop's letters (*litteris*) (Ms.19DR m.4d)

> For William see appendix 4 no.22.

257. Thirty seven shillings and ten pence spent at Farnham on W(illiam) de Vad' and J. Sarace(n) staying there for nine weeks *pbe* (Ms.19DR m.4d)

258. Fifteen shillings spent at Farnham in carrying the bishop's letters (*litteris*) to London, Wargrave and Winchester (*Lond' Weregrav' et Wint'*) (Ms.19DR m.4d)

259. Eight pence spent at Farnham in carrying the bishop's letters (*litteris*) to (Bishop's) Sutton, Herriard (*Herierd*) and Stapely (*Stapelegh*) (Ms.19DR m.4d)

260. £10 released from Witney to the abbot of Hales *pbe* (Ms.19DR m.7d)

261. Fifty shillings released from Witney to master H(enry) de Ho *pbe* (Ms.19DR m.7d)

For master Henry see below no.287.

262. Ten pence spent at Witney on the abbot of Hales *pbe* (Ms.19DR m.7d)

263. Six shillings and nine pence spent at Witney on two men of P(eter) de Rivall(is) *pbe*, staying there four weeks (Ms.19DR m.7d)

For Peter see appendix 4 no.41.

264. Fifty six cheeses released at Adderbury to the abbot of Hales by the bishop's letters (*litteras*) (Ms.19DR m.8)

265. Fifteen shillings spent at Fareham in buying 1000 cuttle-fish (*siccis*) *pbe* (Ms.19DR m.9d)

266. Twenty shillings spent at Fareham in summoning a ship to take the men of Philip de Aubeni to the Isle (?of Wight) (*ad Insula*) *pbe* (Ms.19DR m.9d)

267. Thirty four shillings and six pence spent at Taunton on T(homas) de Flury staying there for sixty nine days *pbe* (Ms.19DR m.11)

For Thomas see below no.304.

268. Sixty six shillings and eight pence of arrears pardoned to Jordan the reeve (*preposito*) at Rimpton *pbe* (Ms.19DR m.11)

1224/25

269. Fifteen shillings and eleven pence spent at Downton on Henry Bacun and two men staying there for four weeks with the bishop's horse, and in oats *pbe* (Ms.20DR m.1)

270. Fifteen shillings and two pence spent at Downton on Henry de Lortih', Roger Alis and Henry de Champar' with their companions and seventeen horses staying two nights *pbe* (Ms.20DR m.1)

For Roger Aliz see appendix 4 no.37.

271. Two shillings and six pence spent at Downton on John de Herierd' and Roger de Rouen (*Rothomago*) staying two nights *pbe* (Ms.20DR m.1)

For Roger as des Roches' clerk see *RLC* i 557; *Cal.Chart.R.1226–57* 183. For John of Herriard, active as des Roches' attorney for much of the 1220s, see *CRR* viii 234, xii nos.1427, 1545, 2294, xiii no.551; below nos.278, 305, 307, 310.

272. Eighteen pence spent at Downton on Peter Russinol going to the archbishop (of Canterbury) and returning *pbe* (Ms.20DR m.1)

For Peter see appendix 4 no.20.

273. Fifteen pence spent at Downton on William Woodcock returning from Taunton with thirty two hounds *pbe* (Ms.20DR m.1)

William was the bp's principal huntsman.

274. £6, thirteen shillings and four pence released at Taunton to Thomas de Cnoel *pbe* (Ms.20DR m.3d)

275. Eleven shillings and three pence spent at Taunton in buying pigs to present (*ad exsennia facienda*) to the justic(?iar) when he was at Ilchester (*Ivelcest'*) *pbe* (Ms.20DR m.3d)

276. Sixteen shillings and a penny spent at Taunton on Hugh de Weingham who stayed there for seven days whilst imposing (the king's tax of) a fifteenth *pbe* (Ms.20DR m.3d)

277. Six shillings and ten pence spent at (Bishop's) Sutton on W(illiam) Wdecoc and three companions for six days with nine greyhounds (*leporar'*) and thirty two brachet hounds (*brachet'*) around 11 November (1224) *pbe* (Ms.20DR m.6)

278. Four shillings, six and a half pence spent at (Bishop's) Sutton on Hugh de Wingeh(am), John de Herierd and W. the clerk assessing the (king's tax of) a fifteenth by the bishop's letters patent (*litteras patentes*) (Ms.20DR m.6)

For John see above no.271.

279. Forty two pence released at (Bishop's) Sutton to six serjeants (*servient'*) for seven days keeping the ?roads during the fair (*custodiend' iternera in feria*) by the bishop's letters (*litteras*) (Ms.20DR m.6)

280. £7, fourteen shillings and eight pence spent at Farnham in rebuilding the chimney of the bishop's room (*in camino thalami episcopi struendo et reficiendo de novo*) *pbe* (Ms.20DR m.9d)

281. Twenty one shillings spent at Farnham in carrying ten loads of white glass and four of coloured glass (*summis vitri albi et . . . colarati*) from Bromley (*Bromleg'*) to Witney (*Witenay*) *pbe* (Ms.20DR m.9d)

The glass had been bought at a cost of seventy two shillings. It was destined for the bp's foundation at Hales; below no.290.

282. Five shillings and seven pence spent at Farnham in carrying the bishop's writs to Winchester, Meon, Wargrave, Overton, (Bishop's) Waltham, Ropley, Herriard and St Albans (*Winton', Menes, Wrgrave, Overton', Wattham, Repeleg', Herird', Sanctum Albanum*) (Ms.20DR m.9d)

283. Six shillings and ten pence spent at Overton on two huntsmen and eleven men keeping the bishop's hounds for four days *pbe* (Ms.20DR m.11)

The account refers to 26 greyhounds and 6 brachet hounds.

PART 3: WRITS CITED IN THE WINCHESTER PIPE ROLLS 119

284. Six pence spent at North Waltham in the wages of a man who carried the bishop's writ (*breve domini*) to Hugh de Mortimer (*Mort'Mar'*) by the bishop's order (*preceptum*) (Ms.20DR m.11)

285. Two shillings and six pence spent at Highclere in wine and fish for G. the mar(shal) *pbe* (Ms.20DR m.11d)

286. £14 released at Burghclere to the dean of Basingstoke (*decan' de Basingestok'*) in tithes collected at Kingsclere (*ad decimas de Kingescler' colligendas*) *pbe* (Ms.20DR m.12)

287. £17 released at Witney to master Henry de Ho by the bishop's letters (*litteras*) (Ms.20DR m.14)
   See above no.261.

288. Forty shillings released at Witney to William fitz Humphrey by the bishop's letters (*litteras*) (Ms.20DR m.14)
   For William see appendix 4 no.32.

289. £13 released at Witney to master William de Monteacuto by the bishop's letters (*litteras*), and one hundred and sixteen shillings and eight pence to master H. and master W(illiam) de Monteacuto by the bishop's letters for robes, and fifty seven shillings and six pence to the same towards furs for capes and supertunics (*fururas caparum et supertunicarum*) for Pentecost (18 May 1225) by the bishop's letters (Ms.20DR m.14)
   These men are otherwise unrecorded in des Roches' household.

290. Twenty seven shillings and two pence spent at Witney in carrying glass (*vitro*) to Hales by the bishop's letters (*litteras*) (Ms.20DR m.14)
   See no.281 above.

291. Six pence given at Witney to Robert Kene for carrying the bishop's letters (*litteris*) to Taunton, and a further six pence for carrying the bishop's letters to Hales (Ms.20DR m.14)

292. Four shillings spent at Adderbury in buying hay (*in feno appreciato et empto*) by the bishop's letters (*litteras*) (Ms.20DR m.14d)

293. Four pence spent at Southwark in carrying the bishop's letters (*litteris*) to Ivinghoe (*Ivingeho*) (Ms.20DR m.15)

1225/26

294. £80 released at Taunton for twelve marks of gold of musc (*xii marc' auri de muce*) *pbde* (Ms.21DR m.3d)
   This and other purchases of gold recorded at Taunton between 1225 and 1227 were presumably intended to provide des Roches with specie for his forthcoming crusade.

295. £4 and nineteen shillings released at Farnham to John de Culesdun' *pbe* (Ms.21DR m.5)

296. Twenty one shillings and nine pence released at Farnham to master John of Dartford (*Derteford'*) *pbe* (Ms.21DR m.5)

For master John see above no.192.

297. Thirteen shillings and four pence released at Farnham to J. the Roman clerk (*I. clerico Romano*) *pbe* (Ms.21DR m.5)

298. £10 of the arrears owed by the bailiffs of Farnham respited to 2 February (?1227) *pbe* (Ms.21DR m.5)

299. Six pence spent at Alresford on William de Vad' on 26 May 1226, and three shillings and eight pence spent at Alresford on four fishermen and the man of William de Vad' there for four days, all *pbe* (Ms.21DR m.8d)

300. Seven shillings and four pence spent at (Bishop's) Sutton on (William) Wdec(oc), his four companions, ten greyhounds and twenty five brachet hounds there for four days *pbe* (Ms.21DR m.10)

301. Twenty four pence spent at (Bishop's) Sutton on (William) Wdecoc, three companions, ten greyhounds and twenty five brachet hounds for one day *pbe* (Ms.21DR m.10)

302. One hundred and eighteen shillings and two and a half pence spent at Fareham on J. de Pont' and P(eter) Russinol staying there from 18 September to 6 October (1225/6) *pbe* (Ms.21DR m.11d)

For Peter see appendix 4 no.20.

303. Thirty three shillings and four pence spent at (Bishop's) Waltham on H(enry) Bacun and W(illiam) Batill(y) when they came for thirty two days to guard the warren (*custodire warennam*) *pbe* (Ms.21DR m.12d)

For William see appendix 4 no.22.

304. Twenty six shillings and eight pence of the scutage of Montgomery (*scutag' de Mungumer'*) pardoned Thomas de Flury *pbe* (Ms.21DR m.14d)

Thomas de Flury held three knights' fees of the bpric at Taunton (*Reg.Pont.* 390)

1226/27

305. Three shillings and eight pence spent at Downton on J(ohn) de Herriard going to the justices (*iustic'*) in Somerset and returning *pbe* (Ms.22DR m.1)

For John see above no.271.

306. Twenty four shillings, six and a half pence spent at Downton on H(enry) Bacun and 5 boys staying there with horses for thirty one days *pbe* (Ms.22DR m.1)

307. Six shillings spent at Knoyle on J(ohn) de Herierd *pbe* (Ms.22DR m.1d)
For John see above no.271.

308. Eighty three quarters and two bushels of rye at Taunton given in the bishop's alms *per breve eius* (Ms.22DR m.3d)

309. Eight quarters of grain given at ?Rimpton in the bishop's alms *per breve eius* (Ms.22DR m.4)

310. Twelve pence spent at Brightwell on J(ohn) de Herierd *pbe* (Ms.22DR m.6)
For John see above no.271.

311. Eighteen quarters of corn given at Brightwell in the bishop's alms *per breve eius* (Ms.22DR m.6)

312. Sixty shillings of debt respited at Harwell by the bishop, which the lord Aimery (?de St Amand) owes for twenty quarters of corn which he had by the bishop's letters (*litteras*) (Ms.22DR m.6)

313. £4, ten shillings and five pence spent at Wycombe in carrying 8700 pieces of firewood (*busc'*) to (be transported by) water *pbe* (Ms.22DR m.12)

314. Four pence spent at Southwark in carrying the bishop's letters to Wycombe (*Wicumb'*) (Ms.22DR m.13)

1231/32

315. Five shillings, three and a half pence released at (Bishop's) Waltham to Walter, a man of G(eoffrey) de Lucy, there for five weeks and six days *pbe* (Ms.27DR m.1d)
The same man was paid six pence to take his hounds home (*ad canes suas versus patriam suam*)

316. £11, fifteen and a half pence spent at Adderbury in building a hall (*aula de novo facto*) by the bishop's order (*precepto*) and *pbe*, as was shown by the parties at the (bishop's) exchequer, saving the plaster work (*sicut probatur per partes ad scaccar' excepta dealbatione*) (Ms.27DR m.5)

317. Thirteen shillings and eleven pence spent at Taunton in fishing towards Christmas *pbe* (Ms.27DR m.7d)

318. Money (sum illegible) spent at Taunton in fishing for a bream (*brestius*) to carry to Winchester by the bishop's writ (*pbe*) which Thomas Harang' ... (illegible) (Ms.27DR m.7d)

319. Six shillings, four and a half pence spent at Taunton in fishing for the feast of Palm Sunday (4 April 1232) by the bishop's writ (*pbe*) which William Russel brought (Ms.27DR m.7d)

320. Money (sum illegible) spent at Taunton on the earl of ... (illegible) by the bishop's writ (*pbe*) which Roger Alis brought back from ... (illegible) (Ms.27DR m.7d)

For Roger see appendix 4 no.37.

321. Two shillings and ten pence spent at Farnham on Lindsey, Godfrey and ten boys with eight destriers (*dextrariis*) going to the bishop on the way towards Lond(on) *pbe* (Ms.27DR m.8d)

322. Thirteen shillings and six pence spent at Farnham on Reginald de Cundi, Colin de Veniuz and four boys with the greyhounds of P(eter) de Rivall(is) hunting the park for three days *pbe* (Ms.27DR m.8d)

323. Five pence spent at Farnham on the five greyhounds of the said (Peter de Rivallis) for two days *pbe*, and sixteen pence spent on the five greyhounds and eleven brachet hounds of Colin (?de Venuz) for two days *pbe* (Ms.27DR m.8d)

324. Seven shillings and eight pence spent at Farnham on P(eter) Saracen *pbe* (Ms.27DR m.8d)

325. Eight shilllings accounted (*allocantur*) to the reeve (*preposito*) of Farnham *pbe*, which the reeve spent in hay bought for the land of John Blund which (sum) had not previously been accounted to him (Ms.27DR m.8d)

326. Seventeen pence spent at Calbourne on master R. the mason (*cementar'*) coming twice with two men and the bishop's letters (*litteris*) for the ?granary ($g^a oraria$) (Ms.27DR m.9)

The same account mentions repairs to a granary (*granar' emendand' ad tasch'*).

327. Thirteen shillings and nine pence spent at Downton in carrying marble for an altar, with sixteen columns, capitals and bases from Salisbury to Southampton (*petra marmorea ad altare cariand' cum xvi. columpnis capitroll' et balsis de Sar' usque Suhanton'*) *pbe* (Ms.27DR m.12)

Possibly intended for the bp's new foundation at Titchfield. Noticed in Colvin *White Canons* 185n.

328. Forty two shillings and six pence spent at Southwark on W(illiam) fitz

PART 3: WRITS CITED IN THE WINCHESTER PIPE ROLLS 123

Humphrey for four days from 29 October (1231) on the bishop's business ... (illegible) and of the master of the Temple *pbe* (Ms.27DR m.12)

For William see above no.219.

1232/33

329. Four shillings and eight pence spent at Farnham on bishop John (?of Ardfert) on the Wednesday before mid-Lent (9 March 1233) returning from *Kenitun'* towards Winchester *pbe* (Ms.28DR m.8)

For bp John see above pp.xxxiv–v

330. £8, seven shillings and four pence spent at Wycombe buying eighty five pigs *pbe* (Ms.28DR m.9)

331. Forty four shillings and seven pence spent at Downton on Stephen the huntsman (*venatoris*), five men, eleven greyhounds (*leporariorum*) and twenty five hounds (*canum*) for three weeks and two days *pbe* (Ms.28DR m.15)

332. £66, thirteen shillings and four pence (100 marks) released from Taunton to the abbot of Forde (*Forda*) *pbe* (Ms.28DR m.18)

There is nothing in the Forde cartulary to explain this payment, although it is substantial enough to suggest a property purchase by des Roches.

333. £46 released from Taunton to dom. Robert de Vallibus *pbe* (Ms.28DR m.18)

334. Thirty eight shillings released from Taunton to six serjeants, to garrison the castle (of Taunton) for the thirty eight days from 2 August to 8 September (1233) by order of the bishop and the advice of P(eter) de Russ' and by the bishop's letters (*In liberatioo' [sic] vi. servient' ad munitionem castri per precept' domini episcopi et per consilium P. de Russ' et per litteras episcopi a crastino sancti Petri ad vincula usque nativitatem beate Marie per xxxviii dies*) (Ms.28DR m.18)

Reflects the tensions in the west country during the opening stages of the rebellion of Richard Marshal.

335. Twelve shillings spent at Taunton on William de Mulef(ord) and the ?companions (*[so]ciorum eius*) he brought with him, and his other men and hounds taking game (*ad venationem capiend'*) for twelve days *pbe* (Ms.28DR m.18)

1235/36

336. £4, twelve shillings of arrears owed at Bentley by Eimeric de Chanceaux

(*Cancell'*) and Matthew Bezill (*Bastal'* corrected elsewhere) *pbe* for forty small quarters of oats they received, over which Eimeric gave his letters patent (Ms.31DR m.10d)

> Eimeric and Matthew were kinsmen of the bp's alien familiars Peter and Andrew de Chanceaux.

337. Twelve shillings and nine pence spent at Southwark on fuel (*carbon'*) bought against the bishop's arrival *per litteras eius* (Ms.31DR m.15)

338. Fifty shillings released at Southwark to master E(gidius?) de Brideport by the bishop's letters (Ms.31DR m.15)

> Master Giles was also entertained, together with others of the bp's officers, as a nuncio to the archbp (of Canterbury) on 28 March 1236. In 1256 he was to be elected bp of Salisbury.

339. Forty four shillings and four pence spent on three butts of ginger (*butezeles zinzibr'*); seven shillings spent on one pound of cloves (*gariofil'*); six shillings and seven pence spent on two pounds of mace (*macis*); forty two pence spent on one pound of nutmeg (*nuc' musc'*); and four shillings spent on four pounds of cinnamon (*canell'*), at Southwark: *omnia ista empt(a) sunt per litteras episcopi* (Ms.31DR m.15)

> Interesting as an example of luxury goods purchased by the bp.

1236/37

340. Ten shillings spent at Meon in summoning two carts to carry two tuns (*dolia*) of wine to London *pbe* (Ms.32DR m.10)

341. £25 of the arrears owing at Taunton accounted against *dom.* Henry de Ortiaco for grain supplied to him by the bishop's letters (Ms.32DR m.12)

342. Four shillings spent at Taunton for two men of the lord (bishop) of Durham (*Dulmon'* corrected elsewhere) fishing for him for four days by the bishop's letters (Ms.32DR m.13)

343. Fifteen shillings spent at Taunton on the lord Henry de Erlegh' and John de Aure hunting in the park for two days by the bishop's letters (*litteras*) (Ms.32DR m.13)

APPENDIX 1

THE RECORDS OF THE BISHOP'S YEARS AS CHIEF JUSTICIAR AND REGENT 1213–1215

Between *c.*February 1214 and *c.*June 1215 des Roches served as the king's chief justiciar, acting as regent in the absence of the king from February to October 1214. The vast majority of his surviving writs as justiciar were issued in this latter capacity as regent. They are to be found in the following sources:

1. A ROLL OF LETTERS CLOSE recording some 235 writs issued by des Roches as regent between May and October 1214: PRO C54/11, printed *RLC* i 204-13b. Made up of six membranes, 34 cm. wide and varying in length between 42 and 59 cm., written in several chancery hands, with occasional entries on the dorse. Sown together chancery fashion, head to foot. Referred to in a contemporary fine roll (*Rot.Ob.* 547) as a *rotulus litterarum clausarum*. The dating clauses and addresses are more heavily abbreviated than those in rolls of royal letters close, but in all other respects the present record is drawn up in identical form to that of the contemporary royal close rolls. Entries are copied in roughly chronological order. The earliest dated entry, for May 23 (1214), comes on what is now the inner membrane (m.6), which bears no title or any indication that it was originally the first in a series which may originally have stretched back to include the first four months of des Roches' regency, from February to May 1214. There are two holes in the top of membrane 6 by which it may once have been sown to earlier membranes, now lost. The latest dated entry is of a writ issued on October 9 (*ix*) 1214, wrongly transcribed in the printed version as October 20 (*xx*). This suggests that the enrolment of the justiciar's letters may have ceased with the king's return from France. Altogether, the present roll appears to form slightly more than half of a chancery roll, which originally covered the entire period of des Roches' regency from February to October 1214. Even within these confines, it is by no means a comprehensive record of des Roches' correspondence. Numerous writs from the period May–October 1214 are preserved or mentioned in other sources but not enrolled here. Leaps within the chronology of those letters that are preserved imply that many other writs were excluded either by accident or design. Almost all of the 235 individual entries on the roll relate to administrative instructions to sheriffs, constables, bailiffs or others of the

king's officers from whom an account was anticipated. Many such entries are duly marked *com(putate)* in the margin. The roll seems therefore to have served principally as an aid in the compilation of the exchequer pipe roll, representing a development from the *liberate* rolls of John's early years, assisting in the verification of ministers' expenses. As such it was invaluable as a communication between chancery and exchequer, but at the same time deliberately excluded all but a handful of the justiciar's political or general correspondence. Those writs which would be of most interest to the historian; instructions to the Bench, and letters addressed to such figures as the papal legate, the archbishop and the king in France, were never recorded here.

2. THE PIPE ROLLS of the royal exchequer mention various writs issued by des Roches as justiciar (*PR 16 John* 2,20,28,68,79,83–4,109,126–7; *PR 17 John* 14,32,33,58; *PR 3 Henry III* 154; *PR 4 Henry III* 71,111, and probably *PR 2 Henry III* 16), some of which are preserved in full in the justiciar's close roll above . It is likely that many other entries which refer to expenses incurred *per breve* without citing the details of the writ, or *per breve domini regis*, are actually related to writs of the justiciar, since several of the writs preserved in the justiciar's close roll are described, erroneously, as royal writs in the pipe rolls. See for example *PR 16 John* 126–7, and writs to Hampshire in *RLC* i 206b–7, 208b, 212; similarly a writ to Somerset (*RLC* i 210b) entered on the pipe roll as a writ of the king (*PR 16 John* 104).

3. THE PLEA ROLLS OF THE BENCH refer to various writs received from des Roches as justiciar (*CRR* vii 115,125,132–3,149,161,195–6,268,299–300). The compilers of the plea rolls appear deliberately to have drawn a distinction between such written instructions, marked *per breve* or *per litteras*, and instructions issued by word of mouth, marked *per preceptum*. For entries relating to des Roches marked *per preceptum* see *CRR* vii 51–2,113,137,177, 189,209,244,306, and in general see West *The Justiciarship* 155 n.1.

4. A SCUTAGE ROLL, edited by J.C. Holt in *PR 17 John* 105–8 from PRO C72/1, a single membrane measuring 485 × 320 mm. Drawn up in England, the roll records abstracts of 53 licences *de scutagio habendo*, followed by an order in the justiciar's name providing for a general collection of scutage by the sheriffs of every county. Since the present membrane lacks any title and covers by no means all licences to collect scutage known to have been issued by the king in France, its editor suggests that it may represent only one part of a more comprehensive roll, now lost (*PR 17 John* 82). Set against this, there is no indication that the surviving membrane was ever sewn to others, whilst the only external references to the scutage roll, to be found in the justiciar's close

roll, correspond to entries in the surviving ms., described in these instances as a fine roll (*RLC* i 211b) or as *rotulus de summonicione* (*RLC* i 212).

5. FINE or ORIGINALIA ROLLS drawn up in England during the course of 1214, of which various membranes survive: PRO C60/5c mm.5,6, duplicated in PRO C60/5d m.2 and printed *Rot.Ob.* 540–50; PRO C60/5c m. 4 (printed *Rot.Ob.* 533–40); PRO C60/5d m.3 (unprinted). The entries printed in *Rot.Ob.* 540–50 refer to some seventy writs, mostly of a legal nature; writs of entry, *pone, mort d'ancestor* etc., purchased by litigants, as well as memoranda of orders for the renewal of food and supplies at Brill, Windsor, Hertford and Silverstone, a note of arrows and grain supplied by Engelard de Cigogné and two letters issued in the justiciar's name, cancelled, one of them being entered instead in the justiciar's close roll for *c*.June 1214 (*Rot.Ob.* 545, 547; *RLC* i 206b, 207). The membrane printed in *Rot.Ob.* 533–40, recording a further fifty or so fines for the issue of writs, has been assumed to date from after the king's return from France in October 1214 since it carries a note on the dorse stating that des Roches took custody of it at the treasury on 27 November 1214. However, a fine from this membrane was included amongst the entries on the justiciar's close roll for *c*.August 1214, where it is cancelled under a note that it has been entered in the roll of fines (*RLC* i 210b; *Rot.Ob.* 537). This suggests that the membrane itself covers writs issued by des Roches in England, before the king's return. All of the surviving membranes appear to have been drawn up in the chancery for transmission to the treasury, like the justiciar's close roll to assist in the compilation of accounts. Individual fines are duly marked *in rotulo* to show that they had been entered in the *nova oblata* section of the pipe roll.

6. FINAL CONCORDS and FEET OF FINES drawn up under des Roches' presidency of the Bench. The surviving Feet of Fines are fully listed in *PBK* iii pp.ccxc–ccxciv, many of them being printed in the various local record society editions of feet of fines. In addition numerous final concords survive in charter collections or cartularies, occasionally in the original (Westminster Abbey Muniments no.3062; BL Add. Ch. 47958), but more often as copies: *Luffield Charters* i no.14a; *Eynsham Cart.* i no.230; BL mss. Cotton Claudius D xiii (Binham cartulary) fo.56r; Egerton 3033 (Canons Ashby cartulary) fo.98v; Sloane 986 (Braybrook cartulary) fo.42r; Bodl. ms. Rawlinson B142 (Sawtry charters) fo.23v; CUL mss. D.& C. Peterborough 1 (Peterborough cartulary) fo.172r-v, also in Ibid. D.& C. Peterborough 5 fos.125v–126r; Taunton, Somerset County Record Office ms. DD/545 H/348 (Hungerford cartulary)

fo.96v, also in Trowbridge, Wiltshire County Record Office ms. 490/1470 (Hungerford cartulary) fo.146r; Westminster Abbey ms. Domesday fo.241r.

7. MISCELLANEOUS mentions or copies of writs, including a letter to Stephen archbishop of Canterbury preserved in a Lichfield cartulary (printed below), and a mandate to Philip of Burnham and John de Cornard to take custody of the abbey and lands of Bury St Edmunds *c*. June 1214 (printed *Chronicle of Hugh* 80–3, 180 no.19; *Memorials of St Edmund's Abbey*, ed. T. Arnold, 3 vols., Rolls Series (London 1890–96) ii 73–4). For various mentions of writs or awards, now lost, see *Rot.Chart.* 196b; *RLC* i 176b. In addition *RLP* 148 records that 66 sacks containing 9900 marks in coin, sealed with the seal of the bishop of Winchester, were held at Corfe castle in July 1215, the money perhaps forming part of a treasure which des Roches as justiciar had stored at the Tower of London, presumably evacuated before the rebel seizure of London in May 1215. For the treasure at the Tower see *RLC* i 192b; *RLP* 134; *PR 17 John* 33.

Beyond these surviving records of des Roches' justiciarship, there are various references to documents now lost, including:

8. THE MEMORANDA ROLL 16 JOHN (1214–15) (Lost), apparently used to record a judgment delivered by des Roches at the exchequer on 25 November 1214 exempting the vill of Cirencester from tallage taken by the king's officers; *Cirencester Cart.* i no.39 and see no.17, said to have been written in the *magnum rotulum* for the sixteenth year of king John. No such entry survives in *PR 16 John* which nonetheless (p.56) notes a related quittance in relation to Cirencester *per breve regis*. The memoranda roll may also be one of the treasury rolls referred to in the justiciar's close roll, where a pension paid to a mercenary captain was to be recorded (*RLC* i 205b, not in the pipe roll, although see *PR 16 John* 25,145).

9. A ROLL OF FINES (Lost) imposed by des Roches during itinerant legal hearings in the spring of 1214, delivered by des Roches to the exchequer and used in the compilation of *PR 16 John* 47,121,165. For the circumstances of the hearings themselves see *PR 16 John* pp.xxv–vi.

10. AN ESTREAT (Lost) delivered by Richard Marsh, apparently listing

prests made to those serving in Poitou, referred to by the king in issuing instructions to des Roches in June 1214 (*RLC* i 201).

Amongst the surviving, official records of des Roches' justiciarship there is evidence of codification, for example in the cancellation of various writs from the close roll because they were copied elsewhere on the fine or scutage rolls and vice versa. The vast majority of the records drawn up during des Roches' time as justiciar, the close, pipe, memoranda, originalia, fine and scutage rolls, were either productions of the exchequer or chancery productions intended to assist in the compilation of the exchequer pipe roll. The only enrolled records not associated with the exchequer were the plea rolls and concords of the Bench and the various schedules requested by the king in France.

THE JUSTICIAR'S CHANCERY: Although the sources detailed above record copies or mentions of more than four hundred texts, this is by no means the full tally of writs issued by des Roches as justiciar. As we have seen, the pipe rolls disguise many of the justiciar's writs as writs of the king, whilst the justiciar's close roll, with its 235 texts, omits the vast majority of letters which did not call for an account at the exchequer and covers only six of the eighteen months during which des Roches held office as justiciar. In addition, many of the texts recorded in the close roll would have involved the dispatch of up to thirty or more individual writs, to the sheriff of every county, to the bailiffs of a dozen ports, and so on. In all it seems likely that as justiciar des Roches presided over the issue of something well in excess of one thousand individual writs, a rate of production which argues the support of a chancery or writing office of some considerable size and efficiency, beyond the capacity of des Roches' own episcopal chancery. In this respect there seems no reason to question the conclusion reached by H.G. Richardson, that successive justiciars depended upon a branch of the royal chancery, separate from that which accompanied the itinerant king. Already, before his appointment as justiciar des Roches had established some degree of control over the royal chancery. Between August 1213 and February 1214 the majority of royal charters are said to have been issued by the hand (*per manum*) of the bishop of Winchester, a phrase normally reserved for the chancellor or the chief chancery clerk. There is no evidence that des Roches had actually assumed the office of chancellor. Walter de Gray, the titular chancellor, was pre-occupied for much of this period with missions to Flanders, whilst Richard Marsh, the senior chancery clerk, was absent in Rome defending his role in government during the papal interdict. In their absence, des Roches appears to have assumed *de facto* custody of the king's seal, which in December 1213 was released under his custody to Ralph de Neville. Whilst Neville and the great seal accompanied the king to Poitou in February 1214, des Roches continued to be served by members of the royal chancery in his responsibilities as justiciar and regent (above pp. xxix, xli, xlv).

SEALING AND ISSUE: None of des Roches' writs as justiciar survives in the original, and we have only a few examples of writs issued by his predecessors in the office to provide a diplomatic comparison. (For Geoffrey fitz Peter see *PBK* i 5-33; H.G. Richardson introduction to *Memoranda 1 John*. For Hubert Walter see *EEA* iii appendix 1 pp.297-304.) It seems likely that the justiciarship possessed no special seal of office. Des Roches' writs were issued under his title as bishop of Winchester, with no reference to his titles as regent or justiciar, and were probably sealed with his own episcopal seal. At the exchequer he would have had access to a smaller copy of the great seal of the king. It may well have been this so-called exchequer seal that was set to writs issued by the justiciar in the king's name, for example the writs of *mort d'ancestor* and *pone* purchased from des Roches in England and referred to in the fine and originalia rolls. Despite the king's absence overseas, it is almost certain that such routine legal writs would have continued to employ the royal title. Without resort to the exchequer seal, it is difficult to imagine how the Bench and the administrative offices at Westminster could have continued to issue writs in the absence both of the king in Poitou and of des Roches and his episcopal seal, on tour around England. Later, during the reign of Henry III the exchequer seal would occasionally tour England in the regent's custody, as a seal of absence. There is no evidence of this happening in 1214.

DATING AND WITNESSING: Des Roches writs appear to have adopted the standard practices of the royal chancery in respect to dating clauses by day, month and regnal year. Fifty eight of his texts preserve traces of a date or place clause, of which fifty seven are drawn from the justiciar's close roll. Thirty two of these texts are issued *Teste me ipso* of which two carry the further warranty clause *per eundem*, and one is warranted *per* master Eustace de Fauconberg. Eustace witnesses a further twenty one texts, of which two are marked *per eundem*. Two texts are witnessed by William Brewer and one by William Marshal earl of Pembroke. There is no apparent association between individual witnesses and specific administrative concerns. Arguably the form *Teste me ipso* is employed for writs of the greatest political significance, although politically sensitive instructions involving the imposition of scutage on the church were issued *Teste* and *per* master Eustace de Fauconberg (*RLC* i 210). The suggestion that master Eustace was especially associated with routine payments (West *The Justiciarship* 210) does not stand up under scrutiny. Almost all of the surviving texts, with the sole exception of the mandate to Langton printed below, appear to have been issued as letters close.

AUTHORIZATION: More than half of des Roches' writs as justiciar can be linked directly to instructions issued by the king, recorded in the surviving royal patent, fine and close rolls drawn up in France (*RLP* 110b–122, 139–40b; *RLC* i 141b–143b, 160–172b, 200b–202b; *Rot.Ob.* 523–33; PRO C60/5c m.3; C60/5d m.3). In two cases des Roches recites royal mandates in full. Four of his writs are warranted *per breve regis de ultra mare*, and a further ten refer specifically to possession of a royal warrant. The king's enrolled correspondence for 1214 is by no means complete. Various of the justiciar's writs recite or claim the authority of royal letters recorded nowhere else, whilst at least two royal letters to des Roches are preserved only in unofficial sources (*Chronicle of Hugh* 108; *Southwick Cart.* vol.1 part i no.165). In more than a dozen cases it is possible to assess the lapse in time between the issue of royal instructions and their translation into writs by des Roches, varying between just over a fortnight (*RLC* i 170, 212 on Thomas de Erdington) and nearly two months (*RLP* 113b; *RLC* i 208b on Robert de Béthune; *RLP* 118; *RLC* 211b on Peter Saracen). Clearly the speed with which the king's instructions could be implemented depended on the ease of communications between England and Poitou. It seems possible that unfavourable winds compounded the problems of the royal administration in the summer of 1214. The Winchester pipe roll for 1213–14 records the entertainment at des Roches' manor of Fareham near Portsmouth of various nuncios to the king, one of whom, Nicholas de Hattingel, was detained at Fareham for over a month awaiting a crossing (Ms. 9DR mm.2d,3; *RLC* i 207). Besides his correspondence with the king, des Roches' writs record the receipt of letters and the implementation of requests from the papal legate, Nicholas of Tusculum, king William I of Scotland, Geoffrey de Mandeville and possibly archbishop Langton (*RLC* i 208b, 210, 211b). However, it should be emphasised that the most important of the bishop's political letters were almost certainly excluded from the justiciar's close roll as being irrelevant to the compilation of accounts. Of this type of letter we have only three examples, to Stephen Langton preserved in an unofficial source (below appendix 1), to the archbishop of Dublin preserved on the close roll because it involved an account (*RLC* i 205) and to David earl of Huntingdon, copied onto the dorse of the close roll (*RLC* i 213). There is nothing here to match the rich archive of correspondence between the leading figures of the minority of Henry III, preserved in the royal chancery. Above all, whilst there survive numerous of the king's instructions to des Roches, and

132   APPENDIX I

whilst the justiciar's close roll provides for the dispatch of various messengers from des Roches to the king in Poitou, we can only guess at the letters or reports that such nuncios carried with them.

1. *Notification addressed to S(tephen) archbishop of Canterbury that the monks of Coventry and the canons of Lichfield have unanimously elected W(illiam of Cornhill) archdeacon of Huntingdon as their bishop. The bishop of Winchester, William Brewer, and the abbots of St Mary's York and Selby, acting on behalf of the king, have granted assent to this election, which the archbishop is hereby ordered to confirm.*   (Bishop's) Sutton, 13 July 1214

> B = Lichfield Cathedral, Dean and Chapter Library ms.28 (Magnum Registrum Album) fo. 184r-v. s.xiv in.
> Pd (calendar) from B in *Magnum Reg. Album* no. 367. Referred to *Ibid.* no. 464; *Acta Langton* no. 61 p.80.

Venerabili patri in Cristo S(tephano) Dei gratia Cant(uariensi) archiepiscopo, totius Anglie primati et sancte Romane ecclesie cardinali, P(etrus) divina permissione Wintoniensis ecclesie minister humilis salutem et debitam patri reverentiam. Noverit paternitas vestra quod prior et monachi Coventr' et canonici Lich', gratia spiritus sancti invocata, unanimiter elegerunt in pastorem et pontificem[a] ecclesie sue W(illelmum) archidiaconum Huntingdon'. Nos vero et dominus Willelmus Briewerr' et sancte Marie de Ebor' et de Seleby abbates, qui domini regis vices in hac parte gerimus, electioni illi prebuimus assensum, eam ex parte domini regis gratam et ratam habentes. Unde vobis mandamus quatinus munus vestre confirmationis electioni illi apponere velitis, et in huius rei testimonium has litteras nostras patentes vobis mittimus. Teste me ipso apud Sutton' xiii die Iulii, anno regni domini regis Iohannis sextodecimo.

> [a] B fo.184v

> For the background see Cheney *Innocent III and England* 131–2. Langton had earlier quashed the election to Coventry of another curialist, Walter de Gray. In 1214 it was des Roches who suggested to the monks that they elect Cornhill (*Monasticon* vi 1243). The king knew of Cornhill's promotion by 9 July (*RLP* 117b–118). On 6 August he reiterated des Roches' instructions to Langton to receive the elect (*RLC* i 169b; *RLP* 120). On 9 October des Roches ordered the sheriffs of four counties to grant Cornhill seisin of the see's temporalities, although it is by no means clear that the archbp had as yet confirmed the election (*RLC* i 213, dated 20 (xx) October in the printed version, by mistake for 9 (ix) in the ms. (PRO C54/11, m.1)). At some time in the year Michaelmas 1213–14 letters were sent from des Roches to Taunton, *pro morte W. clerici, nuncii directi ad Couvtre* (Ms. 9DR m.9d). Cornhill was eventually consecrated in January 1215. The peculiarity of des Roches' standing in respect to his archbp is reflected in the phraseology of the present letters, in particular the clause 'We order you that you may wish (to confirm)', modelled on royal mandates but here allowing for some degree of choice by Langton.

APPENDIX 2

REFERENCES TO USES OF THE BISHOP'S SEAL AND TO ACTIONS WHICH MAY HAVE INVOLVED THE ISSUE OF EPISCOPAL LETTERS

1. *Collation of William de Sowy, clerk, to the church of Hinton (Ampner) with its land at Bere (in Soberton)* (?1206 × 1225, ?c.1206) (*CRR* xii no. 1545)

   From the context (a suit by William for recovery of land at Bere, supposedly annexed to his church) it is apparent that William had been granted the church by bp Godfrey de Lucy, before 1204, saving an annual pension of 44 shillings to the hospital of St Cross, Winchester. Des Roches' attorney, whilst specifically acknowledging a collation by des Roches to the church of Hinton, was unwilling to warrant the land at Bere because the bp acted towards William only in the latter's capacity as parson and clerk, not as landholder.

2. *Mandate of king John to Hugh de Chaurc' ordering him to release John of Monmouth's son, John, to his father to whom the king has committed him in custody, sealed with the seal of des Roches* (Oxford, 20 October 1208) (*RLP* 87)

3. *Admission of Geoffrey de Caux to the church of Dogmersfield at the presentation of Robert prior of Bath* (c.1210) (*Bath Cartulary* part ii no. 75)

   Geoffrey de Caux was a member of the bp's household, see appendix 4 no. 29.

4. *Foundation of the Domus Dei Portsmouth* (c.1214) (Paris *CM* iii 490)

   Despite Paris' attribution of the foundation to des Roches, it seems more likely that the house came into being through the joint efforts of William of Wrotham and Nicholas de Kivilly, both of them associates of des Roches, at some time before November 1214 (*Rot.Chart.* 202; *CRR* xi nos. 1383,2265,2763). Des Roches undoubtedly sponsored building work at the hospital in the 1230s, whilst in 1229 his official, master Alan of Stokes, effected an important settlement between the hospital and Southwick priory (*CPR 1232–47* 14; *Cl.R.1231–4* 202,242,279; *Southwick Cart.* ii pp.321–3 no. 801).

5. *Charters relinquishing any claim to ancient liberties condemned by the king, issued by the bishop of Winchester, William Brewer and the earl of Chester* (c.January 1215) (*DD* no. 19; *Foedera* 120)

   The existence of such charters or at least of a willingness on des Roches' part to issue them, is recorded in a report from the Roman curia by Walter Mauclerk, setting out the grievances laid before the pope by representatives of the English barons. *Walter of Coventry* ii 218 and Paris *CM* ii 584 confirm that an oath along these lines was demanded by the king during a parley with the barons in January 1215, a meeting at which des Roches stood as one of the guarantors for the safety of the baronial nuncios (*RLP* 126b). See in general Cheney *Innocent III and England* 367–8; Painter *King John* 299–300.

APPENDIX 2

6. *Suspension of Stephen Langton, archbishop of Canterbury* (*c.* 5 September 1215) (*Coggeshall* 174; Paris *CM* ii 629-30)

After no. 100 above. Carried out by des Roches, Pandulph and the abbot of Reading as Langton was waiting on board ship, ready to set sail for the Lateran Council. Confirmed by pope Innocent III, 4 November 1215 (*Letters of Innocent III* nos. 1026-7).

7. *Excommunication of 31 named individuals who had rebelled against king John* (5 September × late November 1215) (Paris, *CM* ii 630, 642-4; *Letters of Innocent III* no. 1028)

After no. 100 above, and in accordance with the papal mandate contained therein. Carried out by des Roches, Pandulph and the abbot of Reading. Confirmed by pope Innocent III, 16 December 1215, who deputed the enforcement of the sentence to William archdn of Poitou, Ranulph of Warham official of Norwich and Hugh abbot of Abingdon, apparently acting as proxies of the three original commissioners. See also *RLP* 170; *RLC* i 269.

7a. *Excommunication of Louis, son of the king of France, and all of his supporters* (Winchester, 29 May 1216) (*AM* ii (Winchester) 82)

8. *Collation of John de la Herce, the bishop's clerk, to the church of Stanford on Soar (Notts.)* (*c.*1216 × 1217) (*CRR* xii no. 379)

9. *Licence from king Henry III to Pain de Chaorciis to return to his homeland to obtain his ransom, sealed by P(eter) bishop of Winchester, W(illiam) Marshal earl of Pembroke rector of the king and his realm, R(anulph) earl of Chester and W(illiam) earl of Derby* (Oxford, 16 January 1217) (*Pat.R.1216-25* 24)

10. *Grant by king Henry III to Henry fitz Count, that he may hold the county of Cornwall on the same terms as Reginald earl of Cornwall, his father. He is not to be disseised save by consideration and judgment of the king's court. Sealed by Guala cardinal priest of St Martin (in Montibus), legate of the apostolic see, W(alter) archbishop of York, H(enry) archbishop of Dublin, P(eter) bishop of Winchester and W(illiam) Marshal rector of the king and his realm* (Gloucester, 7 February 1217) (*Pat.R.1216-25* 30)

11. *Ordination of Jordan de Herst as deacon, at Winchester* (*c.*1217) (*Reg.St Osmund* i 305)

Said to have taken place six years before 1222, presumably after the bp's return to Winchester following the civil war.

12. *An assignment of land in the New Forest made to Beaulieu abbey on behalf of king Henry III by Peter bishop of Winchester and William Brewer, during the king's minority* (1216 × 1224, ?1217 × 1221) (*Cal.Chart.R.1226-57* 216)

Confirmed, with a detailed recital of the bounds, by the king in January 1236. To be dated after the civil war and before des Roches' decline in favour.

13. *Settlement devised and sealed by Guala cardinal priest of St Martin (in Montibus), legate of the apostolic see, P(eter) bishop of Winchester, W(illiam) bishop of Worcester, W(illiam) Marshal earl of Pembroke, rector of the king and his realm, William Brewer, Fawkes de Bréauté and others of the king's council, between Engelard de Cigogné and the crown, by which Engelard receives the manor of Benson with the vill of Henley to keep him in the king's service, with protection against any claim there by John de Harcourt, by reason of Harcourt's status as a crusader. Engelard is to have £100 of land in the first escheats to fall vacant in the bailiwicks of P(eter) bishop of Winchester, W(illiam) Marshal earl (of Pembroke) or Fawkes de Bréauté, and in the meantime an annual pension of £50. The escheats and the pension are to revert to the crown should the king restore the county of Surrey to Engelard* (c.October 1218) (*RLC* i 403–403b)

For the background see Carpenter *The Minority* 66, 96, 122; Vincent *Guala* no. 26.

14. *Seals the will of William Marshal (I) earl of Pembroke* (April 1219) (*Hist.Maréchal* lines 18034–42)

15. *An intention expressed by the bishop to transfer a community of nuns (?Nunnaminster) from the centre of the city of Winchester, where they are exposed to prying eyes, to the church of St Cross, and to install secular canons at the site of the nunnery* (1219) (*Cal.Pap.Reg.* i 66; *Reg.Hon.III* no. 2049)

Mentioned in papal letters of 7 May 1219 in which Honorius III ordered the legate Pandulph to assist des Roches in his scheme. No such transfer is known to have taken place. Nunnaminster is the only nunnery known to have been sited in the city centre.

16. *Letters of king Henry III and the mayor and commune of London, guaranteeing repayment of 2000 marks to be loaned by the mayors and communes of La Rochelle and Bordeaux to Hugh de Lusignan for the defence (of Poitou and Gascony), mentioning similar guarantees by P(eter) bishop of Winchester and H(ubert) de Burgh the justiciar* (c.24 July 1219) (*Foedera* 155–6)

17. *Concord between William bishop of London and various parties, by licence of the bishop of Winchester.* (October 1219) (*CRR* viii 138–9)

18. *Oaths sworn by (Stephen) archbishop of Canterbury, (Peter) bishop of Winchester, Hubert de Burgh the justiciar, (William) earl of Warenne and (William Longespée) earl of Salisbury, promising to observe and uphold the terms of the treaty agreed between the kings of France and England.* (March 1220) (*Foedera* 158–9; *Layettes* i no. 1387)

Although no specific mention is made of letters, letters to this effect survive from the earls of Warenne and Salisbury and from Hubert de Burgh (*Layettes* i nos. 1388-9; *Foedera* 159).

19. *Presentation of Denis (de Bourgueil) to the church of Witney, following the resignation of master Humphrey de Millières.* (1219 × 1220) (*Rot.Hugh Welles* ii 3)

Dated to the 11th year of bp Hugh of Lincoln.

20. *Dedication of the church of* 'Novus Locus', *founded by the bishop* (27 September 1220) (*AM* iv (Worcester) 413: *Item ecclesia Novi Loci, a Petro episcopo Wintoniensi fundata, ab eodem dedicata est v. kal. Octobris*)

Problematical. This particular entry in the Worcester annals is taken *verbatim* from the annals in BL ms. Cotton Vespasian E iv (Pd F.Liebermann, *Ungedruckte anglo-normannische Geschichtsquellen* (Strasbourg 1879) 189), a Winchester-Waverley chronicle whose sources and influence are set out by N. Denholm-Young, 'The Winchester-Hyde Chronicle', *Collected Papers* (Cardiff 1969) 236, 238. The bp was in Winchester on the days surrounding the supposed date of the dedication. It is therefore unlikely that the church in question should be identified with that of Newark priory near Guildford, refounded by bp Godfrey de Lucy. John, first prior of Newark (d.1226) occurs as prior *de Novo Loco* as early as 1207 (PRO CP25(1) 203/17; *AM* ii 302). It is possible, but by no means certain, that the entry refers to some sort of refoundation or rebuilding of the conventual church of Hyde abbey, known in the twelfth century as *Novus Locus*, which had suffered damage during the civil war 1215 × 1217.

21. *Presentation of master Humphrey de Millières to the church of Ivinghoe.* (1220 × 1221) (*Rot.Hugh Welles* ii 54)

Dated to the 12th year of bp Hugh of Lincoln. The register notes the absence or loss of letters of presentation or inquisition, implying perhaps that such letters were taken for granted in other cases.

22. *Establishment of the Dominican house in Winchester* (1221 × 1226)

Traditionally ascribed to des Roches, who had accompanied the first band of Dominicans to reach England on his return from pilgrimage to Compostela in 1221 (*Trivet* 209). By Michaelmas 1225 he had purchased houses in Winchester originally granted to the Domus Dei, Portsmouth by its first master, Nicholas de Kivilly. These he transferred to the Winchester Dominicans before Michaelmas 1226 (Ms.20DR m.7d: *In liberatis fratribus de Portesmues pro vendicione domorum Nicholai de Kivil' datarum fratribus ordinis predicatorum per dominum episcopum c. s(olidos)*; and see Ms.19DR m.10 *c solidos sunt super fratres hospitalis de Portesm' et sunt allocati eis in vendicione domorum Nicholai de Kivilly apud Wint'*). For the background to the Winchester house see Keene *Survey* 822-5, and C.F.R. Palmer, 'The Friars Preachers or Black Friars of Winchester', *The Reliquary*, new series iii (1889) 207-15, which misreads the evidence from the Winchester pipe rolls to imply that des Roches originally established the Dominicans at Portsmouth.

23. *Benediction of Richard of Barking as abbot of Westminster.* (18 September 1222) (Paris *CM*, iii 74; *Flores.Hist.* ii 176)

Follows the settlement no. 73 above, by which the Westminster monks were permitted to seek benediction for their abbot from any bp; a demonstration of their new won freedom from the jurisdiction of the bps of London.

24. *Mediation in a dispute between William (of Merton) archdeacon of Berkshire and Beaulieu abbey over tithes at Faringdon.* (1222 × 1238, ?c.1224) (*Beaulieu cartulary* no. 9)
Probably soon after the appointment of William as archdn (*Fasti* iv 30).

25. *Excommunication in full synod of all who attack the right of the church.* (March 1224 × March 1225) (*AM* ii (Winchester) 84)
The Winchester annals are dated according to the year beginning on 25 March. Cheney (*C. & S.*125) suggested that this excommunication may have marked the issue of des Roches' diocesan legislation. More likely, it was a result of the reprisals against des Roches that sprang from the bp's political disgrace in 1223-4.

26. *Settlement between W(alter) prior of St Swithun's Winchester and Aimery de Sacy, lord of Barton (Stacey), over common pasture in Widemore and the customs owed by the prior's men of Bransbury, made in the presence and through the mediation* (procurante) *of Peter des Roches bishop of Winchester, guardian* (custode) *of the king. Dated 9 Henry III.* (October 1224 × October 1225, ?Spurious) (Winchester Cathedral Library mss. Alchin Scrapbook II/ Henry III no. 6; Libellus Basyng fo.3r-v)
The appearance of des Roches with his title as guardian of the king (forfeit 1221 × 1223), combined with the fact that the witness list includes Henry de Farlee sheriff of Hampshire (otherwise unrecorded) and Eustace de Fauconberg without his title as bp of London, all suggest that the text is a forgery, judging from its script, of the 1220s or 1230s.

27. *Institution of Hugh des Roches as vicar of the church of Ditton at the presentation of Merton priory* (1206 × 1223, ?1222) (*CRR* xi 417; *BNB* iii 467)

28. *Settlement between G(iles) prior of Merton and the priory of St Mary Overy, Southwark, made in the bishop's presence and sealed with his seal, concerning tithe of corn in 'Langecroft'.* (1222 × 1227) (BL mss. Add. 6040 (Southwark cartulary) fo.2r; Cotton Cleopatra C vii (Merton cartulary) fo.95v; and see Heales *Records of Merton* 81-2)
Dates to the time of Giles, prior of Merton (d.1231) and hence before the bp's departure for crusade in 1227.

29. *Licence by Humphrey prior of St Mary Overy, Southwark, to the hospital of St Thomas' Southwark, at the petition of Peter bishop of Winchester and others, that the hospital may have a cemetery for the burial of its master and brothers, and any others who may obtain the permission of their parish priests; the hospital may also have two bells weighing no more than 100 pounds. In* *compensation the master and brothers of the hospital have assigned an annual rent to the priory of six shillings and sixpence in* Trevetlane, *from the land of Walter the thatcher* (tector); *and an annual rent of twelvepence at Easter to the vicar of the church of St Mary Magdalene who has granted his assent to the present agreement. Sealed with the seals of the bishop of Winchester, the prior, the hospital and the vicar. Witnesses: master Alan of Stokes official of the bishop, master Robert Basset, master Humphrey de Millières, Robert de Clinchamps, Roger dean of Southwark, John of Southwark, William fitz Humphrey, Ralph vicar of Wandsworth, John de Grisele and others* (St Thomas' Cartulary no. 2, also in Bodl. ms. Rawlinson D763 pp.10–11; Winchester, Hampshire Record Office ms. 21M65/A/1/5 (Register of bp Stratford) fo. 170r) (1225 × 1234, ?1225 × 1227)

>Dates between the election of prior Humphrey (c.1225) and the retirement of master Alan of Stokes as bp's official. The present agreement follows a settlement devised by master Alan, acting as papal commissioner, in the lifetime of archbp Stephen Langton and before the death of Benedict bp of Rochester (d. December 1226), granting St Mary's priory compensation of two shillings a year in return for the consecration of a cemetery (*St Thomas' Cartulary* no. 5).

30. *Refoundation of the hospital of St Thomas at Acre.* (c.1227 × 1229) (*AM* iii (Dunstable) 126; *Cal.Pap.Reg.* i 150; Paris *CM* iii 490)

>Carried out whilst des Roches was on crusade in the Holy Land, and almost certainly before the refoundation or establishment of a London sister house which had taken place by the autumn of 1228. For the background see A.J. Forey, 'The Military Order of St Thomas of Acre', *EHR* xcii (1977) 481–9. The Dunstable annalist states that des Roches removed the hospital to a less crowded site, replete with a supply of running water, and that he placed it under the custody of the order of Santiago (*Spatae Hispaniensis*). By 1236, when Gregory IX came to confirm the bp's actions, the hospital had been transferred to the rule of the Teutonic knights. Paris claims that des Roches had the assistance of the patriarch of Jerusalem, and that the house was granted to the Templars. Des Roches is said to have left 500 marks to St Thomas' in his will, the least of his benefactions (Paris *CM* iii 490).

31. *Notification addressed to bishop Robert of Salisbury, informing him of the death of master Bartholomew (des Roches) archdeacon of Winchester.* (12 December 1230 × 13 February 1231) (*Reg.St Osmund* i 388)

>Between the death of Bartholomew and the bp of Salisbury's presentation of master Luke des Roches to Bartholomew's former church of Burbage. Master Bartholomew may well have died at the papal curia (PRO SC1/6/96). Bp des Roches was in Italy or France during the months in question.

32. *Presentation of Peter de Chanceaux to the church of Adderbury.* (May 1231 × 1232) (*Rot.Hugh Welles* ii 39)

>Dated to the 23rd year of bp Hugh of Lincoln. Peter was admitted via master Ralph de Appelb(y) (for whom see above actum no. 18) following dispensation from pope Gregory IX, granted in reward for his service to bp des Roches overseas (*Reg.Greg.IX* no. 638, dated 6 May 1231).

33. *Safe conducts to be issued by P(eter) bishop of Winchester, H(ubert) de Burgh, S(tephen) of Seagrave and R(alph) fitz Nicholas to nuncios of L(lywelyn) prince of Aberffraw, coming to Shrewsbury on 23 May 1232 to discuss terms with king Henry (III)* (c.3 May 1232) (*Pat.R.1225–32* 472)

> From the context it is unclear whether the safe conducts were to be made in writing, or by the guarantors personally conducting the Welsh envoys to their meeting.

34. *Tithe settlement by (William) prior of St Bartholomew's London and the dean of St Mary Arches, acting as papal judges delegate in pursuit of a mandate of pope Gregory (IX) to the said prior and dean, and the dean of St Martin's, London, following complaints by (Luke des Roches) archdeacon of Surrey against the prior and convent of Mottisfont and others of the dioceses of Salisbury, Winchester and London concerning tithe, dated at Perugia, 26 January 1232. The judges determine a settlement of tithes in Burbage between Luke archdeacon of Surrey, acting via master Ralph of York, his attorney, and the prior and canons of Mottisfont, in the land in Burbage granted to Mottisfont priory by William Brewer and the land of Penceley* (Pennardeshull') *held by G(eoffrey?) Sturmy. Sealed as a cyrograph by (Peter) bishop of Winchester and the dean and chapter of Salisbury.* (c.1232) (Winchester, Hampshire Record Office ms. 13M63/1 (Mottisfont cartulary I) fo.45v)

35. *Cyrograph between the abbot and convent of Reading and the friars minor, by which the abbot and convent grant the friars land in the culture, called* Vastern, *next to the high road, so that the friars may build and dwell there; the friars undertaking to respect the rights of the abbey with regard to burials, offerings and legacies. The king stands guarantor. One part of the cyrograph, remaining with the abbot and convent, is sealed with the common seal of the friars minor in England and the seals of the king, the archbishop of York and the bishops of Winchester, Coventry and Worcester.* (Reading, 14 July 12[33]) (*Cal.Chart.R. 1226–57* 187)

> For the background to this award and the origins of the Franciscan house at Reading see *Reading Cartularies* nos.1024–5. According to the Dunstable annalist, the establishment of the Reading house followed the dispatch of a papal mandate to des Roches, ordering him to protect the Franciscans against the enemies of Christ (*AM* iii 134).

36. *Letters requested by pope Gregory IX informing the pope of obedience to papal instructions that the bishop secure the release from captivity of Hubert de Burgh and (Margaret) his wife.* (November 1233) (*Reg.Greg.IX* no. 1562)

> Solicited in papal letters of 17 October 1233. Probably never sent. Not only was des Roches opposed to the release of de Burgh, but de Burgh's escape from Devizes late in October 1233 rendered the papal instructions redundant.

37. *Guarantees for the repayment of 800 marks to be loaned by merchants to John de Caen, Master Simon the Norman and Peter Saracen, appointed king's proctors in the Roman curia.* (c. November 1233) (*Cal.Lib.R.1226–40* 267–8)

Mentioned in royal letters of May 1237, when des Roches loans were offset against his obligations to the crown still owing from the tax of a fortieth. For the date of the Roman mission for which the guarantees were made, see *CPR 1232–47* 32–3.

38. *Agreement between the abbot of Waverley and R(obert de Ferles) rector of Witley, made before P(eter) bishop of Winchester, dated 1233.* (1233 × 25 March 1234) (*WC* p.244)

Mentioned as part of one of the lost quires of the Winchester cartulary, apparently confirmed by bp Rigaud de Assier in 1321.

39. *Excommunication of Richard Siward and others, following the seizure of horses belonging to the bishop and the prior of St Swithun's; with an interdict placed on the church and city of Winchester.* (Winchester, 2 July 1234) (*AM* ii (Winchester) 86–7)

The attack upon des Roches' goods followed the bp's dismissal from court and the pursuit by Siward and his supporters of the bp's kinsman, Peter de Rivallis. The attackers did penance on the morning after the excommunication, and were absolved. On the following day the church and city of Winchester were reconciled, implying that the citizens may have seized the opportunity to carry out reprisals of their own against St Swithun's.

40. *Presentation of master William de St Mère Église to the church of Witney.* (June × November 1236) (*Rot.Grosseteste* 450).

Dated to the second year of bp Grosseteste. William was admitted via a proctor named John Parcsiens' and instituted on 28 November 1236 (*Ibid.* 451–2).

41. *Presentation of Peter Russinol (II) to the church of Adderbury.* (June 1236 × 1 June 1237) (*Rot. Grosseteste* 449).

Occured in the second year of bp Grosseteste. Peter (who is not to be confused with an earlier namesake) was admitted via a proctor named Thomas of Winchester, and instituted on 1 June 1237 (*Ibid.* 452). For Peter, see appendix 4 no. 20.

42. *Grant to Michael May, the bishop's butler, of land at West Wycombe* (1206 × 1238, ?1231 × 1238) (*Cl.R.1237–42* 86)

The land was restored in August 1238, following its temporary seizure by the royal keepers of the vacancy. Michael *pincerna* occurs on the Winchester pipe rolls after 1231, supervising the delivery of wine across the bp's estates (Mss. 27DR mm.2,3,7; 28DR mm.3,5; 32DR m.11). He may well have succeeded David *pincerna* who occurs, similarly engaged in the transport of the bishop's wines, between 1223 and 1227 (Mss. 19DR mm.1d,3d,4d,9d; 20DR m.10; 21DR m.10; 22DR mm.5d,8,11d). After 1240 Michael appears to have transferred briefly to the king's service in a similar capacity, as purveyor of wines (*Cl.R.1237–42* 202).

43. *Presentation of R. de Watervill to the church of Patney.* (c.1231 × 1238, ?1238) (*CPR 1232–47* 271)

Upheld by the king in February 1242, despite an earlier attempt to intrude a royal presentee, before the king was aware of des Roches' presentation.

44. *Notification by Luke (des Roches) archdeacon of Surrey and papal subdeacon, that the master and brethren of the hospital of St Thomas, Southwark, have granted him life tenancy in a hall, chapel and stable within the hospital's grounds. Sealed by Luke and with the seal of P(eter) bishop of Winchester.* (May 1238) (Bodleian ms. Rawlinson D763 (St Thomas' cartulary) pp. 8–9; BL ms. Stowe 942 (Ibid.) fo. 3r–v; whence *St Thomas' Cartulary* no. 8)

45. *Grant to William de Cantiloupe, with the assent of Robert Marmiun and Philip Marmiun his son, of all the lands and fees of the said Robert Marmiun in England, with the marriage of the said Philip for a term of three years.* (July 1233 × June 1238, probably made posthumously by the bp's executors June × September 1238) (*Cal.Chart.R.1226–57* 248)

Confirmed by the king in November 1239 as a grant by bishop Peter. Follows the conveyance by Robert Marmiun, son of Robert Marmiun, to the bp, of all his lands and fees together with the custody and marriage of Philip Marmiun, his son, to hold for a term of seven years from the feast of St M(ichael?) 1233, with reversion thereafter to Robert, confirmed by the king in July 1233 (*Ibid.* 186). Before 1221 des Roches had held the wardship of Robert Marmiun, Philip's father. Probably related to the complicated attempt by the Marmiun family to retain their lands in both Normandy and England, for which see Sanders *Baronies* 145. Des Roches appears to have retained possession of Tamworth castle and the rest of the Marmiun barony, up to the time of his death (*Cl.R.1237–42* 62). The present award is probably that confirmed in September 1238 by which William Cantiloupe fined with the bp's executors for possession of Tamworth and the Marmiun lands (*CPR 1232–47* 231)

46. *Appointment of Gilbert le Pestur as usher in the conventual kitchen of the priory of St Swithun, Winchester.* (1206 × 1238) (*Cal.Inq.Misc.*, i no. 261)

47. *Appointment of Ranulph de Mandatis as chief serjeant of the almonry of the priory of St Swithun, Winchester.* (1206 × 1238) (*Cal.Inq.Misc.* i no. 261)

48. *Appointment of Geoffrey le Lavander as washer to the priory of St Swithun, Winchester.* (1206 × 1238) (*Cal.Inq.Misc.* i no. 261)

49. *Appointment of Richard Pruz as chief cook in the conventual kitchen of the priory of St Swithun, Winchester.* (1206 × 1238, ? 1223 × 1238) (*Cal.Inq.Misc.* i no. 261)

A cook known as (Richard) Pruz or Probo appears regularly in the bp's service after Michaelmas 1223 (Mss. 19DR m.11; 27DR mm.2,5d; 28DR mm.3,5,9d,14; 32DR mm.1,2d, 4d,6, 7,8,9d,11,11d,14d). For the possibility that this and the preceding three appointments were made in writing see above actum no. 41.

APPENDIX 3

THE BISHOP'S ITINERARY

As might be expected of a bp pre-occupied with affairs of state, it has been possible to assemble an exceptionally full itinerary for des Roches. The vast majority of the references given below involve attestations to royal charters or letters, most often given in the form *Testibus* or *Teste*, but occasionally, particulary in the 1230s, in the form *coram*. Such references should be treated in the light of reservations expressed by C.R. Cheney in *EEA* iii 308. There are many anomalies in the enrolment of royal letters and charters, and our understanding of chancery operations is at present far from complete. Scribal error, the use of subordinate seals, and the time-lapse between issue, sealing and enrolment, mean that place/date clauses in the chancery rolls, and even in original engrossments, cannot be taken at face value. Enrolments are often scantily dated, lengthy series of letters being entered under the formula *ut supra*, implying that all were issued under a single place/date clause, when often it seems unlikely that this was so. As will be apparent below, enrolments may suggest that des Roches was present on the same day at two or more widely separated locations, or that the court had travelled an impossible distance in the course of a single day. Where such problems arise I have attempted to indicate the more likely location, unlikely or impossible locations being given in brackets. Beyond those formulae such as *testibus* or *coram* which carry a reasonable degree of probability that the bp was present on any given occasion, a large number of royal letters were issued under the clause *per* des Roches, implying that des Roches was the authority who warranted the issue of letters, without necessarily being present at their issue. Warranty could in theory be by the bp's own letters or by verbal instructions sent at a considerable distance from the court, although in practice, as is apparent below, des Roches was generally in the vicinity of the court whenever royal letters were issued under his warranty. Dates for which the formula *per* des Roches is the only key to the bp's whereabouts are given in round brackets below. Likewise, uncertainties of time and place are distinguished by question marks or brackets. Dates joined together by hyphens imply that the bp was present on every intervening day between the outside dating limits, as shown by the accompanying references. Relatively few entries are provided from the bishop's own account rolls, although the accounts themselves frequently note

144   APPENDIX 3

the cost of entertainment supplied to the bishop without giving the precise dates of his visits. For commentary on this, and on the misleading impression given by the present itinerary that des Roches was absent from his diocese for months or years on end, see above pp. xxxi–xxxiii

1205
Feb. 5	Winchester (as bishop elect)	*RLC* i 18b,19
*c.*April 1205–Jan. 1206	Rome (election confirmed 21 June)	*PL* ccxv cols.671–3
September 25	St Peter's Rome (consecrated)	*Flores Hist.* ii 129; *AM* ii 79
October 15	Rome	Actum no.81

1206
March 20	?Windsor	*Rot.Ob.* 370
March 21	Mortlake	*RLP* 60b
March 23,25	Lambeth	*Foedera* 94; *RLP* 60b; *RLC* i 67b
March 26	Winchester (enthroned)	*AM* ii 79 (for 12 ides read 7 kalends)
April 4	Dover	*Rot.Chart.* 164b
April 4,5	Romney	*Rot.Chart.* 163b, 164b
April 23	Dogmersfield	*HMC Wells* i 63; ii 554; iii 33
May 1	Windsor	*RLP* 63
May 9	Southwick	*RLP* 64
May 24	Portchester	*HMC Wells* i 528
May 27,(28)	Bishopstoke	*RLC* i 71b, 84
May 28,(29),30 June 1	Yarmouth	BL Cotton Ch. viii 25; *RLP* 65b, 66; *RLC* i 72b
June (8),9	La Rochelle	*RLC* i 72b; *RLP* 66
June 14	Niort	*RLP* 66
	St-Maixent-l'École	*RLP* 66; *RLC* i 72b
June 15	Châtel-Bruges	*RLP* 66
June 16	St-Maixent-l'École	*RLP* 66
June 19	Niort	*RLP* 67b–68
June 25	St-Jean-d'Angély	*RLP* 66, 66b; *RLC* i 73
June 27	Jarnac	*RLP* 66b
July 18,20, Aug.2	Bourg-Charente	*RLC* i 73, 73b
Aug. 13	St-Émilion	*RLP* 66b
Aug. 25	Niort	*RLP* 67
Aug. 30	Clisson	*RLP* 67,72b
Sept. 9,13	Angers	*RLP* 67; *RLC* i 74
Sept. 16	'*Qulla Episcopi*'	*RLP* 67
Sept. 20	Angers	*RLP* 67
Sept. 23	Le Coudray-Macouard	*RLC* 74b
Sept. 24	St-Amand-sur-Sèvre	*RLP* 67b
Sept.30, Oct. 1	Bressuire	*RLP* 67b; *RLC* i 74b
Oct. 3, 8	Thouars	*Foedera* 95; *RLC* i 74b
Oct. 13	Niort	*RLC* i 74b

Oct. 26	Thouars	*Foedera* 95
Nov. 4-6	La Rochelle	*RLP* 67b-68b; *RLC* i 75
Nov. 8,9,14	St-Martin de Ré	*RLP* 68,68b; *RLC* i 75
Nov. 25	*Insula de Saim*	*RLP* 68
Dec. 26	Farnham	*RLC* i 75b
Dec. 28	Guildford	*RLC* i 75b

1207

Jan. 3,5	Canterbury	*Foedera* 95; PRO C52/24 no.28; C52/26 no.8
Jan. 8,10	Lambeth	*RLC* i 76; *Foedera* 94
Jan. 12	Reading	*RLC* i 76
Jan. 13	Freemantle	*RLC* i 76
Jan. 16,18	Clarendon	*RLC* i 76b
Feb. 8	Marlborough	*RLC* i 77
Feb. 10	Rockingham	*Foedera* 95
Feb. 12	Woodstock	*RLC* i 77b
Feb. 20,22	Rockingham	*RLP* 69; *RLC* i 78; PRO C47/12/6 no.3
Feb. 25	Peterborough	*RLC* i 79
March 3	Geddington	*HMC Wells* i 305-306; *CAR* no. 289
March 10	(Oxford)	*CPR 1348-50* 487
	?Hallingbury	*Foedera* 96; *CAR* no. 300
March 13	Lambeth	*CAR* nos. 224, 284
	Westminster	*RLC* i 103
March 15	Farnham	*RLC* i 79b
March 20,22	Winchester	*RLP* 69b; *RLC* i 79b
March 23,25	Clarendon	*RLP* 70; *RLC* i 80; PRO C47/12/6 no.4
March 27	Cranborne	*RLC* i 80
March 28	Bere Regis	*RLC* i 80
March 31, April 1	Exeter	*RLP* 70,70b; *RLC* i 80b; PRO C52/24 no.25
April 10	Freemantle	*RLC* i 81
April 14	Windsor	*RLP* 70b; *RLC* i 81,81b; PRO SC1/1/4
April 15,21,27,30, May 1-4,7,8	Lambeth	*RLP* 71,71b; *RLC* i 81b-82b; PRO E401/3a m.2d; C52/24 no.31; BL Add.Ch.44694
May 9	Westminster	*RLC* i 82b
May (28)	Bishopstoke	*RLC* i 84
June 16	Ludgershall	*Rot.Chart.* 167,170b
June 19,20	Winchester	*Rot.Chart.* 166b,167; *RLC* i 86
June 29	Lambeth	*Rot.Chart.* 182
June 30	Westminster	*RLP* 73b; PRO E401/3a m.2d
July 2	Winchester	*Rot.Chart.* 167
July 6	Southampton	*RLC* i 86b
July 7,8	Winchester	*RLC* i 87; PRO E401/3a m.2d
July 13,14	Lambeth	*RLC* i 87b-88b
July 14	Westminster	*RLC* i 87b; PRO E401/3a m.2d
July (19)	Devizes	*RLC* i 88
Aug. 4	Woolley	*RLP* 75
Aug. 5,7,8	Woodstock	*Rot.Chart.* 168,171; *RLP* 75; *RLC* i 89b
Aug. 28,30	Winchester	*Rot.Chart.* 169; *RLC* i 91
Aug. 30	Clarendon	*RLP* 75b; *RLC* i 98
Sept. 28	Devizes	*Rot.Chart.* 170b
Oct. 1,3	Winchester	*RLC* i 92b,93
Oct. 5	Lambeth	*Rot.Chart.* 171

Oct. 7,(8)	Westminster	*RLP* 76; *RLC* i 93
Oct. 18,19	Winchester	*Rot.Chart.* 171b; *RLC* i 94,94b; *Reg.Walter Gray* 50
Oct. (28),29	Westminster	*RLC* i 95
Dec. 9	Clarendon	*Foedera* 99; *RLP* 77b; *RLC* i 98
Dec. 15	Egbury	*Rot.Chart.* 173b; *RLP* 77b
Dec. 28	Guildford	*RLC* i 99
Dec. 28,29,31	Farnham	*RLC* i 99b,100

1208

Jan. 1,2	Winchester	*Rot.Chart.* 174; *RLP* 78; *RLC* i 99b
Jan. 3	Ashley Salisbury	*Rot.Chart.* 205b *RLC* i 100
Jan. 4	Salisbury	*RLP* 78b
Jan. 6	Burbage	*Rot.Chart.* 174b
Jan. 22	Lambeth	*RLC* i 100b,101
Jan. 23	Westminster	*RLC* i 101
Jan. 25	Guildford	*Rot.Chart.* 175; *RLC* i 101
c.Jan. × Feb.	?France	*RLC* i 101
Feb. 18,19	Havering	*Rot.Chart.* 176b; *RLP* 79; *RLC* i 102b
Feb. (23)	Lambeth	*RLC* i 103b
Feb. 23,24,26, March 12	Winchester	*Rot.Chart.* 175,175b; *RLP* 79b; *RLC* 104, 104b, 105b; *Reading Cartularies* no.50
March 12	Ludgershall	*Rot.Chart.* 175b
March 15,17,18	Marlborough	*Foedera* 100; *Rot.Chart.* 175b–176b; *RLC* i 105b,106
March 19,20,22	Clarendon	*Rot.Chart.* 176b,178,180; *RLC* i 106b: PRO ms. PRO 31/8/104a no.192
March 23	Southampton	*Rot.Chart.* 178b
March 26	Portchester	*Rot.Chart.* 176b
March 28	Marlborough	*Rot.Chart.* 176
April (2)	Bishop's Waltham	*RLC* i 108
April 2,4	Waverley	*RLC* i 108,108b
April 8,9	Ludgershall	*RLP* 81; *RLC* i 110b
April 10	Marlborough	*RLC* i 111
April 12,13	Woodstock	*RLC* i 111,111b
April 15	Northampton	*RLC* i 112; *Memoranda Roll 10 John* 34
April 22	Clarendon	*RLP* 82
May 4,6,8,18	Lambeth	*Rot.Chart.* 177b–8,179b; *RLC* i 114b; *Beauchamp Cart.* no. 343
May 7	Southwark	*Actum* no.34
May 19	Westminster	*RLP* 83b
May 28	Southampton Portchester	*RLP* 84 *Rot.Chart.* 178b
June 10	Winchester	*Rot.Chart.* 179b
July 14	Windsor	*RLP* 85
July 16	Exchequer	*Memoranda 10 John* 64
Sept. 13	Calstone Wellington	*Rot.Chart.* 182b, 185b
Sept. 21,23	Taunton	*Foedera* 102; *Rot.Chart.* 182b; *RLP* 86b; *PBK* iv no. 3416
Sept. 31	Guildford	*RLP* 86b
Oct. 3	Tewkesbury	*RLP* 86b

Oct. 8,9	Shrewsbury	*Foedera* 101; *RLP* 87
Oct. 16	Witney	*PR4DR* 18
Oct. 20	?Oxford	*RLP* 87
Oct. 24, 25	Westminster	*Rot.Chart.* 183; *RLP* 87
Nov. 6	Downton	*PR4DR* 23
Nov. 8	Clarendon	*RLP* 87
Nov. 11	Downton	*PR4DR* 23
Nov. 11,12	Christchurch	*Rot.Chart.* 183b–184; *RLP* 87b; *Reg. Antiq. Lincs.* i no.204
Nov. 14	Gillingham	*Rot.Chart.* 183b
Nov. 30	Rockingham	*Rot.Chart.* 183b; *Beauchamp Cart.* nos.288–9
Dec. 3	Nottingham	*RLP* 87b
Dec. 14	Burton	*RLP* 88b
	(Gillingham)	*Rot.Chart.* 184b (?*recte* 14 Nov. 1208 or 14 Dec. 1207)
Dec. 16	Haywood	*RLP* 88
Dec. 24	Bristol	*Foedera* 102; *RLP* 88
Dec. 29	Ludgershall	*Rot.Chart.* 184

1209

Jan. 1	Winchester	*RLP* 91b
Jan. 5	Marwell	*PR4DR* 5
c.Jan. 21,22	Witney	*PR4DR* 18
Feb. (3)	Worcester	*RLP* 89
Feb. 4	Prestbury	*Rot.Chart.* 185b
Feb. 8	Ludgershall	*Rot.Chart.* 184; *Reg.Hamonis Hethe* i 36
Feb. 16	Lambeth	*Rot.Chart.* 184b
March 5	Reading	*Rot.Chart.* 184b
March 13	Witney	*RLP* 88
March 16,17	Woodstock	*RLP* 89b
March 23,24	London	*Foedera* 103; *RLP* 90,92
March 29	Bishop's Waltham	*PR4DR* 3,5
April 1	Adderbury	*PR4DR* 59
April 7	Nottingham	*RLP* 91
April 12	Doncaster	*Rot.Chart.* 185
April 14	Tadcaster	*Rot.Chart.* 185–185b
May 10	Bristol	*Misae Roll 11 John* 109
May 14	Bath	*Ibid.* 109
May 16,18	Marlborough	*Ibid.* 110; *PR4DR* 9
May 22	Winchester	*Misae Roll 11 John* 111–12
	Bitterne	*PR4DR* 3
May 24,25	Portchester	*Misae Roll 11 John* 112; *Mon.Exoniensis* 360 (misread as Dorchester: see Bodl. ms Top. Devon d.5 fo.15v)
May 28	Meon	Actum no.21
June 7	Bishop's Sutton	*PR4DR* 4–5
c.June 11	Bitterne	*PR4DR* 31
June 21	Westminster	*Misae Roll 11 John* 115–16
June 22	Chertsey	*Ibid.* 116
June 28,29	Downton	*Ibid.* 117; *PR4DR* 23
June 29	Cranborne	*Misae Roll 11 John* 117
July 2	Milton	*Ibid.* 118

148 APPENDIX 3

July 4	Yarlington	*CAR* no. 321
July 6	Wells	*Misae Roll 11 John* 119
July 7,8	Bristol	Ibid. 119; *God's House Cart.* no.140; *PBK* iii p.cclxxi
July 8	Gloucester	*PBK* iii p.cclxxi; *Norfolk Fines* nos.239–40
July 10	Horton	*Misae Roll 11 John* 119
July 11,13	Gloucester	*Ibid.* 120
July 14	Hanley Castle	*Ibid.* 120
July 16	Tewkesbury	*Ibid.* 121
July 17	Evesham	*Ibid.* 122
	Kenilworth	*Ibid.* 122
July 20,21	Adderbury	*PR4DR* 59
July 24	Bishop's Waltham	*PR4DR* 3
c.July 26–Aug.9	Dover	*Misae Roll 11 John* 123; *Gervase* ii pp.c–civ
August 19	Lound	*Misae Roll 11 John* 128
August 22,?23	Nottingham	*Ibid.* 128–9; *Gervase* ii p.ciii
August 29, c.Sept.1	Witney	*PR4DR* 18
Sept. 1	Adderbury	*Ibid.* 59
Sept. 7–10	Bishop's Waltham	*Ibid.* 3
Sept. 13	Marlborough	*Misae Roll 11 John* 130
Sept. 15	Downton	*PR4DR* 23
c.Sept. 20	Bristol	*Misae Roll 11 John* 131
Sept. 26,29	Knoyle	*PR4DR* 23,76
Oct. 1	Freemantle	*Bradenstoke Cart.* no.554
c.Oct. 6	Dover	*Gervase* ii pp.cvi–vii, cxi–xii
Oct. 14	Brill	*Misae Roll 11 John* 134
Oct. 19	Woodstock	*Ibid.* 134

1210

March 14, May 1	St Bride's London	*Ibid.* 156; *CPR 1396–99* 385
May 4	Dover	*Misae Roll 11 John* 165
June 23	Winchester	BL ms. Add. 29436, fo.32r
Aug. 3	Westminster	*Reading Cartularies* no.51; *Memoranda 1 John* p. lxxv
Summer × Autumn	Wales	*AM* iii 32
Sept. 17	Bristol	*Cart. Gyseburne* ii 98
Oct.1	Bitterne	*PR6DR* 7
Oct. 3	Downton	*PR6DR* 35
Nov. 1	Melkesham	*CAR* no.239

1211

Feb. 5	Downton	*PR6DR* 35
Feb. 20	Fareham	*Ibid.* 110
Feb. 24	Bishop's Sutton	*Ibid.* 142
March 24	Newport (Pagnell)	BL ms. Sloane 986 fo.71r
April 1, May 17,29	Bishop's Sutton	*PR6DR* 142
Late August	?Northampton	Cheney, *Innocent III and England* 324
Nov. 1	Reading	BL ms. Harley 3640 fos. 128r, 152r
Dec. 18	Freemantle	*Monasticon* iii 247; *CAR* no. 212
Dec. 21,22	Winchester	Ms.7DR m.6
Dec. 23	Bishop's Sutton	Ms.7DR m.6

1212

Jan. 2	Woodstock	*CRR* vi 189
May 4	Lambeth	*Foedera* 105; *Rot.Chart.* 186,186b; *RLC* i 116; BL Add.Ch. 11235/9
May (9)	Winchester	*RLC* i 118
May 16,17	Lambeth	*Foedera* 105; *Rot.Chart.* 186b,191; *Misae Roll 14 John* 232
May 18–20	Tower of London	*Rot.Chart.* 186b; *CPR 1301-7* 362; *RLC* i 117b
May 27	'Meves'	Actum no.94; *Misae Roll 14 John* 232
May 29	Winchester (Westminster)	*Misae Roll 14 John* 232 *RLP* 92b (possibly 1211, or by mistake for Winchester)
May 30,31	Odiham	*Misae Roll 14 John* 232-3
June 2	Tower of London	*Ibid.* 233
June 4	Hertford	*Ibid.* 233
June (5),24	Westminster	*Rot.Chart.* 187; *RLP* 93b; *RLC* i 118b
June 27,28	Bishop's Sutton	Ms.7DR m.6
July 3	Westminster	*RLC* i 119
July 6	Nottingham	*RLC* i 119b
July 21	Woodstock	*RLC* i 120b
c.Aug.4	Dover	*Misae Roll 14 John* 237
Sept. 5	Durham	*Charters of Finchale* 47-8
Sept. 23	Havering	*Rot.Chart.* 188,188b; BL ms. Cotton Nero Evi fo.290r
Sept. 26	Woodham	*Rot.Chart.* 190
Oct. 2	Lambeth	*RLC* i 132b
Oct. 29,30	Southwark	*Rot.Chart.* 188b; *Misae Roll 14 John* 245: *RLC* i 126
Nov. 3	Windsor	*Rot.Chart.* 189; *RLC* i 126b
Nov. 7,30, Dec.2, 3,8	Westminster	*Rot.Chart.* 189b; *CAR* no. 227; *RLP* 95b; *RLC* i 127,127b
Dec. 19	Winchester	*Misae Roll 14 John* 249

1213

March 4	Farnham	*RLC* i 147
March (5)	Windsor	*RLP* 97b
March 8	Winchester	*RLP* 97b
March 29	New Temple London	*Foedera* 110; *Rot.Chart.* 190b,191
April 7,8	Portchester	*Misae Roll 14 John* 258
April 10	Ditton	*Ibid.* 258
	Bishop's Sutton	*Ibid.* 258
April (11),22	(Rochester)	*Rot.Chart.* 191; *RLP* 97b (?misdated)
April 22	Winchester	*Misae Roll 14 John* 259
April 24	Lewes	*Ibid.* 260
April 30	Winchelsea	*Ibid.* 261
May 24,25	Ewell	*Foedera* 112,113; *RLP* 99; *Reg.Antiq. Lincs.* i no.207; Lambeth Cart.Misc. XI 7
May 27	Wingham	*Rot.Chart.* 193; *RLP* 99
May 29	Dover	*RLP* 99,99b; *Rot.Chart.* 193b
May 29–31, June 1,3	Wingham	*Rot.Chart.* 192b,193; *RLP* 99b; *RLC* i 134b

APPENDIX 3

June (14)	Aldingbourne	*RLC* i 138
June 15	Bedhampton	*Rot.Ob.* 467; *RLC* i 138
June 18	Bishopstoke	*RLC* i 136
June 25,26	Canford	*RLC* i 137
June 26-8	Bere Regis	*Rot.Chart.* 193; *RLP* 101; *Rot.Ob.* 468; *RLC* i 143b; *Reg.Antiq. Lincs.* i no.208
June 27	Cranborne	*Rot.Ob.* 469-70
c.June 28	Dorchester	*Rot.Chart.* 193
June 29	Corfe	*RLC* i 144
June 30, July 1	Bishopstoke	*Rot.Ob.* 466,470-1; *RLC* i 144
July 4	Bere Regis	*RLP* 102
July 6-8	Cranborne	*Rot.Chart.* 193; *Rot.Ob.* 471-2
July 8	Gillingham	*Rot.Chart.* 194; *Rot.Ob.* 473
July 11	Clarendon	*Rot.Ob.* 475
July 14	Ashley	*Rot.Chart.* 193b
July 15	Corfe	*Ibid.* 194
July 18	Portchester	Lambeth Palace ms. 1212 fo.106v
July 20,21	Winchester	*Rot.Chart.* 194b; 'Annals of Southwark' 48
July 23	Studland	*RLC* i 146
July 24	Corfe	*Rot.Chart.* 194
July 27	Bindon	*Ibid.* 194
July 28	Dorchester	*Ibid.* 194b
July (29),30	Powerstock	*Ibid.* 194b; *RLP* 102b; *RLC* i 147
Aug. 3	Corfe	*RLC* i 150b
Aug. 15	Winchester	*RLP* 103; *RLC* i 148
Aug. 16,(?17)	Ludgershall	*Rot.Chart.* 196b,197; *RLP* 103,105; *Rot.Ob.* 491 (?misdated 17 Sept.)
Aug. 23	Freemantle	*Rot.Chart.* 194b
Sept. 1	Northampton	Norwich Charters i no.39
Sept. 8,(10)	York	*Rot.Chart.* 194b; *RLC* i 150
Sept. 17	Allerton (Ludgershall)	*Rot.Ob.* 492 *Ibid.* 491 (?mistake for 17 Aug.)
Sept. 19	Pontefract	*RLC* i 151b
Sept. 20	Tickhill	*Ibid.* 151b; *Rot.Ob.* 493
Oct. 3	St Paul's London	*Rot.Chart.* 195-195b; BL mss. Harley 1005 fo.66v; Cotton Ch. 58 viii 24
Oct. (4,5)	New Temple London	*RLP* 104b; *RLC* i 152b
Oct. 14,15	Westminster	*Rot.Ob.* 497-8
Oct. 18	Freemantle	*Ibid.* 498
Oct. 19	Winchester	*Ibid.* 499
Oct. 21	Clarendon	*Ibid.* 499; *RLC* i 155
Oct. 31, Nov. 1	Wallingford	*Rot.Chart.* 195; *Rot.Ob.*500-1; *RLC* i 155
Nov. 3,4	Woodstock	*Rot.Ob.* 501-3
Nov. 5	Witney	*Rot.Chart.* 195
Nov. 6	Woodstock	*Rot.Ob.* 503
Nov. 7	Witney	*Ibid.* 504
Nov. 13,19	Woodstock	*Ibid.* 505-6
Nov. 21	Tewkesbury	*Ibid.* 507
Nov. 23,24	Hanley Castle	*Ibid.* 508-9
Nov. 25	Hereford	*Rot.Chart.* 195b
Dec. 10,12,(14)	Reading	*Ibid.* 195b,196; *Rot.Ob.* 512

THE BISHOP'S ITINERARY

Dec. 15	Guildford	*Rot.Chart.* 195b
Dec. 17	St Albans	*Rot.Ob.* 513
Dec. 20	Waltham	*Ibid.* 513
Dec. 22	?Windsor	*RLP* 107; *RLC* i 158b
Dec. 28	Tower of London	*Rot.Chart.* 196b

1214

Jan. 3,(13)	Tower of London	*Ibid.* 196; *RLC* i 160b
Feb. 1	Portsmouth	*Foedera* 118
Feb. 8,9	Yarmouth	*Rot.Chart.* 200,206b (misdated 1215); *CAR* no. 362
Feb. 10	?Westminster	*Norfolk Fines* no. 563; *PBK* iii p.ccxc
Spring	?Exeter	*CRR* vii 158–9
	?Staffordshire	*PR 16 John* p. xxv
	?Shropshire	*Ibid.* p. xxv
March	Northampton	*Chronicle of Hugh* 48–50; *C. & S.* p.21 n.6
April 13,20,27	?Westminster	*PBK* iii p.ccxci
May 4,7,9	?Westminster	*PBK* iii p.ccxci
May 19	Chichester	*RLC* i 204b
May 23	Winchester	*RLC* i 205
May 26, June 1, 8,15,22,25	?Westminster	*PBK* iii p.ccxcii
June 23,26	Tower of London	*RLC* i 207
June 26	Chelmsford	*RLC* i 207
June 28	Bury St Edmunds	*Chronicle of Hugh* 83
July 1,5,8	?Westminster	*PBK* iii p.ccxcii
July 8	Kingston upon Thames	*RLC* i 208
July 10	Waverley	*AM* ii 282
July 13	Bishop's Sutton	Appendix i no.1
	Merdon	*RLC* i 208b
July 15,18	?Westminster	*PBK* iii p.ccxcii; *RLC* i 208b,209
July 20,21	Tower of London	*RLC* i 209
July 26,27	St Albans	*Chronicle of Hugh* 93; *RLC* i 209
July 28	Berkhamsted	*RLC* i 209,209b
Aug. 2	Winchester	*Ibid.* 209b
Aug. 5	Guildford	*Ibid.* 209b
Aug. 8–10	Westminster	*Ibid.* 209b,210
Aug. 13–14	Dover	*Ibid.* 210,210b
Aug. 18	Tower of London	*Ibid.* 210b
c.Aug. 22	London	*Ibid.* 213
Aug. 24	Westminster	*Ibid.* 211
Aug. 25	Tower of London	*Ibid.* 211
	Merton	*Ibid.* 211b
Aug. 26	Westminster	*Ibid.* 211
Sept. 3	Marwell	*Ibid.* 211b
Sept. 4	Beaulieu	*Ibid.* 211b
Sept. 5–7	Downton	*Ibid.* 211b,212
Sept. 9	Winchester	*Ibid.* 212
Sept. 10,11	Bishop's Waltham	*Ibid.* 212,212b
Sept. 13	Winchester	*Ibid.* 212b
Sept. 16	Merdon	*Ibid.* 212b

APPENDIX 3

Sept. 19,25	Winchester	*Ibid.* 212b
Sept. 25	Tower of London	*Ibid.* 212b
Sept. 26	Chelmsford	*Ibid.* 212b
Sept. 29,30	Bury St Edmunds	*Chronicle of Hugh* 109–11
Oct. 5	Canterbury	*Cant.Professions* no.148
Oct. 7	Treasury	*Rot.Ob.* 532
Oct. 9	Westminster	*RLC* i 213 (mistranscribed as 20 Oct.)
c.Oct. 17	Corfe	*Chronicle of Hugh* 113; *RLC* i 173
Oct. 22	Clarendon	*RLC* i 173b
Oct. 23,24	Freemantle	*RLP* 122b; *RLC* i 173
Oct. 26	Reading	*RLP* 123; *RLC* i 175
Oct. 28	Westminster	*RLP* 123
Oct. 29	Tower of London	*Rot.Chart.* 202
Nov. 2	Havering	*Ibid.* 202; *RLP* 122b
Nov. 3,4,?12,14	Westminster	*RLP* 122b; *RLC* i 174b–176b; *PBK* iii p.ccxciii
Nov. 21,22	New Temple London	*Rot.Chart.* 202–202b; *Reg.Antiq.Lincs.* i no.206; *RLP* 124; Lambeth ms. 1212 fos. 27r,105v–106r; *C.& S.* 40–1
Nov. 22	Westminster	*RLC* i 179
Nov. 25	Exchequer	*Cirencester Cart.* i no.39
Nov. 27	Treasury	*Rot.Ob.* 540
Nov. 28	Westminster	*RLC* i 180
Dec. (6)	Gillingham	*RLP* 125; *RLC* i 181
Dec. 27	Worcester	*Rot.Chart.* 206

1215

Jan. 9–11,14	New Temple London	*Rot.Chart.* 203,203b,205b; *HMC Wells* i 310, iii 154; *Domerham* 449; *Glastonbury Cart.* 90; *Cart. St Werburgh* i no.129
Jan. 14	Guildford	*Rot.Chart.* 203–203b
Jan. 15	New Temple London	*Ibid.* 204–204b; *Foedera* 126–7; Lambeth ms. 1212 fo.22v; Maidstone, Kent Record Office ms. D.& C. Rochester, Registrum Temporalium fo. 113v; BL ms. Cotton Claudius B iii fo. 128r–v
Jan. 17,18	Guildford	*Rot.Chart.* 203–205
Jan. 21	New Temple London	*Rot.Chart.* 204; *Reg.Antiq. Lincs.* i no.205
Jan. 22	Knepp Castle	*RLC* i 183
Jan. 20,28	Winchester	*Rot.Chart.* 209–9b; *Acta Langton* no.11; Lambeth ms. 1212 fos.103v–104r
Jan. 31	Christchurch	*Rot.Chart.* 205b; BL ms. Cotton Tiberius D vi part ii fo.9r (misdated 1 Jan. 1215)
March 1	Windsor	*Holm Cultram Cart.* no.217
March 2,8	?Westminster	*PBK* iii p.ccxciii
March (13,14)	Rochester	*RLC* i 189,190b
March 15,16	Tower of London	*RLC* i 189b,191; *HMC 14th Report* part iii 194
April 6,(7),10	Oxford	*Chronicle of Hugh* 165; *RLP* 132; *RLC* i 193
April 14	Wallingford	*RLP* 133
April 16	Oxford	CUL ms. D.& C. Peterborough 1 fo.67r
April 22	New Temple London	*Rot.Chart.* 206b
May 2	Wallingford	*RLC* i 204

THE BISHOP'S ITINERARY 153

May 3	?Westminster	*PBK* iii p.ccxciv
	?New Temple London	*Ibid.* iii p.ccxciv
May (3),5	Reading	*Rot.Chart.* 206b; *RLC* i 198b
May 7,9	New Temple London	*Rot.Chart.* 206b–207; BL ms. Harley 3640 fos. 121r,152r; *Cal.Chart.R. 1427–1516* 285
May 30	Odiham	*Rot.Chart.* 209b
c.June 14–22	Runnymede/ Windsor	*Foedera* 131–2 (the making of Magna Carta)
July (15)	Clarendon	*RLC* i 220b
July 16,18–20, 22	Oxford	*Rot.Chart.* 213b–214b (misdated 22 April 1215), 217b; *CAR* no.485; *RLP* 149; *Reg. Antiq.Lincs* i no.211; *Furness Coucher Bk* ii 573
Sept. 1,2,4,5 13,(17,18)	Dover	*Foedera* 137; *Rot.Chart.* 218b–219b; *RLP* 155b,182; *Magnum Reg. Album* no.688; Actum no.100
Nov. (5,6)	Rochester	*RLP* 158; *RLC* i 234b,235
c.Nov. 12	?Erith	*RLP* 158
1216		
Feb. 23	Bishop's Sutton	Ms.11DR m.7
Feb. 26	Bishop's Waltham	*Ibid.* m.2
c.April	France	*Coggeshall* 180–1
May 12,14	Folkestone	*RLP* 180
May 28,29	Winchester	*Rot.Chart.* 222; *AM* ii 82
June 13	Ebbesbourne	Ms.11DR m.4
June 19	Bere Regis	*Rot.Chart.* 223
June 23,27,30, July 4,14	Corfe	*Ibid.* 223,223b; *RLP* 188b,191
Oct. 18	Newark (death of King John)	*Hist.Maréchal*, l.15154
Oct. 28	Gloucester	*AM* ii 286, iv 409; *Walter of Coventry* ii 233
Nov. 12,(18), Dec. 2	Bristol	*Pat.R.1216–25* 4,9; Stubbs *Select Charters* 336
c.Dec. 21–8	Gloucester	*Pat.R.1216–25* 13
1217		
Jan. 7	Nottingham	*Pat.R.1216–25* 21
Jan. 16,(17)	Oxford	*Ibid.* 24; *RLC* i 296
c.Jan. 23, Feb. 5,7	Gloucester	*Pat.R.1216–25* 26,29,30
Feb. (22)	Oxford	*RLC* i 298b
Feb. (25)	Chertsey	*Pat.R.1216–25* 34
March (7)	Farnham	*Ibid.* 35
March (15,18), April (12,27)	Winchester	*Ibid.* 41,56,112; *RLC* i 300
May 17,18	Newark	*Hist.Maréchal* lines 16225–37
May 19,20	Torksey	*Ibid.* lines 16267–8
May 20,(23)	Lincoln	*Ibid.* lines 16467–16642; Wendover *Flores* iv 19; *Pat.R.1216–25* 64
May (26)	Woodstock	PRO E368/1 m.7d
c.June 6	Reading	*RLC* i 310
June (11)	Chertsey	*Pat.R.1216–25* 68
June (13)	Stanwell	*RLC* i 336

APPENDIX 3

Aug. 10,(11,12)	Oxford	*Pat.R.1216–25* 84,113; *RLC* i 319
Aug. 24	Sandwich	Paris *CM* iii 28
c.Sept. 22	Tower of London	'Annals of Southwark' 53
Oct. 22	Holy Trinity Aldgate	BL ms. Lansdowne 448 fo.5
c.Nov. 8	London	Actum no.107
Nov. 12	Exchequer	PRO E368/1 m.4
Nov. 20	Downton	Ms.13DR m.1
Nov. 23	Bishop's Sutton	*Ibid.* m.11d
Nov. 26	Westminster	*RLC* i 344

1218

Jan. 2	Downton	Ms.13DR m.1
Jan.23,(24)	Westminster	*Pat.R.1216–25* 209; *RLC* i 349b
March (14),?16	Worcester	*RLC* i 355,379
March (27)	Oxford	*Pat.R.1216–25* 144
April 18	Hampstead	*Pat.R.1216–25* 149; *RLC* i 359
May (2,8),9,10	Westminster	*Pat.R.1216–25* 151; *RLC* i 361,361b
May (20)	Woodstock	*RLC* i 319
May (23)	Westminster	PRO E368/1 m.7d
May (26)	Woodstock	*RLC* i 363; PRO E368/1 m.7d
May (29)	Tower of London	*RLC* i 363
May (30)	Westminster	PRO E368/1 m.7d
May (31)	Tower of London	*RLC* i 363
c.May 31, June 4	Downton	Ms.13DR m.1
June 7	Worcester	*AM* iv 409; *Walter of Coventry* ii 240
June (21)	Tower of London	*RLC* i 364
June (22)	Westminster	PRO E368/1 m.8
June (23),25	Tower of London	*Pat.R.1216–25* 160; *RLC* i 364
June 25	Westminster	PRO E368/1 m.7d
July (1)	Tower of London	*RLC* i 364b
July 6	Brightwell	Ms.13DR m.4d
July (23)	Nottingham	*RLC* i 366
Aug. 10	Winchester	Actum no.82
Aug. (19)	Wallingford	*Pat.R.1216–25* 166
Aug. (23)	Winchester	PRO C60/9 m.3
Sept. 1	East Meon	Ms.13DR m.2d
Sept. 3	Wargrave	*Ibid.* m.8
c.Oct. 24, Nov. 2–5,(6), 7–9,(10),12,13	Westminster	*Pat.R.1216–25* 178–81,207; *RLC* i 380b–83, 403b; *Foedera* 152; PRO C60/11 mm.11,12
Dec. 2	Southwark	*RLC* i 383; *E.Rot.Fin.* i 22; PRO C60/11 m.11
Dec. 5,6	St Paul's London	*RLC* i 383b; PRO C60/11 m.10
Dec. 7	Westminster	*RLC* i 383b; *Pat.R.1216–25* 182; *E.Rot.Fin.* i 23; PRO C60/11 m.10
Dec. 8,9,(10),11	Tower of London	*RLC* i 383b,404; *Pat.R.1216–25* 182; *E.Rot.Fin.* i 23; PRO C60/11 m.10
Dec. 11	Westminster	*RLC* i 384,384b; PRO C60/11 m.10
Dec. 12	Tower of London	*RLC* i 384
Dec.13–16	Westminster	*Ibid.* 384b–85,404b;*Pat.R.1216–25* 183–4; PRO C60/11 m.10; *E.Rot.Fin.* i 23
Dec. 25	Winchester	Paris *CM* iii 43

Dec. (31)	Marlborough	*Pat.R.1216–25* 184; PRO C60/11 m.10

1219

Jan. 10	Reading	*Pat.R.1216–25* 185; *RLC* i 385; PRO C60/11 m.10
Jan (14)	Stanwell	*RLC* i 385b
Jan. 16–20,(22), 23,24,26–31; Feb. 1,4,(7),9–11, (13),14,15, (16,17), 18,(19)	Westminster	*Pat.R.1216–25* 185–8,209–10; *RLC* i 385b–89b,404b,405; *Foedera* 154; *E.Rot.Fin.* i 26; PRO C60/11 mm.8–9; E159/2 m.5
March 4–6	Rochester	*Pat.R.1216–25* 188; *RLC* i 388b–89,405; PRO C60/11 m.7; *E.Rot.Fin.* i 28
March 7	Dartford	*RLC* i 405b; PRO C60/11 m.7
	Tower of London	*RLC* i 389; PRO C60/11 m.7
March 9	New Temple London	*RLC* i 389; PRO C60/11 m.7
March 10	Westminster	BL ms. Cotton Cleopatra C vii fo.93v
March 11	New Temple London	*Pat.R.1216–25* 189; *RLC* i 389; *E.Rot.Fin.* i 28; PRO C60/11 m.7
March (13,15)	Tower of London	*Pat.R.1216–25* 189; *RLC* i 389b
March 24	Reading	*RL* i no.xi
March 24, April 2,(3),4	Caversham	*Pat.R.1216–25* 190; *RLC* i 389b–90; PRO C60/11 m.7
April 10–13	Reading	*RLC* i 390, 405b; PRO C60/11 m.6–7
April 14	Oxford	*RLC* i 391b
April 15	Wallingford	*Pat.R.1216–25* 190
April 18,21,(22)	Oxford	*Ibid.* 191–2; *RLC* i 390–90b
April 23	Witney	*Pat.R.1216–25* 191; *RLC* i 390b–91b; PRO C60/11 m.6
April (27,28), 29,(30), May 1,2,(8),?16	Westminster	*Pat.R.1216–25* 192–3; *RLC* i 391–91b; PRO CP25(1) 262/13 no.7
May	Downton	Ms.14DR m.7d
May 25	Gloucester	*RLC* i 393
May 28	London	*Pat.R.1216–25* 194
June ?7,(9)	Westminster	*Ibid.* 194; *RLC* i 392; PRO C60/11 m.5
June 16	?Wallingford	*RLC* i 393
June 16,17,(18), 19,20,(21),22	Westminster	*Ibid.* 392b–94; PRO C60/11 m.5; E159/2 m.2d; CP25(1) 35/2 no.1; CP25(1) 196/3 no.36
June (24)	Oxford	PRO C60/11 m.5
June (25)	Gloucester	*Ibid.* m.5; *RLC* i 394b
June 26–29, July 1–3	Hereford	*Pat.R.1216–25* 195; *RLC* i 394b; PRO C60/11 mm. 4,4d,5
July 4,5	Gloucester	*Pat.R.1216–25* 196,212; *RLC* i 394b
July 6	Cirencester	*RLC* i 395
July (6),7	Witney	*Ibid.* 395
July 8	Oxford	*Ibid.* 395; *Pat.R.1216–25* 196
July 9	Reading	*RLC* i 395
July 16–20,(21),22, 24,25, Aug. 2–5	Westminster	*Ibid.* 395b–97b,406; *Pat.R.1216–25* 196–9,212; *E.Rot.Fin.* i 35; PRO 25, C60/11 m.4
Aug. 9,10	Southwark	*RLC* i 397b
Aug. 10	Lambeth	PRO C60/11 m.3

Aug. (17)	Northampton	*Pat.R.1216–25* 200
Aug. (18)	Oakham	*Ibid.* 200; *RLC* i 398
Aug. (20),21, (22),23	Lincoln	*Pat.R.1216–25* 200,206; *RLC* i 398b–99b
Aug. 24	Grantham	*Pat.R.1216–25* 201; *RLC* i 399–99b
Aug. 29	Ilsley	Ms.14DR m.6d
Sept. 8	Highclere	*Ibid.* m.4d
Sept. 16–24,27	New Temple London	*Foedera* 157; *Pat.R.1216–25* 202–6; *RLC* i 399b–401b; PRO C60/11 mm.2,3
Oct. 6,8,(10,13,17, 18,19,21,23–26, 29,30), Nov. ?3, (4),8,9,10,13,(15)	Westminster	*Pat.R.1216–25* 206,221–3; *RLC* i 401b–403,406b–8b; PRO C60/11 mm.1,2; C60/12 m.9; E159/3 m.1; CP25(1) 172/16 no.60
Nov. (19)	Oxford	*RLC* i 408b
Nov. 20	Westminster	*Pat.R.1216–25* 260
Dec. 12,(14),16,20	Winchester	*RLC* i 409–9b; PRO C60/12 m.8
Dec. 25	Marlborough	Paris *CM* iii 58
Dec. 28	Downton	PRO SC1/1/184; Ms.15DR mm.7d,8
Dec. 29,31	Salisbury	Actum no.112; *CACW* 2
1220		
Jan. (30), Feb. (1, 9,10,14,15), 16–20,23,25, March (4–7),8,(9–10)	Westminster	*Pat.R.1216–25* 229; *RLC* i 410b–14,435b; *DD* no.71; PRO C60/12 m.6
March 3	London	*Layettes* i no.1387; Bouquet *Recueil* xvii 772–4 (misdated 13 March)
April (12,15,16,18, 21,22,24–27)	Westminster	*Pat.R.1216–25* 230–1; *RLC* i 414b–17; PRO C60/12 m.5
May ?4,7	Shrewsbury	*Pat.R.1216–25* 260; *RLC* i 417b; *Foedera* 159
May 21	Westminster	PRO E401/3b m.1
June (12),?15, (16)	York	*Pat.R.1216–25* 234–5; *RLC* i 420b–21; *Foedera* 160
June (23)	Leicester	*Pat.R.1216–25* 238
June (26)	Northampton	*Ibid.* 239
June (26,28,29)	Rockingham	*Ibid.* 239; *RLC* i 422–22b
July (1)	Ware	*RLC* i 422
July (2)	Westminster	*RLC* i 422b
July 3	Southwark	*Ibid.* 422b
July (5)	Westminster	*Ibid.* 422b
July (7)	Canterbury	*Ibid.* 422b
July (28)	Westminster	*Ibid.* 425
?Aug.3	Wallingford	*Ibid.* 425b (given as 3 July)
Aug. (6,10,11)	Oxford	*Pat.R.1216–25* 246; *RLC* i 426b; PRO C60/12 m.3
Aug. 13	Abingdon	*RLC* i 426b
Aug. 30	Winchester	Actum no.119
Sept. (2)	Exeter	*RLC* i 428b
Sept. (16,17),18, 26,28	Winchester	*Pat.R.1216–25* 249,253; *RLC* i 430
Sept. 27	*Novus Locus*	Appendix 2 no.20
Oct. (5),6–14,16–19, 22–24,26,27,29, 30, Nov. (2,3), 4,5,7–9,12,13, (14,15,16),20, (22,23),24,25, Dec. 7,(8,10,12),14	Westminster	*Pat.R.1216–25* 255–7,263,268–70,272,274–5,305; *RLC* i 432–33b,437b,438b–44,473b; *E.Rot.Fin.* i 55; PRO C60/12 mm.1,2; C60/14 m.7; E368/3 m.3(2)

THE BISHOP'S ITINERARY

Dec. 25	Winchester	Paris *CM* iii 67

1221

Jan. (5,6)	Tower of London	*Pat.R.1216–25* 307; *RLC* i 445
Jan. 9	Westminster	*Pat.R.1216–25* 277
Jan. (14,21)	Tower of London	*RLC* i 445b–46,448
Jan. (24,27,28)	Westminster	*RLC* i 446–47,448
c.Feb. 18,19	Bytham	*Pat.R.1216–25* 284; *RLC* i 475
Feb. 26	Adderbury	Ms.16DR m.3
April 11	Fareham	*Ibid.* m.6d
c.April 16–July 16	Pilgrimage to Compostela	*AM* ii 84, iii 68, iv 414; *Walter of Coventry* ii 259–60
c.Early July	Dover, Canterbury	*Trivet* 209
July 19,(20),21, 24,(28)	Westminster	*Pat.R.1216–25* 311; *RLC* i 465b–66b; *Flores Hist.* ii 172–3
Sept. 19	Winchester	*AM* ii 295; *Walter of Coventry* ii 250
Sept. 28, Oct. 6,8, Nov. 3,21(22), 23, Dec. 8	Westminster	*Pat.R.1216–25* 322; *RLC* i 470b,472,478,482–82b; PRO C60/17 mm.7,8
Dec. 30	Newbury	*RLC* i 485

1222

Jan. (26,28), March (17)	Westminster Winchester	*RLC* i 486 *RLC* i 490b
April (29,30), May (1,2),7,(9,10,13, 16,17) June (3, 12,13,14),19,(21, 24), July (15)	Westminster	*Pat.R.1216–25* 331–3; *RLC* i 490b,494–94b,495b–97,498b,499b–501,504
July (29),30	Gloucester	*Pat.R.1216–25* 338; *RLC* i 517
Aug. 24	Havering	*Pat.R.1216–25* 339; Lambeth ms. Register of Abp Warham fo.134r
Sept.18, Oct. 7	Westminster	Paris *CM* iii 74; PRO E368/4 m.5
Oct. 25	Merton	BL ms. Cotton Cleopatra C vii fo.178r
Oct. (25),26, Nov. 3,(8,19,21), Dec. (2,7),9	Westminster	*RLC* i 515b,518b,522b,524b,525b,526b; PRO C60/18 m.10

1223

Jan. 15	Winchester	*RLC* i 530b
Jan. 18	Guildford	*Ibid.* 530b
Feb. (5,9,12,22), March 3,5, April 5, May (23)	Westminster	*Ibid.* 532b, 535b,540,547b; *Pat.R.1216–25* 365; PRO E368/5 m.13(1)d
Aug. (3)	Tower of London	*Pat.R. 1216–25* 380
Aug. 19	Winchester	*RLC* i 561b
Oct. 8	Montgomery	*Pat.R. 1216–25* 411

APPENDIX 3

Oct. 27	Westminster	*RLC* i 573
Dec. (8),*c*.12	London	*AM* iii 83–4; *RLC* i 578
Dec. 25	?Leicester	*AM* iii 84
Dec. 29	?Northampton	*AM* iii 84

1224

c.June/July	Bedford	*AM* iii 87; CUL ms. D.& C. Peterborough 1 fo.270r

1225

Feb. 11	Westminster	Stubbs *Select Charters* 350–1
May 16	London	*Pat.R. 1216–25* 543; *Foreign Accounts* 57
May 18	Witney	Ms.20DR m.14
July 14–Aug. 16	Taunton	*Ibid.* m.3d
c.August 6	Portsmouth	Actum no.127
c.Aug. 29	Bishop's Waltham	Ms.20DR m.7d
c.Dec. 25 × 31	Bitterne	Ms.21DR m.13

1226

Jan. 30,31	Waverley	*AM* ii 301
May 7,12	Westminster	*Pat.R.1225–32* 76,78; *RLC* ii 110b
c.Aug. 29	Wargrave	Ms.21DR m.5d
Dec. 18	Westminster	*Foedera* 183; *Pat.R. 1225–32* 98–100,102
Dec. 20	London	Actum no.128

1227

Jan. 20,30, Feb. 1,3–5,9–14,16–19, March 2	Westminster	PRO C53/18 mm.22,26–8,33,35–6; *HMC 9th Report* 353; *HMC Rutland* iv 83; *Tropenell Cartulary* i 208; *Dunstable Cartulary* no.955D; *CAR* 377; BL Harley Chs. 58H38–40; PRO E315/61 fo.5r
March 7	Marwell	Actum no.31
March 16–18, 20,22–26,28, April 5,6,25, 26,28, May 1	Westminster	PRO C53/18 mm.7–19; *Chichester Cart.* 146,148; *Records of St Barth.* i 484; BL mss. Cotton Ch.VIII.8; Claudius D xiii fo.107r; Vitellius D ix fo.26v; Bodl. ms. Rawlinson B461 fo.38r
May 9	Canterbury	*Flores Hist.* ii 190
May 24, June 8,10,20	Westminster	PRO C53/18 mm.1–3; *Chichester Cart.* 142
c.Mid-August	Brindisi	Start of the August passage to the Holy Land
c.October 28	Acre	Actum no.129

1228–9

	Joppa,Ascalon, Sidon, Acre	*AM* iii 126

1229

March 18,19	Jerusalem	*HDF* iii 109
c.April–May	Acre	Paris *CM* iii 185
c.May	?Cyprus	PRO SC1/6/148
c.July 22	Rome	*AM* ii 308 (although the papal court was in exile at Perugia, being re-admitted to Rome only late in February 1230, whereafter it travelled to Grottaferrata and Anagni; *Reg.Greg.IX* iii 271)
October	San Germano	*San Germano* vii part 2 p.163

1230

Aug. 28	St Justa nr. Ceprano	Actum no.130

1231

c.early Summer	France	Reg.Greg.IX i no.1311; AM iii 127
c.July 25	England	AM ii 310
Aug. 1	Winchester	AM ii 86
Aug. 10,28, Sept. 20	Painscastle	PRO C53/25 mm.3,4
Oct. 26	Westminster	PRO C53/25 mm.1-2
Dec. 20	?Marwell	Actum no. 75 (more likely 1232)
Dec. 22,25	Winchester	PRO C53/26 m.19; AM iii 127; Paris CM iii 211
Dec. 28	Ashley	PRO C53/26 m.19

1232

Jan. 8	Windsor	PRO C53/6 m.17
Jan. 10	Kempton	Ibid. m.17
Jan. 14,16,17,20	Lambeth	Ibid. mm. 16,18
Jan. 20,21	Westminster	PRO C53/26 mm.15,16; BL mss. Cotton Vespasian E xx fo.41r; Harley Ch.58H43
Jan. 23	Lambeth	PRO C53/26 m.15
Jan. 24	Westminster	Ibid. m.15
Jan. 27	Havering	Ibid. m. 15
Jan. 27-29, Feb. 4-6	Westminster	Ibid. m. 14; E159/11 m.12
Feb. 21	?Chippenham	BL ms. Cotton Nero C ix fo.50r (more likely 1234)
April 4	Winchester	PRO C53/26 m.12
May 5,10	Westminster	PRO C53/26 mm.11,12
May 22	Bridgnorth	Ibid. m.10; Worcester Cartulary no.328
May 27,28	Shrewsbury	PRO C53/26 m.10
June 11	Oddington	Ibid. m.9
?June 11,12	Highclere	Ms.27DR m. 11d
June 15	Woodstock	PRO C53/26 m.7
July 15,16	Lambeth	Ibid. mm.3,4; BL ms. Lansdowne 402 fo.107r
July 17,18	Westminster	PRO C53/26 m.3
July 20,21	Lambeth	Ibid. mm.3,4
July 28	Oxford	Ibid. m.3
	Woodstock	Ibid. m.3
Aug. 2	Worcester	Ibid. m.2
Aug. 11	Wenlock	Ibid. m.3; Haughmond Cart. no.424
Sept. (8)	Canterbury	PRO C60/31 m.3
c.Sept. 20	Southwark	Paris CM iii 224
Sept. 20,23	Lambeth	PRO C53/26 mm.1,2; BL ms. Egerton 3316 fo.92v
Sept. 23, Oct. 2	Westminster	PRO C53/26 m.1; C60/31 m.2
Oct. 9,12	Abingdon	Pat.R.1225-32 505; PRO C53/26 m.1
Oct. 25	Reading	PRO C53/26 m.1; Chart. Dieulacres 364
Oct. 29	Wycombe	PRO C53/27 m.15
Nov. 3	Lambeth	PRO C60/32 m.10
Nov. 4,7	Westminster	PRO C53/27 mm.12,15; Cal.Chart.R. 1327-41 88
Nov. 12	Lambeth	PRO C53/27 m.15
Nov. 13-15	Westminster	Cl.R.1231-4 167,287; E.Rot.Fin. i 234; PRO C60/32 m.10

APPENDIX 3

Nov. 18,21	Northampton	PRO C53/27 m.15; C60/32 m.10
Dec. (14)	Hereford	Cl.R.1231-4 290
Dec. 20	Marwell	Actum no.75 (possibly 1231)

1233

Jan. 8,10–13	Woodstock	Cl.R.1231-4 180,182; Cal.Lib.R.1226-40 194; PRO C53/27 m.12; Actum no.
Jan. 18–20,22–24	Westminster	CPR 1232-47 8; Cl.R.1231-4 182-3; Cal.Lib.R.1226-40 196; PRO C60/32 m.8; C66/43 m.7d
Jan. 27	Windsor	CPR 1232-47 9; Cal.Lib.R.1226-40 197
Jan. 29–31	Westminster	Cl.R. 1231-4 185-6,292; Cal.Lib.R. 1226-40 197; E.Rot.Fin. i 236; PRO C60/32 m.8
Jan. (30)	Windsor	Cl.R. 1231-4 186
Feb. 1,3,4,6–11,13–18,20,21	Westminster	CPR 1232-47 10-12; Cl.R.1231-4 186-7, 189-92; Cal.Lib.R. 1226-40 198,200-3; PRO C53/27 m.11; C60/33 mm.7,8; E159/12 m.10; E372/76 rot.4a; C66/43 m.7d
Feb. 21,22	Lambeth	Cl.R. 1231-4 192; Cal.Lib.R. 1226-40 200
Feb. 23–29, March 1,2	Westminster	CPR 1232-47 12,13; Cl.R. 1231-4 192-7; Cal.Lib.R. 1226-40 201-3; PRO C53/27 m.11; C60/32 mm.6,7
March 4–6,(7),8, 9,(10),11	Kempton	CPR 1232-47 13; Cl.R. 1231-4 198; Cal.Lib.R.1226-40 204,206; PRO C53/27 m.11; C60/32 m.6
March 18	Merton	PRO C53/27 m.10
April 15,16	Croydon	Ibid. m.10; Cal.Lib.R. 1226-40 207
April (18)	Stratford-le-Bow	Cl.R.1231-4 209; PRO C66/43 m.7d
April 23,26	Lambeth	PRO C53/27 m.10
April 27,28, May (2),4–7	Westminster	Ibid. mm.5-10; Cl.R. 1231-4 215-6; Cal.Lib.R.1226-40 211-13,215; Chichester Cart. nos. 139,767; PRO C66/43 mm.5d,7d; BL ms. Harley 61 fo.29v
May (25)	Tewkesbury	CPR 1232-47 16; Cal.Lib.R. 1226-40 217
May 28,30	Worcester	PRO C53/27 m.4
June 7	Wenlock	PRO C60/32 m.5
June 11,13,14	Worcester	CPR 1232-47 18; Cl.R. 1231-4 227,312
June 15,(16)	Evesham	Cl.R. 1231-4 230; PRO C53/27 m.4
June 17	Oddington	PRO C53/27 m.4
June 20,22,(23), 24,25	Woodstock	Ibid. m.4; Cal.Lib.R. 1226-40 219-20; PRO C60/32 m.4
June 30, July 1	Oxford	Cl.R. 1231-4 234-5; PRO C60/32 m.4
July 3	Wallingford	CAR 14; Ramsey Cartulary i 115,117; Pinchbeck Register i 439; PRO C53/27 mm.2,3; BL ms. Harley 230 fo.125r
July (6)	Marlborough	PRO C60/32 m.4
July (8)	Sandford	Cl.R. 1231-4 237
July 10	Windsor	Mss. Windsor 13
July 11,13,14	Westminster	Cal.Lib.R. 1226-40 222; PRO C53/27 mm.2,3
July 14	?Reading	Cal.Chart.R. 1226-57 187
July 17,18,(19),28, Aug. 3	Westminster	Cal.Lib.R. 1226-40 224-5; PRO C53/27 mm.2,3

Aug. (5),6	Windsor	*CPR 1232–47* 22; *Cl.R.1231–4* 243,319
Aug. (9)	Reading	*Cal.Lib.R. 1226–40* 227
Aug. (12)	Woodstock	*Cl.R. 1231–4* 246
	Oddington	*Ibid.* 247
Aug. 16	Tewkesbury	PRO C53/27 m.2
Aug. (17,18)	Worcester	*Cl.R. 1231–4* 249; *Cal.Lib.R. 1226–40* 228; PRO C60/32 m.3
Aug. (20,24,27),28	Hereford	*CPR 1232–47* 24; *Cl.R. 1231–4* 252,254,256,322
Sept. 1	Hay-on-Wye	PRO C53/27 m.2
Sept. (7)	Abergavenny	*Cl.R. 1231–4* 265
Sept. 10	Hereford	PRO C53/27 m.1
Sept. 16	Montgomery	*Cl.R. 1231–4* 266
Sept. (17,19)	Shrewsbury	*Ibid.* 267–8; PRO C60/32 m.2
Sept. 21	Kidderminster	*Cl.R. 1231–4* 270
Sept. (23)	Evesham	*Ibid.* 272
c.Oct. 9	Westminster	Wendover *Flores* iv 276
Oct. (13),15,19, (20,21)	Westminster	*CPR 1232–47* 27; *Cl.R. 1231–4* 279,283; *Cal.Lib.R.1226–40* 239; PRO C53/27 m.1
Oct. (29)	Gloucester	*Cl.R. 1231–4* 328
Nov. 2,7,(10)	Hereford	*CAR* 333; *CPR 1429–36* 64; *Cl.R. 1231–4* 333, 337; *Berkeley Charters* no.231
c.Nov. 13	Grosmont	Wendover *Flores* iv 273 (and see *CPR 1232–47* 32)
Nov. (18,20,23, 26,27),28,(30), Dec. (2,3)	Hereford	*CPR 1232–47* 33; *Cl.R. 1231–4* 339–40, 343,345,349; PRO C60/33 mm.10,11; Westminster Abbey mss. Muniments nos. 15163, 15208; Domesday fo.272v
Dec. 15	Ledbury	*Cl.R. 1231–4* 352
Dec. (16),17,(18)	Worcester	*CPR 1232–47* 34; *Cl.R. 1231–4* 353,547; PRO C60/33 m.10
Dec. (20)	Tewkesbury	*Cl.R. 1231–4* 354
Dec. (23,24),27,28	Gloucester	*Ibid.* 356; *CPR 1232–47* 35–6; Oxford, Magdalen College Muniments, Chalgrove Deed 55b

1234

Jan. (4),5,(10)	Gloucester	*CPR 1232–47* 37; PRO C60/33 m.9; *Derbyshire Charters* I 161,1162
Jan. 13	Marlborough	Stratford upon Avon, Shakespeare Birthplace Library ms. E1/66 fo.17v no.489
Jan. 15	Wolvesey	*Actum* no.48
Jan. (16),17	Marlborough	*Cl.R. 1231–4* 368
Jan. (18)	Ludgershall	*Ibid.* 369
Jan. 20,22,?24	Wolvesey	*Acta* nos.48–50
Jan. (25)	Winchester	*Cl.R. 1231–4* 372
Feb. 1	Westminster	*Ibid.* 373–4
Feb. 16	Bromholm	*CAR* no.17
Feb. (18)	Walsingham	*Cl.R. 1231–4* 379
Feb. 21	Chippenham	BL ms. Cotton Nero C ix fo.50r (?misdated 1232)
Feb. (28)	Peterborough	*Cl.R. 1231–4* 383
March 2	Fotheringhay	*Archaeologia* xv (1806) pp.209–10

APPENDIX 3

March 6–8	Northampton	*CPR 1232–47* 40; *Cl.R. 1231–4* 385; *Cal.Chart.R. 1300–1326* 199
March 14,(15,17)	Woodstock	*Cl.R. 1231–4* 391,554; *Cal.Chart.R. 1257–1300* 115; PRO E36/57 fo.69r-v
March 20	Reading	*Oseney Cartulary* vi no.987
March 27,?26	Westminster	*HMC Wells* i 17–18; *Reg.Antiq. Lincs.* i no.239; *Chichester Cart.* no. 808; *Sarum Charters* no. 206; *Magnum Reg. Album* no.220 (dated 26 March); Bodl. ms. Ashmole 1527 fo.96r (the same charter, dated 27 March); CUL Ely Diocesan Records ms. G/3/28 p.111
April 2,3	Canterbury	*St Paul's Charters* no.182; Lambeth ms. 1212 fos.32v–33r
June 28, July 2,3	Winchester	*AM* ii 86–7

1235

Feb. 22	Wolvesey	*Actum* no.70
Feb. 22	?Sets out from Winchester	*AM* ii 87
March 4 or *c*.April 8	Crosses overseas	*Ibid.* ii 87; Wendover *Flores* iv 327
Summer 1235–Spring 1236	With the papal curia, which was in Perugia, Assisi, Foligno, Spoleto, Terni and Viterbo	Paris *CM* iii 309; *RL* ii 12; and for the pope's movements see *Reg.Greg.IX* iii 282–3

1236

c.Sept. 29	England	Paris *CM* iii 378
Oct. 31	Rochester	*CPR 1232–47* 166
Nov. 30	Winchester	*AM* ii 87

1237

Jan. 28,30	Westminster	PRO C53/30 m.7; *Statutes* i 28
March 20	Westminster	PRO C53/30 m.6
c.April 19	Tarrant	Ms.32DR m.7
May 13–15	Selborne	*Ibid.* m.11
June 4	?Farnham	*Actum* no.54 (?rightly 1238)
Dec. 3	Southwark	*Actum* no.3
c.Dec.	Exchequer	*Cl.R. 1234–7* 9–10

1238

Feb. 16	Kingston upon Thames	*Actum* no.51
March 3	Faversham	*Ibid.* no.52
March 24	Wolvesey	*Ibid.* no.40
c.April 21	Overseas (?France)	*CPR 1232–47* 217
c.April 29, May 2	Oxford	*CPR 1232–47* 217; *Flores Hist.* ii 225
May 11	Winchester	*Cal.Chart.R.1257–1300* 227
June 4	Farnham	Acta nos.53,54,84
June 9	Dies at Farnham	Paris *CM* iii 489–90; *AM* i 108, ii 319

APPENDIX 4

THE BISHOP'S HOUSEHOLD

OFFICIALS (in chronological order):

1. Master John of LONDON, served as official probably from the time of des Roches' election in 1205 until his death, 5 September 1211. Son of a Londoner named Geoffrey Lutre and related to an important clerical dynasty within the chapter of St Paul's London. Inherited property in London and promoted to the London prebend of Ealdland c.1192 × 1204 (*Fasti* i 46, 67; *St Paul's Charters* nos. 138–41). Occurs 1196 × 1208 as witness to more than twenty charters of Philip de Poitiers, bp of Durham, perhaps as *de facto* official (e.g. *Charters of Finchale* 167–8; *Reg.Pal.Dun.* ii 1159,1300–1; *Feod. Dunelmensis* 109). Promoted to the Durham living of Kirby Sigston and served alongside bp Philip as papal judge delegate (*Feod.Dunelmensis* 251; *Letters of Innocent III* nos. 362,421). Associated with des Roches from the time of the latter's election, being found in Rome alongside the bp in 1205 (Migne *PL* ccxv cols. 740–1; *SLI* 42 n.4). Thereafter appears as witness to three of the bp's acta (Acta nos. 11,32, and see *Bath Cart.* part ii no.75), once with title as official (Actum no.37). Addressed as official in a letter from Thomas of Chobham c.1208 × 1211, requesting his intervention in a matrimonial suit (Chobham *Summa* p.xl). Present in Winchester cathedral 23 June 1210 at the grant of a rent to the monks of St Swithun's (BL ms. Add.29436 fos. 31v–32r). Recorded in the year Michaelmas 1210–11 delivering livestock to the bp's manor of Twyford and at Bishop's Sutton together with master Elias of Dereham (*PR 6DR* 16,142). Probably served as master of St Cross Winchester, since at Michaelmas 1212 rents were said to be owing from the hospital *de tempore magistri I(ohannis) de London* (Ms. 7DR m.9d). Dead by May 1212, probably 5 September (1211), the date of his obit celebrations at St Paul's cathedral (*St Paul's Charters* no.139; CUL ms. Ee.v.21 fo.86v). Possibly to be identified with a namesake, knowledgeable in natural philosophy, who is said to have taught master John of Garland the grammarian, associated with but not necessarily resident at the schools of Oxford (Emden *Biographical Register* 1157). For further details see N.C. Vincent 'Master John of London (d.1211): Master John of Garland's Teacher?' (forthcoming).

2. Master Alan of STOKES, known also as Master Alan of ST CROSS, official 1211 × 1219– c.1234, the most important of des Roches' lieutenants and the bp's principal agent in the management of spiritualities. Family background entirely obscure. Perhaps a native of the Winchester manor of Bishopstoke, but known to have held no lay feoff in Hampshire (*CRR* xiii no.428). Possibly related to master Ralph of Stokes, royal justice and rector of the Hampshire churches of Somborne and Farlington, but this connection entirely conjectural (*Southwick Cart.* vol.2 iii no.220; Winchester, Hampshire Record Office ms. 13M63/2 (Mottisfont cartulary II), fo.106v). Only known preferment outside des Roches' service, as rector of the churches of Little Yarmouth, Gorleston, Lowestoft and Belton (diocese of Norwich), valued at 100 marks a year, granted by the king in April 1209 by which time he was already established at Winchester (*RLP* 90b; *RLC* i 159b; *BF* 282,392,403). First certain appearance in the household of bp des Roches as witness to a charter of May 1208 (Actum no.34 and see nos. 1,11,14,21,32,38,64). Thereafter found regularly in the early Winchester pipe rolls (*PR 4DR* 2,3,5,17,18; *PR 6DR* 99; Ms. 7DR m.10). Apparently promoted bp's official and master of St Cross hospital in succession to master John of London, d.5 September 1211. Occurs without title 1212–1215 (Acta nos. 13,56). First specifically described as official March 1218 × March 1219 (*Cart. Saint-Pierre* 193) but probably official for some time previously (Acta nos. 26,83). Appointed keeper of St Cross c.1212 × 1217, but already owing pannage for the hospital's pigs at Michaelmas 1214 (Actum no.80; Ms. 9DR m.4). Henceforth occasionally described as master Alan of St Cross (Acta nos. 70,71). Not to be identified with a brother Alan prior of St Cross Winchester who occurs as early as January 1189/90 (Hereford Cathedral Library, Muniments no.486). In des Roches' frequent absences carried a heavy burden of responsibilty for the management of the see. Between c.1215 and 1231 he is to be found supervising clerical taxation (Actum no.46; *Reg. St Osmund* ii 76), passing sentence of excommunication (*CRR* xi no.1823), hearing a case involving an advowson (*CRR* xii no.1483), commissioned at least three times as papal judge delegate (*Cart. Saint-Pierre* 193; *Glastonbury Cart.* i no.xxii; PRO C115/K1/6681 (Lanthony cartulary) fo.4r-v) and on a further five occasions as arbiter in ecclesiastical suits in which context he claimed variously to be acting as ordinary or delegate of bp des Roches (*St Thomas' Cartulary* no.5; *God's House Cart.* i no.120; *Southwick Cart.* vol.2 iii no.801; Winchester, Hampshire Record Office ms. 13M63/1 (Mottisfont cartulary I) fos.69v, 70v; Oxford, Christ Church Archives ms. Notley charter roll m.12). Before 1224 said to have descended in dramatic fashion upon St Thomas' Hospital Southwark, demanding access to the muniment room and carrying

off charters amidst dire threats (*CRR* xi no.2499). Throughout this period, found delivering quantities of stock to the bp's manors, and perhaps himself donating 20 quarters of grain at Fareham in 1217–18 (Mss. 13DR m.8d; 20DR m.9). Served as one of the principal keepers of the see during des Roches' absence on crusade 1227–31. Apparantly vested with powers both to admit and to institute clergy (*CRR* xiii nos. 373,428, 2196; PRO SC1/6/95–6). Required by the crown to intervene in monastic elections at Christchurch Twinham and Romsey (*Cl.R. 1227–31* 146; *Pat.R.1225–32* 420) and to bring clerks before the royal courts (*CRR* xiii nos. 1560,2124 and see no. 1195). His responsibilities presumably greatly augmented following the death of master Bartholomew des Roches, archdn of Winchester, in 1230. May even have served as *de facto* archdn until the bp could nominate a successor. Described (erroneously) as archdn in a court case of the 1240s whereby a later archdn's official sought the right to present to a church in Winchester whose presentation he claimed had previously lain with master Alan as official of the diocese (*CRR* xvi no.1521). One of those who answered to the crown for the bp's temporalities 1228–31 (PRO E159/9 m.1d) and in the same year pursuing a former episcopal bailiff for arrears (*CRR* xiii no.1106). Summoned as des Roches' representative to a church council on Welsh affairs in June 1231 (*Cl.R.1227–31* 590). Still witnesses as official after the bp's return from crusade. Last occurs with title 20 December 1231/2 (Actum no.75). Thereafter continued prominent in the bp's household, witnessing episcopal charters ahead of all other ecclesiastical dignitaries, save heads of religious houses, including his successor as official (Acta nos. 48–50). Prominent in the establishment of Titchfield abbey, visiting the abbey's site together with master Elias of Dereham in 1232/3 and again in 1235/6 (Acta nos. 70,71; Mss. 28DR m.3; 31DR m.11). Apparently resigned as master of St Cross before the bp's death in 1238, since the office conferred upon master Humphrey de Millières. Died 14 January 1239 × 1248. In emulation of the obits decreed for bp Henry de Blois, his executors assigned 24 marks from his estate to endow anniversary celebrations at St Cross on 14 January each year, with the feeding of 100 paupers, the distribution of eight pence to each of the four chaplains of St Cross and of thirteen pence to each of the hospital's thirteen clerks (Winchester, St Cross Hospital ms. Liber Secundus fo.51r, in a confirmation dated December 1248). Presumably a trained canonist. His importance in episcopal administration demonstrated by the prominence assigned to the official in des Roches' diocesan statutes and in the bp's rule for Marwell (Actum no.31).

3. Master William de STE-MÈRE-ÉGLISE, succeeded master Alan of

Stokes as bp's official *c*. 1234– *c*.1238, appearing first with title in January 1234 (Acta nos. 48–50). Presumably a kinsman, perhaps a nephew of his namesake, bp of London (1199–1221). Identified as a canon of St Paul's from before 1231, transferring *c*.1238 to the prebend of Caddington Minor. Resident canon *c*.1238 until his promotion as dean, *c*. December 1241. Died 11 March 1243 (*Fasti* i 6,35,91; Cambridge, St John's College ms.S25 (Cartulary of the almoner of St Paul's) fos.41v–42r). No known connection with des Roches prior to 1234, although des Roches closely associated with bps William and Eustace of London. Occurs without title as witness to deeds of Selborne priory (*Selborne Charters* i 6, ii 64). As official, remained in England during the bp's exile 1235–6, supervising various ecclesiastical matters in the courts and in September 1236 involved in the excommunication of royal bailiffs who had impounded a great fish claimed by the prior of St Swithun's (*Surrey Eyre* ii no.57; *Cl.R.1234–7* 378–9). Provided episcopal assent to the election of an abbess of Wherwell *c*. June 1236 (*CPR 1232–47* 149). Presented by Walden abbey to the church of Amersham after March 1235 but resigned before 7 December in the same year, at which time he was in deacon's orders. Subsequently presented by des Roches to the episcopal living of Witney and instituted *c*. 28 November 1237 (*Rot. Grosseteste* 341–2, 451–2). Presumably to be identified as the un-named official who served as papal judge delegate in a case involving the church of Idbury, 29 July 1237 (Salisbury, D.& C. Library Muniments mss. Press IV Box E4 Idbury 2 and 6). Replaced as bp's official by 16 February 1238. His church of Witney vacant by 6 April 1243, confirming his identification with the dean of London d.11 March 1243 (*Rot.Grosseteste* 480; *CPR 1232–47* 371). As canon of London much in demand as papal judge delegate, arguing skill in canon law (*Malmesbury Cart.* ii 60–64; Sayers *Papal Judges Delegate* 65,131,273).

4. Master Humphrey de MILLIÈRES, official 1238, son of William de Millières, descended from a Norman family (Millières, Manche) holding East and West Preston in Sussex from the honour of Arundel from the early twelfth century (Farrer *Honors* iii 97–8), and Overstone, Northants., of the Avranches fee from at least 1166 (*Red Book* 193; Hatton *Seals* no.343). Almost certainly related to another family, holding land at Happisburgh and elsewhere in Norfolk from the same honour of Arundel (Farrer *Honors* iii 153–4; *Red Book* 399). Humphrey's father forced to fine £100 in 1207 for the release from captivity of Humphrey's brother William junior. William, junior or elder, captured at Northampton in 1215 as an adherent of the barons and released into the custody of the earl of Arundel (*RLP* 71,157b). Humphrey first appears on the Winchester estates, Michaelmas 1217 × 1218, at Highclere and

at Knoyle in company with William archdn of Poitiers (Ms. 13DR mm.3d,5d). As early as June 1217 presided over an exchange of prisoners taken captive at Winchester (*Pat.R.1216-25* 75). Thereafter appears as a regular guest, in 1218-19 receiving over £12 at Adderbury and Ivinghoe (Ms.14DR mm.6,10d). Under des Roches' patronage served on several occasions as royal ambassador, to France in March 1220, Rome *c*.October 1220-*c*.June 1221, and to France again in October 1227 (*RLC* i 414, 477b, ii 215; *Foedera* 159; *Pat.R. 1216-25* 257,267-8). In 1224-5 attended discussions in London with archbp Langton and in the same year found at Witney returning from des Roches' foundation at Halesowen (Ms. 20DR mm.9d,14,15). Promoted to a stall at Chichester by the king *sede vacante* in 1222, perhaps in reward for his diplomatic work (*Pat.R.1216-25* 340). Presented by his brother to the church of Overstone (*Rot. Hugh Welles* ii 107-8). Presented by des Roches to the episcopal living of Witney before 1220, but resigned in order to take up another Winchester peculiar, the church of Ivinghoe (*Rot.Hugh Welles* ii 3,54). Served in 1229 as arbiter in an ecclesiastical suit at Oxford (*HMC Wells* i 478). Clearly a man of some learning, in 1231 planning to renew his studies at Oxford where he owned a house, previously let with royal approval to Pontius de Pons, son of the seneschal of Poitou (*Cl.R.1227-31* 552; Emden *Biographical Register* 1496-7). In 1230 succeeded to his family's estate in Sussex and Northants. following the death of his brother, William, former constable of Dover and Windsor castles, and hence presumably an adherent of Hubert de Burgh (*BF* 498,501; *CRR* xv no.183; *Memoranda 14 Henry III* 78,80,85; *VCH Northants.* iv 96; *Cal.Lib.R.1226-40* 51,72,174). In Hampshire held the manor of Swanwick and part of Portchester by purchase from William bp of Avranches. Sold this portion of his estate to des Roches for 400 marks, the land to be put towards the bp's foundation at Titchfield *c*.August 1231. The sale probably to avoid the land escheating to the crown as *Terra Normannorum* after master Humphrey's death (*Plac.Quo.Warrant.* 767; *Cal.Chart.R.1226-57* 140; Actum no.67a). Clearly anticipated problems with inheritance after his death, perhaps because his nearest heir was a Norman. Part of his brother's estate, at Littlehampton, escheated to the crown as Norman land and in 1233 granted to Aimery des Roches, the bp's nephew, but recovered by master Humphrey in 1236 from a man named Engelram de Vuill (*Sussex Fines* no.331; *Cl.R.1231-4* 239). Engelram possibly Humphrey's brother-in-law, a Norman, since Humphrey's nephew known as master Gilbert de Wiarvill (?St Germain-de-Varreville, Vierville or Varouville, all dep. Manche) (*Cal.Chart.R.1226-57* 324). In 1237 Humphrey confirmed a grant made by William his brother to Tortington priory of land in West Preston; Humphrey retaining life possession of the land and securing obit celebrations at Tortington both for himself and for William (*Sussex Fines* no.339). Witnesses numerous Winchester acta before 1238 (Acta nos. 14,35,49,50,61,65,74). Succeeded as bp's official before 16 February 1238 and thereafter witnesses with or without title up to the time of the bp's death in June (Acta nos.51-3, 84). Required to surrender Farnham castle to the king (*CPR 1232-47* 224). Involved in the establishment of des Roches' posthumous foundation at Netley (PRO E210/120). Without title but probably acting as bp's official, issued letters instructing the dean of Basing to induct master Adam de Esseby, chancellor of Salisbury into possession of the church of Odiham (*c.*1237 × 1241) (Salisbury, D.& C. Library Muniments ms. Press III Chancellor's Box; *Fasti* iv 19). Succeeded master Alan of Stokes as keeper of St Cross Hospital Winchester, probably before June 1238. No letters of appointment obtained from the king as would have been necessary had Humphrey been promoted *sede vacante*. Dead by 13 April 1241 (*CPR 1232-47* 249-50, 252). Endowed obituary celebrations at Chichester cathedral (*Chichester Cart.* no.558). These later confirmed by master John Mansel who had acquired an interest in Humphrey's lands at Preston (*Chichester Cart.* no.457; *CPR 1232-47* 253). Preston and the Overstone estate initially seized by the crown but subsequently granted for life to Felicia, Humphrey's sister and heir. Overstone passed in 1247 to her son, master Gilbert de Millières or de Wiarvill, a royal clerk active in Anglo-French diplomacy (*E.Rot.Fin.* i 363, ii 13; *Cal.Chart.R. 1226-57* 324; *Cal.Inq.Misc.* no.843), but after 1271 escheated as Norman land (*VCH Northants.* iv 96). Preston demised by Felicia to John Mansel, descending thereafter to Mansel's sister and her assigns (*Cal.- Chart.R.1258-1300* 114,166). Humphrey not to be confused with a namesake, a Norfolk knight sprung from the ?related Millières family of Happisburgh; this namesake appointed to keep Hubert de Burgh at Devizes castle after 1232, and with his ?son, William de Millières, instrumental in Hubert's escape in 1233 (*Cal.Lib.R.1226-40* 190; *Cl.R.1231-4* 180,545; *CPR 1232-47* 27; *CRR* vi 172). Following an attack against des Roches' baggage in 1233, master Humphrey actively engaged in pursuit of Richard Siward and the rebels who later abducted de Burgh from Devizes (*CRR* xv no. 214). Nonetheless intriguing to note the close relations between des Roches' clerk and the circle of Hubert de Burgh, des Roches' most bitter political opponent.

ARCHDEACONS OF WINCHESTER (in chronological order):

5. ROGER (I), archdn *c.*1180-1207/8, said to have been promoted rector of Alton, Hants., by Richard of Ilchester when the latter was bp-elect of Winchester (May 1173 × October 1174) (*CRR* iii 119), but this must have

predated his appointment as archdn, an office filled by a man named Ralph (I) from before 1154 until at least March 1178 (*Fasti* ii 92). As archdn, Roger witnesses several charters of bp Richard (e.g. *WC* nos. 34,72; PRO E40/6112) and continued active under bp Godfrey (Actum no.75; PRO C115/K1/6681 (Lanthony cartulary) fo.3r). Served alongside bp Godfrey and G(uy) prior of Southwick as papal judge delegate in a suit between Tavistock abbey and the rector of North Petherwin (Exeter, Devon Record Office ms. Duke of Bedford's collection W1258M/D80/3; *EEA* viii no. 246). Perhaps assisted by an archdn's official named master Walter de Langton who occurs 1190 × 1220 (BL Harley Ch. 111C68; Newport, Isle of Wight Record Office, ms. St Helen's/P8). Supported the candidacy of Richard Poer, bp Richard of Ilchester's son, during the election dispute at Winchester 1204-5 (Migne *PL* ccxv col. 672). Found at Abingdon, 6 February 1205, as witness to a charter of bp Herbert of Salisbury, Richard Poer's brother (Oxford, St John's College muniments III.i (Cartulary of St Nicholas' Wallingford) fo.27v). Supervised a settlement concerning the chapel of Shorwell, *sede vacante* (*Carisbrooke Cart.* nos. 174-5). Assented to the consecration of altars at Southwick, November 1204 (*Southwick Cart.* vol. 2 part iii no. 964). Last occurs February 1206, when the papal legate was instructed to ensure that Roger suffered no molestation from the king following Richard Poer's failure to secure Winchester (*Letters of Innocent III* no.680). Died March 1207 × March 1208 (*Fasti* ii 92).

6. RALPH or RANULPH, archdn 1208-(?*c.* 1213), occurs as witness to a Mottisfont charter (*Cart.Treas.York* no.6). Since archdn Roger was dead by March 1208, probably Ralph who witnesses a charter of des Roches in favour of Mottisfont, 7 May 1208 (Actum no.34). A grant by William Campanarius to Quarr abbey, dated on the evidence of its handwriting *c.*1210, was made for the souls of Ralph, late archdn of Winchester, Vincent and William his brothers, and mentions a third brother, master Andrew, then still living, holding lands outside the Winchester Northgate (PRO E326/6250; calendared *Quarr Charters* no.527). An un-named archdn of Winchester continued active throughout the papal interdict after 1208. In 1210-11 his robes were carried from Southwark to Reading and thence to Newark (*PR 6DR* 41-2,57,85-6,155). He was at Wargrave in November 1211, stayed at Clere the following Easter week, and was at Farnham for Pentecost (13 May) 1212 (Ms. 7DR mm.2,8,9).

7. Master Bartholomew des ROCHES (*de Rupibus*), archdn *c.*1213-1230, the bp's nephew (*nepos*), presumably a fellow native of the Touraine (*RLP* 113b). First certain appearance, already styled archdn, 28 May 1213, presented by

the king to the church of North Walsham, Norfolk, during a vacancy at the abbey of St Benet of Hulme (*RLP* 99b). Not to be identified with Bartholomew of the king's chamber, a clerk active *c*.1205–post 1221, whose principal interests lay in Ireland (*RLC* i 78b,174b,296, 296b,451,451b). Despite later contacts with Salisbury and York, unlikely to be identified with master Bartholomew, canon of Salisbury, familiar of Hubert Walter at York and Salisbury, fl. *c*.1188–*c*.1192, or with Bartholomew chaplain of bp Herbert Poer of Salisbury (Cheney *Hubert Walter* 25,41–2; *Sarum Charters* 49–50,64; *Reg. St Osmund* i 264,267,345; *Reading Cartularies* i 169, ii 31–4,363; *RLP* 39b; Colvin *White Canons* 349; *Fasti* iv 118–9). Bartholomew des Roches promoted canon of Salisbury, prebend of Burbage, Wilts., in succession to Richard Barre (d. before 1215) (*Fasti* iv 79). As canon and archdn of Winchester witnesses ordinances of Richard Poer, dean of Salisbury, *c*.1214. Before January 1215, without title as archdn, granted an oratory to Henry Esturmy (*Reg. St Osmund* i 250–1,374–80; *Sarum Charters* 76–7). In April 1214 postulated dean of York by the king, as part of bp des Roches' bid for the see of York. In return, asked to relinquish various unspecified rents granted through royal favour (*RLP* 113b). Appears with title as dean of York on various of the Winchester manors in 1214 (Ms. 9DR mm.5d,8d), but promotion as dean quashed September 1214 in the face of local opposition (*RLC* i 202b). Retained prebend at York into the 1220s (*York Minster Fasti* i 2; ii 89). Canon of Lichfield, prebend unknown, promoted whilst bp des Roches had custody of the see in 1223 (*Pat.R.1216–25* 382–3). Canon of London, prebend of Ealdstreet *c*.1209–*c*.1230 (*Fasti* i 48). Canon of Exeter (Exeter Cathedral Library ms. 3518 fo.58r). Rector of Martock, Somerset., promoted by the abbey of Mont-St-Michel before 1226 (BL Add. Ch. 19065–7; PRO ms. PRO 31/8/140B part ii 312–3). Dean of the royal free chapel of St Mary Stafford, occurs 1226 × 1228, promoted 1213 × 1216 (*BF* 385; *VCH Staffordshire* iii 309). Apparently non-resident in all of these churches save York, where witnesses various deeds after 1214 (*York Minster Fasti* ii 89,138; BL ms. Cotton Claudius B iii (York Minster cartulary) fos. 41v,107r). At Salisbury fined for non-residence, 1227 (*Reg. St Osmund* ii 77). Regular appearances on the Winchester estates 1213–19 (Mss. 9DR mm.4,5d,6,7d–9; 13DR mm. 3,3d,4d,5d,7d; 14DR mm.6d,10). Between 1221 and 1227 engaged in a dispute with the prioress of Nuneaton over presentation rights at Catherington and Clanfield churches (*CRR* x 247, xiii no.428; *BNB* no.276; *Pat.R.1225–32* 165). Found making payments of over £50 from the church of North Waltham to the bp's bailiff at Southwark 1224/5, presumably vacancy receipts (Ms. 20DR m.15). In 1223–4 paying a representative of the bp of Exeter at Farnham (Ms. 19DR m.4d and see mm.2,5d,7d,8d,9,10,11).

Deputed by the crown in November 1219 to oversee repayments to Bolognese merchants who claimed to have loaned money to Richard I before 1199 (*Pat.R.1216-25* 260; *RLC* i 408). Appointed papal judge delegate after 1216 in a dispute involving Peter fitz Herbert, a magnate with Hampshire interests, deputing his responsibilties to master Nicholas de Vienne (*Brecon Cartulary* 166-7). Assisted by master Adam of Ebbesbourne, acting as archdn's official (*Reg. St Osmund* ii 76; Winchester, Hampshire Record Office ms. 13M63/1 (Mottisfont cartulary I) fos. 69v-70r). With the bp's departure for crusade in 1227, Bartholomew acted alongside master Luke des Roches to control episcopal patronage, being described as the bp's *procurator* in 1227-8 (*Rot. Hugh Welles* ii 29). 1228 × 1229 sealed an ordinance by master Alan of Stokes, bp's official (*Southwick Cart.* vol.2 iii no. 801). As canon of the Salisbury prebend of Burbage, involved in a prolonged dispute with the crown over right to present to the church of Hurstbourne, and by July 1229 had set out to lodge an appeal with the pope (PRO SC1/6/96; *CRR* xiii nos. 1906,2044,2196; *Pat.R.1225-32* 246). Perhaps joined bp des Roches at the papal court, since it was the bp who, from overseas, reported Bartholomew's death (*Reg. St Osmund* i 388). Died 12 December 1230, when still in the orders of a deacon (Exeter Cathedral Library ms. 3518 fo.58r), his benefices being re-apportioned in February 1231 (*Reg. St Osmund* i 388; *Pat.R.1225-32* 424). Perhaps donated a cope to Salisbury cathedral, before 1214 (*Fasti* iv 79).

8. ROGER (II), archdn c.1230-1232, succeeded Bartholomew, perhaps as one of those who had accompanied bp des Roches on crusade after 1227. Probably to be identified as the un-named archdn of Winchester appointed papal judge delegate alongside master Alan of Stokes, 15 June 1230, and granted papal dispensation to hold two benefices in plurality, 6 May 1231 (*Cal.Pap.Reg.* i 127; *Glastonbury Cart.* i no.xxii). May have returned to England with the bp in August 1231, but died 'a pilgrim' before March 1232 (*AM* ii (Winchester) 85).

9. Master P., occurs once as archdn of Winchester, appointed bp's executor, 4 November 1236. Probably a scribal error for master H(ugh des Roches) (*CPR 1232-47* 166). See below no.10.

10. Master Hugh des ROCHES (*de Rupibus*), archdn c.1232-1253, brother of Geoffrey and Aimery des Roches, the bp's nephews (*Southwick Cart.* vol.2 iii nos. 374-5). First appears 1218-19, receiving twopence from the bp's marshal *ad elemosinas*, perhaps acting as bp's almoner or in some related office (Ms. 14DR m.12d). Presented by Merton priory to the church of Ditton, Surrey, before January 1223 (*CRR* xi no.417; *BNB* iii 467). Canon of Exeter (Exeter

Cathedral Library ms. 3518 fo.56v). Rector of Brading, Isle of Wight, from before 1238, advowson disputed (*Reg. Pont.* 726; Hockey *Insula Vecta* 56-7). During the Welsh campaign of 1223 delivered money from the royal treasury to the garrison of Montgomery (*RLC* i 566, 575, 578; *Pat.R.1216-25* 415,417; Ms. 19DR m. 7d). In the same year delivered 10 marks to the bp's bailiff at Southwark (Ms. 19DR m.11d). In 1224-5 took receipt of large sums at Taunton, Farnham and St Albans, and again delivering money to Southwark, arguing a leading role within the bp's itinerant chamber (Ms. 20DR mm. 3d,9d,15). In 1225-6 he was entertained at Bitterne and at Harwell on his way to the bp's foundation at Halesowen (Ms. 21DR mm.13,14). Succeeded Denis de Bourgueil as *de facto* bp's treasurer in 1226-7, and probably continued to serve as such until the bp's return from crusade in 1231, being described subsequently as *receptor* of Wolvesey (Mss. 22DR *passim*; 27DR m.8d). Probably appointed archdn of Winchester soon after the death of Roger (d. before March 1232). The Winchester pipe roll for 1231-2 mentions an un-named archdn of Winchester present at Farnham for the account of Robert Testard, a former bailiff of the manor who was found to owe £32 in arrears. A further £4 was judged to have been received from Testard at the bp's exchequer when Hugh des Roches was *receptor* (Ms. 27DR m.8d). The archdn's appearance at an account assessing receipts by master Hugh suggests strongly that Hugh and the archdn were one and the same person and that the date of his appointment should be carried back before September 1232. This in turn would force us to dismiss the isolated appearance of master P. archdn of Winchester, appointed bp's executor in November 1236, as a scribal error. Hugh's first certain appearance with title as archdn occurs in February 1237 whereafter he witnesses several of the bp's charters and various early deeds of Selborne priory (*Selborne Charters* i 21,23, ii 65; Acta nos. 51,53,84). A master Roger, official of the archdn of Winchester, occurs in November 1236 as witness to a judges delegate decision concerning Mottisfont priory (Winchester, Hampshire Record Office ms. 13M63/1 (Mottisfont cartulary I) fos. 69v-70r). Following the bp's death Hugh assumed an important role in the administration of the see *sede vacante*. Between 1238 and 1244 it was to Hugh or his official, master Alberic de Vitriaco, that the crown presented clerks for admission (*CPR 1232-47* 224,226,230,239,245-6,249,251,269,421,433). In August 1238 Hugh witnessed two charters of master Luke des Roches, archdn of Surrey and in the following year confirmed a settlement by master Luke and other judges delegate ('Surrey Portion' nos.50,53; BL ms. Cotton Tiberius D vi part i (Christchurch cartulary) fos.62v-3r). Presided over a settlement between Southwick priory and Robert de Watevyle 1239 × 1243 (*Southwick Cart.* vol.2

iii nos. 374–5). In 1240 the king petitioned that Hugh, Luke and Aimery des Roches be excused from a papal tax on alien clergy (*Cl.R.1237–42* 176). Together with Luke des Roches, Hugh served as protagonist of the royal candidates for the see of Winchester, first William bp-elect of Valence and then Boniface of Savoy, claiming a role in the election in opposition to the monks of St Swithun's. Probably travelled to Rome as king's proctor in the election. In March 1243 he and his brother Aimery granted royal letters of protection (*CPR 1232–47* 366). Noted as being absent from Winchester in an institution by his official, master Alberic (Winchester, St Cross Hospital ms. Liber Secundus fo.52r). Appointed by the pope, 23 March 1244, to hear a dispute between the prior of St Martin des Champs, Paris, and H(enry) de Tracy concerning patronage of the alien priory of Barnstaple (*Les Actes Pontificaux originaux des Archives Nationales de Paris: I (1198–1261)*, ed. B. Barbiche (Vatican 1975) no.474). Whereas the election of the monks' candidate, William de Raleigh, led to the almost immediate deposition or resignation of Luke des Roches from the archdeaconry of Surrey, master Hugh managed to remain in office. At some time after 1239 he succeeded Raleigh as rector of the Hampshire church of King's Somborne, and in 1248 secured rights to the patronage of the subject chapel of Little Somborne. Resigned the living by March 1249 when bp Raleigh appropriated it to Mottisfont priory (Winchester, Hampshire Record Office ms. 13M63/2 (Mottisfont cartulary II) fos.106v–7r, 114r–115r,121r–122v). Witnesses occasional charters of bp Raleigh (e.g. Oxford, Queen's College Monks' Sherborne Charters 126–7), and witnesses at least two deeds of Netley abbey, established under the will of bp des Roches (PRO E210/83; E210/120). Served in the 1240s by an archdn's official named master Ralph of Southwick (*Southwick Cart.* vol.2 iii nos. 211,343,379,551,835,844,945 and see nos. 133,541 as official after Hugh's death). Hugh still active into the 1250s as archdn under Aymer de Valence bp-elect. Issued letters as late as February 1250/1 (PRO C85/153 no.5). Apparently still active 9 April 1251, when his arbitration mentioned by master Ralph of S(outhwick) in a dispute between Netley abbey and the vicar of Wellow (PRO E210/230). Died 1253/4 (*AM* ii (Winchester) 94, iv (Worcester) 442), probably 9 October 1253 since a Hugh des Roches, subdeacon and canon of Exeter commemorated on 7 ides October (Exeter Cathedral Library ms. 3518 fo.56v).

ARCHDEACONS OF SURREY (in chronological order):

11. Master AMICIUS (fitz OGER), archdn *c.*1192–*c.*1215, first appears 1177 × 1193 as witness to Peterborough abbey deeds in the time of abbot Benedict

(*Carte Nativorum* nos. 496–7, 512). As Peterborough's clerk presented to rights in the chapel of Paston, Cambridge., pending a better benefice. Subsequently promoted to the church of Peakirk and the chapel of Glinton, Cambridge., reserving rights to Herbert Poer, archdn of Canterbury (*Carte Nativorum* nos. 511,513; CUL ms. D.& C. Peterborough 1 (Reg. Swaffham) fo.249v). First entered service at Winchester via bp Godfrey de Lucy. Almost certainly son of Oger dapifer, steward of Richard de Lucy, bp Godfrey's father. In December 1190, together with a sister named Mirabila, widow of Walter de Mandeville, and a brother named Michael fitz Oger, party to a settlement with Holy Trinity Aldgate over the advowson of Bromfield, Essex, the canons presenting Amicius to the church in return for a quitclaim from Mirabila of all future right in presentation (PRO E40/6943–4). Michael fitz Oger, his brother Alan fitz Oger and an Oger fitz Oger witness charters of bp Godfrey de Lucy concerning the de Lucy family estate in Cornwall and Kent (PRO E40/9460; E40/10846). Oger dapifer, their father, was newly enfeoffed in 1166 with land held of Richard de Lucy in Devon. The descent of his estates in Devon, Cornwall, Essex and Lashbrook, Oxfordshire, can be traced into the 1280s (*Red Book* 352; *FF Henry II* no.1; *FF 9 Richard I* no.130; *CIPM* ii no.248; *BF* 481,829,1463; *CRR* xiii nos.757, 1704, xiv no.910; and for Oger dapifer as witness to Richard de Lucy and his heirs see PRO E40/2325; Bodl. ms. Rawlinson B461 (Lesnes cartulary excerpts) pp.55–6; Thorpe *Reg. Roffense* 327). Master Amicius appears in bp Godfrey's service from the beginning of his episcopate, without title and from 5 January 1192 as archdn of Surrey (PRO C115/K1/6681 (Lanthony cartulary) fo.3r; *EEA* viii *passim*). His brother, Oger fitz Oger served alongside Godfrey as a justice of the Bench. As archdn Amicius reported the findings of a chapter held at Guildford into the rights of Godalming church, and in 1198 was ordered to ignore frustratory appeals by parishioners judged liable to tithes (*Reg. St Osmund* i 300; *Letters of Innocent III* no.3). Witnesses numerous deeds within the diocese, including two of Merton priory (BL ms. Cotton Cleopatra C vii (Merton cartulary) fos. 68v,74v). Represented in Rome during the Winchester election dispute 1204–5 as an adherent of Richard Poer against des Roches (Migne *PL* ccxv col.672), but unlike Roger archdn of Winchester, does not appear to have suffered retribution as a result of des Roches' promotion. Maintained contacts with Richard Poer and his brother, bp Herbert of Salisbury, serving in December 1206 alongside bp Herbert and the prior of Merton in an important papal judges delegate decision relating to Rochester (*Letters of Innocent III* no.729; Maidstone, Centre for Kent Studies ms. D.& C. Rochester DRb/Ar2 (Register Temporalium) fo.132r). Played a major role in the re-foundation of St Thomas' hospital Southwark, following the fire of July 1212, buying out his former tenants in the area in order to supply the hospital's new site, and being appointed the hospital's first patron or master by bp des Roches (Acta nos.56-7; *St Thomas' Cartulary* nos. 17,111,127,140,152,219,492-4; PRO CP25(1) 225/3 nos. 46-7,50). Last occurs as archdn 1215 in a settlement over St Thomas' (*St Thomas' Cartulary* no.17; *CPR 1301-7* 339-40), and apparently replaced by a man named master Peter by August-September of the same year (Actum no.13). Dead by 26 October 1216, when, following a fire at Merton, the land there on which he had built houses given as a garden to his nephew and namesake, master Amicius the younger (Heales *Records of Merton* appendix p.xxix). Master Amicius the younger appears with the title offic(ial) of Surrey as witness to Merton deeds in the time of prior Thomas, 1218 × 1222 (BL ms. Cotton Cleopatra C vii (Merton cartulary) fos. 89r,9or). Possibly Amicius' removal as archdn related to the civil war, in which the de Lucy family sided with the rebels, against des Roches. Two of Amicius' brothers are known to have joined their overlord, the leading rebel baron, Robert fitz Walter, heir to part of the de Lucy estate (*RLC* i 216,243b,264,332).

12. Master PETER, occurs as archdn of Surrey in the witness list to des Roches' foundation charter for Halesowen, *c.* August 1215 (Actum no.13) and to a royal charter at Dover, 2 September 1215 (*RLP* 182; misdated in *Fasti* ii 94). Perhaps intruded into the office against master Amicius the elder. Thereafter, apart from the appearances of master Amicius' nephew as official of Surrey, no archdn recorded until 1224-5 when an un-named individual found with his men at Southwark (Ms. 20DR m.15). The master Geoffrey, listed as archdn of Surrey in *Fasti* ii 94, occurs only in a late forgery and can be dismissed (Actum no.63).

13. Master Luke des ROCHES (*de Rupibus*), archdn *c.*1227-1243, perhaps to be identified as Luke the parson, granted grain at the episcopal manor of West Wycombe in 1213-14 (Ms. 9DR m.4). Churches of West Wycombe and Averingdown later held by Luke des Roches (*Rot.Grosseteste* 347). As master Luke of Winchester, canon of Salisbury by April 1220, prebend of Ruscombe Southbury, transferring to the prebend of Coombe and Harnham by 1227 and finally to Hurstbourne and Burbage in succession to master Bartholomew des Roches, in 1231 (*Fasti* iv 64,79,93). Possibly to be identified as master Luke canon of Chichester, promoted 1215 × 1222 (*Chichester Cart.* nos. 756,1120; *Reg. St Osmund* i 259; *Sarum Charters* 107n; Bodl. ms. Rawlinson B461 (Bradsole cartulary extracts) fo.26r). Canon of London, prebend of Wenlocksbarn, promoted *c.*1222 (*Fasti* i 86). Rector of Epsom by the time of his death (PRO E210/11304). Apparently resident at Salisbury for much of the

1220s (*Reg.St Osmund* i 339, ii 38,109; *Sarum Charters* 91,101,121,174,191). Served as papal judge delegate 1226-7, and in 1226 acted as Salisbury's proctor in negotiations over a clerical subsidy (Oxford, Queen's College Monks Sherborne Charter 2276; Winchester, Hampshire Record Office ms. 21M58/T87; *Reg. St Osmund* ii 62-3). Promoted archdn of Surrey, probably on the eve of bp des Roches' departure for crusade, witnessing with title 1227 × 1231 and serving alongside Bartholomew des Roches as bp's proctor in presentations to vacant livings 1227-8 (Actum no.85; *Rot.Hugh Welles* ii 29). Thereafter witnesses several charters of bp des Roches and deeds associated with Selborne priory (Acta nos. 48-51,53,84; *Selborne Charters* i 21,23-4, ii 65). In January 1232 obtained a hearing before papal judges delegate over his tithes at Burbage and in 1232 his own usurpation of tithes subject to an appeal by master Elias of Dereham (*Sarum Charters* 250; Winchester, Hampshire Record Office ms. 13M63/1 (Mottisfont cartulary I) fo.45v). Probably inherited from his predecessors as archdn tenancy of a chapel and hall within the precinct of St Thomas' hospital at Southwark (*St Thomas' Cart.* nos.8-9). Assumed a role in the administration of the bp's temporalities during the bp's exile 1235-6 (Ms.31DR mm.12d.15). In September 1236 named amongst the bp's executors (*CPR 1232-47* 166) and subsequently played a leading role in the administration of the vacant see. Promoted papal subdeacon by 1249 (*St Thomas' Cart.* no.9). Papal judge delegate in a tithe dispute between Geoffrey de Caux, rector of Ringwood, and Christchurch priory, 1239 (BL ms. Cotton Tiberius D vi part i (Christchurch cartulary), fos.62v-3r). Papal judge delegate in a dispute between the monks of Canterbury and archbp Edmund before November 1240 (Canterbury Cathedral D.& C. Library, Shadwell ms.2), and possibly to be identified as the un-named archdn of Surrey active in the early 1240s as papal judge in a dispute between Christ Church Canterbury and the monks of Dover (Canterbury Cathedral D.& C. Library mss. Cart.Antiq. D69, D73). Admitted Netley abbey, founded posthumously by the executors of bp des Roches, to the church of Esher, January 1240 (PRO E210/5707). Active together with master Hugh des Roches as king's proctor in the Winchester election dispute, claiming right of election in opposition to the monks of St Swithun's. Last appears as archdn July 1243 (*CPR 1266-72* 723; *CRR* xvii no.1102). Probably too closely implicated in the election dispute to survive in office following the promotion of William de Raleigh as bp. Resigned his churches of West Wycombe and Averingdown in 1238-9 (*Rot.Grosseteste* 347). Relinquished his property at Southwark 1249 (*St Thomas' Cartulary* no.9), and in the same year amerced in the Hampshire eyre for non-attendance in the hundred of Odiham, perhaps holding unidentified land at Murrell Green near Hook (PRO JUST1/776 mm.27,41). In August 1250 received papal licence to retire to the abbey of Clarté Dieu in the Touraine, another of the houses founded under the terms of bp des Roches' will (*Cal.Pap.Reg.* i 260). But died *c.*1253, probably at Netley abbey. Hurstbourne church vacant by May 1253 (*CPR 1247-58* 229). Luke's will, proved at Netley, granted a substantial collection of plate and vestments to former clerks and associates including master Peter Russinol II. His books were divided between Clarté Dieu and Netley, with Netley securing the lion's share. Will also endowed obit celebrations at numerous religious houses in Hampshire, bestowing the bulk of his estate, including houses outside the Eastgate of Winchester, upon Netley abbey (PRO E210/11304; Winchester, Hampshire Record Office ms. Eccles.II 159696/15).

TREASURERS (in chronological order):

14. John the DEAN (*Decanus*), appears from 1208, the year of the first surviving Winchester pipe roll, until Michaelmas 1218, taking receipt of the bp's cash income at Wolvesey (*PR 4DR*; *PR 6DR*; Mss. 7DR; 9DR; 11DR; 13DR *passim*). In 1211-12 took tallage at various of the episcopal manors and oversaw repairs at Winchester (Ms. 7DR mm.1, 6d, 7, 8,9d). Present for a hundred court at Swainston, Isle of Wight, in 1215-16 (Ms. 11DR m.3). Took cash release from the majority of manors in the account drawn up at Michaelmas 1219, in company with Denis de Bourgueil. Denis had undoubtedly replaced him as head of the episcopal exchequer for the year Michaelmas 1219-20 (Mss. 14DR; 15DR *passim*). John appears with the title steward at Crawley and Witney in 1218-19 (Ms.14DR mm.1d,5d). Possibly the unnamed *decanus* who delivered 99 pigs at Farnham in 1219-20 (Ms.15DR m.9). In 1220-21 found with his men taking a forest regard at Merdon (Ms. 16DR m.1). Last occurs as John the dean, entertained at Burghclere in 1224-5 (Ms. 20DR m.12) but, like several earlier references, this may in fact refer to a rural dean, in this case the dean of Basingstoke recorded in the same year collecting tithes from Kingsclere. The name John *decanus* is highly problematical. May possibly refer to a specific office. In 1211-12 John is almost certainly the man referred to as dean of Wolvesey, taking tallage at Bitterne (Ms. 7DR m.1 and see *PR 4DR passim*). Presumably the addressee of a writ from the bp *c.*1206 × 1208 as 'keeper of the bishop's house of Wolvesey and the bp's receiver (*receptor*) there' (Actum no.20). According to a late thirteenth-century episcopal register, des Roches married John's niece to Geoffrey de la Grave, in which context John is described as 'treasurer of Wolvesey'

(*Reg.Pont.* 666), but the title treasurer of Wolvesey apparently unused in des Roches' day. John is virtually unrecorded outside the Winchester pipe rolls, despite his evident importance. Given his prominence in the bp's service, possibly to be identified as the John *de Chin'* for whom, in company with Geoffrey de Caux and Robert de Clinchamps, the bp obtained special papal privileges in April 1219, protecting them from harassment in ecclesiastical courts (*Cal.Pap.Reg.* i 65-6). A J(?ohn) *de Chinun* appears only twice in the Winchester pipe rolls, in 1218-19 ill at Witney and going from Wycombe to Oxford (Ms. 14DR mm.5d,7). John the dean was certainly at Witney in the same year. The name Chinun implies association with Chinon, close to the bp's birthplace in the Touraine. Alternatively, given that Robert de Clinchamps continued to be credited with the title *camerarius* as head of the episcopal exchequer, John possibly to be identified with John of Twyford, des Roches' *camerarius* awarded three virgates of land in a manor at Winchester held by the treasurer of St Swithun's priory (*Reg.Pont.* 667). A John of Twyford appears only once in the episcopal pipe rolls, without title, supervising the bp's hounds at Bishop's Waltham in 1223-24 (Ms. 19DR m.2). Unlikely to be the John de Teford who witnesses a charter of the bp as late as March 1238 (Actum no.52).

15. Denis de BOURGUEIL (*de Burgolio*), known also as Denis the clerk and Denis *de camera*, presumably a native of Bourgueil in the Touraine and hence a fellow countryman of des Roches. Appears 1208-1215 receiving and distributing money from the bp's chamber (*PR 4DR* 18,38; *PR 6DR* 6,41-2,60,155; Mss. 7DR m.10; 9DR mm. 3d,8d,9d,10; 11DR mm.2d, 7). During the civil war delivered money to William Brewer intended for William Marshal (*RLP* 156). Apparently succeeded John the dean as treasurer of Wolvesey *c.* Michaelmas 1218. Already receiving cash from the bp's manors as *de facto* treasurer in the first year after the re-establishment of the Wolvesey exchequer following the war. Continues as *de facto* treasurer although accorded no official title, Michaelmas 1219 to Michaelmas 1226 (Mss. 14DR; 15DR; 16DR; 19DR; 20DR; 21DR *passim*). Succeeded master Humphrey de Millières as rector of the episcopal living of Witney, 1219 × 1220, and in 1221 repairing his houses there (*Rot.Hugh Welles* ii 3; *RLC* i 458b). In 1226-7 carried out a perambulation at Overton, but perhaps died Michaelmas 1226 × 1227. By Michaelmas 1227 supplanted as *de facto* treasurer by the bp's nephew, master Hugh des Roches (Ms. 22DR *passim*). Witney vacant before 1228 (*Rot. Hugh Welles* ii 29). Witnesses numerous of des Roches' charters as Denis the clerk (Acta nos. 11,13,21,34,38,45,56,61,83), Denis *de camera* (Actum no.26) or Denis de Bourgueil (Acta nos.14,80).

16. Master Hugh des ROCHES, appointed to head the episcopal exchequer on the eve of the bishop's departure for crusade, acting as *de facto* treasurer in the account taken at Michaelmas 1227. Probably continued in office until the bp's return in 1231, but replaced by the time of the next surviving pipe roll, for 1231-2. Subsequently said to have acted as receiver (*receptor*) at Wolvesey (Mss. 22DR *passim*; 27DR m.8d). For further details of his career see above no. 10 as archdn of Winchester.

17. Robert de CLINCHAMPS, possibly a native of Clinchamps-sur-Orne, south of Caen, dep. Calvados. First appears 1213-14 receiving in the bp's chamber (Ms. 9DR m.7d). Possibly transferred to des Roches from the service of Philip de Poitiers bp of Durham, although unrecorded in bp Philip's charters (*RLC* i 190b,262b). Rector of Enford Wilts., in 1223 granted timber for the repair of his church (*RLC* i 540). Appears 1220 × 1240 as rector of either Wyke or Wherwell, prebendal churches of Wherwell nunnery (BL ms. Egerton 2104A (Wherwell cartulary) fos.24v-25r). After 1213 appears regularly in the Winchester pipe rolls, taking receipt of stock and cash, the latter receipts often said to be *in camera* (Mss. 9DR m.7d; 11DR m.5; 19DR m.5). In 1218-19 with five crossbowmen carried the legate's money from Winchester to Farnham (Ms. 14DR m.9). Sent to prohibit a forest regard at Downton 1223-24, in the aftermath of the bp's political downfall, and found at Southwark returning from his homeland (*de patria sua*), perhaps Normandy (Ms. 19DR mm.5,12). Acted as go-between with archbp Langton in 1224-5, in which year found receiving cash from an unusually large number of manors (Ms.20DR *passim* and for Langton see m.15). Between 1231 and 1237 acted as *de facto* treasurer, although retains the title *camerarius* (Ms.28DR mm.16d,17, and for his acting as chief receiver see Mss. 27DR; 28DR; 31DR; 32DR *passim*). Presumably based at the Wolvesey exchequer. In 1231-2 sent a messenger to des Roches 'against the citizens' of Winchester (Ms.28DR m.17). In October 1233 delivered £100 to the king's mariners at Portsmouth (*Cl.R.1231-4* 329). During the bp's exile, 1235-6, authorised disbursement of alms and complied with a royal writ demanding fish for the king's Christmas feast. Found *c.*1236 going to archbp Edmund in London, and in the same year at Southwark returning from discussions with des Roches, perhaps overseas (Ms.31DR mm.7d,8,8d,15). Granted 8 deer from the royal forest of Aliceholt, 1236 (*Cl.R.1234-7* 300). Witnesses des Roches' charters up to the time of the bp's death (Acta nos. 6,13,49-51,53,61,75,80,84). No recorded appearance after 1238. At some time granted a plot of land in Winchester to the city hospital of St John, the land being transferred *c.*1245 to the Winchester Dominicans (Winchester College Muniments ms. 1146).

APPENDIX 4

CHANCERY CLERKS (in chronological order):

18. Master Peter RUSSINOL (I), first appears July 1200, king's clerk, promised £100 of Anjou pending the provision of a suitable benefice. In the same month served as royal ambassador, perhaps to Rome (*Rot.Chart.* 73b,97b). Presented by the crown to the church of Preston, Lancs., July 1202 (*RLP* 14). Apparently accompanied des Roches to Rome during the Winchester election dispute, since the bp's first recorded charter issued at Rome, 15 October 1205, *per manum P(etri) Russinol sigilli nostri custodis* (Actum no.81). Similar formula appears to a charter of 1206 × 1212 (Actum no.32). Another charter, of 7 May 1208, issued by Russinol *per manum* but merely described as bp's clerk (Actum no.34). Witnesses as clerk, but without any other title, charters of 1206 × 1219 (Acta nos.1,11,37 and see possibly no.80). In 1208-9 found at Clere and Harwell, bringing wine from Southampton to Bitterne etc. (*PR4DR* 7,15,31,59,83). At Brightwell and elsewhere 1210-11, including Adderbury 20 November and *c*. 6 December 1210 (*PR 6DR* 57,70-1,85,95,155 and see Ms. 7DR mm.2,3d; 9DR mm.4,7d,8). Probably perpetual vicar of Niton, Isle of Wight (Actum no.28). In June 1213 promoted by the king to a prebend and the precentorship of York (*RLP* 101). Probably involved in des Roches' bid for the archbpric of York, which had failed by November 1214 when Russinol was dispatched to Rome together with two other canons of York to discuss the election (*RLC* i 180). Returned to England by July 1215 when sent to supervise an election at Norwich, resulting in the promotion of Pandulph as bp on July 25 (*RLP* 149b; Norwich, Norfolk Record Office, ms. D.& C. Norwich Register XII fo.11r, and for the date of the election see PRO E210/3480). Set out for Rome again *c*. September 1215 to attend the Lateran Council, serving as king's proctor together with various others including master Robert de Airaines (*RLP* 182,182b). Returned to England by 20 May 1216 when holding des Roches' hundred court at Downton (Ms. 11DR m.4). Replaced as precentor of York by June 1220 (*York Minster Fasti* i 13). Last occurs, in the bp's pipe roll 1217-18, as judge in a robbery case at Meon and holding the bp's court at Bishop's Sutton (Ms. 13DR mm.2d,11d). Dead by November 1218 × July 1219 (*BF* 268,367). Possibly to be identified with a P. Rocinol, d. July 1218, celebrated with an obit at St Martial Limoges (*Chron. Saint-Martial* 102). The same Poitevin source records a P.Russinol, dean of Angoulême, commissioned by the bp of Limoges to enquire into an election at St Martial in 1214 (*Chron. Saint-Martial* 18). The name Russinol probably a nickname, applied on occasion to priests, presumably of nightingale-like singing voice. Peter's church of Preston passed to Aimery des Roches and then in July 1243 to a Guy de Roussillon,

described as king's kinsman and clerk, an adherent of Henry III's Savoyard uncles (*CPR 1232-47* 387; *Cal.Pap.Reg.* i 387).

19. Peter de CHANCEAUX (*de Cancellis*), native of one of several villages called Chanceaux, near to des Roches' birth-place in the Touraine, the Latin version of his name, *Cancellis*, bearing no relation to his functions as *de facto* bp's chancellor. Difficult to determine whether one of several namesakes, or if all references relate to a single clerk. Brother of Guy de Chanceaux and probably brother or half brother of Andrew, John and Matthew de Chanceaux, all appearing for the first time in October 1200 granted letters of protection from the king (*Rot.Chart.* 98b; *RLP* 108b). The entire clan satellites, probably kinsmen, of Girard de Athée, a low-born Tourangeau castellan raised to high office by king John. Following the loss of Normandy, in 1207 Peter settled with Andrew and Guy de Chanceaux and Engelard de Cigogné on the Hampshire manor of Hurstbourne. Thereafter active in Engelard's service in Gloucestershire and the marches (*RLC* i 79b; *RLP* 135; *PR 12 John* 143-4; *Rot.Lib.* 139). Appears to have held custody of the city of Bath *c.*1211 and subsequently of the lands of the monks of Bath and Strata Florida (*PR 13 John* 222,235; *PR 14 John* 146). Witness to several Bath charters in favour of his kinsmen and their adherents (*Bath Cartulary* part ii nos. 72,78,80). Witnesses with others of his clan and bp des Roches, a grant of land in Hampshire before 1213, and a grant by Gerard de Athée to the canons of St Maurice at Tours (*WCM* nos. 15237-8; Paris, Bibliothèque Nationale, ms. nouv.acq. Latines 1183 p.322). Constable of Bristol, appointed before August 1212 and, although proscribed by name in Magna Carta clause 50, remained in office until 20 July 1215 (*RLC* i 121b-214 *passim*, 221; *RLP* 107,135-45b; *CRR* vii 98; Cole *Documents* 236,240). Found during the civil war at des Roches' manors of Brightwell and Harwell, so perhaps part of the royalist garrison of nearby Wallingford (Mss. 11DR m.2d; 13DR m.4d). Thereafter served as one of the bp's inner household. Although afforded no official title, received stock from most of the episcopal estates 1217-1227, presumably as a member of the bp's chamber, supervising the provision of food and entertainment as the bp travelled the country (Mss. 13DR mm.5d,14; 14DR mm. 1d,4d,5d-7d,9,10; 15DR mm.2d,5,6,6d,7d-8d,10,11: 16DR; 19DR; 20; 21DR *passim*; 22DR mm.6d-9,11,11d,13d). Also took receipt of cash, £7 at Witney in 1220-21 and £30 *in camera* at Taunton in 1223-24 (Mss.16DR m.3; 19DR m.10d). Attended forest regard at Farnham in 1223-24 and in the following year again at Farnham, making payments to a clerk of the papal nuncio Otto (Mss. 19DR m.4d; 20DR m.9d). In 1221 involved in a Herefordshire suit, apparently relating to a gift to the religious

(*CRR* x 170). During the same years a namesake, quite possibly the same man, served as deputy to Engelard de Cigogné, having custody over Windsor forest and in 1230 receiving money on Engelard's behalf (*E.Rot.Fin.* i 73; PRO E368/4 m.4; *CRR* xiv no.487). Des Roches' familiar is said to have accompanied the bp abroad 1227 × 1231, but whether on crusade or merely to Rome is unspecified. Rewarded for this service with papal licence to hold a second benefice with cure of souls. Presented by des Roches to the episcopal living of Adderbury via a proctor named master Ralph de Appleby, December 1231 × 1232 (*Rot.Hugh Welles* ii 39; *Cal.Pap.Reg.* i 127). Appears alongside Appleby as witness to a charter of the bp dated 8 January 1233 (Actum no.18; and see *Cal.Chart.R.1226–57* 183). Between 1231 and 1236–7 active as ever in the bp's service, receiving £270 *in camera* at Taunton, ordering the provision of alms at Bishop's Waltham, glass for the episcopal palace at Marwell in 1231–2 and receiving stock for the bp's Christmas festivities (Mss.27DR mm.2,3,4d–5d,6d; 28DR mm.2d,4d,8d,9d,10d,11,14d,18). Last recorded on the Winchester pipe rolls in 1235–6 issuing instructions at Twyford about the bp's horses (Ms. 31DR m.12d). By May 1237 replaced as rector of Adderbury by master Peter Russinol II (*Rot.Grosseteste* 449,452). From December 1231/2 three of des Roches' charters are issued under the clause *per manum* by Peter de Chanceaux, the last being dated 4 June 1238 (Acta nos.48,52,75). Between 1234 and 1235 he accounts as deputy sheriff of Oxfordshire, and he lived until at least 1240 when he represented Engelard de Cigogné as keeper of Windsor forest (*Cal.Lib.R.1226–40* 437; *Cl.R.1237–42* 166; *Oseney Cartulary* i 365; PRO E372/77 m.17; E372/18 m.5). It is likely though not absolutely certain that Engelard's deputy and des Roches' clerk were one and the same man. Peter or a namesake left £30 of Anjou to the cathedral church of Tours to endow obit celebrations on 30 August each year (*Martyrologe de Tours* 57). Not to be confused with P(?hilip) de Chanceaux, who according to the Tewkesbury annals died November 1236 (*AM* i 101).

20. Master Peter RUSSINOL (II), presumably a kinsman of his namesake with whom he has in the past been confused. First certain occurrence in 1224–5, at Downton, going to speak with archbp Langton, and collecting tithes at Wargrave (Ms.20DR mm.1,10d), but may be the P. Russi' found at Farnham in 1220–21 (Ms.16DR m.5). In 1225–6 at Downton with the bp's household, at Bishop's Waltham and at Fareham, 18 September–6 October 1226 (Ms. 21DR mm.1d,11d,13). Again consulting with the archbp at Southwark in 1226–7 (Ms. 22DR m.13). Promoted by St Swithun's priory to the church of Droxford before 24 June 1231 when granted papal licence to hold in plurality (*Cal.Pap.Reg.* i 128). First occurs as witness to a charter of des Roches 20

December 1231/2 and thereafter witnesses regularly (Acta nos. 48-50, 75; *Cal.Chart.R. 1226-57* 186). Not to be confused with Peter de Russeaus or Russell, a Somerset man, active in the bp's political administration 1232-4. Found at Farnham 28-9 May 1233 with the abbot of *La Peruse* (Ms. 28DR m.8). Too much preoccupied with affairs to reside as rector of Droxford and so granted papal licence to install a vicar, June 1235. Similar licence as rector of Bishopstoke, an episcopal living, April 1236 (*Cal.Pap.Reg.* i 148,152). Perhaps accompanied the bp to Italy 1235-6. Found at Farnham in July 1236 and in the same year at Bishop's Sutton, Highclere and at Bishop's Waltham buying hay for the bp. At Farnham again in January 1237 (Mss. 31DR mm.8d,10,11d,14; 32DR m.9d). Presented by des Roches to the church of Adderbury and instituted via a proctor named Thomas of Winchester, May 1237 (*Rot.Grosseteste* 449,452). Charters of the bp issued by Peter *per manum* 4 June 1238, although appears on the same day without title as witness to an episcopal charter written by William, parson of Baughurst (Acta nos. 53-4, 84). Together with master Humphrey de Millières surrendered Farnham and Taunton castles to the crown following des Roches' death (*CPR 1232-47* 224). Witnesses a charter of Luke des Roches, archdn of Surrey, 11 August 1238 ('Surrey Portion' no.50). Employed by the king to make enquiries of the canons of Wells touching a vacancy in the see of Bath and Wells *c.* June 1243 (*HMC Wells* i 97). Witnesses a grant to bp William de Raleigh, 1 April 1245, and in 1251 served as arbiter in a dispute involving the prior of St Swithun's (*WC* nos. 453,551). Received a decorated silver cup from the will of master Luke des Roches *c.*1253 (PRO E210/11304). Dead by 17 November 1259 when his benefices were re-apportioned (*CPR 1258-66* 74,106,132-3). Recorded in an obit list at Wintney nunnery under 2 Ides November as *Petrus Ruscinale benefactor noster*, so perhaps d. 12 November 1259 (BL ms. Cotton Claudius D iii fo.161r). Also endowed an obit with the fraternity of the Kalendars at Winchester (*CPR 1350-54* 371).

STEWARDS:

21. Master Robert BASSET, one of several namesakes active in the early thirteenth century. First recorded contact with the bp, March 1206, overseeing the delivery to des Roches of chapel furniture seized by the king from the estate of the late archbp Hubert Walter (*RLP* 60-60b). Witnesses from at least May 1209 (Acta nos. 1,11,21,26,32,37-8,45,64,80,83). First recorded appearance with the title bp's steward before 1210 (*Selborne Charters* ii 63). Occurs with title 1213-1216 (Mss. 9DR mm.6d,8,8d; 11DR m.1), but clearly, from the time of the earliest Winchester pipe roll, occupied as *de facto* steward, touring

the episcopal estates, issuing instructions by writ, purchasing land etc. (*PR 4 DR*; *PR 6DR*; Ms.7DR *passim*). In 1210-11 attended forest justices at Wilton and the king at Freemantle (*PR 6DR* 34,96). Imposed a tallage on the episcopal estates 1211-12 (Ms. 7DR mm.4d,5,8d) and in 1213-14 oversaw the installation of glass windows at Farnham (Ms. 9DR m.6). Probably succeeded as steward *c*.1217 by William de Batilly. In 1223-4 found at Southwark (Ms.19DR m.11d), in which year he was active in litigation at the Bench in two cases, both involving property in Yorkshire. In the first he sued John de Aiville for 71 marks, apparently arrears of a 4 mark rent. Case dragged on into the 1230s, with master Robert employing at least two of des Roches' familiars, Edmund the clerk and the bailiff John of Southwark, as attorneys (*CRR* x 313, xi nos.461,1124,2049,2336,2850, xii no. 2664, xiii nos. 1231,2358, xiv nos. 435,554). Second case involved a suit against Maurice of Barnby, canon of York, over the land of Ralph of Thixendale in Fridaythorpe, which master Robert claimed was held from him by military service (*CRR* xi nos. 762,1124; *York Minster Fasti* ii 7). Altogether this suggests that Robert was descended from the Bassets of Yorkshire (*EYC* i 499-501 and see *PR 8 John* 207; *PR 14 John* 17), although it is impossible to assign him an exact position within the Basset lineage. He may possibly be the same Robert Basset who in 1224-5 sought land in Gloucestershire by right of his mother Gunnilla, or the namesake involved in litigation over two bovates in Froggatt Derbs. against Thomas Basset (*CRR* xi no.966, xii no. 1991; *RLC* ii 91). For other possibilities see *CRR* x 255, xiii no.112, xiv no.2276; *BF* 871,887,942,1036,1091; *CIPM* i no.140). In 1224-5 the Winchester master Robert was involved in litigation between des Roches and Margaret de Oseville, appearing in one instance as Maragaret's attorney (*CRR* xii 1102; PRO JUST1/36 m.9; Ms. 20DR m.11d). In 1226-7 conducted a perambulation of the bp's manor of Overton, serving as attorney during the bp's absence on crusade, contracting a final concord on des Roches' behalf in company with Achard the bp's marshal (*CRR* xiii nos. 443,551,1235; Ms. 22DR m.9d). Witnesses two grants associated with the foundation of Selborne priory and an episcopal charter as late as December 1232 (*Selborne Charters* i 6, ii 64; Actum no.75). Last appears 1235-6 as the recipient of a gift of nuts from the bp's manor of Overton (Mss. 27DR mm.1,8d; 31DR m.15d). Canon of Romsey abbey, being described as canon and rector of the nunnery's prebendal chapel of *Cumbe* (?Kimbridge) in a settlement of his right to tithe in Timsbury and *Hulle* drawn up by master Alan of Stokes, *c*.1227 × 1236. Since the church of Timsbury was occupied by a rector named Adam by May 1236, who laid claim to identical tithes, it may be that master Robert was already dead or resigned (Winchester, Hampshire Record Office ms. 13M63/1 (Mottisfont cartulary I) fos. 69v-70v).

His other benefices may have included the church of Shelswell, Oxford, to which a namesake was admitted before 1217 (*Rot.Hugh Welles* i 110).

22. William de BATILLY, first appears on the Winchester estates 1210-11 (*PR 6DR* 128). In 1211-12 at Clere with a nephew and namesake (Ms. 7DR m.7d). His writs and instructions are recorded regularly from 1213-14, implying that he exercised authority as steward, although he is not accorded the title until 1224-5 (Mss. 9DR m.4; 14DR mm.6d,9; 15DR m.10; 16DR mm. 5,6d; 20DR mm.4d,6,7d,8d; Actum no. 61). Occurs without title as witness to charters of the bp dated 1215 and May 1224 (Acta nos. 6,13). Described as steward 1224-27, in which years he had charge of the bp's warrens at Bitterne and Bishop's Waltham and made various gifts to visiting dignitaries including the archbp of York, the bp of Carlisle and Richard of Cornwall (Mss. 20DR m.9; 21DR mm.10,12d; 22DR mm.8,12). Served as a knight in the bp's contingent for the Montgomery campaign of 1223 (PRO C72/3 m.1). Probably a native of Essex, where the family of Bataille or Batilly was long-established. Extremely difficult to distinguish from various namesakes (*Red Book* 350; *CIPM* i no.250; iii no.25; *BF* 121,480,1162,1465; *Essex Fines* i 23,33,98; *Colchester Cart.* i 176-8; *HMC 9th Report* appendix 15b; *CRR* i-vi *passim*). In 1219 held Walton Surrey as a crown escheat through des Roches, as part of the Craon estate held in custody by the bp (*BF* 272-4; *RLC* i 479,505, and see perhaps *CRR* i 166). Hence possibly to be identified with a namesake dead by 1231 when his daughter Alice, wife of Roger de Walcot, sought 44 acres of his land in Mitcham, Surrey, held of the Mauduit barony (*CRR* xiv no.1728; and for a Richard de Batayle of Escote, active in Chobham, Surrey, in the 1250s see *Chertsey Cart.* vol.2 (1958) nos.809-10 and note p.lxxvii). The connection between the Batilly families of Essex and Surrey appears to date back at least as far as the reign of Henry I (*Fitznells Cartulary* pp.xxiv-vi). This in turn would enable us to identify William as an attorney of Robert de Mauduit in 1206 (*CRR* iv 127,214), and with the William de Batilly who held land at Langley, Essex, confirmed to Holy Trinity Aldgate by his daughter Alice and Roger de Walcote, her husband c.1240 (PRO E40/140 and see PRO E40/733,736,739). William may have begun his career as a knight in the service of Robert of Thurnham (*Misae Roll 11 John* 133). Since Thurnham served for many years as seneschal of Poitou, this would help explain the choice of William in 1219-20 as royal envoy to Poitou, presumably at des Roches' bidding, to negotiate the release of the king's sister from the custody of Hugh de Lusignan (*Pat.R.1216-25* 199; *RLC* i 397,411; *DD* nos. 48,76,79; *PR 3 Henry III* 73). A Simon de Batilly appears regularly on the bishopric estates between 1217 and 1237 (Mss. 13DR m.1; 14DR m.5d; 27DR mm.3d,5; 32DR

mm.1d,2,4,4d,9,10,11), possibly to be identified as an Essex man with similar landed interests to those of the Essex William de Batilly (*Essex Fines* i 163,219,253,278; PRO E40/821).

23. William of SHORWELL, son of Robert of Shorwell (d. after 1205), a minor landholder on the Isle of Wight, holding the manor of Shorwell and demesne in Chessell, Clatterford and Wolverton from the lords of Carisbrooke (*Carisbrooke Cart.* nos. 4,55, 57–8,232; *VCH Hampshire* v 279). William's chief landholding remained the manor of Shorwell, besides which he held a tenement and meadow at Atherfield from Adam of Compton, the land of *Rugerig*, probably part of Shorwell, held from Walter de Lisle (*Lacock Charters* nos. 448,466) and meadow in Walpen (*Quarr Charters* no.442). Already in the service of bp des Roches by 1208, accounting as constable and bailiff of the honour of Taunton until 5 January 1214, and for Rimpton from before 1208 until Michaelmas 1209 (*PR 4DR* 63,68,72; *PR 6DR* 162,171; Mss. 7DR m.10; 9DR mm.9,9d). In 1208–9 found together with the bp's steward, master Robert Basset, purchasing land at Knoyle. Thereafter makes regular appearances on the episcopal manors (*PR 4DR* 3,39,75; *PR 6DR* 128; Mss. 7DR m.4; 9DR mm.7,7d,10). Reappears as constable of Taunton during the civil war after 1215 (Ms. 11DR m.7d). From 1217 served under des Roches as *de facto* sheriff of Hampshire, accounting at the royal exchequer until the bp's downfall in 1224 (*PR 2 Henry III* 11; PRO E372/68 m.1). As sheriff supervised the collection of carucage in 1220 (*BF* 1438; *Foreign Accounts* 16–17) and witnesses a large number of Hampshire deeds. Reimbursed a nuncio of the count of St Gilles (*RLC* i 415). Appointed to the Hampshire forest enquiry of 1219 (*Pat.R. 1216–25* 212,217). Throughout this period, remained attached to the bp's household, being described as steward in 1217–8, and found taking tallage at the bp's manors of Meon and Downton, holding the bp's hundred court at Meon and on various commissions elsewhere (Mss. 13DR mm. 1,2,3d,4d,7d,8,11d,13; 14DR mm. 1d,4,11; 19DR mm.2d,3,5d,8d). Perhaps already married by 1213–14 when a woman described as the wife of the constable of Taunton received a gift of wine (Ms.9DR m.10). Married again between 1222 and 1224 to Joan, daughter of Walter Walerand, joint-heiress to the barony of West Dean in Wiltshire. Joan was already twice widowed, having married first William de Neville (d. 1220 × 1222) and then Jordan de St Martin (d. before February 1223). She brought William a splendid estate including eight and half fees of her barony, custody of her step-son, William son of Jordan de St Martin, and kinship to John of Monmouth who had inherited the Walerand custody of the New Forest (Sanders *Baronies* 96–7; PRO CP25(1) 203/5 no.78; 203/7 no.61). However, Shorwell removed as

sheriff of Hampshire following the bp's downfall in 1223-4 and in the autumn of 1224 was forced to surrender custody of William de St Martin, his step-son (*CRR* xi no.2381). May have attended the siege of Bedford in 1224 (Ms.19DR m.9). In the following year attended forest justices at Wilton on des Roches' behalf (Ms. 20DR m.2). Finally forced to render accounts for Hampshire 29 April 1225 (PRO E368/7 m.10d) when found to owe £15 for custody of Aliceholt forest 1221-2, his total debts later raised to £40 (PRO E159/7 m.4d; E159/8 m.2). Returned to royal service in 1225, as collector of the tax of a fifteenth in Hampshire (*RLC* ii 147). Appointed to the south-western circuit of the general eyre, 1227, and served for part of the same year as sub-sheriff of Dorset and Somerset under Jocelin bp of Bath (Crook *Eyre* 81; PRO E159/8 m.2; E368 m.6; E368/9 m.5d; E159/9 m.12). In 1227 appointed joint keeper of the manor of Freshwater, Isle of Wight (*RLC* ii 200). Sat on an assize in September 1227 involving claims by Bartholomew des Roches, archdn of Winchester (*Pat.R.1225-32* 165). Last occurs February 1228, pardoned 100 shillings owing from his time as under-sheriff of Somerset (*Cal.Lib.R.1226-40* 69). Dead by 7 August 1228, when his brother Robert fined £22 for William's outstanding debts as sheriff and began repayment (*E.Rot.Fin.* i 175; PRO E372/73 m.2). By 1230 joint resonsibility for William's debts at the exchequer, some £50 all told, assumed by Robert of Shorwell and Ralph de Meriet (PRO E372/75 m.16; *PR 26 Henry III* 331). Meriet's interest is unclear. In 1228 he was a tenant of the Shorwells and later acquired rights in William's lands at Walpen in the Isle of Wight which he transferred to Quarr abbey for 15 marks (*Quarr Charters* no.442; PRO E315/50/236; CP25(1) 203/5 no.78). William's principal holding at Shorwell passed before 1265 from Robert to a kinsman, Nicholas de Depedene who sold his interests there to Lacock abbey in return for £33 (*Lacock Charters* nos. 446-52, 464-5, 468; *VCH Hampshire* v 279-80). William's widow, Joan, was dowered at Shorwell, Atherfield, Walpen and with a third of the mill of Yafford. In addition, between 1228 and 1241, she acquired a hereditary tenancy in the manor of *La Lee*, paying a consideration of 20 marks and an annual quit-rent of twopence to Robert of Shorwell and his heirs. Granted her land there to Romsey abbey in 1245 in return for a chaplain to pray daily for her soul (PRO CP25(1) 203/5 no.78; 203/7 nos.61,75). After her death in 1263 her rights in the barony of West Dean passed to John de Neville and William de St Martin, offspring of her earlier marriages (*CIPM* i no.549; Sanders *Baronies* 97).

24. Richard of STAPLEFORD, first appears 1202 reimbursed payments made on the orders of Geoffrey fitz Peter (*PR 4 John* 284). Served on the fringes of the court as a clerk of Geoffrey fitz Peter, the justiciar, witnessing

Geoffrey's charters and supervising repairs to the Tower of London 1208–13 (*PR 5 John* 7; *PR 10 John* 166; *RLC* i 152b,153; *CRR* v 198; *CAR* no.363; and see possibly *WAC* no.484). Via Geoffrey acquired a royal wardship (*RLC* i 354,358b). Probably to be identified as an Essex landholder, presumably a native of Stapleford, cited in various Essex pleas (*PR 2 Henry III* 70; *PR 3 Henry III* 108; *PR 4 Henry III* 109; *CRR* v 99). In June 1205 confirmed in possession of unspecified lands (*Rot.Chart.* 153). A namesake called upon to warrant land in Herts. in 1214, appearing in 1220 as brother of Henry of Stapleford (*CRR* vii 196,233,299, viii 261,314, ix 279, x 55). Fitz Peter's clerk probably to be identified as a London property owner, described as clerk, granted tenements in the parish of St Faith by the chapter of St Paul's *c.*1213–16, which passed to his nephew, William of Stapleford goldsmith, son of Ralph (*St Paul's Charters* nos.145,152; PRO DL25/123–5 which preserve impressions of Richard's seal). As Richard clerk of Stapleford witnesses a charter of Geoffrey de Mandeville, son of Geoffrey fitz Peter, *c.*1215 × 1216, arguing adherence to the rebel cause during the civil war (*Cart. St Bartholomew's* no.695). His Mandeville connections may explain Richard's recruitment into the household of des Roches, fitz Peter's successor as justiciar and keeper of the Tower of London. Appears at many of the bp's manors 1217–19, frequently with the title steward (Mss. 13DR;14DR *passim*). Apparently left the bp's service *c.* Michaelmas 1219 whereafter continued to owe half a mark advanced to him at North Waltham, last summoned at Michaelmas 1224 (Mss. 16DR m.4d; 19DR m.6). By March 1219 rector of Walton-on-Thames, Surrey, party to a prolonged dispute with Chertsey abbey, eventually settled in Chertsey's favour in February 1228 (*Chertsey Cart.* I nos.93–4). Lived on into the 1230s. Collector of the tax of a fifteenth in Essex/Herts. 1225, described as clerk (*RLC* ii 147; *Pat.R.1216–25* 563,565). At Easter 1231, as Richard son of William of Stapleford respited a prest made by king John (*Memoranda 14 Henry III* 69,78). Probably dead by Easter 1233 when a case involving his custody of land in Essex proceeded without his being summoned as witness (*CRR* xv nos. 188,604). Besides his nephew named William fitz Ralph, Richard may have left a son named William, holding land of Henry of Barnwell at Stambridge Essex. William's daughter was married before 1255 to Richard Tany (*Cal.Chart.R. 1226–57* 440).

25. Roger WACELIN, first appears 1195 fined 100 marks as son and heir of Sampson Wacelin of Barfleur (dep. Manche). Accounts as royal provost of Barfleur 1203 (Stapleton *MRN* i pp.clxxvii, 288, ii pp.clxxix, ccxxvi). Owned a ship used to transport the king across the channel in 1200 and in 1202 crossing from England to Normandy with the king's wine (*Rot.Chart.* 60; *PR 4 John*

79). Confirmed 1203 in land at *Escayo* (?Esquay-Notre-Dame or Esquay-sur-Seulles, both dep. Calvados), but presumably lost control of family estates following the French invasion (*Rot.Chart.* 112b). Appears in England 1205 granted letters of protection, perhaps to accompany des Roches to Rome for the Winchester election dispute (*RLP* 53b). Retained contacts with Normandy into the 1220s, procuring trading licences for ships belonging to Peter Wacelin, a merchant (*Pat.R.1216-25* 507; *Pat.R. 1225-32* 7). By 1208-9 active in des Roches administration, serving as *de facto* steward, overseeing gifts to the king, found on most of the episcopal estates 1210-11, taking tallage 1211-12 and again 1213-14 (*PR 4DR* 21,39,43,57,75-6; *PR 6DR* ; Mss.7DR; 9DR *passim*). By 1216 acquired a rent in Garstreet Winchester from the bp (Mss. 11DR m.7d; 14DR m.12; 28DR m.16d) and by 1232 granted a major estate at Compton, a manor belonging to St Swithun's priory, the estate comprising 300 acres held of the priory and a mill held of the bp (*Pat.R.1225-32* 519; *VCH Hampshire* iii 406; *CIPM* iv no.78). Also held the manor of Ewhurst, Sussex, acquired by marriage to Mabilia, daughter and heiress of Stephen le Borne c.1211, the land held as one and a half fees of the honour of Hastings (*PR 13 John* 128; *VCH Sussex* ix 266; *BF* 692; *CIPM* iv no.78). Sold homage of a tenant in land held at *Messinges* from Richard de Pellered, to Henry of Braybrooke for 30 shillings (BL Sloane ms. 986 (Braybrooke cartulary) fos.52v-53r). Overseeing ships on the bp's behalf 1215-19 and found on many of the bp's manors to 1224-5 when he first appears with the title steward (Mss. 11DR mm.1d,3,4; 14DR mm.7-9,10; 15DR mm.1d,7,7d; 20DR mm.9d,11d). As steward during the bp's absence on crusade defended actions over land and issued with instructions on prisoners, scutages etc. by the king (*Cl.R.1227-31* 248,264,399; *CRR* xiii nos. 443,864,1235,2250,2252; PRO C60/30 m.4). Defending private actions of his own via Nicholas, son of Achard the bp's marshal (*Cl.R.1227-31* 240). In 1229 deprived of a messuage in Winchester he had attempted to purchase from Henry Britton (*CRR* xiii 2018). Party to a settlement over land at Lovington 1228-9 (Winchester Cathedral Library, ms. Book of John Chase fo.34r). Retained his title as steward 1231-3 after des Roches' return (Mss. 27DR mm.3,8d-10; 28DR mm.3,4,8,16d). Following des Roches' appointment as sheriff of Hampshire 1232, Roger acted as deputy with day to day control over the county much as William of Shorwell served as joint steward and sheriff after 1217. In March 1233 received custody of the royal castles of Winchester and Portchester and from July 1232 to c. May 1234 served as keeper of the king's houses at Clarendon, Gillingham and Ludgershall where major building projects were in progress (*CPR 1232-47* 13; *Cal.Lib.R.1226-40* 205-6,215,233; *Cl.R.1231-4* 369,372-3,411,414,428). As sheriff witnesses many of the early deeds of Selborne priory (*Selborne Charters* i 6,8,14–15, ii 4–5,47,64). However, removed as sheriff following the bp's poltical downfall and subsequently accused of various misdemeanours before the general eyre of 1236 (PRO JUST1/77 mm. 17d,18d). His debts as sheriff, calculated at £10, cleared in 1238 (PRO E372/82 m.7d). Remained active on the Winchester estates until at least 1237 but apparently no longer as steward (Mss. 31DR m.10; 32DR mm.1,5d,6d,7,8,9–10). Witnesses charters of des Roches up to the time of the bp's death (Acta nos. 21,52,80,84). Last appears c.1242 (*BF* 692). Died 1242 × 1250. Bp William de Raleigh seized back his land at Compton, restoring it to Roger's son Nicholas only after the latter had agreed to marry the bp's kinswoman, a daughter of Ralph de Raleigh (*Reg.Pont.* 667,758). Nicholas was succeeded at Ewhurst and Compton by a son named John. John died an idiot c.1302 whereafter a partition was effected amongst his sisters (*CIPM* iv no.78).

26. Robert of WALTHAM, perhaps a native of the manor of Bishop's Waltham, with no proven connection to namesakes active in Essex, Wiltshire and Norfolk. First appears 1213–14 overseeing expenses at Clere (Ms. 9DR m.5d). Taking tallage at Wargrave 1215–6 and himself accounts as bailiff of Bishop's Sutton (Mss. 11DR mm.5d,7; 15DR m.4). Bailiff of Farnham and Bentley from before Michaelmas 1221 until 1226 (Mss. 16DR mm.1,5; 19DR mm.4,4d; 20DR mm.9d,10; 21DR m.5). Owed arrears for Farnham, Michaelmas 1227 (Ms. 22DR m.5). Reappears 1235–6 as steward, purchasing a rent at Alresford and witnessing a sale (Ms. 31DR mm.7d,10). Last appears assessing grain at Farnham 1236–7 and as witness to a charter of des Roches, 3 March 1238 (Ms. 32DR m.9d; Actum no.52). Perhaps dead by May 1238 when land at Basingstoke held by a namesake from Walter son of Alexander fitz Hugh of Basingstoke granted to Walter of Merton (*Cl.R. 1237–42* 132; PRO CP25(1) 203/7 no.10). A Robert son of Robert of Waltham, one of several brothers, confirmed this grant for 40 shillings, Merton promising to use the money to find Robert suitable employment so that he might be trained up in an office in London or elsewhere, the promise to be honoured by 1 November 1241 (Oxford, Merton College Muniments no.1755). Alternatively, des Roches' steward perhaps to be identified as Robert of Waltham, who in the spring of 1241 obtained temporary possession of a moiety of the manor of Upton Grey and an 8 mark rent in lieu of a moiety of the manor of Ludshott, Hants., with reversion in both manors to the heir of William of Arundel when the latter should come of age (PRO CP25(1) 203/7 no.58). A Robert of Waltham and William his son witness a Wargrave deed of c.1230, as does a Reginald son of Robert of Waltham (Reading, Berkshire Record Office ms. D/EN M4/T6). A

master Reginald of Waltham witnesses a charter of des Roches, 4 June 1238 (Actum no.84)

CLERKS:

27. Master Robert de AIRAINES (*de Arenis*) presumably a native of Airaines, near Abbeville, Picardy. First appears April 1202, appointed king's proctor in Rome (*RLP* 9). Found in 1210-11 at the Winchester manors of Meon and Southwark (*PR 6DR* 123,156). Witnesses a charter of des Roches (Actum no.26). Canon of York from before 1215 (*York Minster Fasti* ii 89). In receipt of a 20 mark fee from before January 1215 when serving as royal ambassador alongside the archdn of Airaines (*RLC* i 186). Returned to England before April 1215 when sent by the king to oppose the election of master Robert of York as bp of Ely (*RLP* 132b,133). Set out for Rome again, September 1215, serving as king's proctor to the Lateran Council in company with master Peter Russinol I (*RLP* 182,182b). In April 1218 received 50 marks arrears of his annual pension to cover unspecified expenses in Rome (*RLC* i 358). His pension paid until at least July 1221 (*RLC* i 194,221b,383b, 393b,410b,423b,466b). Perhaps serving as clerk of the papal legate Pandulph, whose legation ceased in July 1221. In July 1219 sent with Stephen of Seagrave to consult with the legate and given two and a half marks to buy a palfrey (*RLC* i 396b). In turn, sent by Pandulph to des Roches with news of a truce with Scotland and duly found on the bp's manor of Highclere, presumably late summer 1219 (*Foedera* 157; Ms.14DR mm.4,4d). Canon of Exeter and in the orders of a subdeacon at the time of his death, 18 June 1224 (Exeter Cathedral LIbrary ms. 3518 fo.28v).

28. Richard of BARKING, a professional attorney at the Bench from 1199 (*Rot.Cur. Reg.* i 246; *CRR* i 337,403, ii 74,182; *PBK* iii no.1155; *Essex Fines* 19) and by 1206 serving as a clerk of the justiciar Geoffrey fitz Peter (*CRR* iv 251,299). Apparently transferred to des Roches' service after Geoffrey's death, still serving in 1214 as an attorney at the Bench (*CRR* vii 268). Perhaps in receipt of robes as a clerk of the exchequer from before 1215. Grant of robes renewed and continued 1219-November 1224 (*RLC* i 402,441b,473,523b,574, ii 8b). In 1220 seeking debts of fifty two shillings from William de Dammartin, one of des Roches' bailiffs (*CRR* viii 379). Possibly presented by Waltham abbey to the church of Lambourne, Essex, 1216 × 1223 (*CRR* xi no.768). Presented by the monks of Ely to the church of Melton, Suffolk, and instituted 1218 × 1221 (CUL, Ely Diocesan Records ms. G/3/28 (Ely Liber M) p.206). Still acting as attorney 1221 (*CRR* x 221; *Essex Fines* 50). Identified as the

clerk of the exchequer responsible for writing the exchequer pipe rolls 1217–1228, and various other exchequer documents, probably as a satellite of the treasurer, Eustace de Fauconberg (*PR 5 Henry III* pp.xxiv–vi). In 1221 given parchment to draw up the rolls of the Jewish exchequer (*RLC* i 517b). Witnessed a grant to Fauconberg as bp of London, and at least one of Eustace's episcopal charters (*St Paul's Charters* no.329; Westminster Abbey muniments no.1265). Served as Fauconberg's attorney 1224 (*CRR* xi no.2756). Named together with Fauconberg as a baron of the exchequer, December 1226 (Vincent 'Chancellorship' 106n.). Last appears *c*. Easter 1228 as an exchequer clerk (*Cal.Chart.R.1226–57* 77). Probably to be distinguished from namesake(s) holding land in Essex and Norfolk (*Pat.R.1225–32* 446; *CRR* x 32, 63; *Essex Fines* i nos. 637, 1023, 1303). Unrecorded on the Winchester pipe rolls but witnesses four of des Roches' charters, 1206 × 1224, mostly in association with Eustace de Fauconberg (Acta nos. 26,38,45,83).

29. Geoffrey de CAUX (*de Cauz* or *de Caleto*) probably descended from the Caux family of Water Eaton, Bucks., royal serjeants by custody of the king's falcons from the reign of Henry I (*Dunstable Cartulary* 294–7 and pedigree 10). Brother of James de Caux, whose relation to the family of Water Eaton is not seriously in doubt (*RLP* 118b; *Dunstable Cartulary* 297). Active in des Roches' service from at least 1208, witnessing the bp's charters, found in 1208–9 at Farnham meeting the expenses of a visit by the king, and at Witney in charge of the king's arms (*PR 4DR* 7,17–18,39,42,60,74; Acta nos. 11,14 and see possibly no.34 where *G. de Caleston'* may be a scribal error for *Caleto*). Appears to have served simultaneously under king John and des Roches, being described in a royal letter to the bp as 'ours and your clerk' (*RLP* 118b). Brother of Roger de Caux, to whom des Roches was ordered to grant a wardship and marriage in 1214. Roger unlikely to be identified with Roger de Caux III of Water Eaton, keeper of the king's falcons, whose daughter was married to John son of Stephen of Seagrave the justice (*RLC* ii 82,174,183; *Dunstable Cartulary* 295). In 1210–11 Geoffrey presented gifts to the king and the papal nuncio Pandulph on des Roches' behalf, and attended the king at Witney and Taunton (*PR 6DR* 65,154,172). Thereafter makes regular appearances in the Winchester pipe rolls, supervising tallages in 1211–12, 1217–18 and 1220–21 (Mss. 7DR mm.1,2,4,4d,5,6d,7,8, 9d-10,11,11d; 13DR m.2; 16DR mm.4,4d,5d,11). During the civil war carried £100 to Taunton and made a journey to Exeter (Ms. 11DR mm.8,8d). Overseeing provision for Henry III at Downton 1218–20 (Mss. 14DR m.7; 15DR m.7d). Served as the bp's messenger, probably to France, November 1212 (*RLP* 95b). Accompanied the king to Poitou 1214, together with his brother James (*RLP* 118b;

Rot.Chart. 199b,201). Ambassador to Rome after October 1218, to deal with the issue of papal census, his expenses being guaranteed by des Roches (*Pat.R.1216-25* 167,181,184,208-9; *RLC* i 370b,387; *DD* no.25). Used this mission to secure papal dispensation to commute a crusading vow in return for paying the wages of four crusaders. Described as deacon. At the same time licensed to hold in plurality the churches of West Camel, Somerset, and 'Wethmenes', possibly Wilmington, Sussex, to which he had been presented by the king in February 1209 (*RLP* 89; *Cal.Pap.Reg.* i 65). May also have negotiated papal letters protecting himself and des Roches' clerks Robert de Clinchamps and John de *Chin'* from harassment in church courts (*Cal.Pap. Reg.* i 65-6). Between 1213 and 1222 held the manor of Bentworth, Hants., a Norman escheat originally granted by the king to des Roches (*RLC* i 100,499b; *Rot.Ob.* 512; *PR 16 John* 132; *PR 3 Henry III* 28; *PR 4 Henry III* 124; *Pat.R.1216-25* 329; PRO C60/17 m.5). In 1214 presented by the king as vicar of Ringwood, a church held by the crown as Breton escheat (*Rot.Chart.* 195b-6). Presented *c.*1210 to Dogmersfield, Hants., supposedly as a clerk of the prior of Bath, but more likely through des Roches' influence (*Bath Cartulary* part ii no.75). May have held land at the bp's manor of Adderbury, 1218 (*RLC* i 348b). Retained the vicarage but resigned the rectory of Ringwood 1219 in favour of a nephew and namesake, Geoffrey de Caux II, rector until at least 1249. Probably Geoffrey II rather than Geoffrey I who quitclaimed rents at Pennington in return for 100 marks to repair the chancel of Ringwood (*Pat.R.1216-25* 205; *BF* 256,1365,1417; PRO E326/4475). Geoffrey I last appears 1220-21 (Ms.16DR mm.4-5d,6d,11,11d). Almost certainly to be distinguished from a knight and namesake, sent abroad in 1209, in 1212-13 paying the expenses of the duke of Limburg and the Scottish princesses in the custody of king John, active into the 1230s. A John de Caux appears on the Winchester estates 1224-27 (Mss. 20DR mm.3d,4,4d,9d; 22DR m.3d,4); a Martin de Caux between 1231 and 1233 (Mss. 27DR mm.9d,10; 28DR m.6d).

30. Master Eustace de FAUCONBERG, bishop of London (1221-1228), younger son of Peter de Fauconberg, a major Yorkshire landowner. Born after 1166. Presented to the family living of Catwick, Yorks. E.R. (*Chart. Pontefract* ii 491-3, 551 no.445; *Lincs. Assize Rolls* p.xxiv; *EYC* iii 48-9; *CP* v 268n). Granted half a carucate of land at Catwick to Nunkeeling priory, a nunnery patronised by other members of Eustace's family including his father Peter (BL ms. Cotton Otho C viii (Nunkeeling cartulary) fo.72v). Held the Southwell prebend of Rampton, Notts. to which he had been presented by Geoffrey archbp of York after 1189 (*CRR* x 219). Probably rector of Kirk-

leatham, Yorks. N.R., land there being the subject of litigation in 1219 (*CRR* viii 67). Witnesses a grant by his brother Walter de Fauconberg to Welbeck abbey which by 1202 was under Fauconberg family patronage (BL ms. Harley 3640 (Welbeck cartulary) fo.47v and see fo.119r; Colvin *White Canons* 248n, 292; *Yorks. Fines* 12–13). Held a moiety of the manor of Kettlewell, Yorks. W.R. at farm from his brother Walter (PRO JUST1/1042 m.27; *CIPM* iii no.284). Perhaps came into contact with Godfrey de Lucy during the latter's time as archdn of Richmond. Witnesses more than a dozen of Godfrey's charters as bp of Winchester, July 1193–May 1203 and probably earlier (*EEA* viii *passim*). Appears as attorney for bp Godfrey, October 1199 (*Rot.Cur.Reg.* ii 61). Followed Godfrey into a career as royal justice. Attorney for the monks of Vaas (dep. Sarthe), Hilary 1199 (*CRR* i 74). Promoted justice of the Bench under king John, serving at Westminster or on eyre in virtually every term to 1209. With the closure of the Bench in 1209 transferred briefly to the court *coram rege* (*PBK* iii pp.cxlix–cl, clxiv–cclxxii *passim*). Influence as a justice probably explains his promotion to various churches including Godington, Oxford, presented by Christina de Camville before 1200 (*CRR* x 250–1; *EEA* iv no.57); unidentified, ?Thurlton, Norfolk (*CRR* viii 321); Compton Chamberlayne and Barford St Martin, Wilts. (PRO DL42/2 (Coucher Book of the Duchy of Lancaster II) fo.186r; BL ms. Stowe 882 (Blaunchard cartulary) fo.39r); and a prebend in the college of St Martin-le-Grand, London, by 1208 (*Essex Fines* 43 no. 241). Land at Bratton, Wilts. from Geoffrey de Mandeville (*CRR* ii 24), and at Clapton, Cambs. (*CRR* iii 332). Royal ambassador to France, August 1204, in the midst of the French invasion of Normandy. Ambassador to Flanders, December 1204 (*RLC* i 16, 32b; *PR 7 John* 112–13). Appears alongside des Roches as witness to a royal charter as early as 1201 (*Rot.Chart.* 86b). Continued in episcopal service at Winchester after des Roches' election, witnessing many of the bp's charters (Acta nos. 1,38,45,56,59,60,83). Served as des Roches' principal lieutenant during the bp's period as regent-justiciar 1214 (Appendix 1) and as proctor during an election at Bury St Edmunds (*Chronicle of Hugh* 110–11). His legal expertise no doubt valuable to the legally inexperienced justiciar. Found during the civil war at the bp's manor of Meon, taking receipt of grain, 1215–16 (Ms. 11DR m.6). Promoted royal treasurer after the re-opening of the exchequer, 1217, presumably through des Roches' influence. Remained in office until his death in 1228. Served in 1218 as arbiter in a dispute between des Roches and William Brewer (Actum no.82), and continued to witness charters associated with des Roches as late as 1218 × 1221 (Actum no. 35n). Granted custody of the lands of John of Boreham, Essex, by the king, October 1220 (*RLC* i 432). Prebendary of Holbourn, London c.1217 × 1221 (*Fasti* i 54). Elected bp of

London 25 February 1221, probably with the assistance of des Roches (*Henry of Avranches* 81-7; Ms. 16DR m.5; Actum no.30). Consecrated 25 April 1221 (*Fasti* i 2). As bp, in July 1223, agreed to pay 5 marks a year to the chapter of St Paul's to endow obit celebrations for kings Henry II and John, perhaps in gratitude for their patronage (CUL ms. Doc.126). Promoted the careers of several associates of des Roches, including Richard of Barking, master Philip de Fauconberg and master Luke des Roches. But rapidly detached from des Roches' circle of political intimates. Sided with archbp Langton and the other English bps against des Roches in 1223 and served as gaoler to Fawkes de Bréauté (Paris *CM* iii 87; *Hist.Ang.Min.* ii 265-6). Perhaps alienated after 1221 by des Roches' role in procuring Westminster's independence from the jurisdiction of the bps of London (Actum no.73; Appendix 2 no.23). Died late October 1228 (*Fasti* i 2)

31. Master Philip de FAUCONBERG, a kinsman, perhaps an illegitimate younger brother or a cousin of master Eustace. His mother, Agnes, commemorated in the obit list of Wherwell abbey on 18 March ('Wherwell Kalendar' 91). Eustace's mother apparently named Beatrice (*Chart. Pontefract* ii 491, 550 no.444). Appears alongside master Eustace in the household of bp Godfrey de Lucy of Winchester, August 1197 – April 1199, and perhaps to be identified with a man named Philip the clerk who witnesses a further half dozen of bp Godfrey's charters May 1197–September 1199, and who is said to have drawn up (*per manum*) a further ten charters, January 1201–August 1204, implying service as the bp's chancellor or scribe (*EEA* viii *passim*). Attorney to the prior of Newark Surrey in 1205 and witnesses at least two Newark deeds (*CRR* iv 29,39; BL mss. Harley Charters 45E32; 47B51). Witnesses three charters of des Roches after 1206, twice in company with master Eustace (Acta nos. 14,38,45). Witnesses a deed associated with master Eustace's settlement in favour of des Roches 1218 (*WC* no.329). Present at Winchester alongside master Alan of Stokes in 1218-19 when both men are described as learned in the law (*Cart. Saint Pierre* 193). Found at the bp's manor of Meon in 1220-1 (Ms. 16DR m.6). Witnesses numerous deeds of Wherwell abbey and on one occasion described as canon of the abbey, although prebendal church unidentified (BL ms. Egerton 2104A (Wherwell cartulary) fos. 32v–33r, 122r–124r, 169r, 176v). Witnesses a deed of the Hospitallers' preceptory of Baddesley alongside master Philip de Lucy (BL ms. Loans 29/57 (Baddesley cartulary) fo.32r). Promoted archdn of Huntingdon, March × August 1223, with an unidentified prebend at Lincoln (*Fasti* iii 28,114). Served by an archdn's official named master Richard of Waltham (BL mss. Add. Charters 33630-1). Appointed *c.*1223 to the prebend of Mora in St Paul's cathedral

following the election of Eustace de Fauconberg as bp of London, transferring after 1225 to the prebend of Caddington Major (*Fasti* i 33,62). Witnesses many of Eustace's acta (e.g. *St Paul's Charters* no.202; Sayers *Papal Judges Delegate* 331; BL mss. Harley 2110 (Castle Acre cartulary) fo.120v; Add. 43972 (Little Wymondley cartulary) fos.17v-18r). Also canon of Hereford, prebend unidentified (Rawlinson *Hereford* appendix p.30). Granted 5 marks in 1226 to sustain in royal service (*RLC* ii 127). Last occurs November 1228 as an executor of bp Eustace (*Cl.R.1227-31* 122). Dead, probably 2 or 3 December 1228 (*Fasti* iii 28; 'Wherwell Kalendar' 92). His own executors included master Philip de Lucy, master Geoffrey de Lucy dean of St Paul's and Reginald archdn of Middlesex. Executors delivered 20 marks to the nuns of Wherwell so that Philip's obit might be commemorated each year. Money used to purchase a rent in Winchester worth 13 shillings a year, of which ten shillings were to go to the pittancer and three shillings for the feeding of 100 paupers (BL ms. Egerton 2104A fo.193v). Will also bequeathed 3 marks and a tooth relic of St Ethelbert to Hereford cathedral (Rawlinson *Hereford* appendix p.30). Later commemorated alongside Eustace de Fauconberg in obit celebrations endowed by master Geoffrey de Lucy at St Paul's (*St Paul's Charters* nos. 299,300). Not to be confused with a namesake active as a Yorkshire property owner into the 1240s.

32. William fitz HUMPHREY, bearer of an extremely common name, can perhaps be identified with a namesake, clerk, granted Winnianton in Gunwalloe and Merther, both in Cornwall, by the king in December 1216 in lieu of clerical preferment, in reward for his services to king John (*RLC* i 294b; *Pat.R.1216-25* 14), and hence, perhaps, with the William fitz Humphrey presented to the church of *Beche* (?Waterbeach, Cambridge) in the diocese of Ely in May 1216 (*RLP* 183b). In September 1219 his presentation by king John to the Cornish churches of Winnianton in Gunwalloe, Cury and Breage was confirmed (*Pat.R.1216-25* 204; O.J. Padel, *A Popular Dictionary of Cornish Place-Names* (Penzance 1988) 58-9,77,120,188-9). The same William fitz Humphrey, former clerk of king John, received £10 of land in the St Pol escheats at White Roding, Essex, during bp des Roches' regime of 1233, retaining them despite an assize of *quo warranto*, until at least 1236 (*CPR 1232-47* 12-13; *CRR* xv no.1090; *BF* 590,1360). Des Roches' clerk of the same name first appears as witness to a charter before January 1217 (Actum no.45) and thereafter in 1219-20, receiving £25 at Adderbury and overseeing threshing and harvesting at Brightwell and Harwell (Ms. 15DR mm.5,5d). One of the most active of the bp's officers, witness to several of the bps' charters (Acta nos. 14,18,45,75), whose identity with the clerk of king John

can perhaps be substantiated by the journeys he made to Cornwall in 1231–2 (Ms. 27DR mm.7d,12). Spent much of 1220–21 at Taunton (Ms.16DR mm.4,9,10–11). Attended a forest regard in 1223–4 (Ms. 19DR m.4d), and in 1223 served as des Roches' attorney at the Bench (*CRR* xi no.451). Before the bp's departure for crusade, attended Martin of Pattishall at Southwark and found carrying gifts to the king at Woodstock and to the earl of Salisbury (Mss.20DR mm.14,15; 21DR m.1d). Appears to have combined a subordinate role within the bp's marshalcy (Ms. 21DR m.14d), with general functions as clerk, messenger and accountant. Following the bp's return in 1231, pardoned £9 received at Taunton and found at several manors, including Southwark, buying jewels (Ms.27DR mm.2,5,5d,6d, 7,8,8d,11d,12). In the next year oversaw the duke of Brittany's expenses at Bishop's Waltham, made a journey to Lincoln and found at Southwark for various pleas (Ms.28DR mm.3,10,16d). May have served as ambassador overseas in April 1233 (*CPR 1232–47* 14). In the bp's absence 1235–6 received 12 carts of timber at Merdon and incurred expenses at the Winchester fair (Ms.31DR mm.6d,15). Last occurs 1236–7 (Ms.32DR mm.9d,11,15). Nicholas prior of Christchurch Twinham, *c*.1242, granted a Stephen son of William fitz Humphrey a ten shilling rent at Whippingham, Isle of Wight, which William had held by grant of prior Roger (*c*.1216–1232) (BL ms. Cotton Tiberius D vi part i (Christchurch cartulary) fos.48v,106v–7r).

33. Master John of LIMOGES (*de Lemovic'*), Poitevin clerk, witness to charters of des Roches dated 1227 × 1231 and January 1233 (Acta nos. 18,85). Found at Southwark in 1223–4 (Ms.19DR m.12). In December 1223 dispatched by des Roches to Rome to protest against the assumption of power at court by archbp Langton and Hubert de Burgh (*Walter of Coventry* ii 263). In June 1224 given permission to cross to England. Appears to have done so, since, in September 1224, his earlier licence overturned and des Roches forbidden to retain him in his household because of the evil he had preached against the king in Rome (*Pat.R.1216–25* 447–8; *RLC* i 632b). Apparently exiled 1224 to 1229. In 1225 in attendance upon the papal legate in France, pleading the cause of Fawkes de Bréauté (*DD* no.182). Forgiven his speeches against the king and granted licence to return to England, February 1229 (*Pat.R.1225–32* 238). Returned by January 1233 at the latest (Actum no.18). Promoted by des Roches to the Winchester living of East Woodhay before June 1233 (*Cal.Pap.Reg.* i 136; *VCH Hampshire* iv 306). As bp's clerk presented by Nuneaton priory to the church of Blendworth, in succession to master R. de Grandon who had been admitted before 1212 (Acta nos.37,39). Papal licence to appoint a vicar to Woodhay in order to return to the schools,

June 1233 (*Cal.Pap.Reg.* i 136). At much the same time appointed king's proctor in Rome, making his last recorded appearance in this capacity in November 1233 (*CPR 1232–47* 14,32). Succeeded as rector of Blendworth before March 1238 by a clerk named Aimery of Limoges, perhaps a kinsman (Actum no.40). Possibly, though unlikely, to be identified as the author of a treatise on kingship, the so-called 'Morals of the Dreams of Pharaoh', dedicated to Theobald I king of Navarre (1234–53), of which at least three mss. survive in England (Oxford, Balliol College ms. 163; Salisbury Cathedral Library ms.8; BL ms. Add. 62132). Elsewhere authorship ascribed, probably erroneously, to a Cistercian namesake active in France *c.* 1270, whose works in turn have been misappropriated by Hungarian commentators and ascribed to an abbot of Zirc (*c.*1208–18) (intro. to *Iohannis Lemovicensis Opera Omnia*, ed. C. Horvath, 3 vols. (Veszprém 1932); partially corrected by P. Glorieux *Répertoire des Maîtres en théologie de Paris au xiiie siècle*, 2 vols. (Paris 1933) ii 252–4).

34. Master Philip de LUCY, probably an illegitimate son of bp Godfrey de Lucy, in 1224 granted papal dispensation for default of birth (M. Cheney 'Master Geoffrey de Lucy' 758–61). Definitely brother of master Stephen de Lucy (Cole *Documents* 272), who in turn is described as brother of bp Godfrey's foster-son, William de Lucy of Charlecote (*E.Rot.Fin.* i 130; *RLC* i 14b). Witnesses several charters of bp Godfrey in August 1204 (*EEA* viii *passim*). Witnesses early charters of Thelsford priory, Warwicks., founded by William de Lucy of Charlecote, foster-son of bishop Godfrey, *c.* 1205 × 1208 (Bodl. ms. Phillipps-Robinson e.77 (Thelsford cartulary abstract) fos. 1r–v,4v,13r). Witnesses a charter of Geoffrey de Lucy relating to family land in Cornwall (BL Harley Ch. 53B42). Presented to the churches of Selborne, Basing and Basingstoke and admitted by bp Godfrey 13 August 1197 (BL Add. Ch.19068). Letters of presentation by Mont-St-Michel dated 25 March 1204, in which Philip is said to have succeeded master William de Ste-Mère-Église, elected bp of London December 1198 × May 1199 (*Selborne Charters* ii 2). Presented by Lire abbey to Newchurch, Isle of Wight, and admitted by bp Godfrey before 11 January 1201 (*Carisbrooke Cart.* no.37; BL ms. Egerton 2104A (Wherwell cartulary) fos. 99r,114v–15r). Also rector of Cuddington, Surrey, presented by Merton priory 1198 × 1206, having previously served as perpetual vicar of the same (BL ms. Cotton Cleopatra C vii (Merton cartulary) fo.66v; Heales *Records* 63; *CRR* xv no.190). Rector of St Mary's Southampton from before March 1224 (*St Denys Cart.* no.283; *WCM* no.10662). In June 1206 presented unsuccessfully by the crown to the treasurership and the prebend of Carlton Kyme at Lincoln (*RLP* 66b; *Fasti* iii

19). Earlier in the same year presented by the crown to the church of Middleton (?Cheney, Northants.), perhaps effectively (*RLP* 59; *RLC* ii 169). In 1205 fined 60 marks for custody of the lands and heir of Richard of Warwick in Yorks., Warwicks. and Lincs. (*EYC* vi 109–10; *RLC* i 21b; *Rot. Ob.* 255). Also held a lay feoff at Micheldever, Hants.; half a hide at farm from Hyde abbey, which Philip restored to the monks after 1222 (BL mss. Cotton Domitian A xiv (Hyde cartulary) fos.54v–55r; Harley 1761 (Hyde cartulary) fo.41v). Farmed the manor of Bishop's Lydeard, Somerset, from the bpric of Bath, 1205 (*RLC* i 54). In the same year his men involved in an affray with followers of the Hampshire baron, Adam de Port (*RLC* i 52b; *RLP* 84b). Served as clerk of the king's chamber March 1205–March 1207, witnessing and authorising payments *in camera* (*RLP* 51,56b,59b–60,61–2,65b,66b; *RLC* i 21b *passim* 79b). Accompanied the king to Poitou 1206 (*RLC* i 74,75). On the king's return to England, problems persuading the treasury at Westminster to release money to Philip (*RLC* i 76–76b). Philip still witnessing chamber writs throughout February 1207, last appears in an official capacity 13 March 1207, sent to the treasury to take receipt of £210 (*RLC* i 77b–79b,103). Dismissed by 23 July 1207 when fined 1000 marks in return for release from all prests, receipts and arrears he owed for the chamber. Payments spread over three years and met in full by Michaelmas 1210 (*RLP* 74b,82; *Rot.Ob.* 382,386–7; *PR 9 John* 149–50; *PR 10 John* 124; *PR 11 John* 168; *PR 12 John* 187). Probably the outcome of a struggle between treasury and chamber, leading to the establishment of new provincial treasures 1207–8, but perhaps some suspicion of sharp practice against Philip (Jolliffe 'The Chamber and the Castle Treasures' 121,128–9). Philip excused usury on his Jewish debts to help meet his fine and for the same reason restored to his goods in April 1208 following temporary seizure at the start of the papal interdict (*RLC* i 89,108b). Letters of protection September 1213 for a pilgrimage to Compostela (*RLP* 104b). Worked alongside des Roches in the king's chamber and continues to witness the bp's charters as late as 1218 (Acta nos. 14,35; *WC* no. 329). Given a cheese at the bp's manor of Adderbury, 1218–19 (Ms. 14DR m.6). Served 1216 × 1221 as papal judge delegate (*Sarum Charters* 111). In 1221–2 his own rights as rector of Selborne subject to papal judges (*Selborne Charters* ii 14). Close association with master Philip and Eustace de Fauconberg, witnessing charters alongside Philip (BL mss. Loans 29/57 (Baddesley cartulary) fo.32r; Harley Ch. 45E32) and several charters of Eustace as bp of London (e.g. Sayers *Papal Judges* 331; BL ms. Add. 43972 (Little Wymondley cartulary) fos. 17v–18r). Witnesses alongside master Philip de Fauconberg and master Geoffrey de Lucy, a grant by Isabella de Lucy relating to Lucy family land in Cornwall (M. Cheney 'Master Geoffrey de Lucy' 763). Served as executor of master Philip in company with master Geoffrey de Lucy, November 1229 × 1232 (BL ms. Egerton 2104A (Wherwell cartulary) fo.193v). In 1224 master Stephen de Lucy, Philip's brother, served as king's proctor in Rome, representing the interests of archbp Langton and Hubert de Burgh against des Roches (*DD* nos.136,149,182–3). Probably master Stephen who obtained papal dispensation for Philip's illegitimacy (*Cal.Pap.Reg.* i 95), perhaps so that Philip could proceed to higher office. A namesake occurs, only once, pre *c*.1230, perhaps 1223 × 1226, as archdn of Wiltshire (PRO E40/4821, temp. Alan Basset and Alice de Gai his wife, lords of Wootton Bassett, Wilts. For possible dates see *Fasti* iv 36 and n.). Granted hunting rights by the crown 1230 (*Cl.R.1227–31* 340). Last occurs after April 1231, subject to papal judge delegacy as rector of Basingstoke, seeking tithes against Monks Sherborne priory (*Selborne Charters* ii 14). Dead by April 1233 (*CRR* xv no.190). Commemorated by an obit at Wintney nunnery 5 kalends October, so presumably d. 27 September 1231/2, more likely 1232 (BL ms. Cotton Claudius D iii fo.158r). His former benefices of Selborne, Basing and Basingstoke, together with the manor of Selborne held previously by his brother, master Stephen, went to form the nucleus of des Roches' foundation at Selborne priory. Not to be confused with a younger namesake, fl.1245–47, rector of Overton, Hants., dispensed for illegitimacy at the prompting of Richard earl of Cornwall (*Cal.Pap.Reg.* i 226,231).

35. Master William de SAINT-MAIXENT (*de Sancto Maxentio*), perhaps a native of St-Maixent, near Le Mans, dep. Sarthe. First appears in royal service, October 1203, mentioned in a writ authorised by des Roches as treasurer of Poitiers (*Rot.Norm.* 108). Occurs from 1205 as a junior clerk in the royal chamber, authorising payments, often in association with des Roches (*RLC* i 18–103b *passim*; Cole *Documents* 261, 271–2,274). Appears on the Winchester estates between 1208 and 1217–18 (*PR 4DR* 59,68; *PR 6DR* 128,179; Mss. 7DR m.4; 13DR m.14). Described as bp's clerk in 1213 when delivering instructions on the release of money from Winchester (*RLP* 103b). During the king's expedition to Poitou in 1206 granted a prebend in the collegiate church of St Martin, Angers, and in 1214 as king's clerk promoted dean of the same, a position held before 1204 by des Roches (*RLP* 67; *Rot.Chart.* 199b). Accompanied king John's expedition to Ireland, 1210 (*Rot.Lib.* 187,227). Rector of Ewhurst, Surrey, promoted 1206 × 1207, the advowson lying with Merton priory (*CRR* iv 206, v 242; *VCH Surrey* iii 101). In 1209 promoted to a prebend in the royal college of Bridgnorth (*RLP* 89). Rector of Mixbury, Oxford., from before December 1216 when engaged with

Osney abbey in a dispute over tithes. Canon of Lincoln, prebend unidentified (?Thame), by May 1218 (*Oseney Cart.* nos.867a,868a; *Fasti* iii 101,149). Appointed to the Hampshire circuit of the 1219 forest eyre alongside William of Shorwell and Maurice de Turville (*Pat.R.1216-25* 212,217; PRO SC1/2/70). Witnesses (?c.1220) a deed of Missenden abbey, Bucks. (*Missenden Cart.* iii no.663). Thereafter unrecorded, perhaps returned to France. His church of Mixbury vacant by December 1222 (*Rot.Hugh Welles* ii 9). Tentatively identified as William archdn of Buckingham 1219-21 (*Fasti* iii 40). Witnesses one charter of des Roches, 1211 × 1224 (Actum no.26).

36. Master Nicholas de VIENNE, a Frenchman of unidentified birth, possibly a Norman. First appears 1218/19 at Farnham in association with master Humphrey de Millières (Ms. 14DR m.9). Served in the autumn of 1220 as attorney of the bp of Avranches, himself a guest at des Roches' manor of Bitterne in 1219-20, in a dispute over the manor of Swanwick. Swanwick later held from the bp of Avranches by master Humphrey and given to Titchfield abbey (*RLC* i 431, 437b-38; Ms.15DR m.2; Actum no.67a). In 1223-4 found at Bishop's Waltham with the archdn of Winchester (Ms.19DR m.2) and in the following year at several manors, including Bishop's Sutton together with master Humphrey de Millières, returning from London where they had attended archbp Stephen Langton (Ms. 20DR mm.1,2,6,9d). In 1226 found for eight days at Southwark around January 13, attending the council convened by the papal nuncio Otto, presumably as des Roches' proctor, to discuss proposed subsidy to Roman clergy (Ms. 21DR m.14; *C.& S.* 155). At court reimbursed 10 marks expenses as a *protégé* of des Roches in April 1223 (*RLC* i 540). Served June 1223 as royal ambassador overseas, possibly to France (*RLC* i 550b,551). Proxy of Bartholomew des Roches archdn of Winchester, serving as papal judge delegate before 1230 (*Brecon Cartulary* 166-7). Witnesses two undated charters of des Roches (Acta nos. 65,80) and a settlement by master Alan of Stokes (Winchester, Hampshire Record Office ms. 13M63/1 (Mottisfont cartulary) fo. 69v). No recorded appearance in des Roches' household after the bp's departure for crusade 1227, but possibly maintained contacts with England. In 1247 a namesake granted robes by the king in company with Roger Aliz, another former familiar of des Roches. In May 1248 pardoned eleven and a half marks amercement from the Hampshire forest eyre and in September granted venison by the king (*Cl.R.1247-51* 15,52,86). In the same year recorded as rector of Donnington (West Sussex), whose advowson lay with Hyde abbey (*Chichester Cart.* no.282). A John, son of master Nicholas *de Vyane*, accused of forest offences within the bailiwick of

APPENDIX 4

Aymer de Valence, bp-elect of Winchester, c.1254 (PRO E32/157 m.6). Presumably not to be identified with a master Nicholas de Vienne granted 3 years' protection overseas, August 1272 (*CPR 1266–72* 672).

KNIGHTS:

37. Roger ALIZ (II), heir to the manor of Allington near Southampton, held by William Aliz as early as 1086 and in 1166 by Roger Aliz I as one fee in chief from the crown. The family perhaps a cadet branch of a baronial family of Breteuil in Normandy (*VCH Hampshire* iii 485; *Red Book* 28,705; Crouch *Beaumont Twins* 106n). Roger's ancestors made grants in the manor to several local religious houses, most prominently to St Denys' Southampton (*St Denys Cart.* pp.6–7; *VCH Hamsphire* ii 221; *WCM* no.10661A). Roger II first appears c.1202 × 1215, selling further land at Allington to God's House Southampton, in return for 19 marks, a silver cup for Agnes his wife and gold rings for his daughters, Agatha, Joan, Petronella and Isabella (*God's House Cart.* 134,161–2). Given that Roger and his son Thomas continued to alienate portions of the manor, and that Allington was their only recorded asset, it seems reasonable to suggest that economic factors encouraged Roger to enter the service of des Roches. By 1230 Roger owed at least 24 marks to the Winchester Jews (*St Denys Cart.* p.7n; *WCM* nos. 1558–9; PRO E326/28; *Cl.R.1227–31* 428; *PR 14 Henry III* 196; PRO E372/79 m.5). First appears in the bp's service 1211–12 at Brightwell and travelling from Farnham to London (Ms. 7DR mm.3,8). Thereafter makes occasional appearances on the bpric estates 1215–1232, representing the bp in an assize at Witney 1226–7 and in 1231–2 carrying a writ from the bp to Taunton (Mss. 11DR m.2; 20DR m.1; 22DR mm.5,13d; 27DR mm.5,7d). Witnesses several of des Roches' charters, 1209–1238 (Acta nos. 9,13,21,51–2). Granted temporary custody of rebel land at Wallop, Hants., and at Tew in Oxfordshire during the civil war 1215–17 (*RLC* i 232b,236b,305). Granted forty shillings by the king in 1223 (*RLC* i 534b). Served as a Hampshire knight of assize 1229 and in the following year accompanied the king to France (*Pat.R.1225–32* 292; *Cl.R.1227–31* 428). During des Roches' political ascendancy after 1232, promoted keeper of Bristol, serving as *de facto* constable under Peter de Rivallis, and having custody of the king's cousin, Eleanor of Brittany, from November 1233 until the bp's downfall in May 1234 (*CPR 1232–47* 43; *Cl.R.1231–4* 331–2,341,364). Rewarded with temporary grant of rebel land at Cheddar, Somerset and a £10 debt transferred to him in Cardiff (*Cl.R.1231–4* 383–4). During the same period witnesses numerous charters associated with des Roches' foundation at Selborne, often in company with Thomas, his son

(*Selborne Charters* i 4,20–22, ii 65; Actum no.52). Father and son continued in association with the see of Winchester into the 1240s, witnessing at least one charter of bp William de Raleigh (*Reg.Common Seal* no.306). In 1242 Roger's manor of Allington described as half a fee held of the crown via Isabella de Mortimer (*BF* 704). In the same year Roger appointed to take the Hampshire assize of arms (*Cl.R.1237–42* 485). Impleaded over custody of a ward 1245 (*Cl.R.1242–7* 365). Thomas, Roger's son acting as an independent landholder by 1256 at the latest (*New Forest Docs.* no.29; *E.Rot.Fin.* ii 233). But Roger apparently still living, though presumably a very old man. Received robes from the crown and gifts of wine 1247–56 (*Cal.Lib.R. 1251–60* 342; *Cl.R.1247–51* 15,56). Witnesses a grant to Aymer bp-elect of Winchester and with the seizure of Aymer's estates after 1258, granted robes as he was wont to receive them from the bpric, until as late as June 1261 (*Cal.Lib.R.1251–60* 478,525; *Cal.Lib.R.1260–7* 41; *Cl.R.1256–9* 442; *Cl.R.1259–61* 335,392,397). The Aliz landholding at Allington survived until the failure of the male line c.1304 (*VCH Hampshire* iii 485).

38. Eustace de GREINVILLE, sprung from an important Anglo-Norman family, native to Grainville-la-Teinturière, Seine-Maritime. One branch of the family tenants of the Giffard honour at Longueville in Normandy and Chilton, Bucks., and hence later of the Marshal earls of Pembroke, heirs to the honour of Giffard (*Early Bucks. Charters* 65–8; Loyd *Anglo-Norman Families* 47–8; *Red Book* 312; 'Nutley Cartulary' *passim*; Bodl. ms. Dugdale 39 (Notley cartulary extracts) fos.68r–73v; *BF* 880,1155; *Longueville Charters* nos. xi,xxii). Another, closely related branch tenants and possibly kinsmen of the earls of Gloucester, holding at Stambourne in Essex and Bideford in Devon (*Early Bucks. Charters* 69–70; *Red Book* 154,291; *Gloucester Charters* nos. 51–2,70–1,95,174,284; *Stoke by Clare Cart.* ii nos.300–307, iii p.13; *Cal.Chart.R.1226–57* 328). The relationship between the various branches in England and Normandy clouded by the appearance of several men with the same Christian name, but clear that family maintained cross-channel connections long after the loss of Normandy in 1204. In the 1220s the Giffard tenancy at Grainville-la-Teinturière was partitioned, following the death of another Eustace de Greinville, between men named Robert de Greinville and Eustace de Greinville, a clerk. Robert's father, Girard, is said to have died in England (*Cart.Normand* no.351). Eustace de Greinville, des Roches' knight, was probably of Norman birth. After his death, his brother and heir, William, was said to be an adherent of the king of France in Normandy (*CRR* xvii no.110). Eustace's lands in England all apparently by new enfeoffment rather than inheritance, so probably a younger son newly introduced from

Normandy; perhaps family lands in Jersey, later passed to a kinsman, Gilbert de Greinville (*RLC* i 349b; *Cal.Chart.R.1226-47* 244; *Plac.Quo Warrant.* 839b). Possibly to be identified with a namesake who appears 1200-2, active in the defence of Normandy (*Rot.Norm.* 44; *Rot.Ob.* 186). Enfeoffed before 1202 by Aimery count of Evreux, heir to the honour of Gloucester, with 100 shillings of land at Sheet in Petersfield, rendering the service of one third of a knight's fee (*Gloucester Charters* no.94; *Red Book* 154; *PR 3 John* 56). By 1208-9 active in the household of bp des Roches (Actum no.21; *PR 4DR* 3-5, 7,17,45,48,75). In 1210, his service of one third of a fee for Sheet transferred from the honour of Gloucester to bp des Roches (Actum no.9). May also have held of the bp elsewhere, since in 1225-6 a moiety of a mill newly built at Meon was said to belong to one of Eustace's men (Ms.21DR m.10). Married c.1213, through des Roches' influence, Isabella daughter of William Painel and widow of William Bastard, heiress to part of the honour of Hooton Pagnell, Yorkshire (Sanders *Baronies* 55; *RLC* i 322,323b,422b; *Rot.Ob.* 494). Served under des Roches, from 1214 until after the fall of London to the rebels in May 1215, as constable of the Tower, held by the bp by virtue of his office as justiciar (*RLC* i 175b,182,198b,450; *Memoranda 10 John* 132). Active on the episcopal estates 1208-1226, regularly respited scutage for his fee at Sheet and in 1213-14 exercising some sort of custody over the the future Henry III, then in the bp's keeping (*PR 6DR* 42,75,96,109,127; Mss. 7DR mm.4d,7; 9DR mm.5d,6d,8,10d; 11DR m.5; 14DR m.12d; 20DR m.5; 21DR mm.8,14d). Appointed steward of the royal household c. September 1217, serving until at least December 1225 (*RLC* i 322,345b,355 *passim* 578, ii 13,22b,25b,36b, 86b,87b). Rewarded with various privileges and grants of land, including custody of Barton, Yorks., originally as part of the Painel inheritance, but seized back after Eustace's dismissal from the stewardship, as a Norman escheat (*RLC* i 322,323b,399, ii 25b,125b,131; *CRR* viii 2,64,95-6,241-2,289-90). Attempted, unsuccessfully, to obtain the church of Barton for R. the clerk, his nephew, probably to be identified as master Robert, Eustace's nephew, presented, again unsuccessfully, to the church of Somerton, Oxfordshire in 1231 (PRO SC1/2/4; *Rot.Hugh Welles* ii 35). First wife dead by 1220, when her lands passed to her aunt, Frethesant widow of Geoffrey Lutterel (Sanders *Baronies* 55). Eustace married again, probably before 1223, Joan daughter of Robert de Arsic, heiress after 1230 to half of the barony of Cogges, with lands in Oxfordshire, Lincs. and at Hallaton in Leics.(Sanders *Baronies* 36-7; *CRR* xi no. 617; *BF* 451,455,546,549,1064; and for Hallaton see *CRR* xiii no.1558; *RLC* i 618b,643b; *BF* 520,524-5; *Pat.R.1225-32* 302; *VCH Leics.* v 124). Eustace seeking warranty of half the manor of Hallaton against Robert de Arsic and Sybilla his wife as early as October 1223 (*CRR* xi no.617).

Even earlier, in January 1223, he obtained a carucate and thirty two acres of land at Dungewood in Shorwell and Wilmington in Freshwater, both in the Isle of Wight, lying within manors where Joan de Arsic may have had some claim via her uncle (d.1205), married to Margaret de Vernon, heiress of Freshwater (*CRR* x 110; PRO CP25(1) 203/5 no.8; Sanders *Baronies* 36). After Eustace's death, Joan and her second husband still active as landholders at Atherfield, within Shorwell, perhaps to be identified with the estate at Dungewood held earlier by Eustace (*VCH Hampshire* v 281). Served on the Welsh campaign of 1231 (*Cl.R.1227-31* 547; PRO C72/2 m.10d). Transferred his land at Sheet, held of the bpric of Winchester, to des Roches' foundation at Selborne 1234 × 1238 (Actum no. 51; *Selborne Charters* ii 65, which preserves Eustace's seal, badge charged with 3 scallop shells; *Reg.Pont.* 389-90), possibly to avoid it escheating as Norman land. Died before January 1241, when his interest in the barony of Cogges reverted to his wife who sold it almost immediately to Walter de Gray, archbp of York (*CRR* xvi nos. 758,1403,1465,1691,2572,2625; *BF* 1064; Sanders *Baronies* 36; *Cal.Chart.R.1226-57*, 264-5,270,285; Eustace had mistakenly been recorded as dead, July 1231: PRO JUST1/1042 m.26d). Eustace's heir and brother, William, said to be a Norman adherent of the French and therefore Eustace's other lands seized as Norman escheats (*CRR* xvii no.110). A claim to his estate at Hallaton, Leics., raised by Gilbert de Greinville, presumably a kinsman, former knight of the royal household, holding at Drayton Beauchamp, Bucks., of the honour of Berkhamsted, probably by new enfeoffment as constable of Berkhamsted, and at Prested, Essex, by right of his wife. Gilbert also held in Jersey by grant of Eustace (*BF* 482,877,892,1152; *CRR* xv no.1252, xvii no.110; *Cal.Chart.R.1226-57* 244; *RLC* i 507b,565b,617). Gilbert's claim to Hallaton apparently upheld, his portion of the manor descending *c*.1248 to a daughter named Joan, married to John de Engaine. Joan later described as a descendant of Eustace de Greinville (BL Add.Ch.26937). Eustace's widow dowered with land at Hallaton, held of Joan, Gilbert's daughter; and also apparently at Atherfield, in the Isle of Wight (*Berkshire Eyre* no.495; *VCH Leics.* v 124; G.F. Farnham, 'Hallaton: Notes on the Descent of the Manor', *Transactions of the Leicestershire Archaeological Society* xiii (1923-4) 149; *VCH Hampshire* v 281). Eustace not to be confused with a namesake, son of Richard de Greinville, granted the Greinville family interest in the manor of Stambourne, Essex, *c*.1244 following the failure or defection to Normandy of the male line there and the land's reversion to the crown as *Terra Normannorum*. The Greinville tenancy at Chilton, Bucks., of the Giffard fee, also suffered seizure as Norman land following the death of Robert de Greinville *c*.1244, being granted to Paulinus

Peyvre by the king (*Cal.Chart.R.1226–57* 328; *Cl.R.1242–7* 192,204; *CPR 1247–58* 4; *BF* 1404,1410). Before 1248 Paulinus Peyvre enfeoffed Gilbert de Greinville, kinsman of Eustace de Greinville the steward, with land at Chesham (*[C]estresham*), Bucks. (BL Harley Ch. 54G10, and see *Early Bucks. Charters* 11,42–3), a further indication of the tangled interweaving between the Norman and English branches of the family. It is intriguing that Eustace's family should have been so closely associated with the Giffard lands in Normandy, and hence with the Norman lordship of Richard Marshal, des Roches' political rival of the 1230s.

39. Geoffrey de LUVEREZ, brother of Michael de LUVEREZ who held Steventon near Basingstoke as half a fee from the crown from before 1194 (*Red Book* 91,148). Related to landholders at Middleton, Wilts. and West Tytherley, Hants., including a namesake, Geoffrey de Luverez (d.1213). These kinsmen held by serjeanty, acting as keepers of the king's wolfhounds (*luverez*) (Round *King's Serjeants* 293–6; *Red Book* 485; *BF* 12; *CRR* vi 8,44,58; *Rot.Ob.* 473,478; *PR 12 John* 38; *VCH Hampshire* iv 522). Kinship proved by the seals of Michael and Geoffrey, both of which depict a wolf's head, with lolling jaw (Oxford, Queen's College Monks Sherborne charters 241–2). Michael appears to have forfeited his lands by remaining in Normandy after 1204. In 1210 Geoffrey fined 50 marks for Steventon (*PR 12 John* 192; *E.Rot.Fin* i 261; *VCH Hampshire* iv 171). Already active as a Hampshire knight 1205 (*CRR* iii 283, iv 292). Between 1220 and 1233 augmented a grant to Monks Sherborne priory in Steventon by Michael de Luverez and Margaret, Michael's wife, to endow Michael's obit celebrations (Oxford, Queen's College Monks Sherborne Charters 241–2). Active in des Roches' household from at least 1210–11, apparently acting within the bp's butlery or as *de facto* household steward, receiving grain and stock for the bp's kitchens and preparing for the bp's reception as he travelled the country (*PR 6DR* 68,91,120,128; Ms. 7DR mm.1,2,4d,5d,7,9,10,11d). In 1211–12 made similar provision for the king's arrival at Marwell, Meon and Witney (Ms. 7DR mm. 1d,7,8d) and in 1213–14 overseeing the expenses of the future king Henry III at Merdon and Bishop's Waltham (Ms. 9DR mm.2d,7). Continued to receive stock and provisions 1215–21, frequently in association with Peter de Chanceaux (Mss. 11DR mm.3,4,6; 13DR; 14DR *passim*; 15DR mm.3,9; 16DR mm.4,6). In 1220–21 held three burgages in the bp's new borough at Overton (Ms.16DR m.4d). Possibly minor land holding at Carisbrooke, Isle of Wight, 1219 (PRO CP25(1) 203/4 no.19). Witnesses two of the bp's charters, 1218 × 1227 (Acta nos. 35,61). Involved in actions over land at Steventon and Well in Long Sutton, 1224 (*CRR* xi nos. 787,1640,2302). In 1231 defending his tenure at

Steventon against a claim from the crown that it should escheat as Norman land (*Cl.R.1227-31* 581). Last appears September 1232, respited debts to the Winchester Jews (*Cl.R.1231-4* 108). Apparently dead by October 1233 when Steventon was granted by the king to Geofrey des Roches, the bp's nephew, as a Norman escheat (*CPR 1232-47* 38; *Cl.R.1231-4* 282). Geoffrey des Roches lost custody of the manor with the bp's political downfall in 1234 and the manor passed to Luverez' kin; his sister Annora, married to Hugh of Wingham, and his nephew Philip de Saunderville who fined £80 for custody (*E.Rot.Fin.* i 261). In the Hampshire eyre of 1236 a Nicholas de Luverez was impleaded over one virgate of land at Steventon on a writ of *mort d'ancestor* by Geoffrey Burel (PRO JUST1/775 m.6). In the 1240s Geoffrey des Roches revived his claim to the manor against Hugh of Wingham and Manasser de Saunderville. Steventon was eventually annexed to the des Roches estate through purchase *c.*1258, when Michael de Luverez (?a Norman), nephew of Michael, Geoffrey's brother, issued a quitclaim in favour of Martin des Roches (*Cl.R.1256-9* 310; *CRR* xvi nos. 1176,1287,2558; *CIPM* ii no.225).

40. Maurice de TURVILLE, probably a younger brother of William de Turville (d.1222) and hence related to the family of Turville of Weston Turville, Bucks., descended from Tourville, Eure, major tenants of the Beaumont honours of Leicester and Warwick (Crouch *Beaumont Twins* 116-20,218-9). William de Turville held a mesne tenancy at Bramshill in Hampshire from the earls of Warwick (*BF* 705; *CIPM* i no.10) and the manor of Lomer from the St John fee of Hyde Abbey (*VCH Hants.* iii 250). After William's death Lomer passed to Maurice de Turville who, in 1222, awarded dower there to William's widow Annora or Amicia, by then remarried to Thomas de Chaucombe (*CRR* x 175; PRO CP25(1) 203/5 no.1). Maurice's other Hampshire tenancies appear to have been newly created, since no record of them is to be found in the returns for 1166. From the St John/Hyde fees he held a virgate in Micheldever (*BF* 46). From the bps of Winchester he held a half fee in Wield near Preston Candover (*Reg.Pont.* 387), paying a quit-rent to Southwick priory for grazing rights at Candover (*Southwick Cart.* vol.1 115). First appears in 1200 as witness to a charter of bp Godfrey de Lucy (*Southwick Cart.* vol.2 no.550). Said to have acted as Godfrey's steward, the bp marrying Maurice's sister to the father of Robert of Morestead (*Reg.Pont.* 666,756). In 1210-11 Maurice was amerced for offences in Cornwall, an account subsequently transferred to Hampshire, but suggesting enfeoffment at a junior level with land in the de Lucys' Cornish estate (*PR 12 John* 41; *PR 13 John* 160,184; *PR 14 John* 96). Appears regularly as a witness to Hampshire deeds 1200-1225 and as a knight of assize (*CRR* passim). Witnesses a charter

of des Roches as early as 1210 (Actum no.9, and see *Selborne Charters* ii 63). Found regularly on the bpric estates 1208–1212 (*PR 4DR* 45; *PR 6DR passim*; Ms. 7DR mm.1d,11). Appointed as the bp's attorney in 1214 (*CRR* vii 275). In the same year respited sixty shillings of scutage owing to the bp (Ms. 9DR m.10d), and regularly respited scutage 1218–26 (Mss. 14DR m.12d; 20DR m.15; 21DR m.14d). Probably constable of the king's castle at Winchester in 1214 (*RLC* i 213) and serving jointly there together with William de Falaise in May 1215 when described as the bp's knight, apparently acting as custodian of the future king Henry III (*RLP* 136). Appointed alongside other Hampshire knights as royal spokesman to the county community, February 1215 (*RLP* 128b). Despite witnessing numerous charters of his lay lords, Adam de Port and William de St John (*Montacute Cart.* nos. 110–11; Oxford, Queen's College Monk's Sherborne Charter 228), did not follow the St Johns into rebellion after 1215. With the recapture of Winchester castle from the rebels in 1217, entrusted with custody of rebel hostages (*Pat.R.1216–25* 79). Served as justice on the southern circuit of the general eyre of 1218–19 (Crook *Eyre* 72). Appointed escheator in Hampshire 1218 and to Hampshire forest enquiries in 1219 and 1223 (*Pat.R.1216–25* 172,212,217,401 and see Actum no.82). He, or more likely a namesake served as coroner for Gloucestershire before 1225 (*RLC* ii 24b–25). Occurs as witness, together with his sons Peter, Henry and William de Turville after 1217 (*WCM* nos. 12096,16609; BL ms. Cotton Domitian A xiv (Hyde cartulary) fos. 48v–49r). Last certain occurrence 1236, when fined half a mark to concord with William and Roger his sons on a plea of warranty (PRO JUST1/775 m.3). Occurs on a schedule of the bp of Winchester's military tenants *c*.1245 (*Reg.Pont.* 387). Apparently succeeded by Peter his son (*VCH Hampshire* iii 250), but by the 1280s his half fee at Wield held of the bpric estate by William de Wyvenhulle (*Reg.Pont.* 593).

KINSMEN:

41. Peter de RIVALLIS or de ORIVALLIS, the bishop's nephew (*CRR* xv no.1031; Ms. 22DR m.5), branded des Roches' son with malicious intent by the St Albans chroniclers (Paris *CM* iii 220,240,245,252,265). Birthplace unknown. Generally described as a Poitevin (?native of Orival, Airvault or Airvau), but could perhaps have been Norman, native of Orival or Roche d'Orival near Rouen. A Peter de Orival occurs 1164 × 1202 as witness to a grant by Geoffrey de Orival in Ste Colombe near Rouen (L. de Glanville, *Histoire du Prieuré de Saint-Lô*, 2 vols. (Rouen 1890–1) ii 316 no.15d). This man is unlikely to be the same Peter, described as son of Roger fitz Odo de Orivalle granting land to Foucarmont abbey in *Busci Fresnai* (?Fresnoy-en-

Val, near Orival, cant. Neufchâtel-en-Bray) before 1178 (Rouen, Bibliothèque Municipale ms. 1224 (Foucarmont cartulary) fo.22r). Possibly the nephew of des Roches captured at Caen in 1204 and ransomed for an important Norman knight (*RLP* 45b). First certain appearance, 9 June 1204, in England, granted all the churches which the late Gilbert de Beseby held in the county of Lincs., perhaps ineffective, as was a subsequent grant of the next vacant prebend at Lincoln cathedral in March 1208 (*RLP* 43,80b,84b). Eventually promoted to the Lincoln prebend of Welton Ryval, to which he gave his name (*Fasti* iii 107). Rector of Trumpington, Cambridge, by 1225 (*Cambridge Portion* nos.42,55). Succeeded Roger I, archdeacon of Winchester (d.1208) as rector of Alton, Hants.(*CRR* iii 119, xvii no.2013; *Cl.R.1227-31* 543; *AM* ii (Waverley) 353). Rector of Mottisfont, Hants., probably presented by William Brewer. Received timber to repair his houses there, recently burned, April 1223 (*RLC* i 543). In October 1228 involved in a dispute with the prior of St Denys' Southampton over the respective rights of the parish churches of Mottisfont and East Tytherley, and in May 1229 sued the prior and convent of Mottisfont before papal judges delegate, for 20 marks claimed in tithes and offerings lost by his parish church, obtaining an annual pension of 1 mark a year henceforth (*St Denys Cartulary* no.341; Winchester, Hampshire Record Office ms. 13M63/2 (Mottisfont cartulary II) fos. 148r-9r). Rector of Claverley, Shrops., promoted 1222 × 1228, probably before 1224 in reward for service at court (*BF* 384, 1342). Promoted dean of the royal college of Bridgnorth, Shrops., May 1223 (*Pat.R.1216-25* 372). Occurs regularly on the Winchester estates from at least 1211-12, and probably to be identified as Peter, the bp's nephew, in 1208-9 travelling to Lincoln and granted firewood whilst staying at Oxford, perhaps at the schools (*PR 4DR* 17-19; Mss. 7DR m.4; 9DR m.8d; 13DR mm.3,5,5d,8; 14DR m.1; 15DR m.6; 16DR mm.2,2d; *WC* no.463). Intriguing that both in royal chancery sources and the Winchester pipe rolls, the spelling of his name changes from Orivallis to Rivallis c.1223. Introduced to royal administration following the end of the civil war, serving as clerk or chamberlain of the king's wardrobe between November 1218 and the downfall of bp des Roches late in 1223 (*RLC* i 383,391b,410b,415b *passim* 574; *Pat.R.1216-25* 329). Accused c.1223 of poaching a clerk named P. Orelluz from the Poitevin magnate Hugh de Lusignan (*DD* nos. 128-9). Granted wardship of the heir of Robert of Bassingham, 1222 (*E.Rot.Fin.* i 91), but dismissed from court following the fall of bp des Roches in 1223-4. Occurs on the Winchester estates, 1224-7 (Mss.20DR mm.7d,10d,11d,13,14d; 21DR mm.2,5,6d,9d,10d; 22DR mm.5-6d,8). Apparently retired thereafter to Poitou. Acting via a proctor for his English affairs, master W. le Bronum, in May 1229 (Winchester, Hampshire Record Office ms. 13M63/2 (Mottisfont cartulary II) fos. 148r-9r). Safe conduct to come to England *c*.February 1230, the terms of the safe-conduct implying an earlier period of disfavour at court (*Pat.R.1225-32* 325). Meanwhile, before 1232, promoted chèvacier (*capicerius*) of an unidentified church in Poitiers, possibly St-Hilaire where des Roches had been treasurer before 1205. Rose to enormous power and authority in England under des Roches' regime 1232-4, being promoted royal treasurer, chamberlain, keeper of escheats, titular sheriff of over twenty counties and in his own right granted custody of wide estates in Sussex and the Welsh marches, some in hereditary fee, others with provision for descent to his heirs should he renounce his clergy, suggesting that he considered abandoning the church for a career as lay magnate. Apparently intended marrying his most important ward, the son of William de Braose, to his niece named Aaleys (*CPR 1232-47* 504; *Cal.Chart.R. 1226-57* 156-85 *passim*). Dismissed from office and stripped of his custodies and lands March 1234, amidst rebellion by the earl Marshal, and accusations of abuse of power. Thereafter subject to recriminations from the new regime at court. Fled to Winchester, but repeatedly ordered to attend court to render accounts and to answer his accusers (*CRR* xv nos.1031,1225,1289; *AM* ii (Winchester) 86-7; *Cl.R.1231-4* 573-4,579-83; *Cl.R.1234-7* 1,163). Gifts of silk and gold to the king, apparently to buy back favour (PRO E372/79 mm.5d,11d). Ordered into exile January 1236 (*Cl.R.1234-7* 332), but orders revoked. Formal pardon May 1236 (*CPR 1232-47* 145) and by April 1237 in receipt of minor favours, timber to keep him at Banstead, Surrey, where he appears to have owned a house (*Cl.R.1234-7* 440). Thereafter regular gifts of timber and venison from the king until his death. Appears on the Winchester estates 1235-7, and witnesses two charters of bp des Roches, 1238 (Acta nos.51,53; Mss. 31DR mm.8d,10d,12; 32DR mm.1,11d). Served after 1238 as one of the bp's executors (*CPR 1232-47* 423). Witnesses grant to Southwick priory by Emma, widow of Geoffrey des Roches, and various Selborne priory charters of the 1250s (*Southwick Cart.* vol.2 iii no.465 [for *Binall'* read *Rivall'*]; *Selborne Charters* i 42,45, ii 29). Himself granted land to Selborne priory *c*.1251 at Shoelands in Puttenham near Guildford, Surrey (*CPR 1338-40* 176). In 1246 supported by the king in enforcing discipline as dean of Bridgnorth (*CPR 1232-47* 495). Possibly to be identified as Peter *de Rimilis*, granted land at Bartley in the New Forest by Walter Marshal, earl of Pembroke (1241-45), in return for an annual rent of a pair of white gloves, a grant witnessed by Simon of Kilkenny, styled 'chancellor', and Geoffrey de Caux II (BL ms. Harley 1240 (Mortimer cartulary) fo. 90r-v). Perhaps to be identified as Peter *de Orivall'* of Bursledon, appeared before the Hampshire eyre of 1249, having found £110 in the sea off the Hamble estuary which he gave to a Jew of Winchester. Both

Peter and the Jew to answer to the crown for the money (PRO JUST1/776 m.28). Returned to court 1250, granted temporary custody of the king's seal and involved in the transport of money from the treasury (*Cl.R.1247-51* 266,285,288; *Cal.Lib.R.1245-51* 288). Witnesses and warrants numerous royal letters 1250-52, royal custodian during a vacancy at St Denys' Southampton 1252, granted houses in Newgate, London, June 1253 (*CPR 1247-58* 86,101,128,151,170,198; *Cl.R.1251-3* 9-50 *passim*; *Cal.Chart.R. 1226-57* 371,407-8; *Cal.Chart.R. 1300-26* 58; *Cal.Chart.R.1327-41* 229; *Cal.Lib.R.1245-51* 293,335,375). Re-admitted as baron of the exchequer on the eve of the king's departure for Gascony, June 1253 (*Cl.R.1251-3* 371; *CPR 1247-58* 326). Treasurer of the king's wardrobe September 1257-June 1258 (Tout *Chapters* i 282). Envoy to France with Aymer, bp-elect of Winchester, January 1257. Witnessed the king's commitment to reform, May 1258 (*CPR 1247-58* 537,626). Granted land in Winchester in hereditary fee, May 1258 (*Cal.Chart.R.1257-1300* 10). Promoted canon of St Paul's London, prebend of Islington *c*.1260 × 62 (*Fasti* i 59). Appointed to take the king's daughter, Beatrice, to Brittany, July 1261 (*CPR 1258-66* 170). His men hunting with huntsmen of the lord Edward in Hampshire, 18 April 1262 (PRO E32/158 m.15d). Peter still alive July 1262 (*Cl.R.1261-4* 69). Dead by 2 November 1262 when his prebend and houses at London re-apportioned (*CPR 1266-72* 739; *CPR 1258-66* 238). Described in a corrupt, fifteenth-century Mottisfont chronicle as brother of the Hampshire baron William Brewer; Peter de Rivallis, known as 'the holy man in the wall', who died on 23 November leaving jewels and cash to Mottisfont priory after his death. Account clearly garbled, but perhaps correct in describing him as benefactor of Mottisfont (*Monasticon* vi 481, taken from Bodl. ms. Dodsworth 97 fo.81r-v).

42. Aimery des ROCHES (*de Rupibus*) brother of Hugh des Roches archdn of Winchester, and of Geoffrey des Roches the knight (*Southwick Cart.* vol 2 iii no.374; *CPR 1232-47* 366). First appears 9 August 1219 as rector of Preston, Lancs., to which he had been presented by the crown in succession to master Peter Russinol I (*Pat.R.1216-25* 199). As rector of Preston 1226 × 1228 described as *nepos* of bp des Roches (*BF* 367). Found at Wargrave in 1224-25 with his men and horses going to Oxford (Ms. 20DR m.10d); at Wargrave on 13 December 1225 and at Highclere in 1226-27 (Mss. 21DR m.5d; 22DR m.8). Plays no discernible role in the bp's administration thereafter. Letters of protection for two years from September 1219 and a further two years from July 1222, so perhaps chiefly resident abroad (*Pat.R.1216-25* 199,336). In May 1228 his proctor and the other keepers of Preston church granted letters of protection for two years (*Pat.R.1225-32* 189). Perhaps the Aimery de

Rivallis, recipient of a grant by the crown of a Norman escheat from the estate of William de Millières, brother of master Humphrey, 1233 (*Cl.R.1231–4* 239), but grant probably ineffective. Returned to England during the vacancy after des Roches' death, witnessing a settlement of August 1239 drawn up before his brother, Hugh the archdn (*Southwick Cart.* vol.2 iii no.374). Together with master Luke and Hugh des Roches described as royal clerk in 1240 in a list of alien clergy whom the king wished exempted from papal tax (*Cl.R.1237–42* 176). In the same year as rector of Preston contested the patronage of a chapel at Chipping (*CRR* xvi nos. 1582,1651). Last occurs 1 March 1243 granted royal letters of protection with his brother Hugh, presumably going overseas (*CPR 1232–47* 366). Preston vacant by 22 July 1243, so Aimery either dead or permanently retired abroad (*CPR 1232–47* 387). Probably the nephew of bp des Roches presented to the church of Newchurch, Isle of Wight, in succession to master Philip de Lucy (d.1231/2) (Actum no.29). By 1244 Newchurch held by master Peter de Gattebrig' (BL mss. Egerton 2104A (Wherwell cartulary) fos.99r–v, 114v; Cotton Tiberius D vi (Christchurch cartulary) fo.103r).

43. Geoffrey des ROCHES (*de Rupibus*) bears a name in common use amongst the des Roches family of Château-du-Loir in the Touraine. A namesake sided with the French after 1203, appointed castellan by William des Roches, seneschal of Anjou, but captured and ransomed by king John (*Rot.Norm.* 62; *RLP* 23,24,25; *Chron. de Touraine* 148). Possibly the same Geoffrey des Roches confirmed in land at *Longi Vadi* in 1215, previously held by Baldwin des Roches (*Act.Phil.Auguste* no.1576). Baldwin was a nephew of William des Roches of Château-du-Loir, the seneschal (*Cart. Château du Loir* 79–80,91,95–7). By a deed drawn up at Château-du-Loir in 1216 Geoffrey gave fifty shillings of rents at Sougé to Holy Trinity Vendôme for the soul of Baldwin des Roches, *dilectus suus* (*Cart. Vendôme* ii 38). Unclear whether the Geoffrey active in the Touraine 1202–16 can be identified with bp des Roches' kinsman who first appears in England in 1220–21 as a guest at the bp's manor of Farnham (Ms.16DR m.5). His brothers Aimery and Hugh active in England since c.1219. Geoffrey spent thirteen days at Ivinghoe in 1224, apparently guarding the bp's horses during the nearby siege of Bedford (Ms. 19DR m.8d). Promoted to wide estates following the death of Roger fitz Henry, one of the bp's principal tenants who had held 4 fees at North Fareham, Hurstbourne and Bradley of the see of Winchester by direct descent from his grandfather, William fitz Roger, enfeoffed before 1130 (*Red Book* 205; *Reg.Pont.* 387; *RLC* ii 3). Roger's son, William, appears to have predeceased him, leaving a daughter and heiress named Emma, held in des

Roches' custody at Farnham from September to November 1224 (Ms.20DR m.9d) but thereafter granted by the king to Hubert Hose, by virtue of a single fee which Roger fitz Henry had held of the crown in chief at Lulworth, Dorset (*Red Book* 149,167; *RLC* ii 4; *BF* 424; *CIPM* i no.482, ii no.297). Despite Hose's intervention, bp des Roches was permitted to buy back Emma's wardship together with custody of Lulworth for a fine of two destriers, eventually pardoned in 1227 (PRO C60/23 m.7; *RLC* ii 16b,189b). At some time between February and September 1225 Emma was married to Geoffrey des Roches, the bp providing for their marriage feast at Farnham (Ms. 20DR mm.9d,10). The marriage probably involved further negotiations with the crown, since Geoffrey is found at Witney taking presents to the king at Woodstock, and then at Brightwell on his way to his wedding feast (Ms. 20DR mm.13,14). In 1229 Geoffrey and Emma defended an action brought by Henry and Alice fitz Roger, probably children of Roger fitz Henry, Emma's grandfather, over land at Ellisfield which Roger had held in chief, having perhaps acquired it through marriage. Settled upon Geoffrey and Emma for a fine of 20 marks (*RLC* ii 3; PRO CP25(1) 203/5 nos. 55,60). However, Henry fitz Roger appears to have ignored this settlement and only withdrew his claims in 1236 following protracted litigation (PRO JUST1/775 m.1d; *CRR* xvi no.144a). In 1229 Emma and Geoffrey launched a suit against Henry of Bradley, again apparently to secure obedience to an earlier fine (*CRR* xiii no.1472). The case appears to have involved three virgates at Bradley, eventually confirmed to Emma and Geoffrey in October 1241 in return for a grant to Henry of 41 acres at Ellisfield, doing three-weekly suit to their court there (PRO CP25(1) 203/7 no.59). In 1236 Emma, Geoffrey and others were accused of disseisin by Roger, chaplain of Bradley, who claimed he had demised two tenements there to Richard his son in the court of Geoffrey des Roches, only to have them seized after Richard's death (PRO JUST1/775 m.3d). Such suits notwithstanding, Geoffrey's must be accounted one of the more successful alien marriages by comparison to the litigation and hostilities which attended the introduction of others of the bp's alien familiars to estates in England. Nonetheless, whilst Geoffrey witnesses various of the bp's charters and deeds, especially those associated with the bp's foundation at Selborne (Acta nos. 51-2, 71; *Selborne Charters* i 4,20,22,31; *Cal.Chart. R.1226-57* 183), it is noticeable that he never once appears as a Hampshire knight of assize, suggesting that he was kept at arm's length by the local gentry. His arrears of scutage allowed to run on from year to year by the bp (Ms.28DR m.17). In October 1233, during the bp's political ascendancy, received Steventon near Basingstoke at the king's pleasure, the manor having been claimed as a Norman escheat following the death of Geoffrey de Luverez

(*Cl.R.1231–4* 282; *CPR 1232–47* 38). Seized back by the crown after the bp's downfall, but Geoffrey appears to have maintained a residual interest there. In 1240 he sued de Luverez' heirs, Hugh of Wingham and Manasser de Saunderville, claiming that he had been unjustly disseised by the king (*CRR* xvi nos. 1176,1287,2558). In the immediate term his claim failed. Wingham and Saunderville were still in possession in 1249 (PRO JUST1/776 m.14d). But between 1258 and 1260 Martin des Roches, Geoffrey's son, secured most of Steventon, paying 100 marks to buy out Wingham's moiety (PRO CP25(1) 204/10 no.31; *Cl.R.1256–9* 310; *CIPM* ii no.225; *VCH Hampshire* iv 171). Geoffrey last appears in October 1241 (PRO CP25(1) 204/7 no.59), although he is named amongst a list of the bpric's knight service *c.*1245 (*Reg.Pont.* 387). His one recorded charter was issued before November 1246, probably some years previously; a grant to Walter of Merton, described as *dilecto socio et amico meo*, of a meadow named *Rughemed* near Basingstoke, witnessed by Silvester de Everedon, like de Merton a rising courtier, future archdn of Chester and bp of Carlisle (Oxford, Merton College muniments no.1730, which preserves Geoffrey's seal, round, device: a shield charged with a single lion or leopard *couchant*. The arms given in *VCH Hampshire* iii 52 are unauthenticated). Geoffrey was certainly dead by October 1253. At some time previously, Emma his widow granted land at West Boarhunt, within her manor of North Fareham, to the canons of Southwick for the soul of her late husband, an award witnessed by master Hugh des Roches, archdn of Winchester (d.1253) and Peter de Rivallis (*Southwick Cart.* vol.2 iii no.465). Emma lived on for several years, receiving a grant of the land of Quob in North Fareham, held previously by her grandfather Roger fitz Henry, in heredity from bp des Roches' foundation at Titchfield (Winchester, Hampshire Record Office ms. 5M53/36), and securing her right to dower at Bradley as late as 1260 (*VCH Hampshire* iv 202). The descent of her inheritance after Geoffrey's death was complicated by Emma's decision to partition the estate between her two sons, Martin and Hugh des Roches. Martin, the elder, received Bradley and Hurstbourne and by 1260 had also secured the family claim to Steventon. Other lands at Candover and Stoke Charity may have come to him by marriage. He was knighted in October 1256, having perhaps served on the Gascon campaign of 1253 (*CPR 1247–58* 232; *Cl.R.1256–9* 18). Sheriff of Hampshire September 1269–October 1270, he died *c.*August 1277, leaving a widow named Lucy, still active in 1286 (*CCR 1279–88* 426; *CFR 1272–1307* 81–2; *CIPM* ii 225). His younger brother, Hugh, received the remainder of Emma's estate, the manor of North Fareham and appurtenant property in Wickham, Titchfield and Whipstrode, to hold of Emma and her heirs for the service of 20 marks a year and one and a quarter knights (BL Add.

Ch. 15692). Inherited the remainder of Emma's estate following Martin's death in 1277 (*CIPM* ii no.225; *Reg. Pont.* 593). Also secured titular lordship over Lulworth, Dorset, a fee held in chief by Emma's ancestors. Emma may have attempted to repudiate the fee, in light of the fact that Lulworth was actually occupied by a sub-tenant and that through its possession Emma and Geoffrey became tenants-in-chief, liable to a series of feudal obligations to the crown (*CIPM* i no.482, ii no.297). Hugh married the daughter and heiress of Roger de Hoo. He died 1295 × 1300 leaving a son named John. The des Roches family interest survived until 1342 when, with the declared idiocy of Hugh's grandson William des Roches, the estate merged via a female line into that of the Brocas family.

44. Peter des ROCHES (*de Rupibus*), a knight, first occurs at Wycombe in 1216, spending most of the early summer there under the care of a doctor (Ms. 11DR m.5). Found on several of the bp's manors 1219-21 (Mss. 15DR mm.6,7d,10,10d; 16DR m.5) and in 1220-21 at Portchester castle (Ms. 16DR m.6d). In November 1220 together with Jocelin de Brissac (?Brissac-Quincé or Brissarthe, both dep. Maine et Loire), granted custody of the manor of Keyston, Huntingdon., as knights of the royal household, bp des Roches authorising and witnessing the award (*RLC* i 442; and for Jocelin see further Ms.16DR m.12). In March 1221 Peter and Jocelin received 20 marks to maintain them in royal service (*RLC* i 450b), but whereas Jocelin continued in receipt of money from the king (*RLC* i 414b,483,487b), Peter makes no further appearance in England. The P. de Rupibus recorded in 1226-7, receiving arrears from the bp's bailiffs of Overton, is almost certainly a scribal error for master H(ugh) des Roches (Ms.22DR m.9d).

INDEX OF PERSONS AND PLACES

Arabic numerals refer to the numbers of the acta in this edition, the capital letter W denoting an appearance as witness. Roman numerals refer to the page numbers of the introduction. The first and third appendices, on the bp's role as justiciar, and his itinerary, are indexed as app.i and app.iii followed by a page number in arabic numerals, within brackets. The second and fourth appendices, on uses of the bp's seal and on the bp's household, are indexed as app.ii and app.iv followed in brackets by an arabic numeral representing the appropriate numbered heading within the text. Where possible place names and toponyms have been identified, for England, Ireland and Wales by county in the old and the new form, for France by *département*. Personal names formed by patronym, as in the name Roger fitz Henry, are indexed throughout as Fitz. Beyond the standard abbreviations for the names of English counties, the letters IOW represent the Isle of Wight.

A., the chaplain, monk of St Swithun's Winchester, 6W and n.
—the clerk, 225-6
—mr, 135
—master of the schools of Winchester, xlii n.
Abbeny, *see* Aubigné
Abbeville (Somme), app.iv (27)
Aberffraw (Anglesey/Gwynedd), prince of, *see* Llywelyn
Abergavenny (Monmouth. / Gwent), app. iii (161)
Abingdon (Berks./Oxon.), app.iii (156, 159), app.iv (5)
—Ben. abbey, abbots of, *see* Hugh, monks of, *see* Richard
—Edmund of, archbp of Canterbury, xxxv, xliii n., 133n., 338n., app.iv (13, 17)
Abrincensis, *see* Avranches
Abshot, Abbechute (Hants.), *villata* of, 67
Acangre, *see* Oakhanger
Achard, the bp's marshal, xxxviii, app.iv (21)
—Nicholas Achard, his son, xxxviii, app.iv (25)
Acre (Holy Land), 129, app.iii (158)
—hospital of St Thomas, lv, lvii, 43, 57n., app.ii (30)
Adam, rector of Timsbury, app.iv (21)
Adderbury (Oxon.), 178-9, 202-3, 219-20, 236-7, 264, 292, 316, app.iii (147-8, 157), app.iv (4, 18, 32, 34)
—church of, app.ii (31, 41), app.iv (19, 20) rectors of, *see* Chanceaux, Russinol
—land at, app.iv (29)
Aigle, Aquila, Gilbert de, 85

Aimery, count of Evreux, earl of Gloucester, 9, app.iv (38)
Airaines, Aren', Arenis (Somme), archdn of, app.iv (27)
—mr Robert de, 26W, app.iv (18)
——biography of, app.iv (27)
Airvau (Deux-Sèvres), app.iv (41)
Airvault (Deux-Sèvres), app.iv (41)
Aiville, John de, app.iv (21)
Alan, abbot of Chertsey, 6
—brother, prior of St Cross Winchester, app.iv (2)
—mr, 75
—official of the bp of Winchester, *see* Stokes
Alard, prior of St Denys' Southampton, 1W
Albinus, bp of Ferns, xxxiv
Aldenham (Herts.), church of, 74n., rector of, *see* Farnham
Aldgate, *see* London, Holy Trinity
Aldingbourne (W.Sussex), app.iii (150)
Alemanno, Alem, William (de), 59W, 60W, 61W
Aleppo (Syria), 129
Alexander, mr, 75W
—the clerk, 100
Alicehoit (Hants.), forest of, app.iv (17, 23)
Aliz, Alis, Alys, Roger (I), app.iv (37)
—Roger (II), 9W, 13W, 21W, 51W, 52W, 270, 320, app.iv (36)
——biography of, app.iv (37)
——Agnes his wife, Agatha. Isabella, Joan and Petronella his daughters, app.iv (37)
—Thomas, 52W, son of Roger (II), app.iv (37)
—William, app.iv (37)

INDEX OF PERSONS AND PLACES

Allerton (Yorks.W.R./W.Yorks.), app.iii (150)
Allington (Hants.), app.iv (37)
Alresford (Hants.), 149, 165, 185–6, 299, app.iv (26)
—bailiffs of, see Martre, Osbert
Alresford Forum (Hants.), 152, bailiff of, see Daniel
Alswynus, 6
Alton, Aulton' (Hants.), 11n.
—church of, 1, app.iv (41)
—manor of, 1n.
—rectors of, 1n., app.iv (5), and see Rivallis, Roger archdn of Winchester
Alured, the clerk, 181
Amersham (Bucks.), church of, app.iv (3)
Amicius, mr, archdn of Surrey, xxxvii, lv, 56–7, 75W, app.iv (12)
——biography of, app.iv (11)
——family of, see Fitz Oger
——master of St Thomas' hospital Southwark, 57
——nephew of, see Amicius the younger
——sister of, see Mandeville
—mr, the younger, official of the archdn of Surrey, xxxvii n., app.iv (11, 12)
Amico, the clerk, 56W
Amo, see Hamo
Amport (Hants.), church of, 7
—persona of, see Hervey
Anagni (Italy), app.iii (158)
—convent of St Mary de Gloria at, 85n.
Anastasius, succentor of Salisbury, 247n.
Andeli, Andely, John de, 10
—Roger de, 172
—Walter de, 9W
Andrew, mr, brother of Ralph archdn of Winchester, app.iv (6)
—mr, master of the schools of Winchester, xlii n.
—prior of St Swithun's Winchester, inspeximus by, 70
Andwell (Hants.), Ben. priory of, xxxiv
Anesy, see Anstey
Angers (Maine-et-Loire), app.iii (144)
—St Martin, collegiate church, dean of, xxvii, app.iv (35)
——canon of, app.iv (35)
—St Nicholas, Ben. abbey of, lii
Angevin kings of England, xxxix, and see Plantagenet
Angoulême (Charente), count of, see Lusignan
—dean of, see Russinol
Anjou (France), seneschal of, see William des Roches
Anstey, Anesy (Hants.), Richard de, 82W
Appleby, Aplubi, mr R(alph) de, 18W, app.ii (32), app.iv (19)

Appleshaw (Hants.), chapel of, 7n.
Aquila, see Aigle
Arceveske, Philip le, 22
—Dionisia, wife of, 22n.
—Isemania, wife of Philip le, 22
Arches, see London, St Mary Arches
Archevêque, William le, 128
Ardfert (Ireland, co. Kerry), bp of, see John
Aren, see Airaines
Arenis, mr Stephen de, canon of Le Mans, 25n.
Arles (Bouche-du-Rhône), archbp of, see Hugh
Arreton (IOW), church of, 27, rector of, see Robert son of Robert
Arrow (Warwicks.), 3n.
Arsic, Robert de, Sybilla his wife and Joan his daughter, wife of Eustace de Greinville, app.iv (38), and see Vernon
Arundel (W.Sussex), earl of, 96, 184, app.iv (4), and see William
—honour of, app.iv (4)
—William of, his heir, app.iv (26)
Ascalon (Holy Land), app.iii (158)
Ashey (IOW), land at, 75n.
Ashley, Essel', Esseleg', (Hants.), church of, 34
—Gilbert de, 82W, and see Bere Ashley
Ashley (Wilts.), app.iii (146, 150, 159)
Assier (Lot-et-Garonne), Rigaud de, bp of Winchester, app.ii (38)
Assisi (Italy), app.iii (162)
Athée(-sur-Cher) (Indre-et-Loire), Girard de, app.iv (19)
Atherfield (IOW), app.iv (23, 38)
Aubigné, St-Aubin d', Abbeni, Aubeni (Ille-et-Vilaine)
—Nicholas de, clerk, 105
—Philip de, 127, men of, 266
——his chaplain, see Ernisius
—Robert de, son of William, 105
—William de, 100, 105, wife of, see Trussebut, and see Aubigny
Aubigny, St Martin d' (Manche), Robert de, 78, and see Aubigné
Aubrey, earl of Oxford, see Vere
Auckland (?Durham), Arnold or Ernold of, 94
—the king's treasurer, 157
Augustine, St, rule of, 34, 49
Aulton, see Alton
Aure, John de, 343
Averingdown (Bucks.), church of, app.iv (13)
Aversa (Italy), Rainald of, sons of, 130
Avranches (Manche), bp of, app.iv (36)
—bp and chapter of, 67a, and see William
—honour of, app.iv (4)
—mr Henry of, xlii, 30n.
Awbridge (Hants.), rents at, 43
Azo, brother, 198

INDEX OF PERSONS AND PLACES

Baalun', Roger de, 82W
Bacun, Henry, 252, 269, 303, 306
—Peter, bailiff of Southwark, clerk of the sheriff of London, 205
Baddesley, *see* North Baddesley
Bagehurst, *see* Baughurst
Balsham (Cambs.), Hugh of, bp of Ely, charter of, 58n.
Banstead (Surrey), app.iv (41)
Barfleur (Manche), app.iv (25)
Barford St Martin (Wilts.), app.iv (30)
Barking, Berching, Berking (Essex), 116n.
—Richard of, abbot of Westminster, 73n., app.ii (23)
—Richard of, xli, xlvi, 26W, clerk, 38W, 45W, 83W, app.iv (30)
——biography of, app.iv (28)
Barnby, Maurice of, canon of York, app.iv (21)
Barnstaple (Devon), Clun. priory of, app.iv (10)
Barnwell, Henry of, app.iv (24)
Barre, Bar', Richard, archdn of Ely, canon of Salisbury, 18, app.iv (7)
Bartholomew, of the king's chamber, app.iv (7)
—mr, canon of Salisbury, app.iv (7)
—chaplain of bp Herbert Poer of Salisbury, app.iv (7)
Bartley (Hants.), app.iv (41)
Barton, Berton', in Titchfield (Hants.), manor of, 67
Barton (Yorks.N.R./N.Yorks.), manor and church of, app.iv (38)
Barton Priors, Barton', Berthone (Hants.), 82n., tithes of, 81
Barton Stacey (Hants.), app.ii (26), *and see* Widemore
—rector of, 76, *and see* Teutonicus
Basing, Basinges, Bassinges, Basynges (Hants.), church of, 32, 48–50, app.iv (34)
——rector of, *see* Lucy, Philip de
—dean of, app.iv (4), *and see* Basingstoke
—lords of, *see* Port, Adam de, St John
Basingstoke, Basingestok', Bassingestok', Basyngstok' (Hants.), app.iv (26, 43)
—church of, 48–50, app.iv (34), rectors of, *see* Lucy, Philip de
—dean of, 286, app.iv (14), *and see* Basing
—Rughemed, meadow near, app.iv (43)
—Walter, son of Alexander fitz Hugh of, app.iv (26)
Basset, Basseth, Alan, app.iv (34), *and see* Gai
—Gunnilla, app.iv (21)
—mr Robert, 1W, 11W, 21W, 26W, 32W, 38W, 45W, 56W, 64W, 75W, 80W, 83W, app.ii (29W)
——bp's clerk 37W

——bp's steward, 9n., app.iv (23)
——biography of, canon of Romsey, app.iv (21)
—Thomas, xlv, app.iv (21)
Bassingbourn (Cambs.), John of, 94
Bassingham (Lincs.), Robert of, heir of, app.iv (41)
Bastal, *see* Bezill
Bastard, William, his widow, *see* Painel
Bataille, Batayle see Batilly
Bath (Somerset/Avon.), app.iii (147)
—Ben. priory of, prior of, app.iv (29), *and see* Robert
——lands of, app.iv (19)
—bps of, *see* Wells, Jocelin of, Savaric
—bpric, app.iv (34), election to, 86–7, app.iv (20)
—city of, app.iv (19)
Batilly, Bataille, Batayle, Batill', Batilli, Richard de, app.iv (22)
—Simon de, app.iv (22)
—William de, 6W, 13W, 256, 304
——bp's steward, 61W, app.iv (21)
——biography of, app.iv (22)
——daughter of, *see* Walcot
Baughurst, Bagehurst (Hants.), *persona* of, *see* William
Beatrice, daughter of king Henry III, app.iv (41)
Beauchamp, Robert de, 75
—William de, and his daughter, 78
Beaulieu, Belli Loci Regis (Hants.), app.iii (151)
—Cist. abbey, 2, app.ii (12, 24)
——abbot of, the bp's executor, 43, *and see* Hugh (I), Hugh (II)
Beaumont, family, honours of, app.iv (40)
Beauvais (Oise), bp of, *see* Miles
Beche, *see* Waterbeach
Becket, Thomas, archbp of Canterbury, 56n.
Beckenham, Begeham, Regeham (Kent/London), Robert of, 61
Beddington, Bedinton', Bedyngton', Bedynton' (Surrey/London), tithe of a mill at, 61
Bedford (Beds.), xlvii, app.iii (158)
—siege of, app.iv (23, 43)
—mr Robert of, bp-elect of Lismore, 110
Bedhampton (Hants.), app.iii (150)
Belton (Norf.), rector of, app.iv (2)
Benedict, abbot of Peterborough, app.iv (11)
—bp of Rochester, *see* Sawston
Bendenges, Gunnora de, wife of John fitz Hugh, 12
Benson (Oxon.), app.ii (13)
Bentley (Hants.), 336, app.iv (26), bailiff of, *see* Waltham, Robert of
Bentworth (Hants.), app.iv (29)
Bera, *see* Bere

INDEX OF PERSONS AND PLACES

Berching', *see* Barking
Bere (Hants.), forest of, 71n., *and see* Bere Ashley
Bere, in Soberton (Hants.), app.ii (1)
Bere, Bera, Berre, Adam de la, 71W
—Richard de, 9W
—Robert de la, John son of, xlviii
Bere Ashley (Hants.), forest of, 82n., *and see* Bere
Bere Regis (Dorset), app.iii (145, 150, 153)
Berengaria, queen, 121
Berghes, Hamo de, 15
Berkhamsted, Burhamstede (Herts.), 238, app.iii (151)
—honour of, app.iv (38), constable of, *see* Greinville, Gilbert de
Berking, Berkyng', Berkyngg', *see* Barking
Berkshire, archdn of, *see* Merton
—sheriff of, *see* Fitz Regis
Bermondsey (Surrey), Clun. abbey, annals of, 57
Bernevall, R. de, 250
—Reginald de, 127n.
Berre, *see* Bere
Berthone, Berton, *see* Barton Priors
Beseby, Gilbert de, app.iv (41)
Béthune (Pas-de-Calais), Robert de, app.i (131)
Bezill, Bastal', Matthew, 336
Bideford (Devon), app.iv (38)
Bindon (Dorset), app.iii (150)
Bingham, mr Robert de, bp of Salisbury, xli, 18, app.ii (31)
Bishop's Lydeard (Somerset.), app.iv (34)
Bishopstoke, Stoke (Hants.), 241, app.iii (144-5, 150), app.iv (2)
—church of, 14, 31, app.iv (20)
Bishop's Sutton (Hants.), 94n., 195, 243, 259, 277-9, 300, 301, app.iii (147-9, 151, 153-4), app.iv (1, 18, 20, 26, 36)
—bp's treasury at, 280, charter dated at, app.i (132)
—bailiff of, *see* Robert of Waltham
Bishop's Waltham, Wattham (Hants.), 2n., 147-8, 188, 252, 282, 303, 315, app.iii (146-8, 151, 158), app.iv (14, 19, 20, 26, 32, 36, 39)
—bp's warrens at, app.iv (22)
Bitterne (Hants.), app.iii (147-8, 158), app.iv (10, 14, 18, 36)
—bp's warrens at, app.iv (22)
Bladon (Oxon.), 112n.
Blendworth, Blaneword', Blaneworthe, Blanewrth', Bleneword', Blenewurth' (Hants.), chapel of, 37-8
—church of, 39, 40, app.iv (33)
—*persona* of, *see* Grandon', rector of, *see* Limoges

Blois, Henry of, bp of Winchester, liv, 66, 80, app.iv (2)
—letters and charters of, 1, 31
—Peter of, xxxv n.
—William of, bp of Worcester, 108, app.ii (13, 35)
—William of, bp of Lincoln, 86-7
Blund, John, 325
—mr John, candidate for the see of Canterbury, xlii n.
Blundeville, Thomas de, bp of Norwich, 128
Boarhunt, *see* West Boarhunt
Bolebec, Hugh de, sheriff of Northumberland, 120
Bologna (Italy), merchants of, app.iv (7)
Bonlieu (Sarthe), Cist. abbey of, lv
Bookham, Bocham (Surrey), *persona* of, *see* Robert
Boothby (Lincs.), Osbert of, William his son, 105
Borbech', *see* Burbage
Bordeaux, Burdegalen' (Gironde), archbp of, 113-14, *and see* William
—dean and chapter of, 113-14
—mayor and commune of, app.ii (16)
Bordesley (Worcs./H. & W.), Cist. abbey, 3
Boreham (Essex), John of, app.iv (30)
Borne, Mabilia daughter of Stephen le, wife of Roger Wacelin, app.iv (25)
Bourg-Charente (Charente-Maritime), app.iii (144)
Bourgueil, Burgalio, Burgoil, Burgolio (Indre-et-Loire), Denis de, xxxix, biography of, app.iv (15)
—chamber clerk and treasurer, xlvi, 14W, 80W, app.iv (10, 14)
—rector of Witney, app.ii (19), *and see* Denis the clerk *and* Denis of the chamber
Bouvines (Nord), campaign of, 101
Brademere, Willam de, 15
Brading (IOW), church of, app.iv (10)
Bradley (Hants.), app.iv (43), chaplain of, *see* Roger
—Henry of, app.iv (43)
Bramshill (Hants.), app.iv (40)
Bransbury (Hants.), men of, app.ii (26)
Braose, Giles de, bp of Hereford, xxxix, 95, 100
—William de, Leuca his grand-daughter, wife of Geoffrey de Camville, 3n.
—William de, another, his heir, app.iv (41)
Bratton (Wilts.), app.iv (30)
Braybrooke (Northants.), Henry of, app.iv (25)
Breage (Cornwall), church of, app.iv (32)
Bréauté (Seine-Maritime), Fawkes de, 108, app.ii (13), app.iv (30, 33)
—William de, marriage of, lii n.

INDEX OF PERSONS AND PLACES

Brébeuf, Brayboef, Breibuf (Calvados), Anketil de, demesne of, 75
—Henry de, 9W
Brembelacr', Brembulaker, *see* Cobham
Bressuire (Deux Sèvres), app.iii (144)
Breteuil (Eure), app.iv (37)
Bretingehirst, Brettyngherst, Brettynghurst, *see* Briddinghurst
Brewer, Briewerr', Briwer', Briwerr', John, 34
—William (the elder), lii, 33–4, 82, 90, 94, 96, 108, 132, app.i (130, 132), app.ii (5, 12, 13, 34), app.iv (15, 30, 41)
——brother of, app.iv (41), wife of, 82
—William (the younger), 82n.
—William, bp of Exeter, 129, 131, app.iv (7), inspeximus by, 86–7
Brian, son of Ralph, *see* Fitz Ralph
Briddinghurst, Bretingehirst, Brettyngherst, Brettynghurst, in Camberwell (Surrey/London), Reginald of, 61
Bridgnorth (Shrops.), app.iii (159)
—royal college at, canon of, app.iv (35)
——dean of, app.iv (41)
Bridport (Dorset), mr Giles of, xli, 76n., 338, bp of Salisbury, 338n.
Brightwell (Oxon.), 172, 175, 221, 231, 310–11, app.iii (154), app.iv (18, 19, 32, 37, 43)
—chapel at, liii n.
Brill (Bucks.), app.i (127), app.iii (148)
Brindisi (Italy), app.iii (158)
Brissac(-Quincé) or Brissarthe (Maine-et-Loire), Jocelin de, app.iv (44)
Bristol (Gloucs./Avon), app.iii (147–8, 153)
—Aug. canons of, 67n., Aug. priory of St Augustine, prior of, 76
—Ben. priory of St James, prior of, 76n.
—constable of, app.iv (37), *and see* Chanceaux, Rivallis
—dean of, 76n.
Brito, Britonis, 142, 194
Britton, Henry, app.iv (25)
Brittany (France), xxx–xxxi, 127, app.iv (41)
—duke of, app.iv (32)
—lands of the Bretons in England, app.iv (29)
—Eleanor of, the king's cousin, app.iv (37)
Briwer', Briwerr', *see* Brewer
Briwes, John de, 21W
Brocas family, app.iv (43)
Broke, *see* Titchfield
Bromholm (Norf.), app.iii (161)
Bromley, Bromleg' (?Kent/London), 281
Bronum, mr. W. le, app.iv (41)
Broomfield (Essex), church of, app.iv (11)
Brun, Nicholas de, 210
Buche, Aimery, 25n.
Buckingham (Bucks.), archdn of, 218, *and see* William
Buluse, John de la, 229

Burbage, Borbech', Burebech' (Wilts.), app.iii (146)
—prebend of, in Salisbury cathedral, 18, app.ii (31), app.iv (7, 13)
——tithes in, app.ii (34), app.iv (13)
——canons of, *see* Barre, Roches, Bartholomew *and* Luke des
Burel, Geoffrey, app.iv (39)
Burgalio, Burgolio, *see* Bourgueil
Burgate, Robert of, 94
Burgh, Burgo (Norf.), Hubert de, xxix–xxxi, lxvi, 85n., app.ii (33, 36), app.iv (4, 33–4)
—earl of Kent and justiciar, 18W
—justiciar, 107–8, 110n., 116, 118–9, 132, app.ii (16, 18)
——letters to, 111–13
—Margaret, wife of, app.ii (36)
Burghclere (Hants.), 196, 286, app.iv (14)
Burhamstede, *see* Berkhamsted
Burierchas, *see* Chobham
Burnham, Philip of, app.i (128)
Bursledon (Hants.), app.iv (41)
Burton on Trent (Staffs.), app.iii (147)
Bury St Edmunds (Suff.), app.iii (151–2)
—Ben. abbey of, app.i (128), app.iv (30)
Busci Fresnai, *see* Fresnoy-en-Val
Butt', Daniel, 21W
Bydun, Peter, gate keeper of St Swithun's priory Winchester, 41
Bytham (Lincs.), app.iii (157)

Caddington Major, prebend in St Paul's cathedral London, app.iv (31)
Caddington Minor, prebend in St Paul's cathedral London, app.iv (3)
Caen (Calvados), app.iv (17, 41)
—John de, app.ii (37)
Caesarea (Holy Land), 129, archbp of, *see* Peter
Cainhoe (Beds.), barony of, 78n.
Cairo (Egypt), sultan of, 129n.
Calbourne (IOW), 326
—church of, ornaments at, lvii
Calce, *see* Caux
Calstone Wellington (Wilts.), app.iii (146)
Caleston', G. de, clerk, 34W, app.iv (29), *and see* Caux
Camberwell (London), 61n.
Camel, *see* West Camel
Camera, *see* Denis, Jordan
Camoys, Ralph, 53n.
Campanarius, William, app.iv (6)
Camville, Christina de, app.iv (30)
—Geoffrey de, son of Albreda de Marmiun, 3n., his wives, *see* Braose, Frankley
—William de, father of Geoffrey, husband of Albreda de Marmiun, 3

INDEX OF PERSONS AND PLACES

—Richard de, son of Geoffrey, 3n.
—Thomas de, 74n.
—William de, the younger, son of Geoffrey, 3n.
Cancell', *see* Chanceaux
Cancie, *see* Kent
Candover (Hants.), 10n., app.iv (40, 43), *and see* Preston Candover
Canford (Dorset), app.iii (150)
Canterbury (Kent), app.iii (145, 152, 156-9, 162)
—archbps of, 56n., app.i (126), *and see* Abingdon, Becket, Blund, Grant, Langton, Pecham, Walter, Hubert, Winchelsea
—archbpric of, xxviii, 43
——election to, lii, 88
——suffragans of, xxviii, lxvii, lxviii n., 88, 100, ranking of, 86n.
——vacancy in, 86-7
—archdn of, *see* Poer, Herbert
—Ben priory of Christ Church, monks of, 92n., app.iv (13)
——prior of, 95
—cathedral church of, lv
—Martin of, official of the archdn of Surrey, xxxvii n.
Cantiloupe, Walter, bishop of Worcester, 3
—William de, 94n., app.ii (45)
Capell', Hugh de, 14W
Capella, H. de, 18W
Capetian kings of France, xxvii, xxxix
Cardevile, Richard de, 71W
Cardiff (Glamorgan / S.Glamorgan), app. iv (37)
Cardinan, Cardin', Robert de, sheriff of Cornwall, 118
Carisbrooke (IOW), app.iv (39)
—Ben. priory of, 27-9
—lords of, app.iv (23)
Caritate, Karitate, John de, 13W, 96b, knight, 83W and see Charite
Carlisle (Cumberland/Cumbria), bps of, *see* Everdon, Hugh, Mauclerk
—Aug. canons of, 104
Carlton Kyme (Lincs.), prebend of in Lincoln cathedral, app.iv (34)
Carrucarium, *see* Reginald, Walter
Cashel (Ireland, co. Tipperary), archbp of, 125, *and see* Donnchad
Caterham (Surrey), church of, 72, rector of, *see* Waltham
—vicar of, *see* Gatton
Catherington, Katerington', Katerinton' (Hants.), chapel of, 38, app.iv (7)
Catwick (Yorks.E.R./Humberside), church of, app.iv (30)

Caune, Herbert de, knight, tithes of, 76n.
Caux, Calce, Calceto, Cauz
—Geoffrey de, 11W, 14W, 108, 151, app.iv (14)
——biography of, app.iv (29)
——clerk of bp and king, xli
——rector of Dogmersfield, app.ii (3)
——rector, later vicar of Ringwood, app.iv (29)
——*and see* Caleston'
—Geoffrey de (II), rector of Ringwood, app.iv (13, 29, 41)
—Geoffrey de, knight, app.iv (29)
—James de, app.iv (29)
—John de, app.iv (29)
—Martin de, app.iv (29)
—Roger de, app.iv (29)
—Roger de (III), of Water Eaton, and his daughter, app.iv (29)
Caversham (Berks.), app.iii (155)
Celano (Italy), counts of, 130
Ceprano (Italy), 130, app.iii (159)
Ceresia, Charles (*Karolus*) de, 146
Cerne, Robert de, clerk, 45W, 83W
Cestretune, *see* Chesterton
Chaliton, mr Thomas de, 75W
Chalton, Chalgton', Chauton' (Hants.), church of, 38, subject chapel of, *see* Idsworth
Champar', Henry de, 270
Chanceaux (-prés-Loches, or -sur-Choisillé), Cancell', Chaunceles, Chanceus (Indre-et-Loire)
—Aimery de, 336
—Andrew de, xl, 209, 336n., app.iv (19)
—Guy de. app.iv (19)
—John de, app. iv (19)
—Matthew de, app.iv (19)
—Peter de, 14W, xxxix, xliv-v, 18W, 336n., app.iv (39)
——biography of, app.iv (19)
——charters issued by, lxiv, lxxiii, 48-9, 52, 75
——constable of Bristol, deputy sheriff of Oxfordshire, app.iv (19)
——rector of Adderbury, app.ii (31)
—Philip de, app.iv (19)
Chaorciis, Pain de, app.ii (9), *and see* Chaurc'
Charite, John de la, 9W, *and see* Caritate
Chark (Hants.), chapel of, 67, tithes of, 70
Charlecote (Warwicks.), app.iv (34)
Charlton-on-Otmoor (Oxon.), church of, 74n., rector of, *see* Farnham
Chase, John, chapter clerk of Winchester, xlix
Château-du-Loir (Sarthe), xxvii, app.iv (43), lord of, *see* William des Roches
Châtel-Bruges (Deux Sèvres), app.iii (144)

INDEX OF PERSONS AND PLACES 223

Chaucombe, Thomas de and his wife, widow of William de Turville, app.iv (40)
Chaurc', Hugh de, app.ii (2), *and see* Chaorciis
Chauton', *see* Chalton
Cheddar (Somerset.), app.iv (37)
Chelewarton', *see* Cholderton
Chelmsford (Essex), app.iii (151–2)
Chenduit, Chesneduit, Julian, 74W, steward and attorney of Westminster abbey, 74n.
Cheney, Christopher, lix
Cherlewode, Chorlwud (unident., ?in Hambledon, Hants.), 71
Chertsey, Certes', Certeseye (Surrey), app.iii (147, 153)
—Ben. abbey of, liii, lxxiv–v, app.iv (24)
——abbot of, *see* Alan, Hugh
——abbot and convent of, 5, 6
——obit celebrations at, 43
——sacrist of, 5, 43
Chesham, Cestresham (Bucks.), app.iv (38)
Chesneduit, *see* Chenduit
Chessel, in Shalfleet (IOW), app.iv (23)
Chester (Chesh.), archdn of, *see* Everdon
—bp of, *see* Cornhill
—constable of, *see* Lacy
—earl of, lii, 78n., 132n., app.ii (5), *and see* Ranulph
Chesterton, Cestretune (Cambs.), church of, 107, vicar of, *see* Wisbech
Chichester, Cicestren', (W.Sussex), app.iii (151)
—bps of, xxxv, *and see* Neville, Poer, Simon, Warham, Wich
—cathedral church of, l, 7
—canons of, app.iv (4), *and see* Luke, Millières, Waltham
—dean of, *see* Peregore
—dean and chapter of, 7n.
—estates of, endowment of, 43
—obit celebrations at, app.iv (4)
Chiddingfold, Chidingesaud' (Surrey), *persona* of, *see* Richard
Chilling, Chullyng, in Hook (Hants.), *villata* of, 67
Chilton (Bucks.), app.iv (38)
Chin', John de, app.iv (14, 29), *and see* Chinon
Chinon, Chinun (Indre-et-Loire), John de, 199, app.iv (14)
—W. de, 252
Chippenham (Cambs.), app.iii (159, 161)
Chipping (Lancs.), chapel of, app.iv (42)
Chobham (Surrey), app.iv (22)
—church of St Lawrence, 5
—parishioners of, 5
—places in, 'Burierchas' and 'Fletlande', 5
—mr Thomas of, 5, app.iv (1), subdean of Salisbury, 5n.

—tithes of, 5
—vicar of, *see* Thomas
Cholderton, Chelewarton' (Hants.), chapel and pasture of, 7n.
—land in, 75
Chorlwud, *see* Cherlewode
Christchurch, formerly Twinham (Hants./Dorset), app.iii (147, 152)
—Ben. priory, xxxiv, 8, app.iv (13)
——election at, app.iv (2)
——obit celebrations at, 43
——prior of, *see* Nicholas, Roger
——refectory and priory gate, 43
Chullyng', *see* Chilling
Cicestren', *see* Chichester
Cigogné (Indre-et-Loire), Engelard de, 94, 123n., app.i (127), app.iv (19)
Cirencester (Gloucs.), app.iii (155), vill of, app.i (128)
Cîteaux (Côte d'Or), Cist. abbey, abbot of, 43
—order of, lv–vi, monks and lay brothers of, lvi, app.iv (33)
Clacton (Essex), manor of, 30
Clanfield, Clanefeld', Clenefeld' (Hants.), chapel of, 38, app.iv (7)
Clapton (Cambs.), app.iv (30)
Clare, Clara (Suff.), earl of, 100
Clarendon (Wilts.), app.iii (145–7, 150, 152–3)
—king's houses at, app.iv (25)
Clarté Dieu (Indre-et-Loire), Cist. abbey, lv, 43, app.iv (13)
Clatterford (IOW), app.iv (23)
Claverley (Shrops.), church of, app.iv (41)
Clenefeld, *see* Clanfield
Clere (Hants.), 153, 166, app.iv (6, 18, 22, 26), *and see* Burghclere, Highclere, Kingsclere
Clifford, Walter de, 132n.
Clinchamps(-sur-Orne), Climchamp, Clincham', Clinchamp, Clinclamp, Clynchamp (Calvados),
—Robert de, clerk, xxxix–xl, 6W, 13W, 49W–51W, 53W, 61W, 75W, 80W, 84W, app.ii (29W), app.iv (14, 29)
——biography of, app.iv (17)
——bp's chamberlain (*camerarius*), app.iv (14)
Clinton, William, lii n.
Clisson (Loire-Atlantique), app.iii (144)
Cnoel, *see* Knoyle
Cobham, Coveham (Surrey), church of, 6
—places in, 'Brembelacr' or 'Brembulaker', 'Hukelescroft', 'Le Hezemore', 'La Rudynge', 'Wolcroft', 6
—rector of, *see* Michael
Cobham, Agnes de, 71n., daughter of Thomas de Venuz
Cogges (Oxon.), barony of, app.iv (38)

INDEX OF PERSONS AND PLACES

Coldred (Kent), church of, 72n., rector of, *see* Waltham
Colesdon', Culesden', *persona* of, *see* John
—John de, 51W, 53W, 295
Colevill', John de, knight, 83W
Compostela, Santiago di (Spain), pilgrimage to, 42, app.ii (22), app.iii (157), app.iv (34), *and see* Santiago
Compton (Hants.), manor of, 82n., app.iv (25)
Compton, in Freshwater (IOW), Adam of, app.iv (23)
Compton Chamberlayne (Wilts.), church of, app.iv (30)
Convers, Roger le, king's serjeant, gate keeper of St Swithun's priory Winchester, 41
Coombe (Bisset) and Harnham (Wilts.), prebend in Salisbury cathedral, app.iv (13)
Corfe Castle (Dorset), app.i (128), app.iii (150, 152-3)
Corhampton, Cornhametone, Cornhampton' (Hants.), manor of, 67, 71
—Adam of, 9W, 71
Cormeilles (Calvados), Ben. abbey, abbot and convent of, 2
—abbots of, *see* Durandus, Philip
Cornard, John de, app.i (128)
Cornevill', Robert de, clerk, 75W
Cornhametone, Cornhampton', *see* Corhampton
Cornhill, William of, archdn of Huntingdon, 34W, 94, app.i (132)
—bp of Coventry, 13W, app.i (132) letters and charters of, 97-9, 107
—*described as* bp of Chester, letters and charters of, 24n.
Cornwall, Cornub', 190, app.ii (10), app.iv (11, 32, 34, 40)
—earls of, *see* Fitz Count, Reginald, Richard
—sheriff of, 116, *and see* Cardinan, Fitz Count
Cosham (Hants.), manor of, 68
Coudray-Macouard, Le (Maine-et-Loire), app.iii (144)
Coveham, *see* Cobham
Coventry, Couvrtre, Coventr', Coventrensis (Warwicks./W.Midlands),
—Ben. priory of, lv, app.i (132)
—bps of, *see* Cornhill, Gray, Muschamp, Stainsby, Weseham
—'Walter of', chronicler, app.ii (5)
Cranborne (Dorset), app.iii (145, 147, 150)
Craon, family estate, app.iv (22)
Craucumb', *see* Crowcombe
Crawley (Hants.), app.iv (14)
Crofton (Hants.), chapel of, 67, tithes of, 70
Crondall, Crundal (Hants.), *persona* of, *see* Simon
Crowcombe, Craucumb' (Somerset), Godfrey of, 18W

Croydon (Surrey/London), app.iii (160)
Crundal, *see* Crondall
Cuddington (Surrey), church of, app.iv (34)
Culesden', *see* Colesdon'
Cumbe, *see* Kimbridge
Cumin', Baldwin de, 134
Cumyn, William, 13W
Culesdun', *see* Colesdon'
Cunde, Cundi, Reginald de, 71W, 322
Curbridge, Curebrigge (Hants.), villata of, 67
Curci, Robert de, 64n.
Cursus Romane curie, *see* Rome
Cury (Cornwall), church of, app.iv (32)
Cyprus, app.iii (158)

Dacy, William, 4n.
Damascus (Syria), Sultan of, 129
Damerham (Hants.), lii n.
Damietta (Egypt), 78n., archbpric of, xxix
Dammartin, mr William de, 85W
—William de, bp's bailiff, app.iv (28)
Daniel, bailiff of Alresford Forum, 152
Dapifer, *see* Oger
Dartford, Derteford' (Kent), app.iii (155)
—mr John of, bp's clerk, 192, 296
Dartmoor (Devon), forest of, 82n.
David, the bp's butler (*pincerna*), app.ii (42)
—mr, canon of Lismore, 110
—earl of Huntingdon, app.i (131)
Denis, the clerk, 11W, 13W, 21W, 34W, 38W, 45W, 56W, 61W, 83W, *and see* Bourgueil
—of the chamber (*de camera*), 26W, *and see* Bourgueil
Denmead, Donemede (Hants.), Matthew of, xlviii n.
Depa, Roger de, 85W
Depedene, Nicholas de, app.iv (23)
Derby (Derbs.), earls of, *see* Ferrers, William
Dereham, Derham (Norf.), mr Elias of, xliii, 51W, 53W, app.iv (1, 2, 13)
—the bp's executor, xliii, 43
Derteford', *see* Dartford
Devizes (Wilts.), app.ii (36), app.iii (145), castle, app.iv (4)
Devon, 118n., app.iv (11)
—sheriff of, 116
Dionisia, wife of Philip le Arceveske, 22n.
Dispensatore, *see* Geoffrey
Ditton (Surrey), app.iii (149)
—church of, app.ii (27), app.iv (10), rector of, *see* Roches, Hugh des
Dogmersfield (Hants.), app.iii (144)
—church of, app.ii (3), app.iv (29), rectors of, *see* Caux
Domesday Book, script of, lxiii

INDEX OF PERSONS AND PLACES

Dominicans, order of friars preachers, lv, lvii, app.ii (22), *and see* Winchester
Doncaster (Yorks.W.R./S.Yorks.), app.iii (147)
Donnchad, archbp of Cashel, 110
Donemede, *see* Denmead
Donnington (W.Sussex), church of, app.iv (36)
Dorchester (Dorset), app.iii (150)
Dorset, sheriff of, *see* Somerset
Douhet, le, Duay (Charente-Maritime), Joldewin de, king's household knight, 231
Dover (Kent), 13n., 91, app.iii (144, 148-9, 151, 153, 157)
—Ben. priory of, monks of, app.iv (13)
—charters dated at, 100, app.iv (12)
—constable of, *see* Millières, William de
Downton (Wilts.), bp of Winchester's manor of, 31n., 112n., 141, 176, 206, 247-8, 269-73, 305-6, 327, 331, app.iii (147-8, 151, 154-6, app.iv (20, 23, 29)
—forest regard at, app.iv (17)
—bp's hundred court at, app.iv (18)
—Wick, Wik', in, 141
Drayton, Draiton', mill of, *see* Winchester
Drayton Beauchamp (Bucks.), app.iv (38)
Drayton Cannes, Drayton (Hants.), tithes in, 75, 76n.
Drogo, Drocon', mr, 84W
Droxford, Drokeneford' (Hants.), church of, app.iv (20)
—rural dean of, *see* Thomas
Duay, *see* Douhet
Dublin (Ireland, co. Dublin), archbp of, app.i (131), *and see* Henry
Dungewood, in Shorwell (IOW), app.iv (38)
Duniun', Lawrence de, 176
Dunstable (Beds.), Aug. priory, annals of, 57n., app.ii (30, 34)
—prior of, *see* Richard
Durandus, abbot of Cormeilles, 2n.
Durham, Dulmon' (Durham), xl, app.iii (149), app.iv (1)
—bp of, 342, *and see* Farnham, Marsh, Poer, Poitiers, Puiset
—see of, xxix, lv
Durngate, mill of, *see* Winchester

Ealdland, prebend in St Paul's cathedral, London, app.iv (1)
Ealdstreet, prebend in St Paul's cathedral, London, app.iv (7)
Earley, Erlegh' (Berks.), Henry of, 343
Eastleigh, Esli (Hants.), Gilbert of, 9W
East Meon, Menes (Hants.), 94n., 184, 214, 244-5, 282, 340, app.iii (147, 154), app.iv (18, 23, 27, 30, 39)
—charter dated at, 21
—hundred court at, app.iv (23)
—mill at, app.iv (38)
—reeve of, *see* Robert
Easton (Hants.), rents at, 43
East Preston (W.Sussex), app.iv (4), *and see* West Preston
East Tytherley (Hants.), church of, app.iv (41)
East Woodhay (Hants.), church of, app.iv (33)
Ebbesbourne, Ebleburn', Eblesbourne, Eppelburn (Wilts.), 10n., p.106, 207, app.iii (153)
—mr Adam of, official of the archdn of Winchester, xxxvii n., app.iv (7)
—mr Thomas of, xliii, 1W and n., 26W, bailiff of Fonthill and subdean of Salisbury, 26n.
Ebor', *see* York
Edessa (Holy Land), 129
Edmund, archbp of Canterbury, *see* Abingdon
—the clerk, bp's familiar, app.iv (21)
Edward I, king of England, 55n.
—son of Henry III, huntsmen of, app.iv (41)
Edward II, king of England, confirmations by, 31, 68, 71
Edward III, king of England, confirmation by, 21, 63, 83-4
Edward IV, king of England, confirmation by, 63
Egbury (Hants.), app.iii (146)
Eling, Elyng (Hants.), church of, 34
Ellisfield (Hants.), app.iv (43)
Elmham, Elmaham, Elman', Helmam (Norf.), Richard of, 26W, 183
—clerk, xlvi, 38W, 45W, 83W
Elsteham (unident., ?Surrey), rents at, 43
Ely, Eliens' (Cambs.), archdns of, *see* Barre
—bps of, *see* Balsham, Eustace, Northwold, York
—Ben. priory of, prior of, *see* Roger
——monks of, app.iv (28)
—diocese of, app.iv (32)
—Nicholas of, bp of Winchester, xlix
——his official, *see* Thomas
—mr Thomas of, 11W
Elyng, *see* Eling
Enford (Wilts.), church of, app.iv (17)
Engaine, John de, Joan his wife, daughter of Gilbert de Greinville, app.iv (38)
England, xxviii, xxx-xxxi, xxxix, 122, app.iii (159, 162), app.iv (8, 18, 25, 33-4, 36, 38, 41-3)
—barons of, xxx, xxxi, lii, 98-100, 132n., app.ii (5), app.iv (4, 11)
—kings of, *see* Edward, Henry, John, Richard
—ports of, 90
Eppelburn, *see* Ebbesbourne
Epsom (Surrey), church of, app.iv (13)
Erdington (Warwicks./W.Midlands), Thomas of, 107, app.i (131)

Erith (Kent), app.iii (153)
Erlegh', *see* Earley
Ernisius, chaplain of Philip de Aubigné, rector of Farlington, 64n.
Escote (unident., ?Surrey), app.iv (22)
Esher (Surrey), church of, app.iv (13)
Esli, *see* Eastleigh
Esquay-Notre-Dame, or Esquay-sur-Seulles, Escayo (Calvados), app.iv (25)
Esseby, mr Adam de, chancellor of Salisbury, app.iv (4)
Essel', Esseleg', *see* Ashley
Essex, 112n., app.iv (11, 22, 24, 26, 28)
—earl of, *see* Fitz Peter
Essex and Hertfordshire, sheriff of, *see* Verdun
Estthorpnatele (unident., in Farnham, Surrey), xlix n.
Esturmy, Henry, app.iv (7), *and see* Sturmy
Eton, Etton', *see* Nuneaton
Eudo, dean of Ewell, official of the archdn of Surrey, xxxvii n.
Eustace, bp of Ely, 86-8, 91-2, 95, 168
—bp of London, *see* Fauconberg
Everdon, Silvester de, archdn of Chester, bp of Carlisle, app.iv (43)
Evesham (Worcs./H.&W.), app.iii (148, 160-1)
Evreux, Evereus (Eure), count of, *see* Aimery
Ewell (Kent), app.iii (149)
Ewell (Surrey), dean of, *see* Eudo
Ewhurst (E.Sussex), app.iv (25)
Ewhurst (Surrey), church of, app.iv (35)
Exchequer, royal, the bp at, xxix, app.iii (146, 152, 154, 162)
—Red Book of, 132n.
Exeter, Exoniensi (Devon), 119, app.iii (145, 151, 156), app.iv (29)
—bps of, letters and charters of, 24n., *and see* Brewer, Marshal, Robert, Simon
—bp-elect of, 92n.
—canons of, *see* Airaines, Roches, Bartholomew and Hugh des
—John of, bp of Winchester, 20n.
Eynsham, Adam of, xxxiii

Fakkenberge, Falkenberge, *see* Fauconberg
Falaise, William de, app.iv (40), bp's knight, 96b
Falelia, *see* Fawley
Fareham (Hants.), xliii, 127n., 145-6, 154-5, 164, 189, 208, 213, 246, 265-6, 302, app.i (131), app.iii (148, 157), app.iv (2, 20)
—bp's harbour of, 67n., *and see* North Fareham
Faringdon (Berks./Oxon.), tithes of, app.ii (24)
Faringdon, Farendon' (Hants.), 11

Farlee (?Hants.), Henry de, sheriff of Hampshire, app.ii (26)
Farlington, Farlyngton (Hants.), church of St Andrew, 64-5, rector of, *see* Ernisius, Stokes, Ralph of
Farnham (Surrey), xxxi, xxxv n., xlviii n., xlix n., 30n., 94n., 136-9, 142, 156-7, 167-8, 169-71, app.iii (145-6, 149, 153, 162), app.iv (6, 10, 14, 17, 19, 20-1, 26, 29, 36-7, 43)
—in the Winchester pipe rolls, 240-42, 253-59, 280-82, 295-8, 321-5, 329,
—bailiff of, *see* Testard, Waltham, Robert of
—castle of, 255, app.iv (4, 20)
—chapel at, ornaments of, lvii
—charters issued at, 53-4, 84
—forest regard at, app.iv (19)
—mr Nicholas of, 74W, royal physician and bp of Durham
Fauconberg, Fakenbergh, Fakkenberge, Falkenberge, Fauc', Faucoberg, Faucunberg, Faucunberge, Faukeberg', Faukebergh, Faukenbergh
—mr Eustace de, xl, xlii, 1W and n., 35n., 38W, 45W, 56W, 59W, 60W, 83W, app.i (130), app.ii (26), app.iv (31, 34)
——bp's clerk, 13n., treasurer of the king, xlvi, 82, 83n., app.iv (28)
——bp of London, xlvi, 30n., 38n., 73, 83n., 126, 128, app.iv (3, 31, 34)
——biography of, xli, app.iv (30), Beatrice mother of, app.iv (31)
—family of, 13n., app.iv (30-1)
—Peter de, app.iv (30)
—mr Philip de, xl, xlii, 14W, 38W, 45W, app.iv (30, 34)
——biography of, app.iv (31), Agnes mother of, app.iv (31)
——archdn of Huntingdon, canon of Hereford, Lincoln and London, app.iv (31)
—Walter de, app.iv (30)
Faversham, Faverisham (Kent), app.iii (162)
—charter witnessed at, 52
Fawley, Falelia (Hants.), 187
Felde, *see* Titchfield
Fereby, John de, 100
Ferentino (Italy), letters dated at, 100
Ferles (?Firl, E.Sussex), Robert de, clerk, rector of Witley, 85, app.ii (38)
Fernham, *see* Farnham
Ferns (Ireland, co. Wexford), bp of, *see* Albinus
Ferrers, William earl, 96, *and see* William earl of Derby
Fitz Adam, Roger, xlviii n.
Fitz Baldwin, Peter, 15
Fitz Count, Henry, son of the earl of Cornwall, 94, app.ii (10)
—sheriff of Cornwall, 116, 118n.

Fitz Henry, Roger, 82W, app.iv (43), *and see* Fitz Roger
—Emma his grandaughter, wife of Geoffrey des Roches, app.iv (43)
Fitz Herbert, Matthew, 64n., his son, *see* Fitz Matthew
—Peter, 21, 94, app.iv (7)
Fitz Hernic', Richard, 84
Fitz Hugh, John, 12, 79n., 94, letters to, 150
Fitz Humphrey, William, xlviii, 14W, 18W, 75W, 219, 221, 288, 328
—clerk, 45W, app.ii (29W)
—biography of, app.iv (32)
—Stephen son of, app.iv (32)
Fitz John, Roger, 116
Fitz Matthew, Herbert, 64n., his father, *see* Fitz Herbert
Fitz Nicholas, Ralph, app.ii (33)
Fitz Oger, Alan, app.iv (11)
—Michael, app.iv (11)
—Oger, royal justice, app.iv (11), *and see* Amicius, archdn of Surrey
Fitz Osbert, Richard, 15
Fitz Peter, Geoffrey, justiciar, xxix n., xli, xlvi, 90, app.i (130), app.iv (24, 28), his son, *see* Mandeville
—earl of Essex, 94
Fitz Philip, John, 18W
Fitz Ralph, Brian, 4
—Walter, 13W
Fitz Regis, Richard, 115
Fitz Reinfrey, Gilbert, 96a
Fitz Richard, Ralph, 82W
Fitz Roger, Alice, app.iv (43)
—Henry, app.iv (43)
—Robert, 120
—William, grandfather of Roger fitz Henry, app.iv (43)
—William, son of Roger fitz Henry, app.iv (43)
Fitz Walter, Robert, 44, app.iv (11), marshal of the army of God, 100, chaplain of, *see* R.
Flanders, xlv n., app.i (129)
—ambassador to, app.iv (30)
—men of, 90
Fletlande, *see* Chobham
Fluri, Flury, N. de, 222, 225
—Thomas de, 267, 304
Foligno (Italy), app.iii (162)
Foliot, Folioth, Hugh, archdn of Shropshire, 13W, bp of Hereford, executors of, 43
Folkestone (Kent), app.iii (153)
Fonte, Reiner de, 15
Fonthill (Wilts.), bailiff of, *see* Ebbesbourne, Thomas of
Forde, Forda (Dorset), Cist. abbey of, abbot of, 332
—mr Robert of, xlii n., 70
Fortin, Walter, 2, *and see* Funten'

Fotheringhay (Northants.), app.iii (161)
Foucarmont (Seine-Maritime), Ben. abbey, app.iv (41)
Foxcott, Foxcote (Hants.), Herbert of, 188
France, xxvii, xxx, xxxi, xxxix, lxvii, 118n., app.i (125-7, 129, 131), app.iii (146, 153, 159, 162), app.iv (35-7)
—ambassadors to, app.iv (4, 29, 30, 41)
—men of, xxxviii, xlv, 21n., 67n., app.iv (43)
——as bps in England, xxxix
—king of, app.ii (18, 38), *and see* Louis, Philip Augustus
—papal legate to, app.iv (33)
—peace negotiations with, lii, 111
—religious corporations of, 43, *and see* Normandy
Franciscans, friars minor, lv n., app.ii (34), *and see* Reading
Frankley (Worcs./W.Midlands)
—Felicia of, sister of Simon, wife of Geoffrey de Camville, 3n., her son, *see* Camville, Richard de
—Simon of, 3n.
Frederick II, emperor of the Romans, king of Sicily, xxx, lii, 47, 129-30
Freemantle (Hants.), app.iii (145, 148, 150, 152), app.iv (21)
Freshwater (IOW), app.iv (23, 38), *and see* Wilmington
Fresnoy-en-Val, Busci Fresnai (Seine-Maritime), app.iv (41)
Friars Minor, *see* Franciscans
Fridaythorpe (Yorks.E.R./Humberside), app. iv (21)
Froggatt (Derbs.), app.iv (21)
Fulflood (Hants.), 82n.
Full', 152
Funten', R. de, 225, *and see* Fonte

G., the huntsman (*venator*), 242, *and see* Gervase
—the marshal, 285
Gai, Alice de, wife of Alan Basset, app.iv (34)
Gal', *see* Guala
Garland, mr John of, xxxv, app.iv (1)
Gascony (France), 127n., app.ii (16), app.iv (41)
—expedition to, app.iv (43)
—natives of, *see* Douhet
Gattebrig', mr Peter de, rector of Newchurch, 29n., app.iv (42)
Gatton, Thomas de, vicar of Caterham, 72
Geddington (Northants.), app.iii (145)
Geiste, Roger son of Everard de, 72n.
Geoffrey, xlii n.
—archdn of Surrey, 63W, app.iv (12)

INDEX OF PERSONS AND PLACES

—bp of Le Mans, 25n.
—the butler (*dispensator*), 18W
—prior of St Swithun's Winchester, 1W
—the swineherd (*porcar*'), 84
Germanus, the baker (*pistor*), 234
—monk of St Swithun's Winchester, 81W
Germany, princes, counts and barons of, 130
Gerold, patriarch of Jerusalem and papal legate, 129, 131
Gervase, 75W
—the huntsman (*venator*), 187, *and see* G.
—monk of St Swithun's Winchester, liv n.
Giffard, honour of, app.iv (38)
—William, bp of Winchester, 66, charters of, 63n.
Giles, the crossbowman (*balistarius*), 225
—prior of Merton, app.ii (28)
Gillingham (Dorset), app.iii (147, 150, 152)
—king's houses at, app.iv (25)
Gimmeg', *see* Gymminges
Gisors (Eure), John de, 67n.
Glanvill, Gilbert, bp of Rochester, 86–8
Glastonbury (Somerset), Ben. abbey of, lii
Gliddon, Gleddun (Hants.), Edith of, 17
Glinton (Northants./Cambs.), chapel of, app.iv (11)
Gloucester, Glocestr' (Gloucs.), 106, 125, app.ii (10), app.iii (148, 153, 155, 157, 161)
—burgesses of, 102
—earls of, app.iv (38), *and see* Aimery, Mandeville
—Ben. abbey, abbot of, xxxiv n.
—mr Robert of, 94
—priory of St Oswald (Aug.), prior of, 125
Gloucestershire, app.iv (19, 21)
—coroner of, *see* Turville, Maurice de
—sheriff of, *see* Musard
Godalming (Surrey), church of, 46n., app.iv (11)
Godfrey, 321
—bp of an Irish see, xxxiv
—bp of Winchester, *see* Lucy
—clerk of bp Godfrey de Lucy, xliii, 75W
Godington (Oxon.), church of, app.iv (30)
Godstow, Godestowe (Oxon.), Ben. abbey, nuns of, 11
Gonville, Edmund, lvi
Goodworth Clatford (Hants.), parson of, *see* Simon
Gorleston (Norf.), rector of, app.iv (2)
Goscelin, mr, 1W
Grai, *see* Gray
Grainville-la-Teinturière (Seine-Maritime), app.iv (38)
—natives of, *see* Greinville
Grandon', mr R. de, *persona* of Blendworth, 37, app.iv (33), *and see* Grendon
Grant, Richard, archbp of Canterbury, xliii n.

Grantham (Lincs.), app.iii (156)
Grave, Geoffrey de la, app.iv (14)
Gray, Grai, John de, bp of Norwich, 86–7, 94–5
—Richard de, 158
—Walter de, 94n.
——archbp of York, 13n., 22n., 104, 106–7, 117, app.ii (10, 34), app.iv (22, 38)
——bp of Worcester, li n., 13, 13W, 97–9
——bp elect of Coventry, app.i (132)
——royal chancellor, xlv n., 93, app.i (129)
Gregory, bp's chaplain xliv, 1W, 11W, 34W, 83W, charter dated by, 21
——*dominus*, 56W
—chaplain of bp Godfrey de Lucy, xliv, 75W
Gregory I, pope, quotation from, lvi, 13n.
Gregory IX, pope, xxx–xxxi, lii, 130
—letters of, lii–iii, 5n., 8n., 23n., 25n., 76n., 77n., 132n., app.ii (30, 32, 34–6), app.iv (8, 19, 20, 33)
—letters to, lxvi, 23, 132
—cardinal legates of, *see* Sabina, St Sabina
—court of, xxxii, app.ii (31, 37), app.iii (158, 162), app.iv (7)
—judges delegate of, 22n., app.iv (8, 10, 13, 34, 41)
—subdeacon of, *see* Roches, Luke des
—tax of, on alien clergy, app.iv (10, 42)
Greinville, Greinvile, Greinvill', *and see* Grainville
—Eustace de, xxxviii–xl, 9, 21W, 51
——biography of, app.iv (38), constable of the Tower of London
——steward of the king's household, xli
——wives of, *see* Arsic, Painel
—Eustace son of Richard de, app.iv (38)
—Eustace de, another, app.iv (38)
—Eustace de, clerk, app.iv (38)
—Gilbert de, app.iv (38), daughter of, *see* Engaine
—Robert de, son of Girard, app.iv (38)
—mr Robert de, Eustace's nephew, app.iv (38)
—William de, brother of Eustace, app.iv (38)
Grendon, Robert de, 37n., *and see* Grandon'
Grisele, John de, app.ii (29W)
Grosmont (Monmouth./Gwent), app.iii (161)
Grosseteste, Richard, archdn of Wiltshire, 5 and n.
—Robert, bp of Lincoln, app.ii (40–1)
Grottaferrata (Italy), app.iii (158)
Guala, Gal', cardinal priest of St Martin in Montibus and papal legate, 82, 104, 107, 110, 111n., app.ii (10)
—letters of, 102, 105–6, app.ii (13)
—money of, app.iv (17)
Guildford, Guldeford (Surrey), app.iii (145–6, 151–2, 157), app.iv (41)
—chapter held at, app.iv (11)

INDEX OF PERSONS AND PLACES

—church of Holy Trinity, vicar of, *see* John
—dean of, *see* Robert
Gunwalloe (Cornwall), app.iv (32)
Gurst, *see* Test
Guy, prior of Southwick, app.iv (5)
Gymminges, Gimmeg', Gumming', Nicholas de, 64n., his wife, *see* Merlay
—Thomas de, 71W
—William de, 21W

Hackel', *see* Oakley
Hale (Lincs.), 105
Hales, *see* Halesowen
Halesowen, Hales (Worcs./W.Midlands)
—church and manor, 13, vicarage in, li
—Premonst. abbey of BVM, l–li, lv, lxxii, 3n, 13, 43, 67n., 281n., 290–1, app.iv (4)
——abbot of, 203, 220, 228, 260, 262, 264
——abbot and convent of, 3, 13
——foundation charter, 13, app.iv (12)
——patrons of, 3n.
Hallaton (Leics.), app.iv (38)
Hallingbury (Essex), app.iii (145)
Hama (Syria), 129
Hamble, Hamele (Hants.), river, estuary of, app.iv (41)
—Ben. priory of St Andrew, 14
Hambledon (Hants.), xlviii n., 71n.
Hameleweye (unident., ?Hants.), ride (*chiminuus*) of, 71n.
Hamo, treasurer of York, 33
Hampshire, xxx, 37n., 44, app.i (126), app.iv (2, 19, 40)
—assize of arms, app.iv (37)
—escheator of, *see* Maurice de Turville
—eyre of, 22n., app.iv (13, 39, 41)
—forest enquiry in, app.iv (23, 40)
—forest eyre in, app.iv (35–6)
—gentry of, xli
—knights of assize, app.iv (37, 39, 40, 43)
—religious houses of, app.iv (13)
—royal bailiffs of, 12n.
—sheriffs of, xli, app.iv (23, 25), *and see* Farlee, Roches, Martin des, Shorwell, Wacelin
Hampstead (Berks.), app.iii (154)
Hanley Castle (Worcs./H. & W.), app.iii (148, 150)
Happisburgh (Norf.), app.iv (4)
Harang', Thomas, 318
Harcourt, John de, app.ii (13)
Haring, Nicholas, 138
Harnham (Wilts.), *see* Coombe
Hart' (unident., ?Hants.), chaplain of, *see* N.
Harwell (Berks./Oxon.), 312, app.iv (10, 17, 19, 32)
Hasting', R. de, 170

Hastings (E.Sussex), church of St Andrew, 72n., rector of, *see* Waltham
—honour of, app.iv (25)
Hattingel, Hattinggeli, Nicholas de, app.i (131)
—Peter de, 9W
Hauville, Hauvill', Geoffrey and Gilbert de, keepers of the king's falcons, 112
—Henry de, 112
—Walter de, 112
Havering (Essex/London), app.iii (146, 149, 152, 157, 159)
Havleherod (unident., ?Hants.), wood of, 71n.
Hay-on-Wye (Brecknock./Powys), app.iii (161)
Haye, Thomas de, 127n.
Haywode, Gilbert de, 6W
Haywood (Staffs.), app.iii (147)
Helmam, *see* Elmham
Henley (Oxon.), app.ii (13)
Henne, William atte, 71W
Henry I, king of England, app.iv (22, 29), grants and confirmations by, 18
Henry II, king of England, xxxix, grants and confirmations by, 18, 38
—obit celebrations for, app.iv (30)
Henry III, king of England, xxx–xxxii, xxxv, xliii, lxix, 18n., 25n., 28n., 43, 64–5, 67a, 69, 78, 85n., 102, 105–6, 112–15, 117–18, 125, 128, app.ii (33), app.iv (7, 25, 29, 31–2, 36, 38–41)
—grants and letters of, lx, 4n., 10n., 18, 30n., 41–2, 42a, 43–4, 46, 48–9, 67n., 67a n., 68–9, 71, 78n., 79n., 82, 104, 107, 109–10, 116–17, 121, 123, app.ii (9, 10, 12, 16, 18, 37, 42–3, 45), app.iv (10, 17, 32–4, 36–8, 42–3)
—letters to, lxxii, 15–17, 127
—brother of, *see* Richard
—carucage of, app.iv (23)
—chamberlain of, *see* Rivallis
—chancellor of, *see* Marsh, Neville, Ralph de
—chancery of, xlvi, xlvii–viii, archives, 25n.
—clerks of, *see* Neville, Nicholas de, Roches, Aimery, Hugh and Luke des
—council of, 110, 113
—court of, 48, 53n., 73n., 206n., app.ii (39), app.iv (33)
—cousin of, *see* Brittany
—daughter of, *see* Beatrice
—exchequer of, xlvi, 74n., 115, app.iv (23, 28, 30)
——seal of, app.i (130)
——barons of, app.iv (28, 41)
——clerks of, *see* Barking
——rolls of, xlvi
—falcons of, keeper of, *see* Hauville
—forest charter of, 82
—foresters of, 82n.

230 INDEX OF PERSONS AND PLACES

—guardian of, xxix, app.ii (26)
—hound of, 201
—household knights of, 109, *and see* Brissac, Greinville, Gilbert de, Roches, Peter des
—house of Jewish converts, xxxiv
—Jewish legislation of, lix
—justices of, lviii, 116, 305, *and see* Pattishall
—justiciar of, 275, *and see* Burgh, Seagrave
—justiciar of Ireland, *see* Marsh
—keeper of escheats of, *see* Rivallis
—kinsmen of, *see* Roussillon, Valence
—knights of, *see* Douhet, Montferrat
—mariners of, app.iv (17)
—minority of, xlvii, liii, lviii, 132, app.i (131), app.ii (12)
—physician of, *see* Farnham
—Poitevin half-brothers of, *see* Valence
—proctors of in Rome, 108, app.ii (37), app.iv (33), *and see* Caen, Caux, Limoges, Lucy, Stephen de, Norman, Richard monk of Abingdon, Saracen, Warham
—proctors in the Winchester election dispute, app.iv (13)
—purveyor of wines of, app.ii (42)
—rector of, *see* William Marshal I
—religious foundations of, lv–vi
—scutage of, *see* Montgomery
—seal of, 48, 132–3, app.ii (35), app.iv (41)
—serjeants of, *see* Convers
—sheriffs and bailiffs of, 42
—sisters of, 117, *and see* Isabella, Joan
—son of, *see* Edward
—stewards of, 206, *and see* Greinville
—tallage of demesne, 102n.
—tax of a fifteenth, 276, 278, app.iv (23, 24)
—tax of a fortieth, app.ii (37)
—treasure ship of, 127
—treasurers of, *see* Fauconberg, Rivallis
—treasury of, app.iv (10)
—vice-chancellor of, *see* Neville, Ralph de
—wardrobe of, clerk, chamberlain or treasurer of, *see* Rivallis
Henry VI, king of England, confirmation by, 63
Henry, archbp of Dublin, 95, 97, 99, 104, 106, app.ii (10) as archdn of Stafford, 94
—the chaplain, 49W, 50W, 59W, 60W, 70, 75W
—the crossbowman (*balistarius*), 211
—mr, 59W, 60W
—mr, the crossbowman (*arbalistarius*), 154
Herbert, bp of Salisbury, *see* Poer
—the clerk, 18W, 217
Herce, John de la, bp's clerk, rector of Stanford on Soar, app.ii (8)
Hereford (Hereford./H. & W.), app.iii (150, 155, 160–1)

—archdn of, *see* William
—bps of, *see* Braose, Foliot
—bpric, lands of, 43
—cathedral church of, app.iv (31)
——canons of, *see* Fauconberg, Philip de
Herefordshire, app.iv (19)
Herlewin, bp of Leighlin, xxxiv
Herriard, Herierd, Herird' (Hants.), 259, 282
—John of, the bp's attorney, 271, 278, 305, 307, 310
Herst, Jordan de, deacon, app.ii (11)
Hertford (Herts.), app.i (127), app.iii (149)
Hertfordshire, app.iv (24)
Hervey, mr, *persona* of Amport, 7n.
Hezemore (le), *see* Cobham
Hibernia, Robert de, clerk, xlii n.
Hida, *see* Hyde
Highclere (Hants.), 222, 230, 285, app.iii (156, 159), app.iv (4, 20, 27, 42)
Hinton Ampner (Hants.), church of, app.ii (1), rectors of, *see* Sowy
Hirm', William de, 61W
Ho, mr. H. de, 261, 287
—Thomas de, 9W
—William de, 53W, knight, xlviii n.
Hod, Lohod, Lood, Joan widow of Robert le, 53–4, *and see* Hot
Holbourn, prebend in St Paul's cathedral London, app.iv (30)
Holy Land, xxx, 47, 130
Hommet, Humeto, mr Nicholas de, 74W and n.
—William de, abbot of Westminster, 73, 74n.
Honorius III, pope, letters and mandates of, xxxiv–v, li–ii, 5, 26n., 30, 36n., 47, 60n., 73, 125, 132, app.ii (15), app.iv (14, 29, 34)
—chamberlain of, *see* Pandulph
—census owed to, 108, app.iv (29)
—judges delegate of, app.iv (7, 11, 34, 36)
—legates of, app.iv (33), *and see* John, Gerold, Guala, Pandulph
—letters to, 104, 121
—nuncio of, *see* Otto
Hoo, Roger de, daughter of, wife of Hugh des Roches, app.iv (43)
Hook (Hants.), app.iv (13)
Hooton Pagnell (Yorks.W.R./S.Yorks.), honour of, app.iv (38)
Hordle (Hants.), chapel of, 8
Horton (Gloucs./Avon), app.iii (148)
Hose, Hubert, app.iv (43)
Hospitallers, military order of St John, 130, master of, 129
—mill of at Winchester, 19, *and see* North Baddesley
Hot, Robert le, 9W, *and see* Hod
Hottot, Hotot, Robert de, 9W, 21W
Hou, Robert le, 4n.

Hoyville, Philip de, 71n.
Hubersete, Huberset', Ailwin de, 15
—William de, 15
Hugh, abbot of Abingdon, app.ii (7)
—(I) abbot of Beaulieu, 1W and n., 2n.
—(II) abbot of Beaulieu, 1n., bp of Carlisle, 1n., 2n., 104n., 110, 117
—abbot of Chertsey, 1W
—archbp of Arles, 130
—archdn of Winchester, *see* Puiset, Roches
—bp of Lincoln, *see* Wells
—the chaplain, 49W, 50W, styled mr, 80W
—viscount of Thouars, 128
Hukelescroft, *see* Cobham
Hull', Richard de la, priest, vicar of Sparsholt, 84
Hulle (unident., ?Hants.), app.iv (21)
Hulle, Odo de la, 6
Hulme, *see* St Benet of Hulme
Humeto, *see* Hommet
Humphrey, Umfrido, chaplain, 26W
—chaplain of bp Godfrey de Lucy, 75W
—the clerk, 74W
—mr, 197, 204, *and see* Millières
—prior of St Mary Overy, Southwark, 62n., app.ii (29)
Huntbourn' (unident., ?Hants.), ride of, 71n.
Huntedon, Huntendun, *see* Huntingdon
Huntifeld', Thomas de, 21W, *and see* Huntinges'
Huntingdon, Huntedon, Huntendun (Hunts./Cambs.), archdns of, *see* Cornhill, Fauconberg, Philip de
——official of, *see* Waltham, Richard of
—earl of, *see* David
Huntinges', T. de, 159, *and see* Huntifeld'
Hurstbourne (Hants.), app.iv (43)
Hurstbourne Tarrant, Hwseburne, Hwusseburne (Hants.), church of, 18, 46n., app.iv (7)
——perpetual vicar of, *see* Peregore
—manor of, app.iv (19)
—prebend of Hurstbourne and Burbage in Salisbury cathedral, app.iv (13)
—rectors of, *see* Neville, Nicholas de, Roches, Luke des
Hyde, Hida, Hyda (Hants.), Ben. abbey of St Peter, liii, 1, 19–21, 31n., 82n., app.iv (34, 36)
——abbot and convent of, 19
——abbots of, *see* Suthill, Walter
——church of, lights in, 21, known as *Novus Locus*, app.ii (20)
——prior of, *see* O.
——St John fee, held of, app.iv (40)
—mr Walter of, 214

Iakel', 248
Idbury (Oxon.), app.iv (3)

Idsworth, Iddesworth', Idesworth' (Hants.), chapel of, 38
Ilchester, Ivelcest' (Somerset), 275
—Richard of, bp of Winchester, xxviii, xxxv, his sons, *see* Poer
——charters of, lxix, 11, 14, seal of, lxi
——bp-elect of Winchester, app.iv (5)
Ilsley (Berks.), app.iii (156)
Inmede, meadow of, *see* Winchester
Innocent III, pope, xxviii, lii, 91, 95–6, app.ii (5–7)
—cardinals of, *see* John, Langton, Tusculum
—court (curia) of, 81n.
—legate of, app.i (126), app.iv (5), *and see* John, Tusculum
—letters of, xlii n., li, lvii, lxvii–viii, 59n., 81n., 100
—letters to, 86, 88
—subdeacon and familiar of, *see* Pandulph
Innocent IV, pope, licence from, app.iv (13, 34)
Insula, *see* Lisle
Insula de Saim (unident., France), app.iii (145)
Ireland, xxxi, xxxiv, app.iv (7)
—expedition to, app.iv (35)
—justiciar of, *see* Marsh
—magnates of, 133
Isabella, queen, widow of king John, 128
—sister of king Henry III, 117
Isemania, wife of Philip le Arceveske, 22
Isle of Wight, *see* Wight
Isles (Scotland), bp of, *see* Nicholas
Islington (London), prebend of in St Paul's cathedral London, app.iv (41)
Italy, communes of, xxxi
—natives of, as bps in England, xxxix
Itchen, river (Hants.), 82n., course of, 19n.
Iussai, William de, 189
Iuvene, *see* Peter
Ivelcest', *see* Ilchester
Ivinghoe, Ivingeho (Bucks.), 163, 204, 293, app.iv (4, 43)
—church of, app.ii (21), app.iv (4), rector of, *see* Millières

J., monk of Winchester, 175
—the Roman clerk (*clericus Romanus*), 297
—the steward, 224
Jarnac (Charente), app.iii (144)
Jersey (Channel Islands), app.iv (38)
Jerusalem (Holy Land), xxx, 129, app.iii (158)
—patriarch of, xxx, app.ii (30), *and see* Gerold
Jews, lix, debts to, lv, lix, 48n., 71n., app.iv (34), *and see* Winchester
—converts, *see* London
—exchequer of, app.iv (28)
Joan, daughter of James, 16

INDEX OF PERSONS AND PLACES

—sister of king Henry III, 117, 118n., app.iv (22)
Jocelin, bp of Bath and Glastonbury, *see* Wells, Jocelin of
—the queen's brother (*frater regine*), mill of, 19, *and see* Goscelin
John, king of England, xxix–xxxiii, lv, lxix, lxxv, 18n., 67a, 69, 74n., 86–9, 92, 94–6, 98–100, 124, app.i (125–32), app.ii (5, 7, 17), app.iv (21, 25, 27, 29, 30, 32, 34–5)
—castles of, 96a
—chamber of, xxviii, app.iv (34), clerks of, *see* Bartholomew, Lucy, Philip de, St Maixent
—chamberlain of, *see* Geoffrey de Neville
—chancellor of, xxix, *and see* Gray, chancery clerk, *see* Marsh
—chancery and exchequer of, lxxv, app.i (125–32)
—clerks of, *see* Caux, Fitz Humphrey, Russinol, Peter I
—court of, xxix
—exchequer of, 89, app.iv (28)
—grants and letters of, xli, 13, 18, 34, 90, 93, 96b, 97, 101, app.ii (2), app.iv (2, 7, 12, 19, 21, 29, 30, 34)
—Henry, son of, 96b, *and see* Henry III
—justices of, *see* Fauconberg, Eustace de
—justiciar of, xxix, app.i (125–32), *and see* Burgh, Fitz Peter Roches, Walter, Hubert
—mercenaries of, 96b
—obit celebrations for, app.iv (30)
—scutage of, app.i (126)
—seal of, app.i (129–30), keeper of, *see* Neville
—treasurer of, *see* Auckland
—treasury of, app.iv (34)
—wife of (queen), 96b, 143, *and see* Isabella
—wine of, app.iv (25)
John, archbp of Tours, 121
—*bartonarius* of St Swithun's Winchester, 81W
—bp of Ardfert, xxxiv–v, xlii, 329
—cardinal bp of Sabina, papal legate, 130
—cardinal deacon of Santa Maria in Via Lata, 87, app.iv (5)
—cardinal priest of St Stephen in Celio Monte, papal legate, 110
—the chaplain, 23
—the dean (*decanus*), bp's treasurer, app.iv (15)
——biography of, app.iv (14)
—*persona* of Colesdon', 6W
—prior of Newark, 35, app.ii (20)
—prior of St Swithun's Winchester, inspeximus by, 56–7, 59–62
—vicar of Holy Trinity Guildford, 84W
Joppa (Holy Land), 129, app.iii (158)
Jordan, *de camera*, 75W

—the reeve (*prepositus*) of Rimpton, 268
Josophat (Holy Land), valley of, 131

Karitate, *see* Caritate
Katerington', Katerinton', *see* Catherington
Kempton (Surrey), app.iii (159–60)
Kene, Robert, 291
Kenilworth (Warwicks.), app.iii (148)
Kenitun' (unident.), xxxiv, 329
Kent, Cancie, app.iv (11)
—earls of, *see* Burgh
Kettlewell (Yorks.W.R./N.Yorks.), app.iv (30)
Keyston (Huntingdon./Cambs.), app.iv (44)
Kidderminster (Worcs./H. & W.), app.iii (161)
Kildare (Ireland, co. Kildare), bp of, 125
Kilham, Nicholas de, 82W
Kilkenny (Ireland, co. Kilkenny), Simon of, chancellor, app.iv (41)
Killaloe (Ireland, co. Clare), bp of, 125
Kimbridge, Cumbe (Hants.), church of, app.iv (21)
Kingsclere, Kingescler', Kingesclera, Kynggesclere (Hants.), xlix n.
—church of, 21
—tithes of, 286, app.iv (14)
Kingsmead (Derbs.), Aug. priory of St Mary (nuns), building of, 24
King's Somborne, Sombourn (Hants.), church of, app.iv (10)
—tithes of assarts at, 34
Kingston Deverill (Wilts.), land of, 25
Kingston upon Thames, Kingeston' (Surrey), app.iii (151, 162)
—charter witnessed at, 51
Kington, *see* Little Kington
Kirby Sigston (Yorks.N.R./N.Yorks.), church of, app.iv (1)
Kirkleatham (Yorks.N.R./Cleveland), church of, app.iv (30)
Kivilly, Nicholas de, app.ii (4)
—master of the Domus Dei, Portsmouth, app.ii (22)
Knepp Castle (W.Sussex), app.iii (152)
Knightsbridge (Middlesex/London), vill of, 73
Knights Templar, 130, 177, app.ii (30), master of, 129, 328, *and see* London, Templo
Knoyle, Cnoel (Wilts.), 307, app.iii (148), app.iv (4, 23)
—Thomas of, 274
Kynest, Nicholas, 15

La Lee (unident., Hants.), app.iv (23)
La Peruse, *see* Peruse

La Rochelle, *see* Rochelle
Laci, *see* Lacy
Lacock (Wilts.), Aug, abbey (nuns), app.iv (23)
Lacy, Laci, Hugh de, 122
—Iseut de, 136
—John de, constable of Chester, 96a, 100
—Walter de, 106
La Marche (France), count of, *see* Lusignan
Lambeth (London), app.iii (144–7, 149, 155, 159–60)
Lambourne (Essex), church of, app.iv (28)
Lancaster (Lancs.), William of, 106
Lanercost (Cumberland/Cumbr.), Aug. priory of, chronicle of, xxxiii
Langecroft (unident., ?Surrey), app.ii (28)
Langley (Essex), app.iv (22)
Langrish, Langeress' (Hants.), Nicholas de, xlviii n.
Langton (Lincs.), mr Simon of, 92n.
—Stephen, archbp of Canterbury, xxviii, xxx, xxxiv-vii, xliii n., lix, 56n., 110, 272, app.i (128, 131), app.ii (6, 18, 29), app.iv (4, 17, 20, 30, 33, 34, 36)
——letters and charters of, liv, 24n., 34n., 73, 97–9, 126, 128
——letters to, lxvi, 95, 100, app.i (132)
—mr Walter de, official of the archdn of Winchester, xxxvii n., app.iv (5)
Lanthony (Gloucs.), Aug. priory of, prior of, 125
Laon', *see* Killaloe
Lateran, Lat' (Rome), council of 1215, lviii–ix, 13n., 125n., app.ii (6), app.iv (18, 27)
—king's proctor at, *see* Airaines, Russinol, Peter I
—papal mandates dated at, 5, 47, 132
Lavender, Geoffrey le, washer of St Swithun's Winchester, app.ii (48)
Lawrence, mr, 232
Ledbury (Hereford./H. & W.), app.iii (161)
Lee Ground, Leghe, in Titchfield (Hants.), 67
Leghia, Henry de, clerk, 13W
Leicester (Leics.), app.iii (156, 158)
—honour of, app.iv (40)
Leighlin (Ireland, co. Carlow), bp of, *see* Herlewin
Le Mans, *see* Mans
Lemovic', Lemovicen', *see* Limoges
Leuca, grand-daughter of William de Braose, wife of Geoffrey de Camville, mother of William de Camville, 3n.
Lewes (E.Sussex), app.iii (149)
—Clun. priory of St Pancras, monks of, li, 26
Lichfield, Lich' (Staffs.), app.i (128), bps of, *see* Coventry
—cathedral church of, canons of, app.i (132)

——canon of, *see* Roches, Bartholomew des
——dean of, *see* Neville, Ralph de
Limburg (Netherlands), duke of, 129, app.iv (29)
Limoges, Lemovic', Lemovicen' (Haute-Vienne)
—bp of, app.iv (18)
—Clun. abbey of St Martial, app.iv (18)
—Aimery de, clerk, rector of Blendworth, 40, app.iv (33)
—mr John of, xxxix, xliii, 18W, 85W, the bp's clerk, rector of Blendworth, 39
——biography of, app.iv (33)
Lincoln (Lincs.), app.iii (153, 156), app.iv (32, 41)
—battle of, xxix
—bps of, *see* Grosseteste, St Hugh, Wells
——officials of, xxxvi
—cathedral church of, lxix
——canons of, *see* de Fauconberg, Philip de, Lucy, Philip de, Rivallis, St Maixent
——prebends at, app.iv (41), *and see* Carlton Kyme, Welton Ryval
——precentor of, xxvii
——treasurer of, *see* Lucy, Philip de
—diocese of, li
——records of, xlviii
Lincolnshire, app.iv (34, 38, 41)
Lindsey, 321
Lire (Eure), Ben. abbey, abbot and convent of, 27–9, app.iv (34)
Lisle, Insula, Brian de, 94, 96a
—Geoffrey de, 10
—Walter de, app.iv (23)
Lismore (Ireland, co. Waterford), bpric of, 110
—bp of, *see* Bedford, canons of, *see* David, Macrobius
Little Kington (Dorset), 25n.
Little Somborne, Parva Sombourn (Hants.), church of, 34, 46n., app.iv (10)
Little Yarmouth (Norf.), rector of, app.iv (2)
Littlehampton (W.Sussex), app.iv (4)
Llanstephan (Radnor./Powys), 3n.
Llywelyn, prince of Aberffraw, nuncios of, app.ii (33)
Loches (Indre-et-Loire), collegiate church, dean of, xxvii
Locus Sancti Edwardi, 43, *and see* Netley
Lohod, *see* Hod
Loir, river (France), xxvii
Loire, river (France), xxvii
Lombardy, 130, men of, 47
Lomer (Hants.), app.iv (40)
London, Lond', 22n., 56n., 100, 212, 215, 236–7, 258, 321, 340, app.i (128), app.iii (147, 151, 154–6, 158), app.iv (1, 4, 17, 26, 36–7)

INDEX OF PERSONS AND PLACES

—bps of, 73, app.ii (23, 30), *and see* Fauconberg, Eustace de Ste-Mère-Église
—cathedral church of St Paul, app.iii (150, 154)
——canons of, *see* London, John of, Fauconberg, Eustace and Philip de, Rivallis, Roches, Bartholomew and Luke des, Ste-Mère-Église
——chapter of, app.iv (1, 24, 30)
——dean of, *see* Lucy, Geoffrey de, Ste-Mère-Église
——masses and candles in, 73
——obit celebrations at, app.iv (1, 30-1)
——prebends of, *see* Caddington Major, Caddington Minor, Ealdland, Ealdstreet, Holbourn, Islington, Mora, Wenlocksbarn
—city and citizens of, 100, *and see* Lutre
—diocese of, li, app.ii (34)
—Holy Trinity Aldgate, Aug. priory, app.iii (154), app.iv (11, 22)
—House of Jewish converts, xxxv
—mayor of, 171
—mayor and commune of, 123, app.ii (16)
—Newgate, houses in, app.iv (41)
—New Temple at, 78, 123, app.iii (149-50, 152-3, 155-7)
—sheriff of, 109, clerk of, *see* Bacun
—St Bartholomew's, Aug. priory, priors of, *see* William
—St Bride's, app.iii (148)
—St Faith, parish, tenements in, app.iv (24)
—St Martin-le-Grand, college of, dean of, app.ii (34)
——canon of, app.iv (30)
—St Mary Arches, dean of, app.ii (34)
—Tower of, app.i (128), app. iii (149, 151-2, 154-5), app.iv (24, 38)
——constable of, *see* Greinville, Eustace de
—mr John of, xxxix, xliii, lxxii n., 11W, 32W, 136
——bp's official, xxxv-vii, 5n., 21W, 26n., 37W, 80n., 83n.
——biography of, app.iv (1)
—William of, rector of Sparsholt, 83-4
Longespée, William, earl of Salisbury, 94, 127n., app.ii (18), app.iv (32)
Longi Vadi (unident., France), app.iv (43)
Longstock, Stok', Stokes (Hants.), church of, 34
Long Sutton (Hants.), app.iv (39)
Longueville (Seine-Maritime), app.iv (38)
Lood, *see* Hod
Lortih', Henry de, 270
Louis, son of the king of France, app.ii (7a)
Lound (Notts.), app.iii (148)
Loveriis, Lovers, *see* Luverez
Lovington (Somerset.), app.iv (25)

Low Countries, 101n.
Lowestoft (Suff.), rector of, app.iv (2)
Lucy, Luci, family, lands of, app.iv (11, 34)
—Geoffrey de, 67n., 315, app.iv (34), man of, *see* Walter
—mr Geoffrey de, app.iv (34), chancellor of Oxford, xlii
——dean of St Paul's London, app.iv (31)
—Godfrey de, as archdn of Richmond, app. iv (30)
——as bishop of Winchester, xxvii, liv, 1n., 72n., app.iv (11)
——grants and confirmations by, xliii, lxviii-lxx, lxxiii, 11, 18-19, 32, 34, 36, 49n., 75, 83-4, app.ii (1, 20), app.iv (5, 11, 29-31, 34, 40)
——attorney of, *see* Fauconberg, Eustace de
——chancery of, xliii
——chaplains of, *see* Gregory, Humphrey, Terricus
——clerks of, *see* Fauconberg, Eustace and Philip de, Godfrey, Philip, Reginald, Roger, William
——family of, app.iv (34)
——household of, xli, seal of, lxi
——steward of, *see* Turville, Maurice de
—Isabella de, app.iv (34)
—mr Philip de, xl-xlii, 14W, 35W, 67n., app.iv (31)
——biography of, app.iv (34)
——archdn of Wiltshire, app.iv (34)
——canon and treasurer of Lincoln, app.iv (34)
——rector of Basing, Basingstoke and Selborne, 48n.
——rector of Newchurch, 29n, app.iv (42) steward of, *see* Oger
—mr Stephen de, brother of mr Philip de, 48n., 67n., app.iv (34)
—Stephen de, prior of St Swithun's Winchester, xlii n., 81
—William de, foster-son of Godfrey de, app.iv (34)
Ludgershall (Wilts.), app.iii (145-7, 150, 161)
—king's houses at, app.iv (25)
Ludshott (Hants.), app.iv (26)
Luke, archdn of Surrey, *see* Roches, Luke des
—chaplain, 85W
—mr, canon of Chichester, app.iv (13)
—the parson, app.iv (13)
Lulworth (Dorset), app.iv (43)
Lurdun, Lurd', Lurdon', Robert de, 140, 195, 250
Lusignan (Vienne), Hugh de, 118n., 126, app.ii (16), app.iv (22, 41)
—count of La Marche and Angoulême, 128
Lutre, Geoffrey, citizen of London, app.iv (1)

INDEX OF PERSONS AND PLACES

Lutterel, Geoffrey, his widow Frethesant, aunt of Isabella Painel, app.iv (38)
Luverez, Loveriis, Lovers, Geoffrey de, xxxix, 35W, 61W, app.iv (43)
——biography of, app.iv (39)
——nephew of, *see* Saunderville, sister of, *see* Wingham
—Geoffrey, another, app.iv (39)
—Michael de, Margaret his wife and Michael his nephew, app.iv (39)
—Nicholas de, app.iv (39)

Macrobius, mr, canon of Lismore, 110
Magna Carta, xxxi, liii, 97, 100n., app.iv (19)
Maine (France), 121n.
Maleth, Hugh, 13W
Malo Lacu, *see* Maulay
Mandatis, Ranulph de, app.ii (47)
Mandeville, Geoffrey de, app.i (131), app.iv (24, 30)
——earl of Gloucester, 100, *and see* Fitz Peter
—Walter de, Mirabila his widow, sister of mr Amicius archdn of Surrey, app.iv (11)
Mans, Le (Sarthe), app.iv (35)
—bp, dean and chapter of, 25,
—bps of, *see* Geoffrey, Maurice
—canon of, *see* Arenis
—dean of, *see* R.
——land of in England, *see* Kingston Deverill
—St Pierre de la Cour, collegiate church of, 121
Mansel, mr John, app.iv (4)
Mapledurham, Mapelderham, Mapeldreham, Mapulderham (Hants.), 9, 51
Mara, John de, 21W, 34W
March, the (Italy), 130
Marche, *see* La Marche
Marisco, *see* Marsh
Mark, Philip, 105
Marlborough (Wilts.), app.iii (145-8, 155-6, 160-1)
Marlow, Merlave (Bucks.), 9n., 215-16
Marmiun, Albreda de, 3n., her son, *see* Camville, Geoffrey de
—Geoffrey de, father of Albreda, 3n.
—Philip son of Robert the younger, app.ii (45)
—Robert de, heir and barony of 3n.
—Robert, son and heir of Robert, 53W, 78n., app.ii (45)
Marmoutier (Indre-et-Loire), Ben. abbey of, abbot of, *see* Roches, Hugh des
Marsh, Marisco, Geoffrey, justiciar of Ireland, 110
—Richard, xlv n., 96b, app.i (128-9)
——archdn of Northumberland, 94, bp of Durham, 107, 110

——chancellor of the king and bp of Durham, 117
Marshal, family, earls of Pembroke, app.iv (38)
—Gilbert, earl of Pembroke, 85n.
—Henry, bp of Exeter, 86-7
—John, 107
—Richard, 133, 334n., earl of Pembroke, xxxi, app.iv (38, 41)
—Walter, earl of Pembroke, app.iv (41)
—William (I), earl of Pembroke, xliii n., 79, 96, 105, 107, app.i (130), app.ii (14), app.iv (15)
——rector of the king and his realm, app.ii (9, 10, 13)
—William (II) junior, 79
—*and see* G.
Martigny, Martino, Geoffrey de, constable of Northampton, 103
Martin, prior of St Mary Overy, Southwark, 57
Martin IV, pope, xxviii n.
Martock (Somerset), church of, app.iv (7)
Martre, Henry la, bailiff of Alresford, 149
Marwell, Merewell', Merewelle (Hants.), app.iii (147, 151, 158-60), app.iv (39)
—bp of Winchester's palace at, 31n., app.iv (19)
—chaplain of, *see* William
—charters dated at, 31, 75
—college of, xxxvi, l, lv-vi, 31, app.iv (2)
Massingham (Norf.), church of, lii n.
Mathilda, abbess of Wherwell, 75
Mauclerk, Walter, app.ii (5), bp of Carlisle, 128, app.iv (22)
Mauduit, Robert de, barony of, app.iv (22)
Mauger, bp of Worcester, 86-7, 91-2
Maulay, Malo Lacu (Vienne), Peter de, 94, 105-6, 193
Mauléon, Mauliun (Deux-Sèvres), Savaric de, 96b, clerk of, 139
Maurice, bp of Le Mans, 121
Mautravers, W., 191, John his companion, 191
May, Michael, bp's butler, app.ii (42), *and see* Michael the butler
Meath (Ireland, co. Meath), bp of, 125
Melksham (Wilts.), app.iii (148)
Melton (Suff.), church of, app.iv (28)
Menesfeld', *see* Shenfield
Menes, Meon, *see* East Meon
Meon, Mune, in Titchfield (Hants.), villata of, 67
Meon Church (Hants.), bp of Winchester's manor of, 31n., 194
—bailiffs of, *see* William
Merdon, Meredon' (Hants.), 137, 229, app.iii (151), app.iv (32, 39)
—forest regard at, app.iv (14)
Meriet, Ralph de, app.iv (23)

Merlay, Roger de, 64, Agnes his daughter, 64n., *and see* Gymminges
Merewell', Merewelle, *see* Marwell
Merther (Cornwall), app.iv (32)
Merton (Surrey), app.iii (151, 157, 160)
—Aug. priory of, app.ii (27), app.iv (10, 11, 34–5)
——fire at, app.iv (11)
——priors of, app.iv (11), *and see* Giles, Thomas
—Walter of, app.iv (26, 43), bp of Rochester, lvi
—mr Walter of, archdn of Berkshire, app.ii (24)
Messinges (unident.), app.iv (25)
Meves (unident., ?Hants.), app.iii (149), charter dated at, 94
Michael, the bp's butler (*pincerna*), app.ii (42)
—rector of Cobham, 6
Micheldever (Hants.), app.iv (34, 40)
Middlesex, archdn of, *see* Reginald
Middleton (Wilts.), app.iv (39)
Middleton Cheney (Northants.), church of, app.iv (34)
Miles, bp of Beauvais, 130
Milford-on-Sea (Hants.), church of, 8
Millières, Milers, Milleiis, Miller', Milleriis, Millers, Milliers, Mulers (Manche)
—Felicia de, app.iv (4)
—mr Gilbert de, alias de Wiarvill, app.iv (4)
—Humphrey de, knight, and William his son, app.iv (4)
—mr Humphrey de, xxxix, xlii, 6W, 14W, 35W, 49W, 50W, 52W, 61W, 65W, 67a, 197n., app.ii (29W), app.iv (20, 36, 42)
——biography of, app.iv (4)
——bp's official, xxxvi, 51W, 53W, 84W
——master of St Cross Hospital, Winchester, app.iv (2, 4)
——rector of Ivinghoe, app.ii (21), Witney, app.ii (21), app.iv (15)
—William de, the elder and the younger, app.iv (4, 42)
Milton Abbas (Dorset), app.iii (147)
Missenden (Bucks.), Aug. abbey, app.iv (35)
Mitcham (Surrey/London), app.iv (22)
Mixbury (Oxon.), church of, app.iv (35)
Moine, Monachus, Ralph le, 44
Molendinarium, *see* Reginald
Molendinis, Roger de, 20, mill of, 19
Mombray, *see* Mowbray
Monachus, *see* Moine
Monks Sherborne (Hants.), Ben. priory of, app.iv (34, 39)
Monmouth (Monmouth./Gwent), John of, 106, app.iv (23)
—and his son John, app.ii (2)

Montacute, Monteacuto, mr H. de, 289
—mr William de, 289
Monte Morelli, Adam de, 113
Montferrat, Montferrant (Var), Imbert de, 48
Montgomery (Montgomery/Powys), xl n., app.iii (157, 161), app.iv (22)
—garrison of, app.iv (10)
—scutage of, 304
Mont-Saint-Michel, Sancti Michaelis de Periculo Maris (Manche), Ben. abbey of, 32, 48n., app.iv (7, 34)
Mora, prebend in St Paul's cathedral London, app.iv (31)
Morden, Mordon (Surrey/London), church of St Lawrence, 74, rector of, *see* Shenfield
Morestead (Hants.), Robert of, married sister of Maurice de Turville, app.iv (40)
Mortimer, Mort'Mar', Hugh de, 106, 284
—Isabella de, app.iv (37)
Mortlake (Surrey/London), app.iii (144)
Mottisfont, Motesfont', Motesfonte, Motesfunt', Mottesfont (Hants.), Aug. priory of Holy Trinity, lxii, lxvii, lxx–lxxi, 33–4, 48n., 49n., 82n., app.ii (34), app.iv (6, 10, 41)
——cemetery of, lxix, 33
——chronicle of, app.iv (41)
——prior of, 49n., *and see* Stephen
————attorney of, *see* Wich
—land at, 34
—parish church of, 33, app.iv (41), rectors of, *see* Rivallis
Mowbray, William de, 100
Muleford, William de, 335
Mune, *see* Meon
Murrell Green (Hants.), app.iv (13)
Musard, Ralph, sheriff of Gloucestershire, 106
Muschamp, Geoffrey, bp of Coventry, 86–7
Mussel, Hugh, notarial inspection by, 32, 50, 54

N., the chaplain of Hart', 239
Narbonne (Aude), archbp of, *see* Peter
Navar', Ralph de, 61W
Navarre (Spain), Theobald I, king of, app.iv (33)
Nazareth (Holy Land), archbp of, *see* Nicholas
Neatham, Netham (Hants.), tithes and community of, 1
Neovila, *see* Neville
Netley (Hants.), Cist. abbey of, xxxvii, lv, 25n., app.iv (4, 10, 13)
—originally founded at Locus Sancti Edwardi, 43
Neufchâtel-en-Bray (Seine-Maritime), app.iv (41)

INDEX OF PERSONS AND PLACES

Neville, Neovila, Geoffrey de, 127n., chamberlain, 96a, 113
—Henry de of Hale, William his son, 105
—Hugh de, 89, 94, 96b
—John de, app.iv (23)
—Nicholas de, king's clerk 18 and n.
—Ralph de, bp of Chichester, royal chancellor, xli–ii, 18W, 128, 133n.
——bp-elect of Chichester, vice-chancellor of the king, 132
——dean of Lichfield, 110, 116
——the bp's executor, 43, correspondence of, p.105
——keeper of the seal of king John, xli, app.i (129)
——the bp's steward, xli
—William de, app.iv (23)
Newark, Novus Locus (Surrey), Aug. priory of, l, 35, 36, app.ii (20), app.iv (6, 31)
—prior of, *see* John
Newark-on-Trent (Notts.), app.iii (153)
Newburn (Northumberland/Tyne and Wear), 120
Newbury (Berks.), app.iii (157)
Newchurch, Niwechirche, Nywecherch' (IOW), 75, church of, 29, 75, app.iv (34, 42)
—rectors of, *see* Gattebrig', Lucy, Philip de, Roches, Aimery des
Newport Pagnell (Bucks.), app.iii (148)
New Forest (Hants.), app.ii (12), app.iv (23, 41)
Nicholas, archbp of Nazareth, 129
—bp of the Isles, xxxiv
—bp of Winchester, *see* Ely
—chaplain, attorney of Wherwell, 76n.
—prior of Christchurch Twinham, app.iv (32)
Niort (Deux-Sèvres), app.iii (144)
Niton (IOW), church of St Michael, 28, app.iv (18), perpetual vicar of, *see* Russinol
—rectors of, *see* Wautham,
Niwechirche, *see* Newchurch
Niwel, Walter, 152
Norfolk, app.iv (4, 26, 28)
Norman, mr Simon the, app.ii (37)
Normandy, xxxix, 127, app.ii (45), app.iv (17, 19, 25, 38)
—duke of, writing of, xlix n.
—French invasion of, xxxix, app.iv (25, 30, 38)
—land of the Normans (*Terra Normannorum*), xxxix, lv, 67n., app.iv (4, 29, 38–9, 42–3)
—natives of and Anglo-Normans, xxxviii–ix, app.iv (38–9, 41)
—religious houses of, lv–vi
Nornis, 142, *and see* Norwich
North Baddesley (Hants.), preceptory of Hospitallers at, app.iv (31)
North Fareham (Hants.), app.iv (43)
North Petherwin (Cornwall), church of, app.iv (5)
North Walsham (Norf.), church of, app.iv (7)
North Waltham (Hants.), 239, 284, app.iv (7, 24)
Northampton (Northants), app.iii (146, 148, 150–1, 156, 158, 160, 162), app.iv (4)
—burgesses of, 103, constable of, *see* Martigny
Northumberland, archdns of, *see* Marsh
—sheriff of, *see* Bolebec
Northwold (Norf.), Hugh of, bp of Ely, 132
Norton, Northton', Nortone (Hants.), James of, 49, 71W
Norwich, ?Nornis, Norwycensis (Norf.),
—bishops of, 142n., *and see* Blundeville, Gray, Pandulph, Raleigh
—official of, *see* Warham
—election at, app.iv (18)
Notehat, Thomas, wife of, 196
Notley (Bucks.), Aug. abbey of, canons of, xlii n.
Nottingham (Notts.), app.iii (147–9, 153–4)
Novus Locus (unident.), church of, app.ii (20), app.iii (156), *and see* Hyde, Newark
Nuneaton, Eton, Etton' (Warwicks.), Ben. priory, prioress and nuns of, 37–40, app.iv (7, 33)
Nunkeeling (Yorks.E.R./Humberside), Ben. priory (nuns), app.iv (30)
Nywecherch', *see* Newchurch

O., prior of Hyde, xlii n.
Oakham (Rutland/Leics.), app.iii (156)
Oakhanger, Acangre (Hants.), James of, 49
Oakley, Hackel' (Hants.), tithes of land and assarts at, 34
Oddington (Gloucs.), app.iii (159–61)
Odiham (Hants.), app.iii (148, 153), church of, 46n., app.iv (4)
Oger, the steward (*dapifer*) of Richard de Lucy, app.iv (11), *and see* Fitz Oger
Oldcotes (Notts.), Philip of, 94, 96a, seneschal of Poitou, 118n.
Orelluz, P., clerk, app.iv (41)
Orival (Charente), app.iv (41)
Orival, Orivalle (Seine-Maritime), Geoffrey de, app.iv (41)
—Peter de, app.iv (41)
—Roger fitz Odo de, app.iv (41), *and see* Rivallis
Orivallis, *see* Rivallis
Orleton, Adam de, bp of Worcester, inspeximus by, 13
—bp of Winchester, xlix n.
Ortiaco, Henry de, 341
Osbert, mr, bp's official, 63W

—the reeve of Alresford, 149
Oseville, Margaret de, app.iv (21)
Osney (Oxon.), Aug. abbey of, 33n., app.iv (35)
Ospringe (Kent), Domus Dei at, 58n.
Ossory (Ireland, co. Queen's), bp of, 125
Otto, nuncio of pope Honorius III, app.iv (19, 36)
Overstone (Northants.), manor and church, app.iv (4)
Overton (Hants.), 223, 282–3, app.iv (15, 21)
—borough of, app.iv (39)
—church of, app.iv (34, 44)
Over Wallop (Hants.), rector of, 77
Oxford (Oxon.), xlii–iii, 182, 234, app.iii (145, 147, 152, 154–6, 159, 160, 162), app.iv (4, 14, 42)
—earls of, *see* Vere
—schools of, xxxvi, xlii, app.iv (4, 41), chancellor of, *see* Lucy, Geoffrey de
Oxfordshire, xxxii, app.iv (11, 38)
—deputy sheriff of, *see* Chanceaux

P., mr, archdn of Winchester, the bp's executor, 43, app.iv (9, 10)
Paddington (Middlesex/London), vill of, 73
Painel, Thomas, 155
—William, his daughter Isabella, widow of William Bastard, wife of Eustace de Greinville, app.iv (38), *and see* Lutterel
Painscastle (Radnor./Powys), app.iii (159)
Pamesbirg', *see* Rammesbiryg
Pandulph, app.ii (6, 7)
—bp-elect of Norwich, 13W, 121, app.iv (18)
——papal chamberlain and legate, 35, 118–119
——papal legate, 110–112, 245, app.ii (15), app.iv (27)
——letters to, lxiv
—papal subdeacon, 97, 99, 100, and nuncio, app.iv (29)
Parcsiens', John, proctor of William de Ste-Mère-Église, app.ii (40)
Paris (France), xlii
—St Martin des Champs, Clun. priory, app.iv (10)
Paris, Matthew, chronicler, xxxv, 43, 131n., 133n., app.ii (4, 5, 30)
Passelewe, Robert, 101n.
Paston (Northants./Cambs.), chapel of, app.iv (11)
Patney (Wilts.), church of, app.ii (43), rectors of, *see* Watervill
Patterson, Robert, lxii n.
Pattishall, Patishill' (Northants.), Martin of, king's justice, 22n., 116, app.iv (32)

Pavilly, Pavill', Pavilli, J. de, 250
—mr Richard de, 135
—mr R(obert) de, 11W, 21W, 32W, 34W
Peakirk (Northants./Cambs.), church of, app.iv (11)
Pecham, John, archbp of Canterbury, xxviii n.
Peckham (London), 61
Pellered, Richard de, app.iv (25)
Pembroke (Pembroke./Dyfed), earls of, *see* Marshal
Penceley, Pennardeshull' (Wilts.), app.ii (34)
Pennington (Hants.), app.iv (29)
Percy, Richard de, 100, chaplain of, *see* Seamer
Peregore, mr Simon de, dean of Chichester, 18
Perray Neuf (Sarthe), Premonst. abbey of, lv
Perugia (Italy), app.ii (34), app.iii (158, 162)
Peruse, La (unident.), abbot of, app.iv (20)
Pestur, Gilbert le, app.ii (46)
Peter, archbp of Caesarea, 129
—archbp of Narbonne, 129
—archdn of Surrey, 13W, app.iv (11), biography of, app.iv (12)
—the bp's nephew, app.iv (41), *and see* Rivallis, Roches
—the clerk, 75W
—the doorkeeper (*portarius*), 6W
—the younger (*iuvene*), 21W
Peterborough (Northants./Cambs.), app.iii (145, 161)
—Ben. abbey, app.iv (11)
Petersfield (Hants.), 51n., app.iv (38), *and see* Sheet
Peter's Pence, xxviii
Petherwin, *see* North Petherwin
Peyvre, Paulinus, app.iv (38)
Pharaoh, Morals of the Dreams of, app.iv (33)
Philip, abbot of Cormeilles, 2n.
—clerk of bp Godfrey de Lucy, xliii, charter issued by, 75, app.iv (31)
—dean of Poitiers, rector of Wearmouth, 118n.
Philip Augustus, king of France, xxvii, lxviii n., 121
Picardy (France), app.iv (27)
Plantagenet, family, xxxi, *and see* Edward, John, Henry, Richard
—lands, xxxix
—Geoffrey, archbp of York, app.iv (30)
Pleystowe, in Titchfield (Hants.), villata of, 67
Poer, Herbert, archdn of Canterbury, app.iv (11)
——as bp of Salisbury, 18, 45, 86, 87, app.iv (5, 11)
——chaplain of, *see* Bartholomew
—Richard, as dean of Salisbury, xxviii, xxxvii, lix, app.iv (5, 7, 11)
——as bp of Chichester, 99

——as bp of Salisbury, xxxvi, 7n., 31n., 53n., 82
——letters and charters of, 73, 107, 126, 128
——letters to, lxx, lxxii, 47
——as bp of Durham, xliii n., 342
Poitiers (Vienne), archdn of, lxviii n., *and see* William
—dean of, 118, 119, *and see* Philip
—St Hilaire, collegiate church
——chèvacier (*capicerius*) of, app.iv (41)
——treasurer of, xxvii, 121n, app.iv (35, 41), *and see* Thouars
—Philip of, bp of Durham, xxxv, xxxix, app.iv (1, 17)
——official of, *see* London, John of
——household of, *see* Chanceaux, Andrew de, Clinchamps
Poitou (France), xxx–xxxi, 113n., 118–19, 123n., 192n., app.i (128–32), app.ii (16), app.iv (22, 41)
—expeditions to, xxix, xxxi, xli n., 127n., app.iv (29, 34–5)
—mercenaries of, 96b
—natives of, xxxviii, *and see* Douhet, Limoges, Lusignan, Rivallis
—seneschal of, *see* Oldcotes, Pons, Thurnham
Pons (Charente-Maritime), Poncius de, son of the seneschal of Poitou, app.iv (4)
Ponsout, Robert de, 21W
Pont', J. de, 302
Pontearche, Pondelarche, William de, 79, 82W
Pontefract (Yorks.W.R./W.Yorks.), app.iii (150)
Pontoise, Pontissara (Val d'Oise), John of, bp of Winchester, xxviii n., 20n., 36n., 55n.
—charters and confirmations by, 39
Porcar', *see* Geoffrey
Porcestr', Porcestre, *see* Portchester
Port, Adam de, 7n., app.iv (34, 40), son of, *see* William de St John
—Agnes de, 71n., daughter of Thomas de Venuz
Porta, Richard de, 6W
Portario, *see* Peter
Portchester, Porcestr', Porcestre (Hants.), 67, 67a, 68, 145, app.iii (144, 146, 149–50), app.iv (4)
—castle of, 246, app.iv (25, 44)
——clerks imprisoned at, 145, constable of, 71n., *and see* Wacelin
Portsmouth (Hants.), 127, app.i (131), app.ii (22), app.iii (151, 158), app.iv (17)
—Domus Dei at, lv, lvii, app.ii (4, 22), master of, *see* Kivilly
Poteria, Matthew de, 75
Potterne (Wilts.), Nicholas of, 46
Powerstock (Dorset), app.iii (150)

Premonstratensian, order, canons of, lv, 13, 43, 69
Prestbury (Gloucs.), app.iii (147)
Prested (Essex), app.iv (38)
Preston (Lancs.), church of, app.iv (18, 42)
—rectors of, *see* Roches, Aimery des, Roussillon, Russinol, Peter I
Preston, East and West (W.Sussex), *see* East and West Preston
Preston Candover (Hants.), app.iv (40)
Pruz, Probo, Richard, app.ii (49)
Puiset, Hugh du, archdn of Winchester, bp of Durham, 1W and n.
Puntyngton', Robert de, 6W
Puttenham (Surrey), *see* Shoelands

Quarr (IOW), Cist. abbey of, lii n., app.iv (6, 23)
Quincy, Saher de, earl of Winchester, 96, 100, 107
Qulla Episcopi (unident., France), app.iii (144)
Quob, Quabbe, in Titchfield (Hants.), app.iv (43), villata of, 67

R., archdn of Winchester, 34W *and see* Ralph
—chaplain of Robert fitz Walter, 100
—clerk, nephew of Eustace de Greinville, app.iv (38)
—dean of Le Mans, 25n.
—dean of Winchester, 32W
—the mason (*cementarius*), 326
—rector of Witley, 85n.
—succentor of Wells, 247
Raleigh, daughter of Ralph de, kinswoman of William de, app.iv (25)
—William de, bp of Winchester, xxxvii, lxiii, app.iv (10, 13, 20, 25)
——charters of, 21, 36, 59, 60, 63n., 65, app.iv (10, 37)
——bp of Norwich, inspeximus by, 86
Ralph (I), archdn of Winchester, app.iv (4)
—(II), or Ranulph, archdn of Winchester, biography of, app.iv (6)
——mr Andrew, Vincent and William brothers of, app.iv (6)
—bp of Chichester, *see* Neville
—the bp's chaplain, 70
—vicar of Wandsworth, app.ii (29W)
Rammesbiryg, Ramesbir', Pamesbirg', mr J. de, 34W, 37W
Rampton (Notts.), prebend of, app.iv (30)
Ranulph, archdn of Winchester, *see* Ralph (II)
—earl of Chester, 94, 96, 106, 122, app.ii (9), *and see* Chester, earl of

Ranvill', Germanus de, 9W
—William de, xlviii n.
Reading (Berks.), app.iii (145, 147-8, 150, 152-3, 155, 159-62), app.iv (6)
—charter dated at, app.ii (35)
—Ben. abbey, abbot of, app.ii (6, 7), *and see* Simon
——abbot and convent of, app.ii (35)
—Franciscan house at, app.ii (35)
—Vastern, in, app.ii (35)
Reginald, archdn of Middlesex, app.iv (31)
—clerk of bp Godfrey de Lucy, xliii, *and see* Regny'
—earl of Cornwall, app.ii (10)
—the miller (*molendinarius*), 15
—the ploughman (*carrucarius*), 15
Regny', clerk, 75W, *and see* Reginald
Reimund, 152
Repeleg', *see* Ropley
Richard I, king of England, xxvii, lv, lxix, 43, 67a, 69, 129n., app.iv (7)
—grants and confirmations by, lxxiii, 18, *and see* Berengaria
Richard II, king of England, confirmation by, 63
Richard, abbot of Titchfield, 52W
—the bp's chaplain, 13W, 38W, 45W, 56W, 83W
—the clerk, 243
—earl of Cornwall, brother of king Henry III, 127n., app.iv (22, 34)
—monk of Abingdon, 108
—*persona* of Chiddingfold, 85W
—prior of Dunstable, 73
—the weaver (*telarius*), 6
Richmond (Yorks.N.R./N.Yorks.), archdn of, *see* Lucy, Godfrey de
Rimilis, Peter de, app.iv (41)
Rimpton (Somerset.), app.iv (23)
—reeve of, *see* Jordan
Ringwood (Hants.), church of, app.iv (29), rector of, *see* Caux
Rivallis, Orivallis, Rivall', Aimery de, app.iv (42), *and see* Roches
—Peter de, the bp's nephew or son, xxx-xxxi, xliii, 48, 51W, 53W, 133, app.ii (39), app.iv (43)
——biography of, app.iv (41)
——constable of Bristol, app.iv (37)
——greyhounds of, 322-323
——men of, 263
——niece of, Aaleys, app.iv (41)
——clerk and treasurer of the king, xli
——rector of Alton, 1n., rector of Mottisfont, 33n.,
——the bp's executor, 43, the bp's kinsman, xxx, 85n.
Robert, bp-elect of Ely, *see* York

—bp of Exeter, charter of, 11
—bp of Salisbury, *see* Bingham
—bp of Waterford, 110
—the chaplain, 59W, 60W
—the clerk, xlviii n.
—dean of Guildford, 6W
—master of St Thomas' Southwark, 26n.
—*persona* of Bookham, 6W
—prior of Bath, app.ii (3)
—reeve (*prepositus*) of Meon, 244
—son of Robert, rector of Arreton, 27
Roche d'Orival (Seine-Maritime), app.iv (41)
Rochelle, La (Charente-Maritime), app.iii (144-5)
—mayor and commune of, 123, app.ii (16)
Roches, Rupibus, Aimery des, the bp's nephew (*nepos*), rector of Preston and Newchurch, 29n., app.iv (4, 10, 18, 43)
——biography of, app.iv (42)
—Baldwin, nephew of William des, app.iv (43)
—mr Bartholomew des, the bp's nephew (*nepos*), 18n.
——archdn of Winchester, xxxvii, 85n., app.iv (2, 13, 23, 36)
——biography of, app.iv (7)
——prebendary of Burbage, app.ii (31)
——dean and canon of York, canon of Exeter, Lichfield and London, app.iv (7)
—Geoffrey des, the bp's nephew, xl, 51W, 52W, 71W, app.iv (10, 39, 42)
——biography of, app.iv (43)
—Emma, his wife, app.iv (41), *and see* Fitz Henry
—Hugh des, abbot of Marmoutier, 121n.
—Hugh des, son of Geoffrey, app.iv (43), wife of, *see* Hoo
—mr Hugh des, archdn of Winchester, xxxvii, 51W, 53W, 84W, app.iv (13, 42-4)
——biography of, app.iv (10, 16)
——the bp's executor, 43, rector of Ditton, app.ii (27)
——the bp's treasurer, app.iv (15, 16)
——orders of, lvii, canon of Exeter, *receptor* of Wolvesey, app.iv (10)
—John, son of Hugh des, app.iv (43)
—mr Luke des, archdn of Surrey, xxxvii, xliii, 18n., 26 and n., 48W-51W, 53W, 56n., 84W, 85W, app.ii (34), app.iv (7, 10, 20, 30, 42)
——biography of, app.iv (13)
——alias Luke of Winchester, app.iv (13)
——the bp's executor, 43, prebendary of Burbage, app.ii (31)
——canon of London and Salisbury, app.iv (13)
——orders of, lvii n.
——papal subdeacon, app.ii (44)
——**attorney of**, *see* York, Ralph of

—Martin, son of Geoffrey des, and Lucy his wife, app.iv (39, 43)
—Peter des, knight, 225-7, 246, biography of, app.iv (44)
—Peter des, bp of Winchester, almoner of, app. iv (10)
——alms of, 31n.
——as regent-justiciar, app.i (125-32)
——attorney of, app.ii (1), *and see* Herriard
——baggage of, app.iv (4)
——bailiff of the liberties of, xxxviii
——bailiff of, *see* Southwark, John of
——baker of, *see* Germanus
——butler of, *see* David, May, Michael
——chamber of, xlvi, app.iv (15, 17, 19) *and see* Bourgueil
——chamberlain (*camerarius*) of, xxxviii, *and see* Clinchamps, Twyford
——chancery of, xxxix, xliii-viii, lxix-lxxv, clerks of, app.iv (18-20), *and see* Chanceaux, Russinol
——chaplains of, lviii, lxxii, *and see* Gregory, Henry, Hugh, Humphrey, Luke, Ralph, Richard, Robert, Simon, Stephen, William
——clerks of, xxxviii, xl, lxxii, app.iv (27-36), *and see* Airaines, Amico, Barking, Basset, Bourgueil, Caleston', Caux, Cerne, Clinchamps, Dartford, Denis, Edmund, Elmham, Fauconberg, Eustace and Philip de, Fitz Humphrey, Herbert, Herce, Leghia, Limoges, Lucy, Philip de, Rouen, Russinol, St Maixent, Stephen, Thomas, Vienne
——courts of, app.iv (18), *and see* hundred court
——crusade of, xxx, xxxii, xxxvi, 8n., 10n., 14n., 15n., 22n., 28n., 29n., 41n., 42, 47, 57n., 61n., 62n., 65n., 67n., 72n., 77n., 78n., 80n., 85n., 113n., 132n., 294, app.ii (28, 30), app.iv (2, 7, 8, 10, 13, 16, 19, 21, 25, 32, 36)
——*custos sigilli* of, xliv-v, *and see* Russinol
——diocesan legislation of, xxxvi, li, lvii, 58n., app.ii (25)
——election of, xxvii-ix, app.iv (1, 5, 18, 25)
——endowment of vicarages, xxxvi
——estates of, app.ii (42), endowment of stock, 43
——exchequer (*scaccarium*) of, xxxiii, xlvi, 20n., 31 and n., 316, app.iv (10, 16), *and see* Wolvesey
——executors of, 43, app.ii (45), app.iv (13), *and see* Beaulieu, Dereham, Neville, P., Rivallis, Roches, Hugh and Luke des
——following of (*familiarite*), 118
——guardian (*custos*) of the king, app.ii (26)
——horses and stables of, xxxviii n., app.iv (19)
——hounds of, app.iv (14)
——household of, xxxvii-xliii, 9n., 233 (*domus*), 118, app.iv (1-44)
——huntsmen of, xxxiii, 283, *and see* Gervase, Stephen, Woodcock
——itinerary of, xxxi-ii, app.iii
——keeper of the see of Lichfield, app.iv (7)
——kinsmen of (*nepotes*), xxxii, 29, 118, *and see* Rivallis, Roches
——knights of, xl, lxxiii, 118, 161, app.iv (22, 37-40), *and see* Aliz, Falaise, Greinville, Luverez, Turville
——mandate to preach the crusade, 47
——markets of, liberty of, 119, *and see* Winchester, fair of
——marshal of, xxxviii, app.iv (10), *and see* Achard, G.
——marshalcy of, app.iv (32)
——men of, 82
——messenger to France, app.iv (29)
——obit celebrations for, 43
——officers of, 26n.
——officials of, xxxv-vii, xlii n., liii, lxvi, lxxii, 22n., 31, 84, biographies of, app.iv (1-4), *and see* London, John of, Millières, Osbert, Ste-Mère-Église, Stokes
——orders of, lvii
——ordinary and delegate of, app.iv (2)
——as patron of the religious, liv-vi
——penitentiary of, lviii
——*procurator* of, app.iv (7, 13)
——ransoms taken by, 44, 79
——*receptor* of, 20
——and reform, lvi-ix
——registers of, xlviii-ix
——scribes of, xlv, lx-lxiv
——seal of, lx-lxii, lxxi, app.i (130), app.ii (2, 9, 10, 13, 14, 28, 30, 34, 36, 44), keeper of, *see* Russinol
——son of, *see* Rivallis
——spigurnel of, xlviii
——stewards of, xxxviii, app.iv (21-6), *and see* Basset, Batilly, J., John the dean, Neville, Shorwell, Wacelin, Waltham
——suffragans of, xxxiv-v
——synod of, 63
——temporalities of, app.iv (13), *and see* Winchester, diocese
——treasurer of, xxxviii, xlvi, lxii-iii, biographies of, app.iv (14-17), *and see* Bourgueil, Clinchamps, John the dean, Roches, Hugh des
——wards of, 78n., *and see* Aubigné, Robert de, Fitz Henry
——will of, lii-iii, lv, 42, 42a, 43, app.ii (30), app.iv (10)
——wines of, app.ii (42), *and see* Winchester, bps of

—William des, lord of Château-du-Loir, xxvii
—— seneschal of Anjou, lv, app.iv (43)
—William des, grandson of Hugh, son of John, app.iv (43)
Rochester (Kent), 43, app.iii (148, 152–3, 155, 162)
—bps of, xxxv, *and see* Glanvill, Sawston
—church of, app.iv (11)
Rocinol, P., app.iv (18), *and see* Russinol
Rockingham (Northants.), app.iii (145, 147, 156)
Roger, (I) archdn of Winchester, xxxvii, 75W, app.iv (6, 11, 41), rector of Alton, 1n.
—— biography of, app.iv (5)
—(II), archdn of Winchester, biography of, app.iv (8)
—chaplain of Bradley, and Richard his son, app.iv (43)
—clerk of bp Godfrey de Lucy, xliii
—dean of Southwark, app.ii (29W)
—mr, official of the archdn of Winchester, xxxvii n., app.iv (10)
—prior of Christchurch Twinham, app.iv (32)
—prior of Ely, 107n.
Roiusinol, *see* Russinol
Romagna (Italy), 130
Rome (Italy), xxviii, xxix, xxxix, xlv, 73, 88n., 100n., 125, app.i (129), app.ii (37), app.iii (144, 158), app.iv (1, 10, 11, 19, 25, 27, 29)
—ambassadors to, app.iv (4, 18, 27, 33, 34)
—cardinals of, *see* Langton
—charter issued at, lxvii, 81, *and see* Lateran
—church of, 100, 113–14
—citizens of, *see* Saracen
—clergy of, lii, subsidy to, app.iv (36)
—clerk of, *see* J.
—court of, app.ii (5), *cursus Romane curie*, lxv
—emperor of, *see* Frederick
—St Peter's, church of, xxviii, app.iii (144) *and see* Lateran *and subjects under* papal, popes
Romsey (Hants.), Ben. abbey (nuns), app.iv (2, 23), canons of, *see* Basset
Ropley, Repeleg', Ropell' (Hants.), 282
—Robert of, 94
Rossinol, *see* Russinol
Rouen, Rothomago (Seine-Maritime), app.iv (41)
—Roger of, 14W, 271, bp's clerk, 271n.
Roussillon (Vaucluse), Guy de, kinsman of king Henry III, rector of Preston, app.iv (18)
Rudynge (la), *see* Cobham
Rufo, Peter, 82W
Rugerig, ?in Shorwell (IOW), app.iv (23)
Rughemed, *see* Basingstoke
Rumyes', Walter de, 82W
Runnymede (Surrey), xxix, 100n., app.iii (153)
Rupibus, *see* Roches

Ruscombe Southbury (Berks.), prebend of, in Salisbury cathedral, app.iv (13)
Russ', Russi', Peter de, 334, app.iv (20), *and see* Russell, Russinol
Russel, William, 319
Russell, Russeaus, Peter, app.iv (20)
Russinol, Roiusinol, Rossinol, Ruscinale, Rusinol, Russig', Russign', Russing', Russingnol, Russinok, Russynol
—P., dean of Angoulême, app.iv (18)
—mr Peter (I), 1W and n., 11W, 80W, app.ii (41), bp's clerk, 37W, app.iv (27, 42)
—— charters issued by, lxxiii, 32, 34, 81
—— biography of, app.iv (18)
—— keeper of the bp's seal, xliv-vi
—— perpetual vicar of Niton, 28
—— canon and precentor of York, app.iv (18)
—mr Peter (II), xliv-v, 48W, 49W, 50W, 75W, 84W, 272, 302, app.iv (13, 19)
—— biography of, app.iv (20)
—— charters issued by, lxiv-v, lxxiv, 53, 54
—— rector of Adderbury, app.ii (41), proctor of, *see* Winchester, Thomas of
—*and see* Rocinol
Ryde (IOW), 75n.

Sabilio, mr Simon de, 85W
Sabina (Italy), cardinal bp of, *see* John
Sace, Sacy, Aimery de, 71W, 123n., lord of Barton Stacey, app.ii (26)
Saher, earl of Winchester, *see* Quincy
St Albans, Sanctum Albanum (Herts.), 282, app.iii (151), app.iv (10)
—Ben. abbey of, xxxv, 131
—— chroniclers of, app.iv (41), *and see* Paris, Wendover
St Amand, S. Amand (?Normandy), Aimery de, 206, 312, steward of the king, 206
St-Amand-sur-Sèvre (Deux-Sèvres), app.iii (144)
St Benet of Hulme (Norf.), Ben. abbey, app.iv (7)
St Birinus, cult of, liv
St Bride's, *see* London
St Cross, *see* Winchester
—mr Alan of, *see* Stokes
St Denys, *see* Southampton
St-Émilion (Gironde), app.iii (144)
St Ethelbert, tooth relic of, app.iv (31)
St-Germain-de-Varreville (Manche), app.iv (4)
St-Gilles(-du-Gard) (Gard), count of, app.iv (23)
St Gregory the Great, quotation from, lvi, lxvii, 13n.

INDEX OF PERSONS AND PLACES

St Hugh, bp of Lincoln, xxxiii
St James outside the Westgate (Hants.), manor of, 82n.
St-Jean-d'Angély (Charente-Maritime), app.iii (144)
St Jerome, quotation from, lxvii
St John, family, lords of Basing, 79n., app.iv (40)
—William de, 7n., 79, app.iv (40)
St Justa (Italy), app.iii (159), charter dated at, 130
St-Maixent, Sancto Maxentio (Sarthe), William de, xli, 26W and n.
—biography of, app.iv (35)
St-Maixent-l'École (Deux Sèvres), app.iii (144)
St Martin, Jordan de, and William his son, app.iv (23)
St-Martin-de-Ré (Charente Maritime), app.iii (145)
St Martin des Champs, see Paris
St Mary Magdalene, see Winchester
St Matthew, quotation from, lxvii, 58
St-Matthieu (Finistère), port of, 127
St Olave, see Southwark
St Oswald's see Gloucester
St Ouen, Walter de, 94
St Paul, quotation from, lxvii-viii
St Peter, patrimony of, 100, 130
St-Pol (Pas-de-Calais), count of, former lands of, app.iv (32)
St Sabina, cardinal priest of, see Thomas
St Swithun, cult of, liv, and see Winchester, St Swithun's
St William of York, xxxiv
Ste-Colombe (Seine-Maritime), app.iv (41)
Ste-Mère-Église, Sancte Marie Ecclesia (Manche), William de, bp of London, 30, 82, 91-2, 95, app.ii (17), app.iv (3, 34)
—letters and charters of, 24n., 86-8, 97-9, 104, 107-8
—William de, official of the bp of Winchester, xxxvi, lxxii, 48W-50W
—biography of, app.iv (3)
—dean of St Paul's London, xxxvi, xli
—rector of Witney, app.ii (40), proctor of, see Parcsiens'
Salisbury, Sar', Saresber', Saresbur', Sarr' (Wilts.), lvi, 112, 327, app.iii (146, 156)
—bps of, 45, letters and charters of, 24n, and see Bingham, Bridport, Poer, Herbert and Richard, Walter, Hubert
—cathedral church of St Mary, lxix, 18, 45, app.iv (7)
—canons of, 45, and see Barre, mr Bartholomew, Roches, Bartholomew and Luke des, Winchester, Luke of
—chancellors of, see Esseby

—dean of, 47n., and see Poer
—dean and chapter of, 46, seal of, app.ii (34)
—office of (Sarum rite), lvi, 31
—prebends of, 46, and see Burbage, Coombe, Hurstbourne, Ruscombe
—precentor of, 77
—proctor of, app.iv (13)
—subdean of, see Chobham, Ebbesbourne, Thomas of
—succentor of, see Anastasius
—diocese of, 47, app.ii (34)
—earls of, see Longespée
Salepsbur', see Shrewsbury
Samara, Sammar', see Seamer
Sampford, Thomas of, 94
Sancta Cruce, mr Alan de, see Stokes
Santa Maria in Via Lata (Rome), cardinal deacon of, see John
Sancte Marie Ecclesia, see Ste-Mère-Église
Sandford (Berks./Oxon.), app.iii (160)
Sandwich (Kent), app.iii (154)
San Germano (Italy), app.iii (158)
Santiago, military order of, app.ii (30), and see Compostela
Sancto Maxentio, see St Maixent
Saracen, J., 256, 257
—Peter, 324, app.i (131), app.ii (37), citizen of Rome, 108
Saracens, xxx, 100, 129
Sarisbury, Sarebury (Hants.), vill of, 67
Sarum rite (officio Saresber'), lvi, 31
Saunderville, Manasser de, app.iv (39, 43)
—Philip de, nephew of Geoffrey de Luverez, app.iv (39)
Savage, Henry le, and his widow, 52, 53n.
Savaric, bp of Bath, 86-7
Savoy (France), Boniface of, app.iv (10)
—Peter of, 85n.
Savoyards, see Montferrat, Roussillon, Vercers
Sawston (Cambs.), Benedict of, bp of Rochester, 30, 97, 110n., app.ii (29)
Sayr, 255
Schirborn', see Sherborne
Schitehagge, see Titchfield
Scot, mr William, 235
Scoteney, Walter de, 82W
Scotland, truce with, app.iv (27)
—Alexander II, king of, 104, 117, Margaret and Isabella, his sisters, 117, app.iv (29)
—William I, king of, app.i (131)
Scriptoria, Walkelin de, xlviii
Scures, Roger de, son of, 71n.
Seagrave, Segrava (Leics.), John, son of Stephen of, and his wife, app.iv (29)
—Stephen of, 18W, app.ii (33), app.iv (27)
—king's justiciar, 48

Seamer, Samara, Sammar' (Yorks.N.R./N.Yorks.), Osbert of, chaplain of Richard de Percy, 100
—Richard of, monk of Whitby, master of Stainfield priory, 100n.
Seckford, Gilbert called Hammergold of, notarial inspection by, 48, 50, 54
Seckington (Warwicks.), 3n.
Segar', lands of, 84
Segrava, *see* Seagrave
Selborne, Seleborne, Selebourn', Selebourne, Seleburn', Seleburne, Selleburne (Hants.), app.iii (162)
—Aug. priory of St Mary, xliii–iv, xlviii n., l, lv–vi, lix–lxii, lxiv–vi, lxviii–lxxi, 34n., 67n., app.iv (3, 10, 13, 21, 25, 34, 37–8, 41, 43)
——foundation charter of, 49
——prior and canons of, 48–54
——priors of, *see* Wich
—church of, 32, 48–50, app.iv (34), rector of, *see* Lucy, Philip de
—king's manor of, 48n., app.iv (34)
—vill of, 49
Selby, Seleby (Yorks.W.R./N.Yorks.), Ben. abbey of, abbot of, app.i (132)
Senesfeld', *see* Shenfield
Sheet, Sithe, in Petersfield (Hants.), land and mill at, 51, app.iv (38)
Shelswell (Oxon.), church of, app.iv (21)
Shenfield, Menesfeld', Senesfeld' (Essex), mr Thomas of, rector of Morden, 74, physician (*medicus*), 74n.
Sherborne, Schirborn' (Dorset), mr Walter of, 75
Shifford, John de, 71W
Shoelands, in Puttenham (Surrey), app.iv (41)
Shorwell, Sorwell (IOW), app.iv (38)
—chapel of, app.iv (5, 23), *and see* Dungewood
—Robert of, father of William, app.iv (23)
—Robert of, brother of William, app.iv (23)
—William of, 190, app.iv (35), biography of, app.iv (23)
——bp's steward, xli
——knight, 83W
——constable of Taunton, 162
——sheriff of Hampshire, xli, 190n., app.iv (25)
——wife of, *see* Walerand
Shrewsbury, Salepsbur' (Shrops.), app.ii (33), app.iii (147, 156, 159, 161)
Shropshire, app.iii (151)
—archdn of, *see* Foliot
Sicily, king of, *see* Frederick
Sidon (Holy Land), app.iii (158)
Silvester, bp of Worcester, 104, 106
Silverstone (Northants.), app.i (127)
Simon, abbot of Reading, 100, 117
—bp of Chichester, 86–7

—bp of Exeter, 7n., 13W, 107
—the chaplain, 56W, vicar of Titchfield, 70
—the clerk, 75W
—parson of Goodworth Clatford, 75
—*persona* of Crondall, 84W
Singe, Ralph, 255
Siward, Richard, app.ii (39), app.iv (4)
Slapton (?Northants.), church of, 72n. rector of, *see* Waltham
Soberton (Hants.), app.ii (1)
Somborne, Sombourn, Sumburn' (Hants.), hundred of, 82,
—Geoffrey de, 82W
—rector of, *see* Stokes, Ralph of
—*and see* King's Somborne, Little Somborne
Somerset, xxxii, 305, app.i (126), app.iv (20)
Somerset and Dorset, sheriff and sub-sheriff of, app.iv (23), *and see* Wells, Jocelin of
Somerton (Oxon.), church of, app.iv (38)
Sorwell, *see* Shorwell
Sougé (Loir-et-Cher), app.iv (43)
Southampton, Suhamton', Suhanton' (Hants.), 82n., 245, 327, app.iii (145–6), app.iv (18, 37)
—burgesses of 55, *and see* Fortin, Isemania, Tyrel, privileges of, 22n.
—English Street 2
—fee of Cormeilles abbey in, 2
—God's House, 55, app.iv (37), master of, *see* Warin
—mayors of, *see* Fortin
—St Denys, Aug. priory, app.iv (37, 41)
——canons of, *see* Warin
——priors of, *see* Alard
—St Mary, church of, app.iv (34)
—suburbs of, 22n.
—water, 43
Southbrook, Southbroke (Hants.), vill of, 67
Southminster (Essex), manor of, 30
Southwark, Southewerk, Suthwerk, Suthewerk', Suwerk (Surrey/London), xlvii, 150, 158–161, 205, 293, 314, 328, 337–9, app.iii (146, 149, 155–6, 159, 162), app.iv (6, 7, 10, 12, 17, 20, 21, 27, 32, 33, 36)
—archdn of Surrey's tenants in, 56, app.iv (11)
—bailiff of, *see* Bacun
—bp of Winchester's exchequer not at, 20n.
—bp of Winchester's lands at, 56–7, meadows at, 60, mills at, 59
—charter witnessed at, 3, 34
—dean of, *see* Roger
—fire at, 56n., 59n., 60n, 61n., 63n., app.iv (11)
—St Margaret, church of, 63
—St Mary Magdalene, chapel or altar of, 63, vicar of, app.ii (29)
—St Mary Overy, Aug. priory, lx, 56n., 63, app.ii (28)
——church of, 63
——infirmary of St Thomas, 56n.

INDEX OF PERSONS AND PLACES

——prior and canons of, 57
——priors of, *see* Martin, Humphrey
—St Olave, sancti Olavi, church of, 26
——vicars of, li, *and see* Southwark, Adam of
—St Thomas', hospital of, xxxiv-v, l, lv, lvii, lxv, lxvii, lxx n., lxxii, 26n., 56-62, 63n., app.ii (29, 44), app.iv (2, 11)
——hall, chapel and stable within the grounds of, app.ii (44), app.iv (13)
——masters of, *see* Amicius, Peter, Robert
——site of, 56-8
—Trevetlane, rent in, app.ii (29)
—Adam of, vicar of St Olave's, 26n.
—John of, 52W, app.ii (29W), app.iv (21)
Southwell (Notts.), prebendal church of, app.iv (30)
——prebends in, *see* Rampton
Southwick, Suthewyk, Suwerk (Hants.), app.iii (144)
——Aug. priory of St Mary, lxv, prior and canons of, 64, app.ii (4), app.iv (5, 10, 40-1, 43)
——infirmary of, l, 65
——prior of, wood of, 71n., *and see* Guy
—pond of, 154
—mr Ralph of, official of the archdn of Winchester, xxxvii n., app.iv (10)
Sowy, William de, clerk, app.ii (1)
Sparsholt, Westpersolte (Hants.), church of St Stephen, 83-4
——rector of, *see* London, William of, vicar of, *see* Hull
—manor of, 82n.
Spoleto (Italy), app.iii (162)
Stafford (Staffs.), archdn of, *see* Henry
—St Mary's, royal free chapel, dean of, app.iv (7)
Staffordshire, app.iii (151)
Staines (Surrey), church of, 73
Stainfield (Lincs.), Ben, priory (nuns), master of, *see* Seamer
Stainsby (Lincs.), Alexander of, bp of Coventry, app.ii (35)
Stambourne (Essex), app.iv (38)
Stambridge (Essex), app.iv (24)
Stanford on Soar (Notts.), church of, app.ii (8), rector of, *see* Herce
Stanham, *see* Winchester, places in
Stanley (Wilts.), Cist. abbey, abbots of, *see* Thomas
Stanwell (Surrey), app.iii (153, 155)
Stapely, Stapelegh (Hants.), 259
Stapleford (Essex), Richard of, xli, biography of, app.iv (24)
—William of, goldsmith, son of Ralph, app.iv (24)
—William son of Richard and William father of Richard, app.iv (24)

Stapleton, Walter, 97
Stef, Adam, 15
Stephen, archbp of Canterbury, *see* Langton
—chaplain, 85W
—clerk, 1W, 21W
—the goldsmith, M. his wife, xlii n.
—the huntsman (*venator*), 33 l
—prior of Mottisfont, 48W, 49W, 50W
Steventon (Hants.), app.iv (39, 43)
Stoke-by-Clare (Suff.), Aug. priory of St John the Baptist, monks of, 35
Stoke, Stokes, *see* Bishopstoke, Longstock
Stoke Charity (Hants.), app.iv (43)
Stokes, Stok', Stoke, mr Alan of, 1W, 11W, 13W, 14W, 21W, 32W, 34W, 38W, 45W, 48W-50W, 56W, 64W, app.iv (8, 21, 31, 36)
——biography of, app.iv (2)
——bp's official, xxxv-vii, lxxii, 26W, 35W, 75W, 83W, app.ii (4, 29W), app.iv (7)
——(mr Alan), mandate to, 46
——master of St Cross Hospital, Winchester, 80, app.iv (4)
——alias mr Alan de Sancta Cruce, 70, 71W
—mr Ralph of, king's justice, app.iv (2)
Strata Florida (Cardigan./Dyfed), Cist. abbey, app.iv (19)
Stratford (Warwicks.), John of, bp of Winchester, inspeximus by, 31, 56-7, 59-62
Stratford-le-Bow (Middlesex/London), app.iii (160)
Studland (Dorset), app.iii (150)
Sturmy, Geoffrey, app.ii (34), *and see* Esturmy
Suhamton, Suhanton, *see* Southampton
Sumborn', Sumburn', *see* Somborne
Sunbury (Surrey), manor and church of, 73
Surrey, Surreye, app.iv (22)
—archdns of, xxviii, xxxvii, lii, lvii, biographies of, app.iv (11-13), *and see* Amicius, Geoffrey, Peter, Roches, Luke des
——officials of, lvii, *and see* Amicius, Canterbury, Eudo
——tenements held of in Southwark, 56, app.iv (12, 13)
—county of, app.ii (13)
Surrey and Sussex, sheriff of, 4n., 85n.
Sussex, app.iv (4, 41)
—sheriff of, *see* Surrey and Sussex
Suthewerk, Suthwark, *see* Southwark
Suthewyk, *see* Southwick
Suthill, John, abbot of Hyde, 19n.
Sutton, *see* Bishop's Sutton, Long Sutton
Suwerk, *see* Southwark
Swainston (IOW), app.iv (14)
Swanwick, Swanewyk', Swanewyke (Hants.), manor of, 67, 67a, 68, app.iv (4, 36)
Swapham, William de, 225, 226
Syria, 129

INDEX OF PERSONS AND PLACES

Tadcaster (Yorks.W.R./N.Yorks.), app.iii (147)
Tamworth (Staffs.), castle and barony of, app.ii (45)
Tany, Richard, app.iv (24)
Tarrant (Dorset), app.iii (162)
Taunton (Somerset), app.i (132), app.iii (146, 158), app.iv (10, 19, 29, 32, 37)
—in the Winchester pipe rolls, 151, 162, 174, 190-3, 209-10, 224- 7, 249-51, 267, 273-6, 291, 294, 304n., 308, 317-20, 332- 5, 341-3
—castle of, 334, app.iv (20)
—constable and bailiff of the honour of, app.iv (23), and see Shorwell
—park of, 251
—Aug. priory of St Andrew, 66
Tavistock (Devon), Ben. abbey, app.iv (5)
Teford, John de, 52W, app.iv (14)
Telarius, see Richard
Temple, Templars, see Knights Templar
—Thomas de, 127n.
Templo, brother Ralph de, 178
Terni (Italy), app.iii (162)
Terra Normannorum, see Normandy
Terricus, chaplain of bp Godfrey de Lucy, 75W
Test, river (Hants.), 82n.
Testard, Robert, bailiff of Farnham, app.iv (10)
Teutonic Knights, military order of, app.ii (30), master of, 129
Teutonicus, G., rector of Barton Stacey, 76n.
Tew (Oxon.), app.iv (37)
Tewkesbury (Gloucs.), app.iii (146, 148, 150, 160-1)
—Ben. abbey of, annals of, app.iv (19)
Thable, Thomas, 71W
Thames, river, xxxii
Thelsford (Warwicks.), Trinitarian priory of, app.iv (34)
Thixendale (Yorks.E.R./N.Yorks.), Ralph of, app.iv (21)
Thomas, abbot of Stanley, 5 and n.
—cardinal priest of St Sabina, papal legate, 130
—the clerk, 56W, 59W, 60W
—the clerk of bp Godfrey de Lucy, 75W
—mr, 59W, 60W
—official of Nicholas bp of Winchester, 70n.
—prior of Merton, 73, app.iv (11)
—rector of Chobham, see Chobham,
—rural dean of Droxford, 39
—vicar of Chobham, 5n.
Thouars, Thuars (Deux-Sèvres), app.iii (144-5)
—viscount of, see Hugh, clerk of, 139
—Geoffrey de, treasurer of St Hilaire, 121n.

Thurlton (Norf.), church of, app.iv (30)
Thurnham (Kent), Robert of, seneschal of Poitou, app.iv (22)
Tichborne, Ticheburne (Hants.), Roger of, 9W
Tichefeld, Tichefelde, see Titchfield
Tickhill (Yorks.W.R./S.Yorks.), app.iii (150)
Timothy, St, 58n.
Timsbury (Hants.), church and tithes in, app.iv (21), rector of, see Adam
Tisted, see West Tisted
Titchfield, Tich', Tichefeld, Tichefelde, Tych', Tychefeld', Tychefelde (Hants.), Premonst. abbey of St Mary, xliii, l, lv, lxx n., 13n., 43, 67, 67a, 68-71, 327n., app.iv (2, 4, 36, 43)
——abbot of, 17n., and see Richard
——abbot and canons of, 67a, 71
——foundation charter, 67
——tenants of, 13n.
—manor of, 67-9
—parish church of St Peter, 67, 69, 70
——chapels of, 69, and see Crofton and Chark
—vicars of, li, 70, and see Simon
—places in, Broke, Felde, Schitehagge, villata of, 67, and see Abshot, Barton, Chilling, Curbridge, Lee Ground, Meon, Pleystowe, Quob, Upton, Southbrook, Warsash
—vill of, 67
Tokynton', see Tufton
Torksey (Lincs.), app.iii (153)
Tortington (W.Sussex), Aug. priory, app.iv (4)
Totford, Tottesford (Hants.), Philip of, 82W
Touraine (France), 121n., 192n., app.iv (7, 13-15)
—the bp's birthplace, 43
—natives of (Tourangeaux), xxvii, xxxviii-xl, app.iv (19, 43)
Tours (Indre-et-Loire), archbp of, see John
—cathedral church of St Maurice, 43, app.iv (19)
—money of, 43
—St Martin, collegiate church of, dean of, 121n.
Tourville (Eure), app.iv (40), and see Turville
Tracy, Henry de, app.iv (10)
Treasury of the king, the bp at, app.iii (152)
Trent, river, xxxii
Trokelawe, Robert de, 120
Trumpington (Cambs.), church of, app.iv (41)
Trussebut, Alice, wife of William de Aubigné, 105
Tufton, Tokynton' (Hants.), assarts in, 75, church of, 75
Turr', mr William de, 75W
Turville, Turevill', Turvill', and see Tourville
—Maurice de, knight, xli, 9W, 82W, 96b, app.iv (35), biography of, app.iv (40)
——sister of, see Morestead

INDEX OF PERSONS AND PLACES 247

——sons of, Henry, Peter, Roger and William, app.iv (40)
—William de, and his widow, Annora or Amicia, app.iv (40), *and see* Chaucombe
Tuscany (Italy), 130
Tusculum (Italy), Nicholas, cardinal bp of, papal legate, app.i (131)
Twinham, *see* Christchurch
Twyford (Hants.), 31n., 69, 140, app.iv (1, 19)
—John of, the bp's chamberlain (*camerarius*), app.iv (14)
Tych', Tychefeld', Tychefelde, *see* Titchfield
Tyrel, Rocelin, fee of in Southampton, 2
Tytherley, *see* East Tytherley, West Tytherley

Umfridus, *see* Humphrey
Upton, Uptone, Uptonne, in Titchfield (Hants.), vill of, 67
Upton Grey (Hants.), manor of, app.iv (26)

Vaas (Sarthe), Ben. abbey, app.iv (30)
Vad', William de, 251, 257, 299
Valence (Charente), bp-elect of, *see* William
—Aymer de, bp elect of Winchester, xxxvii, 41n., app.iv (10, 36-41)
Valentine, prior of St Swithun's Winchester, inspeximus by, 54
Vallibus, Robert de, 333
Valt', John de, 223
Varouville (Manche), app.iv (4)
Vastern, *see* Reading
Vautort, Reginald de and Joan his wife, xlv
Vendôme, Vincestrensis (Loir-et-Cher), 121n.
—Ben. abbey of Holy Trinity, app.iv (43)
Venuz, Veniuz, Colin de, 322-3
—John de, 53W
—Thomas de, 71n., daughters of, *see* Cobham, Port
Vercelli, Vercellensis (Italy), Aug. abbey of S. Andrea, 107
Vercers, Simon de, rector of Witley, 85n.
Verdun, Walter de, sheriff of Essex and Hertfordshire, 202
Vere, Aubrey de, earl of Oxford, 94, letters to, 150
Vernon, Margaret de, wife of the uncle of Joan de Arsic, app.iv (38)
Vescy, Eustace de, 100, clerk of, *see* Fereby
Vienne, Vien', Viene, Vyane (unident., France), mr Nicholas de, 65W, 80W, app.iv (7)
——biography of, app.iv (36)
Vierville (Manche), app.iv (4)
Vieuxpont, Robert de, 96a

Vilers, Roger de, 82W
—W. de, 225
Viterbo (Italy), app.iii (162)
Vitriaco, mr Alberic de, official of the archdn of Winchester, xxxvii n., app.iv (10)
Vuill, Engelram de, app.iv (4)
Vyane, *see* Vienne

W., the clerk, 278, app.i (132)
—mr, the physician (*medicus*), 227
Wacelin, Wascelin, John son of Nicholas, and his sisters, app.iv (25)
—Nicholas son of Roger, app.iv (25), his wife, *see* Raleigh
—Peter, merchant, app.iv (25)
—Roger, 21W, 52W, 80W, 84W
——biography of app.iv (25)
——bp's steward, xxxix-xl, and sheriff of Hampshire, xli
——wife of, *see* Borne
—Sampson, father of Roger, app.iv (25)
Wairvill, mr Gilbert de, app.iv (4), *and see* Millières, Gilbert de
Wakervile, Wakerwyll', Jordan de, 71
Walcot, Walcote, Alice, wife of Roger de, daughter of William de Batilly, app.iv (22)
Walden (Essex), Ben. abbey, app.iv (3)
Walerand, Walter, Joan daughter of, wife of William of Shorwell, widow of William de Neville and Jordan de St Martin, app.iv (23)
Wales, 161, app.iii (148), council on affairs of, app.iv (2)
—marches of, xxxi-ii, app.iv (19, 41)
—campaign in, app.iv (10, 38)
Walesworthe, *see* Wellsworth
Wallingford (Berks./Oxon.), app.iii (150, 152, 154-6, 160)
—garrison of, app.iv (19)
Wallop, Wellop (Hants.), 75, app.iv (37), *and see* Over Wallop
—John of, 71n.
Walpen (IOW), app.iv (23)
Walsall (Staffs./W.Midlands), church of, li n.
Walsingham (Norf.), app.iii (161)
Walter, abbot of Hyde, 48W-50W
—archbp of York, *see* Gray
—bp of Whithorn, xxxiv
—bp of Worcester, *see* Cantiloupe
—the fisherman (*piscator*), 195
—man of Geoffrey de Lucy, 315
—monk of St Swithun's Winchester, 81W
—the ploughman (*carrucarius*), 15
—prior of St Swithun's Winchester, 48W-50W, 82, app.ii (26)
——inspeximus by, 26, 34n., 51n., 52n., 66, 75
—the thatcher (*tector*), app.ii (29)

INDEX OF PERSONS AND PLACES

Walter, Hubert, justiciar, app.i (130), archbp of Canterbury, xliii n., lix, dean of York and bp of Salisbury, app.iv (7)
—chapel furniture of, app.iv (21)
Waltham, *see* Bishop's Waltham, North Waltham
Waltham (Essex), app.iii (151)
—Aug. abbey, abbot and convent of, 72, app.iv (28)
Waltham, mr Reginald of, app.iv (26)
—Reginald son of Robert, app.iv (26)
—mr Richard of, official of the archdn of Huntingdon, app.iv (31)
—Robert of, knight, the bp's steward, xlviii n., biography of, app.iv (26)
—Robert son of Robert of, and his brothers, app.iv (26)
—mr Simon of, rector of Caterham, 72
—William son of Robert, app.iv (26)
Walton-on-Thames (Surrey), app.iv (22)
—church of, app.iv (24)
Walton', mr Henry de, 26W, 160
Wandsworth (Surrey/London), vicar of, *see* Ralph
Wantham, *see* Wautham
Warblington, Warbelton (Hants.), Thomas, lord of, liii n.
—William de, 82W
Ware (Herts.), app.iii (156)
Warenne, earl of, *see* William
Wargrave, Weregrav'. Wrgrave (Berks.), 144, 180–1, 258, 282, app.iii (154, 158), app.iv (6, 20, 26, 42)
Warham (Norf.), Ranulph of, bp of Chichester, 108, 126, will of, 43
—official of Norwich, app.ii (7)
Warin, lands of, 84
—canon of St Denys', master of God's House, Southampton, 55
Warsash, Weresassch, Weresassh' (Hants.), villata of, 67
Warwick (Warwicks.), earls of and honour of, app.iv (40)
—Richard of, app.iv (34)
Warwickshire, app.iv (34)
Wascelin, *see* Wacelin
Waterbeach, ?Beche (Cambs.), church of, app.iv (32)
Water Eaton (Bucks.), app.iv (29)
Waterford, bpric of, 110, bps of, *see* Robert
Watervill, R. de, rector of Patney, app.ii (43), *and see* Watevyle
Watevyle, Robert de, app.iv (10), *and see* Watervill
Wattham, *see* Bishop's Waltham
Wautham, Wantham, Robert de, 52W
—mr Reginald of, 84W

—mr Simon de, rector of Niton, 28, 72n., *and see* Waltham
Waverley, Waverle (Surrey), app.iii (146, 151, 158)
—Cist. abbey of St Mary, xxxi, xxxiv, 1, 85n., app.ii (38)
——annals of, 129n.
Wdecoc, *see* Woodcock
Wearmouth (Durham/Tyne and Wear), church of, 118n.
—rector of, *see* Philip
Wegan, Odo, 110
Weingham, *see* Wingham
Welbeck (Notts.), Premonst. abbey, 13n., app.iv (30)
Well, in Long Sutton (Hants.), app.iv (39)
Wellop, *see* Wallop
Wellow (Hants.), vicar of, app.iv (10)
Wells, Wellens' (Somerset), app.iii (148)
—cathedral church of, 249, canons of, app.iv (20)
——subdean of, 249
——succentor of, *see* R.
—Hugh of, bp of Lincoln, xxxvi, xliii n., 74n., 95, app.ii (19, 21, 32)
——letters and charters of, 24n., 92n., 97–9, 107, 128
——will of, 42a
—Jocelin of, mr, canon of Wells, bp-elect of Bath, 86–8
——as bp of Bath, letters of, 104, 128
——as bp of Bath and Glastonbury, letters and charters of, 97–9, 106–7
——as bp of Bath and Wells, letters and charters of, 24n., 86n., 92n., sheriff of Somerset and Dorset, app.iv (23)
——letters to, 95
——will of, 42a
Wellsworth, Walesworthe (Hants.), manor of, 67–8
Welton Ryval, prebend of in Lincoln cathedral, app.iv (41)
Wendover, Wendovr (Bucks.), Roger of, chronicler, lx, 133
Wenlock (Shrops.), app.iii (159–60)
—Walter of, abbot of Westminster, p.105
Wenlocksbarn, prebend in St Paul's cathedral, London, app.iv (13)
Weregrav', *see* Wargrave
Weresassch, Weressash', *see* Warsash
Wesheham, Roger de, bp of Coventry and Lichfield, li n.
West Boarhunt (Hants.), app.iv (43)
West Camel (Somerset.), church of, app.iv (29)
West Dean (Wilts.), barony of, app.iv (23)
West Preston (W.Sussex), app.iv (4), *and see* East Preston
West Tisted, Westistede, Westistude, Westty-

sted, Westysted, Westystede (Hants.), church of, 54
—land at, 52, vill of 53
West Tytherley (Hants.), app.iv (39)
West Wycombe (Bucks.), app.ii (42), app.iv (13)
—church of, app.iv (13)
Westbourne (Middlesex/London), vill of, 73
Westbury, Westbur' (Hants.), Nicholas of, 9W
Westley, Westeleia, in Sparsholt (Hants.), tithes of, 84
Westminster (Middlesex/London), xxxii, app.i (130), app.iii (145-62)
—Ben. abbey of St Peter, lxii-iv, 73-4, app.ii (23), app.iv (30)
——abbots of, *see* Barking, Hommet, Wenlock
——accounts of, p.105
——attorney and steward of, *see* Chenduit
——king's courts at, app.iv (30)
——king's exchequer at, xlvi, app.iv (34), *and see* Exchequer, Treasury
——parish of, 73
Weston Turville (Bucks.), app.iv (40)
Westpersolte, *see* Sparsholt
Wethmenes, *see* Wilmington
Wherwell, Wher', Wherewell' (Hants.)
—Ben. abbey, app.iv (31)
——abbess and nuns of, 75-7
——abbess of, *see* Mathilda
——attorney and chaplain of, *see* Nicholas
——canons of, *see* Clinchamps, Fauconberg, Philip de
——demesne of, 75
——election of abbess, app.iv (3)
——obit list of, app.iv (31)
——pittancer of, app.iv (31)
——prebendal churches of, *see* Wherwell, Wyke
—church of, app.iv (17)
Whippingham (IOW), app.iv (32)
Whipstrode (Hants.), app.iv (43)
Whitby (Yorks.N.R./N.Yorks.), Ben. abbey, monk of, *see* Seamer
White, Gilbert, xlix n.
White Roding (Essex), app.iv (32)
Whithorn (Scotland), bp of, *see* Walter
Wich, John de, prior of Selborne, attorney of the prior of Mottisfont, 49n.
—Richard de, bp of Chichester, 49n.
Wickham, Wykeham (Hants.), app.iv (43)
—church of, 67, *and see* Wykeham
Wickham St Paul (Essex), manor of, 30
Wicumb', *see* Wycombe
Widemore, in Barton Stacey (Hants.), pasture in, app.ii (26)
Wield (Hants.), app.iv (40)
—church of, 36

Wight, Isle of, 69, 266, app.iv (22)
Wik', *see* Downton, Wick
William, abbot of Westminster, *see* Hommet
—archbp of Bordeaux, 113n., 114n.
—archdn of Buckingham, app.iv (35)
—archdn of Hereford, 100
—archdn of Poitiers, app.ii (7), app.iv (4)
—bp of Avranches, app.iv (4)
—bp of Coventry, *see* Cornhill
—bp of London, *see* Ste-Mère-Église
—bp of Worcester, *see* Blois
—bp-elect of Valence, app.iv (10)
—the chaplain, 14W, 61W
—chaplain of Marwell, bailiff of Meon Church, 31n.
—clerk, xlviii n.
—clerk of bp Godfrey de Lucy, xliii
—clerk, *persona* of Baughurst, xliv-v, charter issued by, 84, app.iv (20)
—constable of Taunton, *see* Shorwell
—earl of Arundel, 94
—earl of Derby, app.ii (9)
—earl of Salisbury, *see* Longespée
—earl of Warenne, app.ii (18)
—the huntsman (*venator*), 201
—prior of St Bartholomew's London, app.ii (34)
—mr, 144, the man of Lawrence de Duniun', 176
Wilmington, ?Wethmenes (E.Sussex), church of, app.iv (29)
Wilmington, in Freshwater (IOW), app.iv (38)
Wilton (Wilts.), forest justices at, app.iv (21, 23)
Wiltshire, xxxii, app.iv (26), archdns of, *see* Grosseteste, Lucy, Philip de
Winchelsea, (E.Sussex), app.iii (149)
—Robert of, archbp of Canterbury, notary of, 32
Winchester, Winton', Wintonensis, Wynton', Wyntoniensis (Hants.), xxxiv, 21n., 70n., 82n., 94n., 118-19, 156, 169, 253-4, 258, 282, 318, 329, app.ii (11, 15, 20, 39), app.iii (144-57, 159, 161-2), app.iv (2, 14, 17, 30, 35, 41)
—annals of, app.ii (25)
—archdns of, xxviii, xxxvii-viii, lii, lvii, 170, app.iv (2), biographies of, app.iv (5-10), *and see* Puiset, P., R., Ralph, Roches, Bartholomew and Hugh des, Roger
——officials of, lvii, app.iv (2), *and see* Ebbesbourne, Langton, Roger, Southwick, Vitriaco
—bps of, lxix, 9, 13 and n., 36n., 43, 48, 59, 60, *and see* Blois, Ely, Exeter, Giffard, Ilchester, Lucy, Orleton, Pontoise, Raleigh, Roches, Peter des, Stratford, Woodlock, Wykeham, bp elect of, *see* Valence
——account rolls of, xxxii n., xxxiii, xxxviii,

xxxviii n., xl–xlii, xliv, xlvi, xlvii–l, lxii–iv, lxxii–iv, lxxvi, 19n., 20n., 31n., 60n., 61n., 69n., 71n., 134–343, app.i (131), app.ii (22, 42), app.iii (143–4), app.iv (10, 14, 17–19, 21, 28–9, 41)
——archives of, xlix, lii
——castles of, *see* Farnham, Taunton, Wolvesey
——harbour of, *see* Fareham
——hundred court of, *see* Downton, East Meon
——manors of, *see* Adderbury, Alresford, Bentley, Bishopstoke, Bishop's Sutton, Bishop's Waltham, Bitterne, Brightwell, Burghclere, Calbourne, Clere, Crawley, Downton, East Meon, Ebbesbourne, Fareham, Farnham, Fonthill, Hambledon, Harwell, Highclere, Ivinghoe, Knoyle, Meon Church, Merdon, North Waltham, Overton, Rimpton, Southwark, Taunton, Twyford, Wargrave, West Wycombe, Wield, Witney, Wycombe
——meadows of, *see* Southwark
——mills of, *see* Southwark
——palaces of, *see* Marwell, Wolvesey
——secular tenants of, xl, lxxii–iii, 9n., *and see* Roches, Peter des, knights of
——treasury of, *see* Bishop's Sutton
——warrens of, *see* Bishop's Waltham, Bitterne
—Blackfriars, *see* Dominican House
—cartulary of, *see* St Swithun's
—castle of, 96b, app.iv (25, 40), constable of, *see* Turville, Maurice de, Wacelin
—cathedral church and chapter of, *see* St Swithun's
—chapter clerk of, *see* Chase
—charters dated at, 75, 79n.
—citizens of, app.ii (39), app.iv (17)
—cloister at, 156
—College of the BVM, 14n.
—deans of, xlii n., *and see* R.
—diocese of, xxxi–ii, xxxiv–v, 9, 11, 50, 58, 81n., 127n., app.ii (34)
——administration of *sede vacante*, app.iv (4, 5, 10)
——vacancy in, 20n., 41n., app.ii (42), app.iv (13)
——Dominicans to act as confessors within, lvii
——interdict within, 33n., 34n.
——ordinations within, lvii–viii
——religious houses of, 34n., visitation of, liii
——parish network, liii
——possessions of church of Salisbury within, 45–6
——preaching within, lviii

——scotales within, lviii
——Sunday markets, lviii–ix
——temporalities of, xxviii–xxix
——Dominican House at, lv, lvii, app.ii (22), app.iv (17)
——earl of, *see* Quincy
——fair (*feria*) of, xxxiii, 119n., 279, app.iv (32)
——forest near, *see* Bere Ashley
——Hyde abbey, *see* Hyde
——Jews of, app.iv (37, 39, 41)
——Kalendars, fraternity of, app.iv (20)
——land in, app.iv (25), *and see* places in
——mills of, 19, 20, *and see* Hospitallers, Winchester, places in, Drayton, Durngate, Stanham, Winchester St Cross
——monks of, *see* J., St Swithun's
——Nunnaminster, Ben. abbey (nuns) of St Mary, 82n., app.ii (15)
————mill of, 19
——ordinations at, app.ii (11)
——pipe rolls of, *see* bp, account rolls of
——places in, Drayton, Draiton', mill of, 19, 20n.
————Durngate, mill of, 19, 20n.
————Eastgate, app.iv (13)
————Garstreet, app.iv (25)
————Inmede, meadow of, 19n.
————Northgate, app.iv (6)
————Stanham, mill of, 20n.
————Westgate, *see* St James
————*and see* Fulflood
——prisoners taken captive at, app.iv (4)
——see of, *see* diocese of
——St Cross, app.ii (15)
————Hospital of, 80, app.ii (1), bailiffs of, 20n.
————masters of, xxxvi, app.iv (1, 2, 4), *and see* London, John of, Millières, Humphrey de, Stokes, Alan of
——————pigs of, 80n., app.iv (2)
——————prior of, *see* Alan
——————mill of, 20n.
——St John, hospital of, app.iv (17)
——St Mary Magdalene, hospital of, 31n.
——St Swithun's, Sancti Swthuni, Ben. priory, liii–v, lxviii n., 81–2, app.ii (39), app.iv (1, 20, 25)
————archive of, xlix, lii, lxii
————*bartonarius* of, *see* John
————cartulary of, xlviii n., lxx n., app.ii (38)
————as cathedral church and chapter of SS Peter, Paul and Swithun, xxix, xxxi, xlvii n., 9, 71n., 81, app.ii (39), app.iv (1)
————authority and dignity of, 34, 49
————benefits and prayers of, 58
————chapter of, lxix–lxx
————charter dated at, 82
————convent of, 49

INDEX OF PERSONS AND PLACES

—cult of saints at, liv
—dignity of, 26, 37
—fee of (*feodum, fundo*), 56–7
—protection of, 34, 49
—records of, xlviii–ix
—retrochoir of, liv
—right and dignity of, 50, 85
——chief cook of, 41n., app.ii (49), *and see* Pruz
——chief serjeant of the almonry, 41n., app.ii (47), *and see* Mandatis
——confraternity with Taunton, 66
——conventual offices, liv, 41n., app.ii (46–9)
——gate keeper of, 41, *and see* Bydun, Convers
——great fish claimed by, app.iv (3)
——kitchen of, *see* chief cook
——manors of, *see* Barton Priors, Bransbury, Compton, Fulflood, St James outside the Westgate, Sparsholt
——men of, 82
——monks of, xxviii, liii, lxvii, lxxii, 6n., app.iv (10, 13), *and see* A. the chaplain, Germanus, Gervase, Walter
——obit celebrations at, xxxiv, 43
——priors of, *see* Andrew, Geoffrey, John, Lucy, Stephen de, Valentine, Walter
——profession of monks at, liv
——treasurer (*hordarius*) of, 148, app.iv (14)
——usher of the conventual kitchen, 41n., app.ii (46), *and see* Pestur
——washer, 41n., app.ii (48), *and see* Lavender
—schools of, xlii, masters of, xlii n., *and see* A., Andrew
—synod of, 1
—Wolvesey, Wlves, Wlvesey, Wolves', Wolveseye, Wolvesye, Wulves', Wulveseya, 186, 208, app.iii (161–2)
——house (*domus*) of, and the bp's *receptor* there, 20, app.iv (14, 16), *and see* John the Dean, Hugh des Roches
——bp's exchequer at, xlvi, lxii, 20n., app.iv (14–17)
————keeper and receiver at, xxxviii, lxvi
————procedure at, xxxiii, liii, lxii–iv, 20n,
————records of, xlix and n., lxii–iv, *and see* Winchester, bp of, accounts rolls of
————treasurer of, xlvi, app.iv (14), *and see* Winchester, bp of, treasurer
——castle or palace of, 96b
——charters dated at, 40, 48–50, 70
——dean of, xxxviii, app.iv (14), *and see* John Winchester, mr Luke of, *alias* Luke des Roches, app.iv (13)
—mr Raymond of, 26W
—Thomas of, proctor of Peter Russinol, app.ii (41), app.iv (20)
Winchcombe, Winchec' (Gloucs.), 198

Windsor (Berks.), app.i (127), app.iii (144–6, 149, 151–3, 159–61)
—charter dated at, 48
—constable of, *see* Millières, William de
—forest of, app.iv (19)
Winemerius, mr, 192
Wingham, Weingham, Wingeham (Kent), app.iii (149)
—Hugh of, 276, 278, app.iv (43)
——his wife Annora, sister of Geoffrey de Luverez, app.iv (39)
Winnianton in Gunwalloe (Cornwall), land and church of, app.iv (32)
Winterburn', Ran(ulph) de, 26W
Wintney, Wynteneia (Hants.), Cist. priory, nuns of, li, 83–4
—kalendar and obits at, 43, 74n., app.iv (20, 34)
Wippeley, Ralph de, xlviii n.
Wisbech (Cambs.), Adam of, vicar of Chesterton, 107n.
Wisley, Wisle (Surrey), Robert of, 85W
Witley, Witle (Surrey), church of, 85, rectors of, *see* Ferles, R., Vercers
Witney, Witenay (Oxon.), app.iii (147–8, 150, 155, 158), app.iv (4, 14, 19, 29, 39, 43)
—in the Winchester pipe rolls, 134–5, 143, 173, 182–3, 197–201, 228, 232–4, 260–3, 281, 287–91
—assize at, app.iv (37)
—burgh of, 235
—church of, app.ii (19, 40), app.iv (3, 4, 14), rectors of, *see* Bourgueil, Millières, Ste-Mère-Église
Wlvem' (unident., ?Hants.), 195
Wodecote, *see* Woodcot
Wodestoc, *see* Woodstock
Woking (Surrey), church of, 35
Wolcroft, *see* Cobham
Wolverton, in Shorwell (IOW), app.iv (23)
Wolvesey, Wulveseya, *see* Winchester
Woodcock, Wdecoc, William, bp's huntsman, 273, 277, 300, 301
Woodcot, Wodecote (Hants.), Henry of, 71W
Woodham (Durham), app.iii (149)
Woodhay, *see* East Woodhay
Woodlock, Henry, bp of Winchester, inspeximus by, 14, 31
Woodstock, Wodestoc (Oxon.), app.iii (145–50, 153–4, 159–62), app.iv (32, 43)
—charter witnessed at, 18
Woolley (Wilts.), app.iii (145)
Woolmer (Hants.), 94n.
Wootton Bassett (Wilts.), app.iv (34)
Worcester, Wigornia, Wygorn' (Worcs./H. & W.), 18, app.iii (160)
—annals of, app.ii (20)
—bishops of, letters and charters of, 24n, *and*

see Blois, Cantiloupe, Gray, Mauger, Orleton, Silvester
—diocese of, 13n.
Wrotham (Kent), William of, lv, app.ii (4)
Wulveseya, *see* Winchester, Wolvesey
Wycombe, Wicumb' (Bucks.), 177, 205, 215–17, 238, 313–14, 330, app.iii (159), app.iv (14, 44), *and see* West Wycombe
Wycombe *Ecclesia* (Bucks.), 218
Wyke (Hants.), church of, app.iv (17)
Wykeham, William of, bishop of Winchester, inspeximus by, 1, 14, *and see* Wickham
Wynton', *see* Winchester
Wyvenhulle, William de, app.iv (40)

Yafford (IOW), mill of, app.iv (23)
Yarlington (Somerset), app.iii (148)
Yarmouth (IOW), app.iii (144, 151), *and see* Little Yarmouth
York, Ebor. (Yorks./N.Yorks.) 33, app.iii (150, 156)
—archbpric of, xxix, 13n., app.iv (7)
—election to of Peter des Roches, xxix, 13n., app.iv (7, 18)
—records of, xlviii
—archbps of, *see* Gray, Plantagenet
—Ben. abbey of St Mary, abbot of, app.i (132)
—cathedral church of, 33, app.iv (7)
—dean of, *see* Roches, Bartholomew des Walter, Hubert
—prebendaries of, app.iv (18), *and see* Airaines, Barnby, Roches, Bartholomew des, Russinol
—precentor of, *see* Russinol
—treasurers of, *see* Hamo
—St William of, shrine of, xxxiv
—mr Ralph of, attorney of Luke des Roches, app.ii (34)
—mr Robert of, bp-elect of Ely, 107, app.iv (27)
Yorkshire, app.iv (21, 30, 31, 34)
Ypres (Belgium), burghers of, 101, 124

Zirc (Hungary), abbot of, app.iv (33)

INDEX OF SUBJECTS

abbot, blessing of, 73, app.ii (23)
account rolls, *see* pipe rolls
accounts, to be rendered, 31
acre, 6, 34, 70
address, lxv–ix
adjudication, 6
admission to churches and chapels, l, 37, 39, 40, 63n., app.ii (3)
advowson, 1n., 7n., 13n., 18n., 21n., 26n., 28n., 33n., 36n., 48, 50, 55n., 64n., 67n., 72n., *and see ius patronatus, patronatus*
agreement, 3, 5, 32n., 78n., 79n., 82n., *and see compositio*
aisles, 63
alder grove, 6
aliens, in the bp's household, xxxviii–xl
almoner, app.iv (10), *and see* pittancer
alms, 13, 58
—gift in, 21
—gift in pure and perpetual, 2, 18, 54, 67, 69, 71, 75
——free, pure and perpetual, 48, 53, 67a
——free and perpetual, 59, 60
——perpetual, 52, 56
—licence to collect, 58n.
—distributed by the bp, 31n., 174, 249, 308–9, 311
altar, xxxiv, 63, 73, 327
—*obventiones* of, 84, *oblationes et obventiones* of, 70
—*and see* consecration
ambassador, xli, app.ii (38), app.iv (4, 18, 27, 29, 30, 32, 36)
—*and see* envoy, nuncio, proctor
animals, feeding of, 81, pasture for, 7n.
—*and see* birds, chickens, fish, geese, hawks, horses, hounds, kids, lambs, pigs, rabbits, sheep
antiphoner, lviin.
apple trees, 56, 215
appropriation of churches, l–li, lxi, 35, 67n., 73
—grants *in proprios usus*, 34, 48–50, 54, 63, 69, 75, 83–4
—licence to appropriate, li, lxi, 26, 36n., 49, 50, 54
—of houses, grants *in proprios usus*, 31, *and see* grants

arbitration, xxxvi, 82, 84, app.iv (2, 4, 10, 20, 30)
archbishops, *see names under* Arles, Bordeaux, Caesarea, Canterbury, Cashel, Damietta, Dublin, Narbonne, Nazareth, Tours, York
archdeacons, *see names under* Airaines, Berkshire, Buckingham, Canterbury, Chester, Ely, Hereford, Huntingdon, Middlesex, Northumberland, Poitiers, Richmond, Shropshire, Stafford, Surrey, Wiltshire, Winchester
archives, ecclesiastical, xlviii–l, lii, lx–lxii, 25n., 64, 65, p.105
—royal, lxii, 25n., 97, 132, app.i (125–9, 131)
arenga, lxiv, lxvii–viii, 48n., 49n.
arrears, xxxiii, 148–9, 162, 173, 232, 268, 298, 336, 341, *and see* debts, loans
assarts, 34, 75, 82n.
attachments, 82n.
attorney, xli, 20, 49n., 74n., 76n., 85n., 101, 124, 271n., app.ii (1n., 34), app.iv (21–2, 28, 30–2, 36, 40), *and see* proctor
authority, 19, 34, 47, 50, 56, 63, 75, 80, 83, 84, 88, 100, 114
—episcopal, 26, 34, 49, 61, 81, 84
—ordinary, 15
—pontifical, 11, 18, 32, 38, 45, 57, 75, 84
auxiliary bishops, *see* suffragan

bacon, *see* pigs
bailiffs, xxxvii–viii, xliv, xlvii, lxxvi, 20n., 26n., 31n., 112n., 149, 152, 173, 205n., 298
—of the king, 12n., 18, 42n., *and see names under* Alresford, Alresford Forum, Bentley, Bishop's Sutton, Fonthill, Farnham, Hampshire, Meon Church, Southwark, Taunton
bailiwick (*ballie*), 82, app.ii (13)
barley, 31, 166
barrels, 165, 185
bartonarius, 81
bells, 62, 69n., 100, app.ii (29)
benefice (*beneficium*), 65, *and see* church
benefits (*beneficia*) of the church of Winchester, 58
birds, 143, 208, *and see* chickens, geese, hawks

254 INDEX OF SUBJECTS

bishops and bishops elect, *see names under* Ardfert, Avranches, Bath, Beauvais, Carlisle, Chester, Chichester, Coventry, Durham, Ely, Exeter, Ferns, Godfrey, Hereford, Isles, Kildare, Killaloe, Leighlin, Le Mans, Lichfield, Limoges, Lincoln, Lismore, London, Meath, Norwich, Ossory, Rochester, Sabina, Salisbury, Tusculum, Valence, Waterford, Whithorn, Winchester, Worcester, *and see* consecration

books, xxxvii, xliii, lvii, app.iv (13, 33), *and see* antiphoner, breviary, gradual, lectionary books, missal, psalter, temporal, troper

borough, *see* Overton, Witney

brachet hounds, *see* hounds

bread, 43n.

bream, *see* fish

breviary, lviin.

building, xxxiii, xliii, liv, 24, 56–8, 62n., 63n., 192n., 280, 316, app.ii (4, 20), app.iv (25)

—of stone, 2, *and see* castle, hall, house

burdens, 63, 84

burgesses and burghers, *see names under* Gloucester, Northampton, Southampton, Ypres, *and see* citizens, commune

burial, app.ii (35), right to chose, 5, app.ii (29)

—in time of interdict, 33n., *and see* cemetery, mortuary legacies

calendar, of the English church, xliv, lxxiii–iv, 40n., 48n., 70n.

—of the Roman church, xliv, lxxiii

candles, xlvii, 71n., 73, 100, *and see* lights, wax

canonical hours, 31, 62, *and see* offices

canons, regular, *see* monastic houses

—of cathedral churches, *see names under* Chichester, Exeter, Hereford, Le Mans, Lichfield, Lincoln, Lismore, London, Salisbury, Wells, York

—of prebendal churches, *see names under* Angers, Bridgnorth, London, St Martin-le-Grand, Southwell

—attached to nunneries, *see names under* Romsey, Wherwell

capes, 151, 289, black, 31

capital lords, 51

captain, 127, app.i (128)

captives, *see* prisoners

cardinals, *see names under* Gregory IX, Honorius III, Innocent III, Rome

carpenters, 154

carriers, 200

carters, 200, 238

carts, 340

castles, xxvii, xxxviii, l, 96a, 96b, 99, 113, 246, 255, 334, app.ii (45) *and see names under* Bedford, Berkhamsted, Bristol, Chester, Corfe, Devizes, Dover, Farnham, Hanley, London, Tower of, Knepp, Northampton, Painscastle, Portchester, Tamworth, Taunton, Winchester, Windsor, Wolvesey

cathedral, *see* archbishops, bishops

cementarius, see mason

cemetery, xxxv, liii, lxix, 5, 33, app.ii (29), *and see* burial, consecration

chalice, of silver, lviin.

chamber, episcopal, xxxviii, xlvi, app.iv (10, 15, 17, 19), *and see names under* Jordan de Camera

—royal, xxvii, xli, app.iv (7, 34–5)

chamberlain, episcopal, xxxviii, lxxii, app.iv (14, 17)

—papal, *see names under* Pandulph

—royal, app.iv (41), *and see names under* Neville, Geoffrey de

chancel, repairs to, app.iv (29)

chancellor, episcopal, xliii, app.iv (31)

—royal, xxix, app.i (129), *and see names under* Gray, Walter de, Marsh, Richard, Neville, Ralph de

—*and see* vice-chancellor, *and names under* Kilkenny, Oxford, Salisbury

chancery, episcopal, xxix, xliii–viii, lxii–iv, lxxi, app.iv (18–20)

—papal, lxvii–viii, lxxiv, 81n., 94n.

—royal, xlv–vii, xlviii–ix, lxii, lxiv–vi, lxviii, lxxi–iv, 25n., 67n., 132n., app.i (125–7, 129–31), app.iv (41)

—of the justiciar, app.i (129–30)

chantry, liii

chapel, 5, 31, 63, app.ii (44)

—of the bishop, xlvii

—of a hospital, 62

—manorial, liii

—subject to another church, 7n., 8, 13, 38, 50, 69, 73

—raised to the status of a church, 5n., *and see* admission, dedication, institution

chaplains, xxxvi, lv–vi, 31

—the bp's, xliv, xlvi–vii, lviii, lxii

—*and see names under* A., Bartholomew, Ernisius, Gregory, Henry, Hugh, Humphrey, John, Luke, N., Nicholas, R., Ralph, Richard, Robert, Roger, Seamer, Osbert of, Simon, Stephen, Terricus, William

chapter, of cathedral, liii, lxvi, lxix–lxx, 7n., 18, 25, 35, 45–6, 48, 67a, 110, 113–14, app.ii (34)

—clerk, *see names under* Chase

charity, lxvii, 11, 26, 32, 46, 48, 49, 51, 53, 54, 58, 62, 65, 71, 81

charter (*carta*), as descriptive term, lxvi, 9, 11, 13, 14, 18, 20, 21, 31, 32, 45, 48, 49, 51–4, 56, 59–61, 65, 67a, 69, 71, 75, 80–2, 85

—*cartula*, 63, *and see* letter, writ, writing

INDEX OF SUBJECTS

cheese, 228, 236, 264, winter, 237
—service of, 82n., tithes of, 84
chickens, service of, 82n.
chimney, 280
chrism, xxxiii, blessing of, 73
church, xxx–xxxi, xxxix, xli–ii, li, lvi–vii, lix, lxv, lxvii, 5, 70, 73, 75, app.ii (15, 29, 31n., 34)
—annexed to cathedral prebend, 18
—annexed to cathedral dignity, 33
—burdens and expenses of, 28, 84
—chapel within the aisles of, 63
—custody of, xxxvi
—election to, 130, app.i (132)
—fabric of, 71n., 249, app.ii (20n.)
—farm of, 26n.
—festivals of, lxxiii
—grants and confirmations of, li, 7n., 13, 32, 34, 38, 45, 48, 63n., 64, 67, 83, *and see* appropriation
—honour of, 118
—intrusion to, 18n.
—language of, lxv
—ornaments, lvii
—patronage of, disputes over, liv
—proceeds of, 69n.
—rights of, app.ii (25)
—settlement involving, 1
—title deeds to, l–li, 63n.
—*and see* admission, appropriation, collation, consecration, custody, dedication, installation to corporal possession, pension, presentation
cinnamon, 339
citizens, *see names under* London, Rome, Winchester, *and see* burgesses, commune
civil war, xxix, l, lii–iii, 4n., 7n., 21n., 100n., 102n., 105n., 108n., 118n., 177n., app.ii (11n., 12n., 20n.), app.iv (11, 15, 19, 23, 24, 29, 30, 37, 41) *and see* rebels
cloister, 156
cloth, 109
—gifts of, 31n.
—*lanecol'*, 207, russet, 179, *and see* vestments, wool
clothing allowance (*vestitum*), 31
cloves, 339
collation, 31, to churches, app.ii (1, 8)
colts, *see* horses
columns, capitals and bases, 327
common pasture, 9n., app.ii (26), *and see* pasture
commune, *see names under* Bordeaux, Italy, La Rochelle, London, *and see* burgesses, citizens
compline, 31
compositio, 1, 82
conduct, *see* safe conduct

confession, lvii, 5n., as a prelude to indulgence, 58, *and see* penitentiary
confirmation, of grants and awards, l, liv, lxvii, lxx, 9, 18, 21, 26, 34, 42, 45, 48n., 49n., 51, 53, 63n., 66, 67a, 68n., 69n., 71n., 75–7, 107n.
—of churches, 32, 34, 38, 45, 64, 83
—of an election, app.i (132)
—of pensions, 11, 14, 21, 35n., 38
—of tithes, liii, 8, 34, 45, 61, 75–7, 81
—registers of, lxix
—by pope, 8, 59n., 60n., 76n., 77n., 107n.
confirmations, religious, xxxiv, 73
consecrated ground, 33n.
consecration, of a bp, xxviii, xxxiv, xlv, lxvii, lxxiv, 75n., 86n., 110, app.i (132), app.iv (30)
—of altars, xxxiv, 73, app.iv (5)
—of cemeteries, 33
—of churches, xxxiii–iv, liii
consent, 1, 18, 19, 48, 100, 107
constable, *see names under* Berkhamsted, Bristol, Dover, London Tower of, Northampton, Portchester, Taunton, Winchester, Windsor
cope, app.iv (7), *and see* capes
cords, used for sealing, lxi
corn, 145, 180, 205, 206, 229, 234n., 311, 312, tithes of, app.ii (28)
—right to grind, 19n., *and see* mancorn
coroner, app.iv (40)
corroboratio, lxxi
council, 129, *and see names under* Lateran
—of king, 82, 110, 113, 118, 119n., 133, app.ii (13)
—*and see* counsel
councillor, 114, 129
counsel (*concilium*), 31, 56, 107, 112, 114, 116, 129
—common, 31
—of prudent men, 1, 57, 58
count, *see names under* Angoulême, Evreux, La Marche, St-Gilles, St-Pol
counterseal, lxi
court, papal, xxx, xxxii, 81n., 108, app.ii (5, 31n., 37), app.iv (7)
—royal, xxviii–xxxii, xxxix–xli, xlvi–vii, lii, lxxiii, 48n., 73n., 94n., 113n., 118n., 119n., 133, 206n., 255n., app.ii (10, 39)
—monastic, 34
courtier, xlv, lxxv, 102n.
court rolls, xlix
cross, 131, *and see* crusade
crossbowmen, xl, *and see names under* Giles, Henry
crown, 100
crusade, to the Holy Land, 100, 106, 113n., 129, app.iv (29)
—preaching of, li–ii, 47,
—against the enemies of God and the king, 106

—*and see* names under Winchester, bp of
cult, of the saints, liv
cultura, app.ii (35)
cultus, 13, 31
cumin, *see* pepper
cure of souls, 69
cursus Romane curie, lxv
custody, of castles and lands, xxxviii, 94, 96b, 132, app.iv (19, 22–5, 30, 34, 37–9, 41, 43–4)
—of churches and religious houses, xxxvi, 34n., 39, 57, 112, app.i (128), app.ii (30n.), app.iv (7, 41)
——right to, 95
—of a charter, 75, of prisoners and hostages, 106, app.ii (2)
—of falcons., 112, app.iv (29)
—of wards and minors, xxx, 3n., 10, 78, 114, 118, app.ii (26, 45), app.iv (22, 23, 29, 34, 37, 38, 40, 43)
—of seals, xli, xliii, 32, 81, app.i (127, 129, 130), app.iv (18, 41)
—*and see* keeper
customs (*consuetudines*), 13, 31, 82, 99, 100, app.ii (26)
cuttle fish, *see* fish
cyrograph, lii, lxi, 10, 73, 82, app.ii (34, 35)

datary. lxiv, lxxiii, *and see per manum*
date and dating elements, xliii–iv, lxxi, lxxiii–v, app.i (130)
dawn, 31
deacons, xxxiv, lvii–viii, 31, app.ii (11), app.iv (3, 7, 29)
deans, of cathedral churches, *see names under* Bordeaux, Chichester, Lichfield, London, Mans, Le, Poitiers, Salisbury, York
—of collegiate churches, *see names under* Angers, Angoulême, Bridgnorth, Loches, Stafford. Tours
—rural (deans of Christianity), *see names under* Basing, Basingstoke, Bristol, Droxford, Ewell, Guildford, London St Mary Arches, Southwark, Winchester
—*and see* names under John the dean, Wolvesey
debts, 44, 48n., 71n., 312, app.iv (23, 25, 28, 34, 37, 39), *and see* arrears, loans
decree (*decretum*), 31
dedication of churches and chapels, 73, app.ii (20)
—of abbeys and priories to the Virgin Mary, 13, 49
——to St Thomas, 56n.
demand, 52, 82
demesne, tithes of, 5, 75, 81

destriers, *see* horses
dignity, 26, 34, 37, 49, 50, 75, 80, 81, 83–5, 88, 100, 114
disafforestation, 82n.
dispensation, *see* papal
dispositio, lxix–lxx
disputes, xxxiv, li–ii, liv, 5, 6, 12n., 13n., 18, 21n., 55n., 74n., 79n., 82, 125n., *and see* elections
disseisin, app.ii (10), app.iv (43), *and see* seisin
ditch, 6, 71n.
Domus Dei, 31, *and see names under* Ospringe, Portsmouth, Southampton
driver, 236
duke, *see names under* Brittany, Limburg, Normandy

earl, *see names under* Arundel, Chester, Clare, Cornwall, Derby, Essex, Gloucester, Huntingdon, Kent, Oxford, Pembroke, Salisbury, Warenne, Warwick, Winchester
eggs, service of, 82n., offered at Easter, 70
elections, xxxvi, lxxiv, 1n., 13n., 30n., 31, 35n., 38n., 83n., 86, 87, 104, 114, 130, 132n.
—dispute over, xxvii–ix, xxxvii, li, 81n., 110, app.i (132)
ells, 179, 207
emperor, *see names under* Frederick
envoy, app.iv (22, 41), *and see* ambassador, nuncio, proctor
eschatocol, lxxi–v
escheats, 41n., 67n., 78, 85n., app.ii (13), app.iv (22, 32), *and see names under* Brittany, Normandy
escheator, app.iv (40, 41)
exactiones, 9, 51, 52, 82
exchequer, episcopal, xxxiii, xxxviii, xlvi, xlix, lii, 20n., 31, 316, app.iv (10, 14–17)
—royal, xxix, xli, xlvi, liii, lxii–iii, 74n., 89, 115, app.i (126, 128–30), app.iv (23, 28, 30)
——barons of, app.iv (28, 41), seal of, app.i (130)
—of the Jews, app.iv (28)
—of Westminster abbey, p.105, *and see* treasury
excommunication, xxviii–xxx, lxvii, 100, 104, 130, app.ii (7, 7a, 25, 39), app.iv (2, 3)
—significations of, lii, lxiv, 15–17
eyre, lviii, 22n., 116n., app.iv (13, 23, 25, 30, 39–41), *and see* forests

facultates, 58
fair, *see names under* Winchester
falcons, *see* hawks

fee, 2, 34n., 56, 57
—lay, 6
—lords of, 52, *and see* capital lords
—hereditary, app.iv (41)
—knight's, xl, 9, 10n., 120, 304n., app.ii (45), app.iv (4, 23, 25, 37–9, 40, 43)
—money, app.iv (27)
feeding, 9, 81, 166, *and see* paupers
fence (*sepes*), 31, 34, 49
field names, *see names under* Chobham, Cobham, Titchfield
final concord, xlix n., 9n., 56n., 57n., app.i (127–9), app.ii (17), app.iv (21, 40)
fire, 56–8, 59n., 60n., 61n., 63n., app.iv (11, 41)
firewood, *see* wood
first person singular, used in charters, lxv
fish, 245, 285, bream, 318, cuttle fish, 265, herrings, 241
fishermen, 195, 299
fishing, 19n., 167, 317–19, 342
foresters, 82n.
forests, 71n., 82, 98, *and see names under* Aliceholt, Bere, New Forest, Windsor
—eyre and regard of, app.iv (14, 17, 19, 23, 32, 35, 36, 40)
—justices of, app.iv (21, 33), *and see* attachments, wood
forgery, lx, 55, 63, 133, app.ii (26)
foundation of religious houses by the bishop, lv–vi, 13, 43, 48, 49, 67, app.ii (4, 30), *and see* foundation charters
foundation charters, 13, 49, 67
frankalmoign, *see* alms
fuel, 337, *and see* wood
fur, 151, 182, 198, 289

game, 335
gardens, 56, app.iv (11), tithes of, 70
gate, 31, 43, *and see names under* Winchester, places in
gate keeper, 41
geese, 166, tithes of, 70, *and see* birds
ginger, 339
glass, 281, 290, app.iv (21)
gold, app.iv (41), marks of musc, 294
goldsmith, *see names under* Stephen
gradual, lvii n.
grain, 189, 220, 309, 341, service of, 82n., tithes of, 5, *and see* barley, corn, mancorn, sheaves
granary, 326
grant, of land, 2, 9, 13, 19n., 34, 49, 51–3, 67, 67a, 68, 69, 100, app.ii (22, 34, 35, 42, 45)
—of pensions, 1n., 14, 19, 21, 31, 36, app.ii (29), *and see* pensions
—of seisin, 4n.

—of office, 41, of money, 43n., of tenancy, app.ii (44)
—of churches, 7n., 11, 13, 26n., 34, 38, 48, 49, 54, 63, 64, 67, 72n., 83, 85n., 107, *and see* appropriation
—of tithes, 59–61, 75, *and see* tithes
—of exemption from tithes, 81
—of liberties, 82, 100, 132n., app.ii (10)
greyhounds, *see* hounds
grove, *see* alder grove
guarantees, *see* pledges

hall, 316, app.ii (44)
hamper, xlix n.
harbour, *see* port
harvest, xxxiii, 69n., 82n., p.105, 223, app.iv (32)
harvesting, service of, 82n.
hauberks, 241
hawks, xxxiii, falcons, 112, app.iv (29)
hay, 31, 240, 292, 325, tithes of, 15, 61, 70, 76n., 81
haymaking, time of (*tempore fenedii*), 31
hedge, *see* fence
herrings, *see* fish
hide, app.iv (34)
homage, 9, app.iv (25)
hordarius, 148, app.iv (14)
horses, xxxiii, 158, 269–70, 306, app.ii (39)
—colts, 140, destriers, 321, app.iv (43), palfreys, 94n., app.iv (27)
hospital, *see names under* Acre; Hospitallers; Ospringe; Portsmouth, Domus Dei; Southampton, Domus Dei; Southwark, St Thomas'; Winchester, St Cross, St John, St Mary Magdalene
hospitality, 13, 26
hostages, 105, app.ii (2), *and see* ransoms
hounds, xxxiii, 172, 191, 201, 273, 283, 331, 335
—brachet hounds, 277, 300, 301, 323
—greyhounds, 147, 277, 300, 301, 322–3, 331
—wolf hounds, app.iv (39)
household, of bp, xxxiii–xlviii, liv, lxxii, lxxvi, 9n., 14n., 39n., 74n., 80n., 108, 118, 233, 289, app.ii (3), app.iv (1–44)
—of king, xlvii, knights of, 48, 109, 231, app.iv (38, 44)
——stewards of, *see names under* Greinville, St Amand
houses (*domus*), 2, 20, 31, 84, app.iv (4, 11, 13, 15, 41), (*mansos*), 31
—of the king, app.iv (25)
—of religion, 20, 26, 31, 34, 49, 50, 54, 61, 62, 80
—*and see names under* Winchester, Wolvesey

hundred and hundred court, xlix n., 82, *and see* names under Downton, Meon, Odiham, Somborne, Swainston
hunting, xxxiii, 191, 251, 322, 343, app.iv (34, 41)
huntsmen, xxxiii, l, 283, *and see names under* G., Gervase, Stephen, William, Woodcock
hurdles, 186, 241

imprisonment, *see* prisoners
induction, *see* installation
indulgences, xxxiv, liii, lviii, lxvii, 24, 58
incontinence, 31
infirmary, l, 56n.,65
injunctio, lxxi
inquisition, 6, 41n., 82n., letters of, app.ii (21)
inscriptio, lxvi
inspeximus, lxx–lxxi, *and see* confirmation
installation to corporal possession of a church, 39
institution to churches and chapels, xxxv–vi, 27–8, 37, 40, 72, 74, 84–5, app.ii (27)
interdict, imposed by bishops and archbishops, 100, 104, 118, 124, app.ii (39)
—papal, xxix, liii, 21n., 33n., 34n., 89, 96, app.i (129), app.iv (6, 34)
intitulatio, lxv
itinerary of bishop, xxxi–iii, app.iii
ius patronatus, 13, 37, 64, *and see* advowson, *patronatus*
ius et proprietatem, 6

jewels, app.iv (32, 41)
judicial work, li–ii
justices, li–ii, lviii, 22n., 305, app.iv (2, 11, 29, 30), *and see* forests
justiciar, xxix, xl–xli, xlvi–vii, l, 13n., 118n., app.i (125–32), app.iv (30, 38), *and see names under* Burgh, Fitz Peter, Marsh, Geoffrey, Seagrave, Walter, Hubert

keepers (*custodes*), 20, 59, 60
kids, 168n.
kings of England, *see names under* Edward I, II, III, Henry I, II, III,VI, John
—of France, *see names under* Louis, Philip Augustus
—of Sicily, *see names under* Frederick
kitchen, app.ii (46, 49), app.iv (39), equipment of, 254
knights, xxxviii–xlii, lxxii–iii, 83, 116, 161, *and see names under* Winchester, bishop of, knights of

knight service, lxxiii, 9, app.iv (21, 22, 38, 43), *and see* fee

lambs, tithes of, 84, *and see* sheep
land, 141, 325
—agreement over, 1, 6, 25, 57, 73, 75, 79, 81, 84, 120, app.ii (26, 29, 34)
—custody of, 10, 78
—grants/confirmations of, 2, 9, 13, 30, 31, 34, 43, 45, 48, 49, 51–3, 56, 66–7, 67a, 68, 70–1, 100, 112, 133, app.ii (12, 13, 35, 42, 45)
lectionary books, lvii n.
legate, papal, *see* nuncio, *and* names under France, Gerold, Guala, John, Pandulph, Sabina, St Sabina, Tusculum
letters, of attorney, 124
—close, lxi, 15, 16, 109, 111, 113, app.i (125–7, 131)
—of credit, 108
—as descriptive term (*littere*), 3, 11, 18, 22, 23, 37, 46, 47, 58n., 62, 64, 100, 102, 107, 110n., 112, 115n., 150, 177, 190, 224, 250–1, 255–6, 258–9, 264, 279, 287–93, 312, 314, 326, 334, 338, 341–3, *and see* letters patent
—newsletter, 127, 129
—papal, *see names under* Gregory IX, Honorius III, Innocent III
—patent, lii, lxi, 12, 20, 33, 40, 50, 90, 92, 95, 101, 110n., 122n., 175, 278, 336, app.i (131–2)
—of presentation and inquisition, app.ii (21n.)
—royal, *see names under* Henry III, John
—testimonial, 86, 87, 98, 99
—*and see* charter, writ, writing
lettering, lxiii–iv, lxxv
liberties, 9, 13, 34, 45, 49, 52, 67a, 68n., 69, 82n., 100, 119, 121, 130, 132n., app.ii (5)
licence, xxxiv, xxxvi, lvi, 23n., 31, 58n., 62, 100, 110, app.i (126), app.ii (9, 17, 29), app.iv (13, 19, 20, 25, 29, 33)
—*and see* appropriation, papal licence
lights, in churches, xlvii, l, 7, 21, 31, *and see* candles, wax
liturgy, *see* offices
loans, 108n., app.ii (16, 37), app.iv (7, 24), *and see* arrears, debts

mace, 339
magistri, xlii, *and see names under* A., Airaines, Alan, Alexander, Amicius, Andrew, Appleby, Arenis, Avranches, Bartholomew, Basset, Bedford, Bingham, Blund, Bridport, Bronum, Chaliton, Chobham, Dammartin, Dartford, David, Dereham, Drogo, Ebbes-

bourne, Ely, Esseby, Farnham, Fauconberg, Forde, Garland, Gattebrig', Gloucester, Goscelin, Grandon', Greinville, Henry, Hervey, Ho, Hommet, Hugh, Humphrey, Hyde, Langton, Lawrence, Limoges, London, Lucy, Luke, Macrobius, Mansel, Merton, Millières, Montacute, Norman, Osbert, P., Pavilly, Peregore, Ralph, Roches, Russinol, Sabilio, St Cross, Scot, Shenfield, Sherborne, Southwick, Stokes, Thomas, Turr', Vienne, Vitriaco, W., Wairvill, Waltham, Walton', Wautham, Wells, William, Winchester, Winemerius, York
mancorn, 180, 243, *and see* corn
mandate, xxxv, l, lxviii, 3, 20, 39, 46, 47, 100, 105, 116, 120, app.i (128, 131–2), app.ii (2), *and see* order, precept
—papal, lii, liv, lxviii, lxx, 5, 47, 95, 100, 104, app.ii (7, 15n., 34–6)
manor, 1n., 2n., 9, 10n., 13, 26n., 30, 31n., 43, 48n., 51, 53n., 67, 67a, 68, 69n., 73, 81, 82n., 94n., 120, 127, 177n., 203n., 205n., app.i (131), app.ii (13)
mansio, *see* house
marble, 327
markets, lix, 119
marriage, grant of, 10, 78, app.ii (45)
—legitimate, l, 12, 22
—settlement, xlv, 117
mason (*cementarius*), 326
mass, lvii, 23n., 62, 73, varieties of, 31, *and see* missal
mayor, *see names under* Bordeaux, La Rochelle, London, Southampton
meadows, 19n., 61, app.iv (23, 43)
—tithes of, 60, 61, 70, *and see names under* Basingstoke, Runnymede, Winchester, places in
mediation, lxix, app.ii (24, 26)
medicine, xlii–iii, lxxii, 26n., 74n., 227, app.iv (44)
merchants, 108, app.ii (37), app.iv (7, 25)
messuage, 84, app.iv (25)
milk, tithes of, 84
millaria (units), xlvii
mills, 9, 19, 20, 51, app.iv (23, 25, 38), tithes of, 15, 59, 61, 81, *and see names under* Molendinarium, Molendinis
missal, lvii n., *and see* mass
monastic houses, building/rebuilding of, xliii, 24, app.ii (20)
—foundation of, l, lv–vi, 33n., *and see* foundation charter
—molestation of, 13, 34, 37, 49
monastic offices, appointment to, liv, 41, app.ii (46–9), *and see* obedientiaries
monks, relations with bishop, liv–vi
—power to appoint and remove, liv, 73

—as witnesses, lxxii, *and see names under* Abingdon, Canterbury Christ Church, Cîteaux, Dover, Ely, Lewes, Stoke-by-Clare, Whitby, Winchester monks of St Swithun's
morning office, 31
mort d'ancestor, writs of, app.i (127, 130), app.iv (39)
mortgage, lv
mortuary legacies, 5, 70, app.ii (35)
mother church, 5, all the sons of holy, 1, 13, 14, 21, 26, 31, 45, 48, 54, 59, 60, 62, 64, 65, 69, 74, 75

narratio, lxix–lxx
newsletter, *see* letter
night, 31, 63
nones (office), 31
notary, inspection by, 32, 38, 48, 50, 54, 107, *and see names under* Mussel, Seckford
notificatio, lxviii–ix
notification, 2, 15–17, 33, 35, 56, 57, 63, 106, 107, 129, app.i (132), app.ii (31, 44)
nuncio, 47, 58, 113n., 338n., app.i (131–2), app.ii (5, 33), app.iv (33)
——*and see* ambassador, envoy, proctor
—papal, *see names under* Otto, Pandulph
nutmeg, 339
nuts, app.iv (21)

oath, 5, 100, 127, app.ii (5, 18)
oats, 31, 146, 153, 155, 164, 205, 213, 252, 269, 336
obedientiaries, *see bartonarius, hordarius,* monastic offices, pittancer, sacrist
obit celebrations, xxxiv, 13, 43, 67a, 69, 83n., app.iv (1, 2, 4, 13, 18–20, 30, 31, 34, 39)
oblationes and *obventiones, see* altar
offerings, 5, 70, app.ii (35), app.iv (41), *and see* pension
office, 75, 81
—divine, 31, 34, 49, 62
—ecclesiastical, 26
—of preaching, 47
—pastoral, lxviii, 34, 48–50, 54
—pontifical, 11, 32, 100
—*and see* canonical hours, compline, mass, morning office, nones, prime, sext, terce, vespers
officials, of bishop, xxxv–vii, 31, *and see names under* Winchester, bishop of, official of
——letter to, 46
—of archdeacons, xxxvii, *and see names under* Surrey, Winchester, archdeacons of

oratories, liii, app.iv (7)
ordeals by fire and water, lviii
order, xlvii, l, lii, 4n., 22n., 25n., 93, 94, 96n., 104, 106, 109, 110, 113, 116, 118, 124, 255, app.i (126–7, 132), app.ii (15n.), app.iv (11, 19, 24, 29, 41), *and see* mandate, precept
orders, ecclesiastical, legitimate, 23, *and see* archbishop, bishop, deacon, subdeacon, ordinations, priest
—religious, *see* names under Augustine, Cîteaux, Dominicans, Franciscans, Hospitallers, Knights Templar, Premonstratensian, Santiago, Teutonic Knights
ordinance (*ordinationem*), 31, 80
ordinary, *ius*, xxxv, 6, authority, 15
ordinations, xxxiv, lvii–viii, 73, app.ii (11), *and see* orders
ornaments, of churches, lvii
oven, 234n.

palfrey, *see* horse
pannage, 80n., app.iv (2)
papal, judges delegate, xxx, xxxv–vii, li–ii, lxi, lxviii, lxix–xx, 2, 5, 22n., 30, 34, 73, 76, 77, 100, 110n., app.ii (29, 34), app.iv (1–3, 5, 7, 8, 10, 11, 13, 34, 36, 41)
—licence and dispensation, xxxiv, 23, app.iv (8, 13, 14, 19, 20, 29, 33, 34)
—register, 47n., 104, 130
—tax, on alien clergy, app.iv (10, 42), *and see* census, chamberlain, chancery, court, *cursus*, legate, letters, mandate, nuncio, subdeacon
parish, li, 1, 73, 84, app.iv (29)
—boundaries of, liii, division of, 5n.
—church, 5, 33, 63n., 67, 69, 75, app.iv (41)
—priest, lvii, app.ii (29)
parishioners, 5, app.iv (11)
parks, xxxiii, 191, 251, 322, 343
parson (*persona*), xliv, li, 6, 7n., 18, 21, 37, 73–5, 84–5, app.ii (13, 20), app.iv (13, 20)
pasture, 7n., 52, *and see* common pasture
patriarch, *see* names under Jerusalem
patristics, *see* quotation
patrons, 3n., 11, 13, 49, 56n., 75, *and see ius patronatus*, *patronatus*
patronatus, 29, *and see ius*
paupers, lxvii, 34, 48–50, 54, 56–60, 62, 80
—feeding of, 43n., app.iv (2, 31)
peace, 56, 75
—form of, xxx, lii, 47, 95, 99, 100, 117, 119n., 128, 130, app.ii (18), *and see* truce
pear trees, 215
penitence, *see* confession
penitentiary, lvii
pension, payable to churches, 1, 21, 71n., 74, 75

—payable from churches, 5, 7, 18, 26n., 27, 28, 31, 35, 36, 63n., 64n., 65, 73, 74
—confirmation of, 11, 14, 21, 28, 32, 37, 38
—grants of, 1, 5, 7, 19, 20, 31, 35, 48, 65, 67, p.105, app.i (128), app.ii (13, 29), app.iv (27, 41)
—augmentation of, 36
pepper, rent of, li, 52
per manum, formula, 18, 21
—formula for datary, xliii–v, 18, 21, 32, 34, 48, 49, 52–4, 75, 81, app.i (129), app.iv (18–20, 31), *and see* datary
perpetual, familiarity, 84, memory, 129n., works, 58, *and see* alms
perpetuity, grants in, lxvi, 5, 6, 9, 13, 18, 20, 26, 31, 34, 38, 43, 45, 48–54, 56, 61, 63, 69, 75, 82, 83, 85, *and see* alms
Peter's Pence, xxviii
petition, lvi, 32, 33, 86, 87, 94, 106, 121, 128, app.ii (29), app.iv (10)
petitorium, 6
pigs, 80n., 174, 275, 330, app.iv (2), for hunting, 191, salted hogs, 212, *and see* pannage
pilgrim, 129, app.iv (8)
pilgrimage, xxxii, 42, app.ii (22), app.iv (34)
pipe rolls, of the royal exchequer, xlvi, xlviii, lxiii, 102n., app.i (126, 129), app.iv (28)
—of the bishops of Winchester, *see names under* Winchester, bishops of, account rolls
pittancer, *see* almoner, *and see names under* Wherwell
plains, 9
plaster work, 316
pledges and guarantees, l, 3, 44, 78–9, 90, 93–6, 105, 116, 126, 130, 133, app.ii (5, 16, 33, 35, 37), app.iv (29), *and see* security
ploughing, service of, 82n.
pond, 154
pone, writs of, app.i (127, 130)
poor, *see* paupers
popes, *see* papal, *and see names under* Gregory IX, Honorius III, Innocent III, IV
ports and harbours, 67n., 90, 127, app.i (129)
pound (weight), xlvii, 5, 21, 339, app.ii (29)
poverty, 63, *and see* paupers
prayers, 13, 26, 58, private, 31
preaching, xxvii, xxxiii, xxxvi, li–ii, lviii, 47, 58n., app.iv (33), *and see* sermon
prebend and prebendaries, 18, 46, 238, *and see names under* Lincoln, London, Romsey, Salisbury, Southwell, York
precedence, lxvi, lxxii, 86n.
precentor, *see names under* Lincoln, Salisbury, Wells
precept, lxxi, 9, 18, 34, 49, 100, 207, 284, 316, 334, app.i (126), p.106, *and see* mandate, order
predia ecclesiastica, 6

presentation to churches, 18, 27–9, 39, 72, 74n., 85, 118n., app.ii (3, 19, 21, 32, 40, 41, 43)
prest, 93
priest, lviii, 18n., 84, app.ii (29)
prime, 31
prisoners, 4n., 17n., 105, 106, 133, 145, app.ii (36), app.iv (4, 25, 41, 43), and *see* custody, hostages
proctor, 25n., app.ii (37, 41n.), app.iv (7, 10, 13, 18–20, 27, 30, 33, 34, 36, 41, 42), *and see* ambassador, attorney, envoy, nuncio
profession of obedience, xxviii, liv, 34, 49
proprietas, see ius
proprios usus, see appropriation
psalter (*salterium*), lvii n.
purprestures, 6
purchases, xlvii, 25, 49, 71, 143, 151, 164, 165, 169, 179, 182, 185, 198, 207, 208, 229, 240, 245, 265, 275, 285, 292, 294, 325, 330, 337, 339, *and see* sales

quarters (weight), 31, 145, 146, 153, 155, 164, 166, 180, 189, 205, 206, 213, 220, 243, 252, 308, 309, 311, 312, 336, app.iv (2)
quitclaims, xlviii n., 3n., 9n., 56n., 71n., 82
quotation, scriptural and patristic, lvi, lxiv, lxvii, 13n., 58n.
quo warranto, assize of, app.iv (32)

rank, *see* precedence
ransoms, lii, 4, 79, app.ii (9), app.iv (41, 43)
rabbits, *see* warrens
rebels and rebellion, xxxi, xxxvii, li, 7n., 44n., 56n., 79n., 96a, 100n., 133n., 177n., 334n., app.i (128), app.ii (7), app.iv (4, 11, 24, 37, 38, 40, 41), *and see* civil war
rector, li, 5, 6, 29n., 31, 40, 48n., 63n., 65, 72n., 76, 77, 83, 84, 85n., app.ii (34, 38), app.iv (2, 5, 7, 10, 13, 15, 17, 19–21, 24, 29, 30, 33–6, 41–2)
—of the king and his realm, *see names under* William Marshal I
reform, of the church, lvi–ix
reformation, 57
regard, *see* forests
relic, app.iv (31)
relict, 3
rents, 19, 22n., 34, 43, 48, 49, 52, 56, 177n., app.ii (29), app.iv (1, 7, 22, 23, 25, 26, 29, 31, 32, 40, 41, 43)
ride (*cheminus*), 71n.
right, *see* ius
rite, *see names under* Sarum

roads and ways, 9, 192, 279, app.ii (35)
robes, 182, 289, app.iv (6, 28, 36, 37)
rumores, 127
russet, *see* cloth
rye, 308

sacrist, *see names under* Chertsey
safe conduct, 89, 91, 92, 106, 122, 126, app.ii (5n., 33), app.iv (41)
saints, *see* cult, relic
sales, 2, 10n., 25, 48n., 71n., 78n., 336, 341, *and see* purchases
salt, 70, *and see* pigs
salutatio, lxvi–vii
sanctio, lxxi
Sarum rite, 31
schools, xlii
scotales, lviii, 56n.
scibes and scripts, lxii–iv, *and see* scriptor
scriptor, xliv, 84
scripture, *see* quotation
scutage, xxxviii, xl, 304, app.i (130), app.iv (25, 38, 40, 43)
—rolls, lxvi, app.i (127, 129)
seal, episcopal, xliii–v, lxvii, l, lx–lxii, lxxi, lxxv, app.i (128, 130)
—of the king, great, xli, xlvii, 48, app.i (130)
—exchequer, app.i (130)
—*and see* chancery, counterseal, spigurnel, wax
security, 44, 56, 78, 79, 99, 116, 123n., 222, *and see* pledge
seisin, 4n., 67n., app.i (132), *and see* disseisin
serjeants (*servientes*), 238, 279, 334
sermon, lviii, lxvii, 58n., *and see* preaching
servants, 34, 49, 62, 100, *and see* serjeants
service, to abbey, 2n., 13n.
—to bishop, 9, 56
—to king, 112, 115, 118, 133, app.ii (13), app.iv (31)
—to secular lord, 51, 52, 82n., 120, *and see* knight service
services, religious, 31, 62, 104, *and see* office
settlement, of disputes, xxxv, xlv, xlviii, l–lii, lxii, 1, 2n., 3, 4n., 5n., 6, 7n., 57n., 62n., 70n., 73n., 76, 77, 79n., 85n., 95, 100n., 117, 128, app.ii (4n., 13, 23n., 26, 28, 29n., 34), app.iv (5, 10, 11, 21, 25, 31, 36, 42, 43)
sext, 31
shearing, 248
sheaves, tithes of, 8
sheep, 239, 248, *and see* lambs
sheriff, *see names under*, Berkshire, Cornwall, Devon, Essex, Gloucestershire, Hampshire, London, Northumberland, Oxfordshire, Somerset, Surrey

ships, 127, 129, 266, app.ii (6), app.iv (25)
siege, xxvii, xlvii, app.iv (23, 43)
significations, *see* excommunication
silk, app.iv (41)
silver, marks of, 67a, 75, *and see* chalice
sowing, 9, 229
spigurnel, xlvii
stabilis, stability, liv n., 21, 48, 51–4, 63–5, 69
stable, app.ii (44)
subdeacons, liv n., lvii–viii, app.iv (10, 27)
—papal, *see names under* Pandulph, Roches, Lukes des
subdean, *see names under* Salisbury, Wells
sub-vicar, li, *and see* vicar
succentor, *see names under* Salisbury, Wells
suffragan bishops, of the province of Canterbury, xxviii, lxvi, 86–8, 100
—within the diocese of Winchester, xxxiv–v, xlii
supertunics, 289
surplus, 31
suspension from office, xxviii, xxx, 100, app.ii (6)
synod, of archdeacon, xxxvii
—of bishop, 1, 63, app.ii (25)

tallage, lix, 102, 120, app.i (128), app.iv (14, 21, 23, 25, 26, 29), *and see* tax
talley sticks, 177, 178
tax, lii, 46, 276, 278, app.ii (37n.), app.iv (2, 10, 23, 24, 42), *and see* tallage
taxation of a vicarage, 70, 84
temporal, lvii n.
tenement, 9, 56, app.iv (23, 24, 43)
terce, 31
thalamus, 280
thorn, letter, lxiv
threshing, 221
tiles, ridge tiles, 169
tithes, xxxv, l–li, liii, lxvii–viii, 5, 34, 45, 49, 76, 77, 286, app.ii (24, 34)
—of the feeding of animals, 81
—of assarts, 34, 75
—of corn, app.ii (28)
—of cheese, 84
—of demesne, 5, 75, 81
—of gardens, 70
—of geese, 70
—of grain, 5
—greater, 70
—of lambs, 84
—of land, 1, 5, 75, 84
—lesser, 70, 84
—of meadows, 60, 61
—of milk, 84
—of mills, 59, 61, 81

—of newly cultivated land, 81
—of sheaves (*garbarum*), 8
—retention of, 15
—exemption from, 81
tittle, lxiii
tournament, 118
travellers (*hospites*), 26, 49, 50, 58
treason, 133
treasure, 127, app.i (128), app.iv (34)
treasurer, episcopal, xxviii, xlvi, lxii–iii, 280, app.iv (10, 14–17)
—royal, *see names under* Auckland, Fauconberg, Eustace de, Rivallis
—*and see* hordarius, *and names under* Lincoln, Poitiers, York
treasury, royal, 93, 97, app.i (127), app.iv (10, 34, 41), *and see* exchequer
treaty, *see* peace
triplex forma pacis, 100
troper, lvii n.
truce, 111, 118, 129, app.iv (27), *and see* truce

vacancy, xxvii–ix, xxxiv, 18, 20n., 35, 41n., 50, 78, 86, 95, 110, app.ii (13, 42), app.iv (3–5, 7, 10, 13, 15, 20, 35, 41, 42)
vespers, 31
vestments, lvii, 31, app.iv (13), painted, lvii n.
—*and see* capes, cope, robes, surplus
vicar, 72, 84, app.iv (10, 18, 20, 30, 33)
—appointment and institution of, 70, 84, app.ii (27)
—perpetual, 5n., 18, 28, 34, 73, 83, 84, 107n., app.iv (34)
—portion of reserved, 26, 34, 35, 49, 50, 54, 69, 73, 83, app.ii (29)
—*and see* sub-vicar
vicarage, 72
—endowment of, xxxvi, l–li, liii, lvii, 26n., 70, 84
vice-chancellor, *see names under* Neville, Ralph de
virgate, 6, 52, 53n., 75, app.iv (14, 39, 40, 43)
viscount, *see names under* Thouars
visitation, liii

war, *see* civil war
wardrobe, xli, app.iv (41)
wardships, xlv, lii, 3n., 10, 41n., 78, app.ii (45), app.iv (22, 24, 29, 37, 38, 40, 41, 43), *and see* custody
warranty, 48, 102n., 109, 122n., app.i (130–1), app.ii (1n.), app.iv (24, 38, 40, 41), pp.105–6, *and see quo warranto*
warrens, xxxiii, 303, app.iv (22)

INDEX OF SUBJECTS

water, 6, 9
—course, 49, 56
—supply of, 57, 58, app.ii (30)
—transport by, 313, *and see* ordeals
wax, for sealing, xlvii, lx–lxi
—pensions of, li, 5, 21, *and see* candles, lights
whore, dirty, xlii n.
windows, app.iv (21)
wine, 127n., 285, 340, app.ii (42), app.iv (18, 23, 25, 37)
witness lists, lxxii–iii
wood, 9, 49, 52, 67, 71, *and see* forest, ride
—firewood, xlii, 216, 313, app.iv (41), *and see* fuel
—sawn planks, 154
—shingles, 156
wool, tithes of, 84
—tied to tag, 49, 50, 52, 54, *and see* cloth
world, 13, 58
writ (*breve*), xxxii, xlvii, xlix–l, iii, 89, 102n., 103, 105n., 116n., 134–343 *passim*, app.i (125–31), app.iv (17, 21, 22, 34, 35, 37, 39), pp.105–6, *and see* charter, letters, writing
writer, *see* scribe
writing (*scriptum*), as descriptive term, lxvi, 1, 6, 13, 14, 19, 21, 26, 32, 34, 38, 45, 47, 49, 51–4, 56–7, 59–61, 64–5, 69–71, 74–5, 83–5, 100